SIGNAL
PROCESSING

McGRAW-HILL BOOK COMPANY

New York
St. Louis
San Francisco
Düsseldorf
Johannesburg
Kuala Lumpur
London
Mexico
Montreal
New Delhi
Panama
Paris
São Paulo
Singapore
Sydney
Tokyo
Toronto

MISCHA SCHWARTZ

Professor of Electrical Engineering and Computer Science
Columbia University

LEONARD SHAW

Professor of Electrical Engineering
Polytechnic Institute of New York

Signal Processing

DISCRETE SPECTRAL ANALYSIS, DETECTION, AND ESTIMATION

This book was set in Times New Roman.
The editors were Kenneth J. Bowman and Madelaine Eichberg;
the cover was designed by Pencils Portfolio, Inc.;
the production supervisor was Charles Hess.
The drawings were done by J & R Services, Inc.
Kingsport Press, Inc., was printer and binder.

Library of Congress Cataloging in Publication Data

Schwartz, Mischa.
Signal processing.

1. Signal theory (Telecommunication) 2. Electronic data processing—Telecommunication. I. Shaw, Leonard, joint author. II. Title.
TK5102.5.S365 621.38′043 74-9850
ISBN 0-07-055662-8

SIGNAL PROCESSING

DISCRETE SPECTRAL ANALYSIS, DETECTION, AND ESTIMATION

7 8 9 10 KP KP 8 6 5 4 3

To
CHARLOTTE and PEEDEE

CONTENTS

PREFACE

The widespread use of computers in thc past decade and the corresponding ability to carry out high-speed calculations on incoming data samples in real time have produced a virtual revolution in the field of signal processing.

Seismic signals, biomedical signals, sonar and radar signals, among many examples, are now routinely processed by computers. A computer may test a signal to see if it possesses characteristics of specific interest, such as evidence in a seismic signal that may indicate the presence of a desired mineral deposit underground, or it may test a child's brain voltage signal for indications of the presence of a congenital learning disability. Other computers process noisy radar signals to provide speed and location estimates to the pilot of an airplane approaching an airport. Radar signal processors also detect the presence or absence of other airplanes along a pilot's intended route. Special-purpose computers designed to rapidly carry out Fourier transform calculations are now available at costs which make them economically justified for many applications. Of particular interest are oscilloscope-minicomputer combinations which, with appropriate software modifications, will carry out signal averaging, fast Fourier transform calculations, spectral analysis, and a host of other signal-processing operations, rapidly displaying the resultant information as curves on the oscilloscope screen, or rapidly printing the curves on paper.

It is our feeling that many of the techniques and algorithms currently employed by engineers and scientists in carrying out signal-processing tasks are widespread and basic enough to warrant introduction into the undergraduate engineering curriculum. We have therefore been developing an upper-level course over the past four years at the Polytechnic Institute of New York that covers some of the more fundamental topics in the area of signal processing. This book is an outgrowth of that course.

The course has been required of all senior students in Electrical Engineering and Computer Science, and has also been selected as an elective by many students in Operations Research and System Science. Junior students in the Honors program at the Polytechnic have taken it as well. The only prerequisites for the course have been a prior course in system analysis and one in probability. (The probability theory could be taken concurrently.) Probability ideas necessary for this course are reviewed in Chapter 3.

As is apparent from the title of the book and from a perusal of the Table of Contents, we have focused on three basic topics in signal processing. These are discrete spectral analysis, discussed in detail in Chapter 4; signal detection, treated in Chapter 5; and signal estimation, covered in Chapters 6 and 7. Chapter 2 on discrete Fourier transforms and Chapter 3 on probability and introductory random processes provide the necessary background for the material on spectral analysis.

Note the emphasis above on *discrete* signal processing. Discrete-time data samples and techniques for handling them form the motivation for this book because of the all-pervasive use of digital computers and digital signal-processing instruments. It is assumed throughout that data *samples* will be available, and hence all techniques discussed are of the discrete-time type. The student is thus exposed immediately to the kinds of computational procedures he will be using on the job or in a research laboratory. He begins to think directly in terms of operations on data samples.

We have noted above that the application of signal-processing procedures involving the use of the high-speed computer has invaded many diverse disciplines. Instrumentation engineers, mechanical engineers, transportation engineers, biomedical engineers, geophysicists, oceanographers, petroleum engineers, etc., as well as electrical engineers and computer specialists, are routinely concerned with signal-processing techniques. Statisticians and economists carrying out time series analysis are, of course, heavily involved as well. Although we have focused on the basic concepts in the three areas noted above, we have drawn freely from a host of applications to demonstrate the widespread utility of the techniques and ideas discussed. We hope the many applications described will whet the reader's appetite, as well as demonstrate to him the widespread usefulness of the material he is learning.

Although the course on which the book is based has been taught principally to students in Electrical Engineering, System Engineering, Operations Research, and Computer Engineering, the diverse applications discussed should make the book useful for courses in other engineering and scientific disciplines which use modern signal-processing techniques. It should also prove useful as a self-study text for practicing engineers and scientists working in some of the many areas encompassed under the broad heading of signal processing.

The book contains somewhat more material than would normally be taught in a one-semester introductory course meeting three hours a week. This allows for some selection in course content, depending on the interests of the teacher and the students. In our own teaching of the material we have experimented with two versions of such a course. In one we stress spectral analysis, in the other we stress recursive estimation. In both versions we have covered essentially all the material in Chapters

1 through 3 (taking time out for only a brief review of the material on probability theory), the introductory sections of Chapter 4, most of Chapters 5 and 6, and the introductory portions of Chapter 7 on discrete recursive estimation. In one offering stressing spectral analysis, we cover the remainder of Chapter 4 as well. In the other version we cover most of the material in Chapter 7 on recursive estimation. Additional selectivity can be made by emphasizing or deemphasizing some of the applications covered, or leaving some for student reading. We have also found it useful to provide computer exercises to be carried out in conjunction with the course. Some of these are included among the problems at the end of each chapter.

The authors would like to thank Prof. Richard Haddad of the Polytechnic Institute of New York, who participated in the teaching of the course, for his many suggestions, help with certain sections of the book, and critique of the course notes on which the book is based. He is responsible for some of the problems as well as some of the application examples included.

We would also like to thank Mrs. Beulah Rudner, who contributed many helpful comments based on her teaching of this material, as well as numerous students whose computer projects provided useful data and motivated some problems included here.

MISCHA SCHWARTZ
LEONARD SHAW

SIGNAL
PROCESSING

1

INTRODUCTION

Our modern world is replete with examples of *signals* available for processing or analysis. Pick up the business section of your daily or weekly paper and note the stock-market averages plotted daily, weekly, or monthly over the past year or two; appear at your doctor's office for your annual checkup and note the printout of the electrocardiogram (EKG) and, in some cases, the electroencephalogram (EEG) taken by the technician or nurse; stand behind a flight controller at an air-traffic-control center and watch the moving radar-screen spots, with position coordinates and velocity superimposed, that represent airplanes moving in the flight corridor to which they are assigned. These and countless other examples of measured data that vary with time represent the signals we shall be discussing in this book. The data in question, besides varying in time, may be random or unpredictable and will generally be carrying some useful information that is to be extracted automatically by computer. This is the *processing* activity mentioned above. The data points may be digital (discrete numbers) or continuously varying (analog). In the latter case a conversion to digital form must be carried out before the processing operation can begin. An example of a time-varying analog signal is shown in Fig. 1.1.

What kinds of processing operations might be carried out? This depends, of course, on the data or signal under consideration and the purpose for which they are

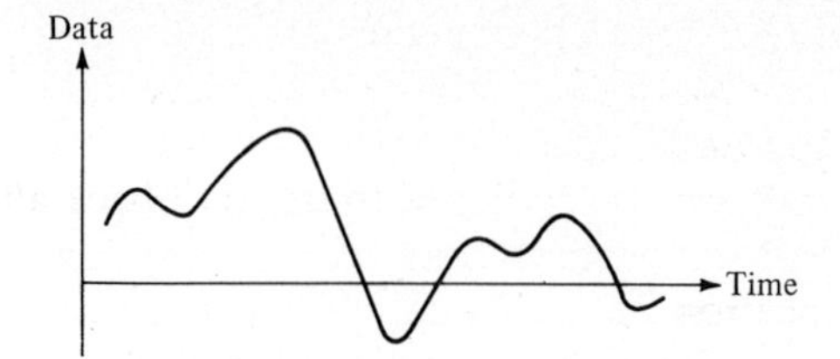

FIGURE 1.1
Typical record of data in question.

intended. An analysis of the stock-market data or other *economic time series*, such as plots of gross national product vs. time, prices, retail sales, employment with time in a given industry, etc., might look for trends: are the data cyclic or periodic? Related to the weather or seasons? Are they generally trending upward? Can they be fitted by straight lines or polynomials? Can one predict or forecast future data points with some confidence? Can one in turn develop models representing any underlying structure and use them for control purposes, i.e., to change the time variation in some desired way?

For the EKG or EEG plots one might ask: What is the specific form of the pattern? How does it deviate from some known classes of so-called "normal" characteristics? For seismic signals traveling through a portion of the earth and picked up by appropriate transducers, can one extract from their particular shape and appearance information concerning the underlying strata? Are there valuable minerals in the earth below? If so, in what quantity and where located? In environmental sensing by satellite, signals reflected from the ground also provide information about the underlying geological structure; the signal and data waveshapes are related in some manner to the reflecting medium, and appropriate signal processing and analysis should provide some clues as to its structure.

One could go on and on describing other examples of signals and processing techniques developed for them—speech signals for automatic voice recognition, automobile velocity and density measurements for highway traffic control, instrument and sensor outputs, etc. In this book we shall stress techniques for the analysis of signals that have broad applicability and serve as a basis for the development of more sophisticated approaches. Some of the problems and some of the approaches are best explained in the context of a simple example. For this purpose we have chosen to discuss signal-processing techniques in the radar surveillance systems of the air-traffic-control system of the United States. The signal-processing problems involved here range from the simplest to the complex. We shall briefly outline some of the approaches used here and return to them in detail in appropriate parts of the book.

Twenty-one radar-equipped air-route traffic-control centers spotted throughout the United States pick up (detect) aircraft and track them to determine bearings, distance, headings, and speed. The purpose of this system is to ensure appropriate separations and control of the aircraft. As individual aircraft reach the vicinity of an airport equipped with radar (there are over 100 such terminal areas in the United States), the airport-surveillance radar takes over the detection and tracking functions. Mounted on each radar is a beacon-interrogator system as well. The beacon interrogators continuously transmit coded signals into space. In aircraft equipped with beacon transponders the transponders are triggered by the ground interrogator and automatically transmit back identification data, e.g., flight numbers. Some aircraft are also equipped to transmit back altitude data as well. The radar and beacon data are processed and combined by computer, on the ground, to provide the aircraft location and speed data to air traffic controllers.

The radar system used in this air-traffic-control system provides very simple examples of the signal-processing functions mentioned earlier. The radar continuously emits high-frequency bursts (pulses) of electromagnetic energy into space. If intercepted by an aircraft, these radar pulses are reflected back to the ground antenna. The round-trip time t taken for the reflected pulse to return provides a measure of the airplane's radial distance r from the radar ($t = 2r/c$, where c is the velocity of light). The radar energy is directed into space by the radar antenna. The antenna beamwidth in these air-traffic-control radars is of the order of 1 to 1.5° in azimuth, i.e., an angle measured in a plane tangent to the earth's surface, so that aircraft location in azimuth is provided by the direction of the antenna beam when the aircraft intercepts the radar pulse. The radar antenna normally turns, or scans, at a fixed rate, so that aircraft can be detected anywhere in azimuth. When the antenna returns after one scan to the vicinity of the azimuth at which an aircraft (target) was detected, another target reflection should take place. Radar-pulse returns from the plane's successive locations can be used to determine the plane's velocity and heading. Each scan of the radar provides another target location point, making aircraft tracking possible.

This in essence summarizes the function of the surveillance radars at both the remote traffic-control centers and at terminals so equipped. But a more intensive look at the aircraft detection and tracking indicates that this simple description is incomplete. For there are significant signal-processing problems involved in detecting and tracking the aircraft; it is the discussion of *these* problems that makes the air-traffic-control example germane to this book. Specifically, as the antenna "looks" into space "searching" for aircraft, it continually picks up electromagnetic energy from the earth's atmosphere, radiation from the earth itself and from the sun if the antenna beam intercepts the sun's rays, and energy from deep in space as well as other sources. This extraneous or interfering energy is highly temperature-sensitive; since it increases with the temperature of the source generating it, it is called *thermal noise*. In addition, the electric circuits in the radar system itself, from antenna to receiver, provide sources of extraneous energy or noise. There is a fundamental physical law that *any* dissipative element in a system (an electric resistor provides an obvious example) is a source of thermal noise at the temperature of that element. The discreteness of electric charges constituting current provides another source of noise, called *shot noise*.

The upshot is that even with no aircraft target present, electric energy at the appropriate frequency is present in the receiver circuits and may occasionally be mistaken for a target. These mistakes give rise to *false alarms*. (Other sources of false alarms, in addition to ever-present noise, include radar reflections from the ground or water, often called *clutter*, and spurious targets picked up by the antenna side

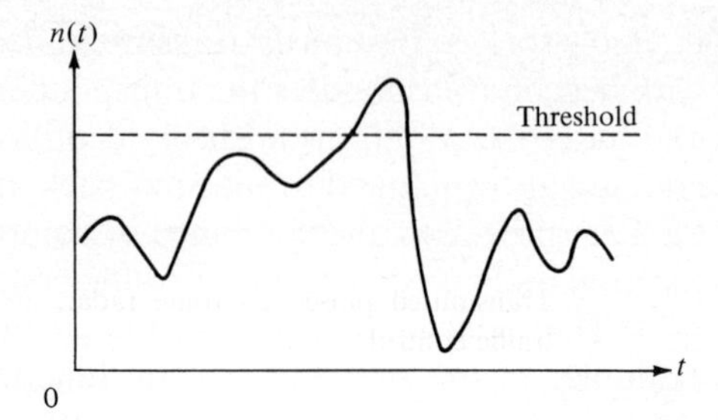

FIGURE 1.2
Typical noise trace. Only positive noise shown.

lobes; no matter how narrow the antenna beam, some energy always spills over into much wider angles, allowing signals to be transmitted and received in directions other than the nominal pointing direction of the antenna.)

One might suggest desensitizing the radar receiver and computer system following it, to ignore the false alarms. This brings up the essence of the signal-processing problem: the extraneous signals or noise picked up are random or statistical in character. An example of a typical noise trace is shown in Fig. 1.2. The exact waveshape can never be predicted in advance, and the extent of the peaks and dips is a function of the intensity of the noise source or sources. One can then talk of desensitizing against noise only in an average sense. As an example, if noise samples below the threshold shown in Fig. 1.2 are rejected from the system, noise in the large peak shown still lies above the threshold.

The problem is that one cannot keep doing this indefinitely. For the signal pulses radiated by the radar are of finite energy and must themselves compete with the noise after reflection by the aircraft. The radar beam spreads steadily as it propagates into space, providing an unavoidable $1/(\text{range})^2$ decrease in power in each direction, or $1/(\text{range})^4$ decrease total. The signal power returned depends on the reflecting properties of the aircraft as well as its distance, and for realistic situations can often be of the order of the noise or even less. An example is shown in Fig. 1.3. If a threshold is used, as in Fig. 1.2, to eliminate some noise, the signal may be eliminated as well.

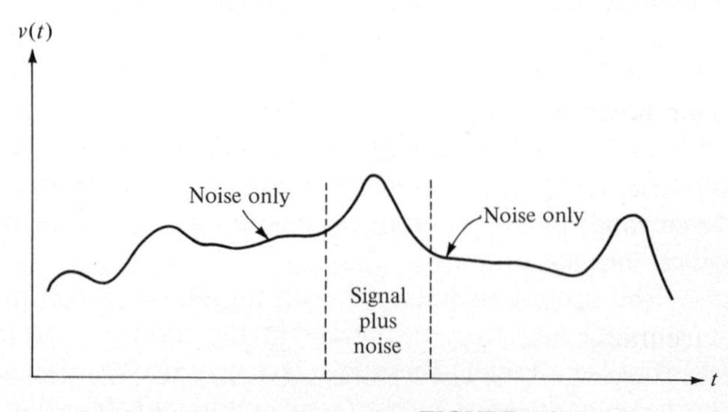

FIGURE 1.3
Signal and noise in a system.

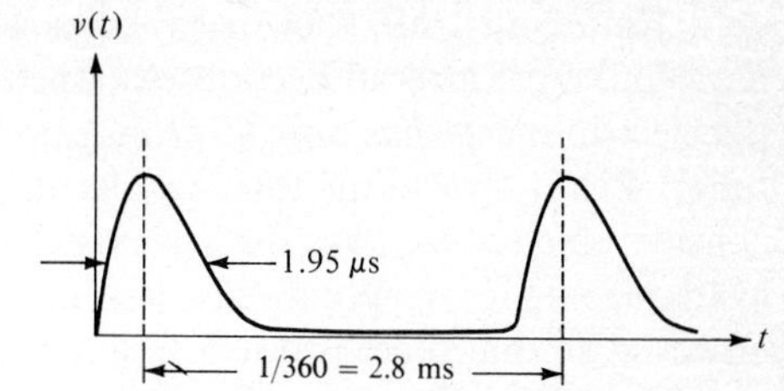

FIGURE 1.4
Transmitted pulses, en route radar, air-traffic control.

This problem of detecting the presence of a signal pulse when it occurs is a critical one in radar and may be considered the simplest example of signal processing: Is there something there or not? We shall consider signal-detection techniques extensively in Chap. 5 and shall come up with quantitative numbers for the *signal-to-noise ratio* (ratio of signal power to noise power) required to detect a target, when it appears, with a certain probability of success (probability of detection), the probability of a false alarm being specified at some tolerable level.

In practice, in many radars, the probability of detecting a signal at one particular range and azimuth setting is enhanced by combining several successive signal-pulse returns from the same target. This is possible if in a single radar scan each location in space is illuminated by several successive pulses. To demonstrate this possibility consider a typical en route surveillance radar with azimuth beamwidth of 1.35°. It rotates at a 6 r/min scan rate, thus covering 360° in azimuth in 10 s. Since the beamwidth is 1.35°, a particular aircraft will be illuminated for $\frac{10}{360}(1.35) = 0.038$ s, or 38 ms. These radars transmit pulses 1.95 μs wide at a rate of 360 pulses per second. This is called the *pulse repetition frequency* (PRF) (Fig. 1.4). During the 38-ms interval in which the beam illuminates the airplane, 13 successive pulses hit the airplane and are reflected from it. Thus these 13 pulses can be used in improving the signal detectability, since the chance of random noise interfering successfully with the signal in all 13 tries will generally be quite small. This is described quantitatively in detail in Chap. 5. The purpose of these radars, however, is not only that of *detecting* the presence of a target but that of tracking it as well. We noted earlier that the two-way time taken for a transmitted pulse to be reflected from an airplane and returned to the radar is a measure of the airplane's radial location ($t = 2r/c$, where r is the radial distance and c the velocity of light). The pulses transmitted are not absolutely rectangular, however (see Fig. 1.4 for an exaggerated example), and noise adding in with the return pulse confuses the decision as to exactly when the pulse begins, i.e., arrives at the antenna. Thus appropriate processing procedures must be used to *estimate* the target range (time of arrival) as accurately as possible. Again, as noted earlier, velocity information may be derived from successive scan-position measurement. Because of inaccuracies in these measurements, however (due to noise, measuring-circuit inaccuracies, wind gusts twisting the antenna, etc.), successive velocity measurements may fluctuate somewhat. Data-smoothing techniques must thus be utilized to derive more reliable estimates. This requires signal filtering, a topic discussed at length in this book.

Future air-traffic-control systems will rely more on beacon inputs than radar as more and more aircraft become equipped with beacon transponders. In these systems the electric energy has only to propagate one way, rather than forward and back as in radar. Beam spreading thus results in a $1/(\text{range})^2$ reduction in signal energy, a considerable saving over the $1/(\text{range})^4$ effect of radar. (The power and antenna size available are correspondingly less on the airplane than with the ground radar, however, so that there is not as much gain as expected.) Therefore the signal-to-noise problem is much less acute. Beacons introduce their own signal-processing problems, however; replies from many other aircraft are often received in addition to the one being tracked, particularly with terminal beacons at busy airports. A different kind of false alarm must now be guarded against. In addition, the beacon reply signal from the aircraft being tracked is omnidirectional and so may find its way back to the ground beacon after reflection from many different points on the ground. This gives rise to another class of interfering signals called *multipath.* The beacon signal-processing system must provide accurate estimates of the aircraft location, bearing, heading, and velocity in the presence of these disturbing influences.

Summarizing, then, in the air-traffic-control radar systems the basic problems are those of first *detecting* an aircraft when it appears and then *estimating* its range, velocity, and azimuth, among other parameters. These two signal-processing functions, detection and estimation, are basic to the processing of all signals and are described in detail in Chaps. 5 to 7. But these are not the only examples of signal-processing techniques. If the signal under investigation is random, one may wish to estimate its average or expected value, its variance and other higher moments, or even its probability distribution if possible. One may look for specific distinguishing characteristics, e.g., peaks, dips, zero crossings, how often they occur, spacings between them, etc. One may wish to "smooth" the signal, to combat against noise and interference in which it is embedded. (We have already alluded to this in discussing the detection problem earlier.) There are a vast number of ways of processing signals. Since many of them are empirically determined, we cannot possibly categorize them all. In this book we shall consider some of the methods of processing signals that have proved most useful in practice.

Basically one would like to take a time-varying signal and extract from it, using various computational techniques, the pertinent parameters for the application at hand. Most often this corresponds to a *reduction* of the data, converting them to another form of much more manageable proportions. (In detection this corresponds to a simple yes-no form: Is the signal there or not? In more complex signal-processing situations, e.g., an EKG, it could be to determine the existence of certain distinct signal patterns. This is the problem of looking for specific distinguishing characteristics noted above.)

One common technique of processing the time-varying signal is to transform it into its equivalent form in the frequency domain. Depending on the type of signal (periodic, aperiodic, or random), this corresponds to finding its Fourier series or Fourier transform representations, discussed in detail in the next few chapters. At this point it suffices simply to point out that these Fourier techniques have been widely used for data-reduction purposes; in many cases the frequency-domain version of the signal is easier to interpret and characterize. Since the frequency-domain approach lends itself readily to explaining the modifications made in signals as they progress through linear systems, it is also useful in that respect. Random signals, described in detail in Chap. 4, are particularly well characterized by the frequency-domain approach, and this so-called spectral analysis of random signals has found widespread application in many areas of science, engineering, economics, social sciences, etc.

In most of the modern signal-processing techniques one works with so-called *discrete* signal samples, either a set of signal values taken at discrete intervals of time or a set formed by *sampling* an analog (continuous) signal waveshape at regular intervals of time. For the essence of modern signal processing is the use of *digital computational techniques*. It is the computer and high-speed digital processing techniques in general that have revolutionized the field of signal processing. In this book we therefore focus almost exclusively on digital processing of signals.

We first consider rather briefly, in Chap. 2, some elementary and yet extremely useful examples of discrete-signal processing. These are subsumed by the description *curve fitting;* given discrete samples of a signal waveshape, how does one fit a straight line, parabola, polynomials, and finally sine waves to the signal? This is obviously a simple example of the search for distinguishing characteristics mentioned earlier. Depending on the application:

1. We may *know* the form of the curve that should be represented by the data samples. Noise and/or interference have perhaps obscured the curve and the object of the curve fitting is to produce the best estimate under the circumstances.
2. We may be interested in seeing which of a set of curves best represents the data points in question. (This is in turn related to the data-reduction question noted earlier. A representation of the signal in terms of a relatively simple mathematical curve or curves requires much less data storage then the original set of data points.)
3. We may hypothesize that the data points in question come from a particular type of signal. Is this hypothesis valid or not? Curve fitting and related approaches help provide the answer. (Applications abound here as well. Are there linear trends in the stock-market averages or in other economic time series? Are there cyclical trends representable by a sinusoid? This may obviously provide very useful information for the stock market investor as well as the economist.)

Curve fitting represents a simple introduction to the problem of signal *estimation*, since we are in essence trying to *estimate* the parameters of some simple curves to which to fit the data. It is no accident that later, when we discuss signal estimation in detail in Chap. 6, some of the results obtained in the curve-fitting examples of Chap. 2 will reappear.

The frequency analysis of signals, on which we shall be spending a great deal of time in the next few chapters, is also in essence an example of curve fitting. Here we are attempting to fit a number of sinusoids to some given data points. This is exactly the approach used in determining the appropriate Fourier series with which to represent a periodic analog signal or one known on a specified time interval. It is less apparent in finding the discrete Fourier transform representation of discrete data points with which we shall be concerned in the next chapter as well. But it is nevertheless a useful way of looking at that representation: What sine waves are needed to represent the data points, and what are their amplitudes and phases?

In order to carry out the discrete processing, discrete samples of the signal must be made available. This is done by sampling the analog waveshape (if that is the original form of the signal) at prescribed time intervals. A discussion of this sampling process, as well as the more complete analog-to-digital (A/D) conversion process necessary for digital processing (sampling is just the first part of the A/D process), is also included in Chap. 2.

Many of the signals encountered in real life are random; they are unpredictable and usually represented by probability distributions. To make this book as self-contained as possible we therefore include in Chap. 3 a review of probability theory and its extension to random-time (stochastic) signals. In Chap. 4 we then discuss the frequency or spectral properties of random signals. We use the discrete Fourier transform approaches of Chap. 2 here but must average appropriately because of the random nature of the signals involved. The spectral analysis of random signals plays a key role in modern data analysis for several reasons:

1. As already pointed out above in discussing data reduction, signal spectra are often much simpler representations of the original signals. This is true of all types of signals, including inherently random signals.
2. We shall show that in explaining the effects of the *linear* processing or filtering of random signals the spectral or frequency representation plays a simplifying role.
3. The representation of random signals in terms of sine waves, which is inherent in spectral analysis, is often physically satisfying, like the curve fitting discussed earlier.

As we progress through the various signal-processing topics of curve fitting, linear filtering, spectral analysis of random signals, detection, and estimation, applications will be described in many fields of engineering and science. As already noted, the air-traffic-control radar system will provide examples for each level of signal processing. The reader is encouraged to search out other applications as well.

2

DISCRETE-TIME SIGNALS

2.1 INTRODUCTION

The air-traffic-control problem mentioned in the first chapter provides numerous examples of signals of various kinds. When a pilot speaks to the control tower, he forms a time-varying air-pressure function near his mouth, a time-varying voltage in his radio transmitter, a time-varying electromagnetic wave between the plane's antenna and the control-tower antenna, etc. These signals are all *analog*, or continuous-time, signals which can be plotted as in Fig. 2.1 with a signal value $s(t)$ for each time t.

An aircraft-identification transponder produces a different kind of signal when it is queried by the control tower. Neglecting the details of how the transponder

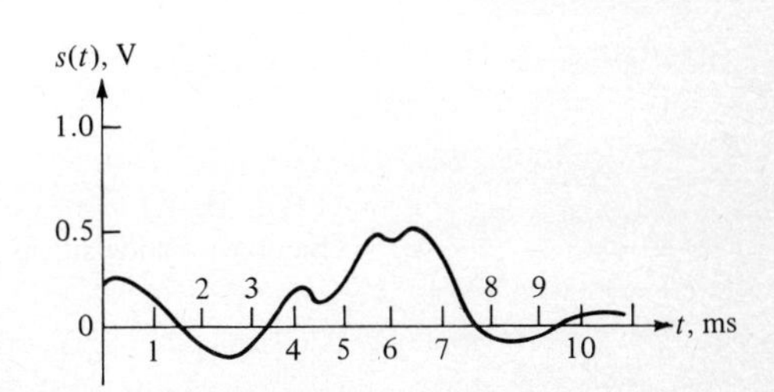

FIGURE 2.1
Graph of an analog signal.

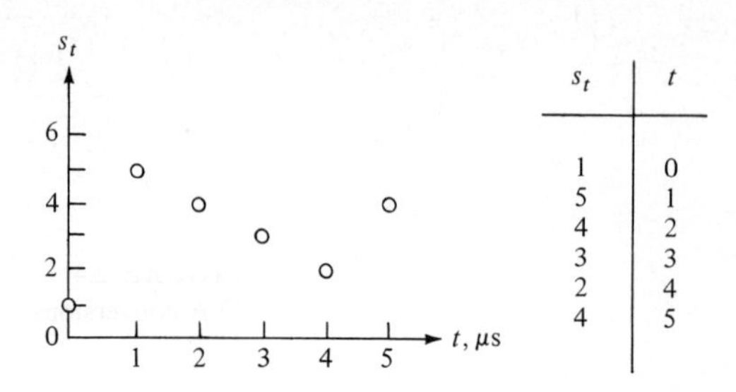

s_t	t
1	0
5	1
4	2
3	3
2	4
4	5

FIGURE 2.2
Digital signal descriptions.

works, we can say that the net result is a sequence of numbers corresponding to the flight number and altitude. When the control computer processes these numbers in order to display them on a radar screen, it is working with a *digital signal*, which has a finite number of possible values (in the binary case only two values) at each of a finite number of specific times. Figure 2.2 shows a digital signal in two ways: as a sequence of numbers s_t at times $t = 0, 1, 2, \ldots$ μs and as a graph of dots at the discrete times. Graphical presentations of digital data often connect the dots with straight lines to make viewing easier.

There are two different kinds of digital signals. In the transponder the time interval between numbers has little significance to the traffic controller and is chosen to satisfy the radio-system design constraints. The sequence spacing conveys important information when an analog signal is *sampled* as part of a data-analysis or data-transmission procedure. For example, if the pilot's spoken message were transmitted via a pulse-code-modulation scheme, the received signal would consist only of the sequence of numbers $s(kT_s)$, at times corresponding to $k = 0, 1, 2, \ldots$, in which the time interval T_s might be 0.5 ms. Figure 2.3 shows the transmitted sequence of numbers as heavy dots on the analog $s(t)$ curve. This process of picking values of an analog signal at a set of discrete times is called *sampling*, and the digital signal is sometimes said to be *sampled data*. Samples are usually equally spaced in time for convenience and analytical simplicity, but some data processors make use of unequally spaced samples. Factors which affect the choice of sample spacing T_s will be discussed later.

Pulse-code modulation involves much more cleverness than simply sampling the analog signals.† Each sample value is rounded off, or restricted to one of a finite set of possible values. This procedure is called *quantization*. Each value in the finite

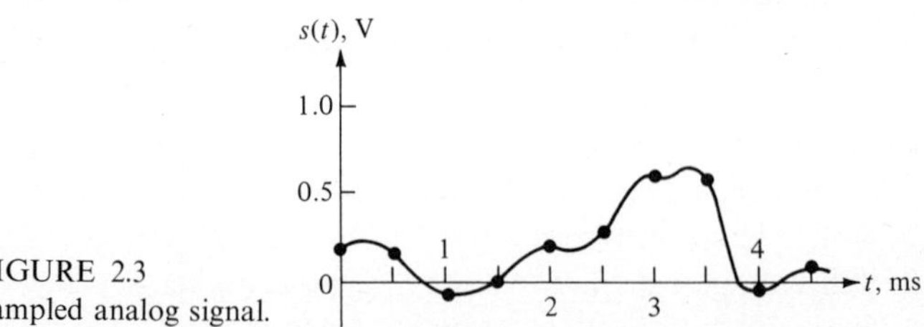

FIGURE 2.3
Sampled analog signal.

†M. Schwartz, "Information, Transmission, Modulation and Noise," p. 138, 2d ed., McGraw-Hill, New York, 1970.

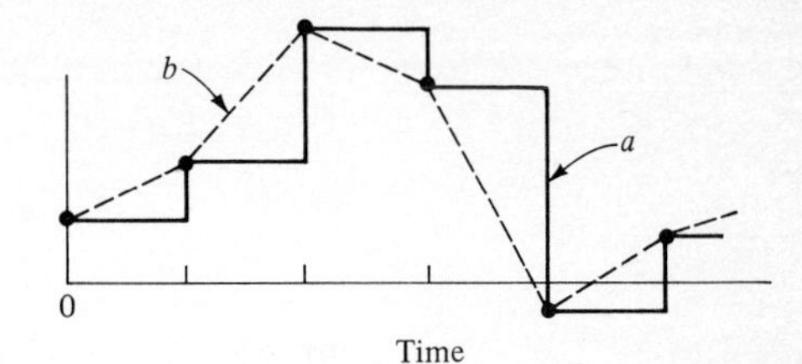

FIGURE 2.4
D/A conversions.

set can then be coded into a sequence of binary numbers for transmission. As an example, $s(kT_s)$ might be restricted to the 16 values $-0.7, -0.6, \ldots, 0, 0.1, \ldots, 0.8$. If $s(5T_s) = 0.567$, the quantized output $s_q(5T_s) = 0.6$, the nearest of the allowable values. Each of the 16 values can be represented by a set of four binary symbols. The pulse-code-modulation output would then be sequences of four-digit binary numbers. Transmission errors are less likely with binary symbols than with the original 16-level quantized signal.

Sampling and quantization occur in all signal-processing work involving digital computers. The combined operations are called *analog-to-digital* (A/D) *conversion*. Digital computers work with sequences of numbers, each expressed as an integer multiple of a fundamental unit, or *quantum*. Quantization produces errors (or destroys information), however, because the quantized $s_q(kT_s)$ is not generally equal to the true $s(kT_s)$ at time kT_s. The trade-off between the cost and complexity of smaller quanta and an increased number of levels on the one hand vs. reduced errors on the other will be discussed later.

The pulse-code receiver must finally convert a sequence of numbers into an analog air-pressure signal which sounds like the pilot's voice. This process is called *digital-to-analog* (D/A) conversion. The basic mechanism of D/A conversion can be quite simple, e.g., forming bar graphs by holding the analog value constant at the previous digital value (Fig. 2.4*a*) or forming straight-line connections between successive sample values (Fig. 2.4*b*). Better D/A procedures yield smoother analog signals and can be described by the filtering operations we shall study later. D/A devices are available for changing digital sequences to ink-drawn graphs on paper, two-dimensional cathode-ray displays, or music, as well as the speech signals in our example. The possible fidelity of A/D and D/A conversion is quite good: most people are unaware that the telephone company is chopping up and reconstructing their voices in this way in some parts of the telephone system.

Figures 2.5 and 2.6 show simplified schematic representations of D/A and A/D converters. The switches in Fig. 2.5 represent electronic switches, e.g., transistors, each controlled by one binary digit in a computer word. As shown, the word is $10 \cdots 0$, with a switch position up (connected to the reference voltage) corresponding to a 1. The operational amplifier with resistor inputs has the effect of adding the switch voltages divided by the corresponding resistance values to produce the output voltage. Each time the computer feeds a new word to the switches, the amplifier output jumps to a new level. The resulting interpolation is a step wave, as in Fig. 2.4*a*.

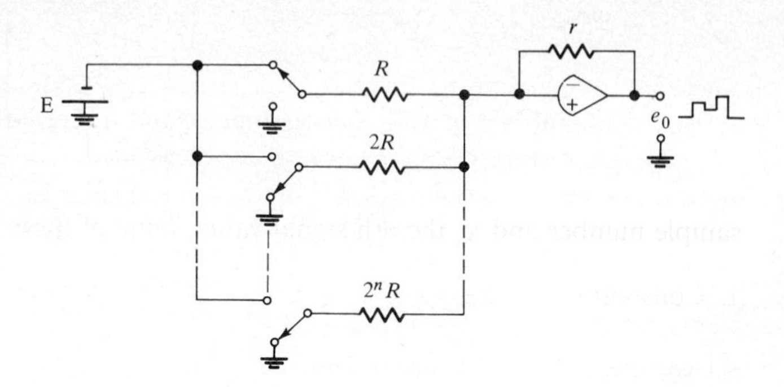

FIGURE 2.5
D/A converter.

The A/D converter in Fig. 2.6 uses the D/A converter as one of its component blocks. The first operation on the analog input is a sample-and-hold, which measures $e_A(t)$ at times spaced by T_s s and maintains a constant analog voltage at the sample value until the next sample is obtained. The digital signal is formed by counting clock pulses until the digital signal is bigger than the current output of the sample-and-hold. This comparison is based on the relative size of analog voltages, and so the digital signal is converted through a D/A operation to provide a comparison voltage. The clock must produce a large number of pulses in T_s s so that the counting can be completed before the next sample value is presented for conversion.

Since the study of quantized signals is more difficult than that of nonquantized sampled data, we shall deal first with the properties of sequences of real (infinitely precise) numbers. Although discrete-time signal and digital signal were used synonymously earlier, we shall henceforth use *discrete time* to indicate the nonquantized case and *digital* to describe the kind of sampled *and* quantized signals which appear in digital computers.

We now proceed to a more quantitative discussion, which uses mathematical symbols to describe simple discrete-time signals.

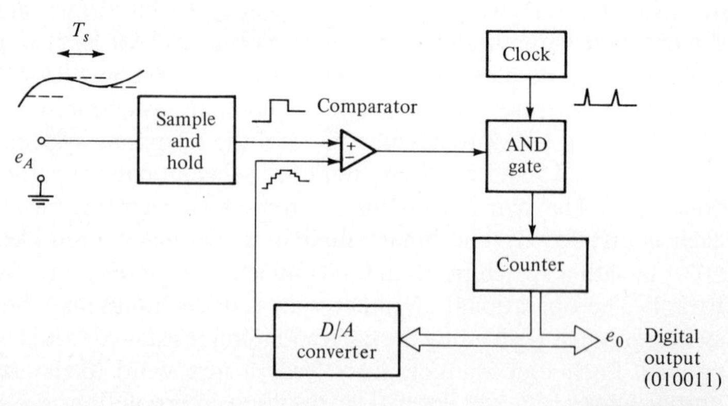

FIGURE 2.6
A/D converter.

2.2 SIMPLE SIGNALS

There are many ways in which signals can be classified or characterized. We have already distinguished among analog, digital, and discrete-time signals.

An interesting class of discrete-time signals consists of those whose sample values can be defined by simple analytical expressions. With n representing the sample number and x_n the nth signal value, some of these simple functions are

1. Constant $\quad x_n = c_0$
2. Straight-line $\quad x_n = c_0 + c_1 n$
3. Sinusoid $\quad x_n = \alpha \cos(\beta n + \gamma)$
4. Exponential $\quad x_n = \mu e^{\lambda n}$

Slightly more complicated signals can be formed by adding or multiplying several of these simple signals. We are not interested in aimless manipulation of simple signals: the goal is to use these nice expressions as *approximations* for real numerical data which, in their original (raw) form, are more difficult to record, transmit, or interpret.

Many similar examples of such signal modeling can be drawn from experimental science. For instance, measurements of current through and voltage across a piece of material can be used to determine the range of values over which the material satisfies Ohm's law, as well as the appropriate resistance constant within that range. Figure 2.7 shows a possible set of measurements. For these data, the range over which

$$e = Ri \tag{2.1}$$

seems approximately valid is

$$-15 \leq i \leq 15 \qquad -30 \leq e \leq 30$$

and the resistance is approximately

$$R = 2\Omega$$

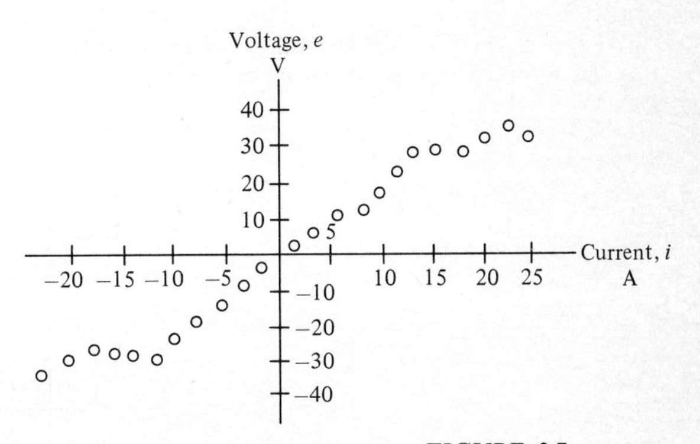

FIGURE 2.7
Voltage vs. current data.

These conclusions follow from passing a straight line near the points which seem to lie approximately along such a line. The many data points can now be well described by the simple relation

$$e = 2i \tag{2.2}$$

(a version of $x_n = c_0 + c_1 n$ with i replacing n) along with a statement of the range of validity.

Fitting a straight line to a set of data points is an example of signal processing often called *data reduction*, for obvious reasons. As another example, the trend in the monthly increase in airport operations (landings and takeoffs) shown later in Fig. 2.33 could be found by fitting a straight line to the data.

One popular measure of the quality of the fit of a straight line

$$\hat{x}_n = c_0 + c_1 n \tag{2.3}$$

to a set of data points

$$x_{-M}, x_{-M+1}, \ldots, x_0, x_1, \ldots, x_M \tag{2.4}$$

is the sum of squared errors (vertical distances between each actual point and the proposed line). The badness of the fit is thus

$$\$(c_0, c_1) = \sum_{n=-M}^{M} (x_n - c_0 - c_1 n)^2 \tag{2.5}$$

When the coefficients are chosen to be the values c_0 and c_1 which minimize \$, the resulting line is called the *least-squares* fit.

An algorithm for least-squares fitting of a straight line will be developed now, both for its own usefulness and for fundamental least-squares concepts, which can be used for fitting other polynomials or sinusoids to data. We might begin by assuring ourselves that (2.5) indeed has a minimum for some values of c_0 and c_1. Figure 2.8 shows the paraboloid shape of $\$(c_0, c_1)$, which can be verified by showing that it intersects any vertical plane along a parabola. For example, if $c_0 = 3$, then

$$\$(3, c_1) = \sum (x_n - 3)^2 - c_1[2 \sum n(x_n - 3)] + c_1^2(\sum n^2) \tag{2.6}$$

which is indeed an upward-opening parabolic function of c_1.

In view of this smooth shape of $\$(c_0, c_1)$, the minimizing coefficient values are

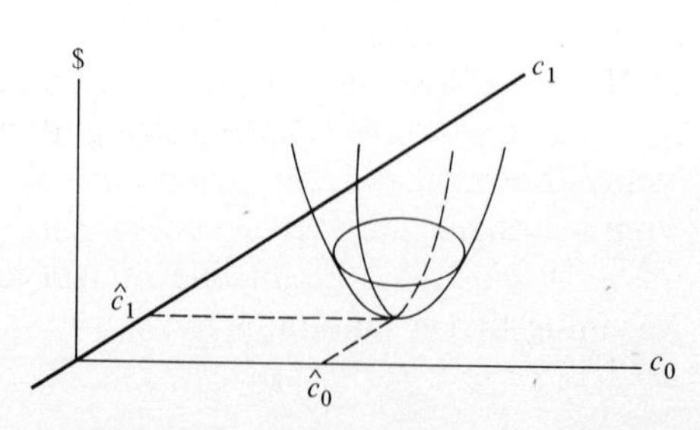

FIGURE 2.8
Paraboloid shape of \$.

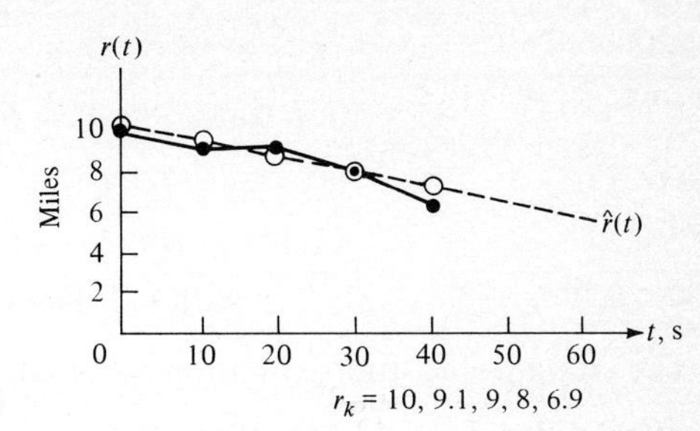

FIGURE 2.9
Sampled range data.

easily found by setting partial derivatives equal to zero:

$$\frac{\partial \$}{\partial c_0} = -2\sum_{-M}^{M} (x_n - \hat{c}_0 - \hat{c}_1 n) = 0 \tag{2.7}$$

$$\frac{\partial \$}{\partial c_1} = -2\sum_{-M}^{M} n(x_n - \hat{c}_0 - \hat{c}_1 n) = 0 \tag{2.8}$$

Rearrangements of these equations, along with the simplifying fact that

$$\sum_{-M}^{M} n = 0 \tag{2.9}$$

lead to the results

$$\hat{c}_0 = \frac{\sum_{-M}^{M} x_n}{2M + 1} \tag{2.10}$$

$$\hat{c}_1 = \frac{\sum_{-M}^{M} n x_n}{\sum_{-M}^{M} n^2} \tag{2.11}$$

We see that $\hat{c}_0$, the intercept, is simply the average of the data values. Furthermore, the slope $\hat{c}_1$ is a weighted sum of the data values, in which the weights are the sample times n divided by the sum of n^2. These expressions owe part of their simplicity to the choice of sample times to be equally spaced and symmetrically arranged about $n = 0$. More general but less neat results, e.g., for data indices running from 1 to N, are explored in some of the problems.

As an example of fitting straight lines, consider the sequence of range measurements shown in Fig. 2.9 for an aircraft approaching an airport. The points deviate from a straight line due to errors in the radar system and speed variations caused by gusts of wind. A straight-line fit will allow prediction of the plane's arrival time, assuming that it maintains the same nominal speed. The preceding formulas can be used if we arbitrarily give the five data points integer time indices from -2 to 2.

Thus,

$$\hat{c}_0 = \frac{10 + 9.1 + 9 + 8 + 6.9}{5} = 8.6$$

$$\hat{c}_1 = \frac{-2(10) - 1(9.1) + 0(9) + 1(8) + 2(6.9)}{4 + 1 + 0 + 1 + 4} = -0.73$$

The straight-line approximation to range vs. time, in terms of time intervals, is

$$\hat{r}_k = 8.6 - 0.73k \qquad k = -2, -1, 0, 1, 2$$

If we now convert to actual time, noting that the spacing between sample times is 10 s and that $k = -2$ corresponds to $t = 0$, we get

$$\hat{r}(t) = 10.06 - 0.073t$$

Details of this calculation are left for the reader to carry out. The arrival time [when $\hat{r}(t) = 0$] is thus 138 s after the first data sample. In actual tracking systems it would be too cumbersome to recompute the straight line and arrival time in this way after each data point is observed. We shall see later how to update the previous estimate after each new observation instead of reprocessing all the observations each time.

Least-squares fitting can be extended in a similar way to parabolas

$$x_n = c_0 + c_1 n + c_2 n^2$$

and higher-order polynomials in n.† At this point we prefer to generalize in the direction of fitting sinusoids to data.

An example of the need for fitting a sine wave to data arises in the study of errors in an inertial navigator. This device enables an airplane or ship to determine its position without external references like stars or radio beacons. The sensors in inertial navigators are gyroscopes, for measuring angular changes, and accelerometers, to measure linear accelerations. Automatic control systems use angular signals to keep the accelerometers fixed in three orthogonal directions: one perpendicular to the earth's surface, one in the north-south direction and one in the east-west direction. The navigation computer integrates the linear accelerations twice to generate the three components of the plane's position changes.

If the accelerometer assembly (sometimes called a *stable platform*) is not tangent to the earth's surface, false linear accelerations enter the computer because, for example, the north-south accelerometer cannot distinguish between a motional acceleration and the component of gravity in the same direction (an accelerometer acts like a mass free to move against a calibrated spring along a straight line). Figure 2.10 shows an exaggerated position error which says that the airplane is at point Q when it really is at P. This results in misalignment of the accelerometers at P, with an erroneous northward acceleration sensed due to gravity. The computer will integrate this acceleration to indicate motion of the apparent position northward from Q. The position error is reduced, but as this process continues, the apparent

†R. W. Hamming, "Numerical Methods for Scientists and Engineers," (Revised Ed.), McGraw-Hill, New York, 1973

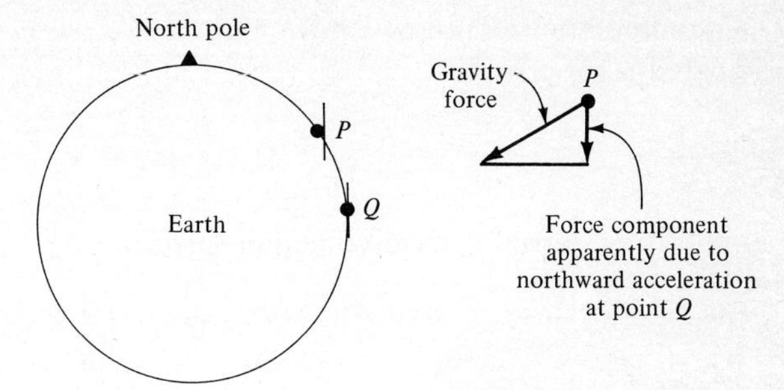

FIGURE 2.10
Navigation error diagram.

position will pass north of the true one and will continue to oscillate. It is fairly easy to show† that an initially small position error will produce an undamped sinusoidally oscillating error with a period of 84 min. This effect is called the *Schuler oscillation* after the man who pointed it out and showed the relation of its period to the earth's radius and gravity force at the earth's surface. (An aircraft at an altitude of a few miles is considered to be effectively at the surface of the earth, since the earth's radius is in excess of 4000 mi.)

Since the Schuler error in the indicated location of a ship does not fade out by itself, it is worthwhile to estimate its amplitude and phase and then to subtract out a corresponding correction oscillation in the navigation computer. Thus, we are led to consider the problem of fitting

$$\hat{x}_n = c \cos\left(\frac{2\pi}{84} n + \varphi\right) \qquad \text{mi} \tag{2.12}$$

a cosine with period $T = 84$, to a set of points like those in Fig. 2.11. These x_n points might represent the north component of a ship's position, at successive minutes, with respect to a fixed navigational reference point. These points differ from a perfect sinusoid due to other error components in gyros, accelerometers, and computer operations, as well as true changes in the ship's position.

Instead of determining c and φ in (2.12), it is better to search for α and β in the equivalent expression.

$$\hat{x}_n = \alpha \cos \frac{\pi n}{42} + \beta \sin \frac{\pi n}{42} \tag{2.13}$$

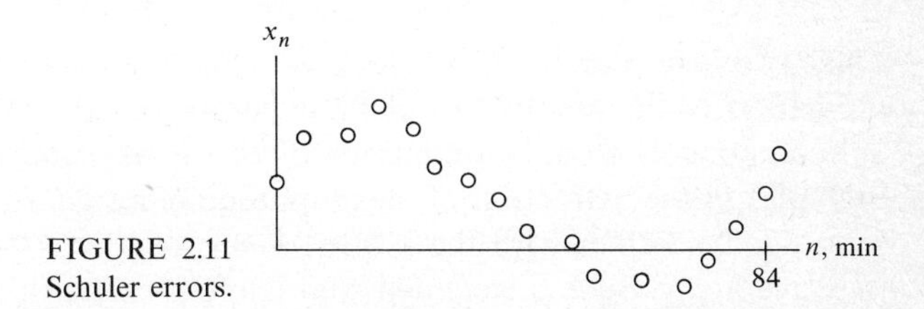

FIGURE 2.11
Schuler errors.

† C. L. McClure, "Theory of Inertial Guidance," p. 182, Prentice-Hall, Englewood Cliffs, N.J., 1960.

(Trigonometric identities show that $\alpha = c \cos \varphi$ and $\beta = -c \sin \varphi$.) Using a squared-error criterion

$$\$ = \sum_{1}^{N} (x_n - \hat{x}_n)^2 \tag{2.14}$$

and setting partial derivatives equal to zero

$$\frac{\partial \$}{\partial \alpha} = 0 \qquad \frac{\partial \$}{\partial \beta} = 0$$

yields the minimization equations

$$\hat{\alpha} = \frac{\sum_{1}^{N} x_n \cos(\pi n/42)}{\sum_{1}^{N} [\cos(\pi n/42)]^2} \tag{2.15}$$

$$\hat{\beta} = \frac{\sum_{1}^{N} x_n \sin(\pi n/42)}{\sum_{1}^{N} [\sin(\pi n/42)]^2} \tag{2.16}$$

These results have assumed for simplicity that N, the number of data points, is a multiple of 84, that is, that the data span an integral number of cycles of the oscillation, so that

$$\sum_{1}^{N} \sin \frac{n\pi}{42} \cos \frac{n\pi}{42} = 0 \tag{2.17}$$

A major observation is that, as in the straight-line-fitting problem, the best coefficient of each function of n is found from a weighted sum of the data values, in which the n-dependence of the weights is precisely the n-dependence of the respective function. That is, the cosine coefficient $\hat{\alpha}$ is a sum in which the data have cosine weights.

With the fitting errors e_n defined by

$$e_n = x_n - \hat{x}_n \tag{2.18}$$

we have viewed the data in Fig. 2.11 as a sum of three terms

$$x_n = \hat{\alpha} \cos \frac{n\pi}{42} + \hat{\beta} \sin \frac{n\pi}{42} + e_n \tag{2.19}$$

a sine, a cosine, and an error. The least-squares choices for $\hat{\alpha}$ and $\hat{\beta}$ can be viewed as estimation of the amplitudes of these components or as extraction of the individual components from observations of the total signal. We shall see many more examples of the extraction of the amplitude of a signal component by weighting the total-signal samples with the known time variation of the desired component and then averaging. Such a weighted sum is often called the *correlation* of the observed data with the desired signal function.

The *matched filter* (discussed in Chaps. 5 and 6) employs the same type of correlation with a specified signal function. The Fourier series signal representation in the next section can be viewed as finding the amplitudes of each of an infinite number of signal components by correlating the total signal with each respective component time function.

2.3 FOURIER SERIES

The rest of this chapter will concentrate on frequency-domain methods for signal characterization. These are, perhaps, most familiar through the specifications of components in music reproduction and transmission systems (radios, phonographs, tape recorders, speakers, etc.) and in the description of filters used in photography. A high-fidelity amplifier should not distort signals with frequency content in the range between 50 Hz (50 c/s) and 20 kHz. An ultraviolet filter has no effect on visible light but blocks ultraviolet frequencies which are present in sunlight and which act like visible colors on the chemicals in the photographic film.

Frequency methods will have many applications in our air-traffic-control problem. Voice communications between pilot and control tower are included in the comments of the previous paragraph. The carrier frequency of a radar (analogous to the labels on the tuning dial of a radio) is a determining factor in the size of the radar antenna. The frequency content of a plane's trajectory corresponds to the sharpness of bends in the plot of this trajectory on a radar screen. The range of possible frequencies in this curve (related to typical airplane maneuvers) will determine how close together in time we must take position samples for a digital-signal processor which predicts the plane's future position.

Frequency methods are also used to describe the random noise which appears in radar receivers. A weak return from a distant airplane may produce a receiver voltage which is no bigger than random voltage produced by electrons in the receiver circuits. The detection of a signal in noise is simplified when one can say something about typical or average properties of the noise. A simple record of noise voltages would appear to be quite a useless mass of wiggles. However, the useful information in such curves can be distilled by looking for the average rates of fluctuation or dominant frequencies. Chapter 4 will study these spectral-analysis techniques.

We have been listing frequency ideas which provide information needed for the signal detection and estimation methods to be developed in Chaps. 5 and 6. Other applications outside the air-traffic realm use frequency analysis as an end in itself. A radio astronomer may be able to deduce the elements present in a planet's atmosphere on the basis of relative strengths of the planet's radiation in various frequency bands. Frequency analysis of irregularities in a runway surface can be used to predict how various planes will bounce as they taxi along the runway. Vibration analysis, speech processing, mass spectroscopy, and crystallographic analysis are among the many areas in which frequency analysis has been used.

In this section we shall begin our frequency-domain discussion by looking at Fourier series for continuous-time functions. Subsequent sections will develop

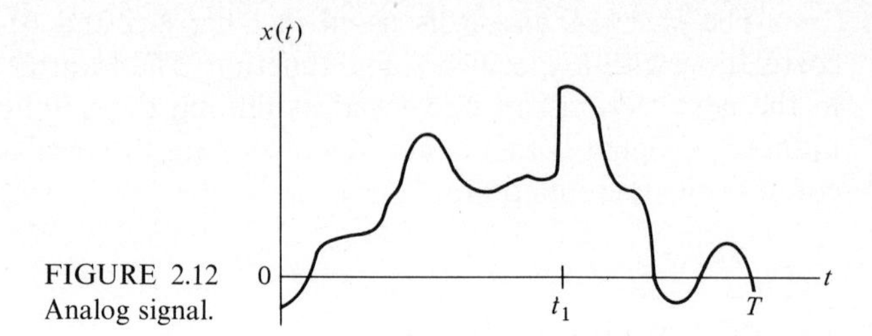

FIGURE 2.12
Analog signal.

frequency methods for discrete-time signals in terms of calculations which are very similar to the more familiar analog-signal Fourier series reviewed here.

We shall use least-squares ideas to lead into Fourier series. The analog function $x(t)$ in Fig. 2.12, defined for $0 \leq t \leq T$, might be approximated by the finite sum

$$x_K(t) = a_0 + \sum_{k=1}^{K} \left(a_k \cos k \frac{2\pi}{T} t + b_k \sin k \frac{2\pi}{T} t \right) \tag{2.20}$$

This sum has a constant term a_0 and sines and cosines at radian frequencies which are integer multiples (harmonics) of $\omega_0 = 2\pi/T$. This ω_0 is the frequency of a sine wave which has exactly one cycle in the time interval between 0 and T. The reason for choosing harmonic sine-wave frequencies will become more evident when we try to adjust the a_k and b_k coefficients to make a good fit to the given $x(t)$.

Again using squared errors, we measure the fitting quality by

$$\$_K = \int_0^T [x(t) - x_K(t)]^2 \, dt \tag{2.21}$$

where the integral over the interval of t-values corresponds to the earlier sums over finite sets of n-values. Substitution of (2.20) into (2.21) and setting derivatives equal to zero yields equations like the following a_3 equation, in which a hat over a coefficient, for example, $\hat{a}_3$, indicates the best coefficient in the least-squares sense:

$$\begin{aligned} \frac{\partial \$_K}{\partial a_3} = -2\Bigg[\int_0^T x(t) \cos 3\omega_0 t \, dt - \hat{a}_0 \int_0^T \cos 3\omega_0 t \, dt \\ - \sum_1^K \hat{a}_k \int_0^T \cos 3\omega_0 t \cos k\omega_0 t \, dt \\ - \sum_1^K \hat{b}_k \int_0^T \cos 3\omega_0 t \sin k\omega_0 t \, dt \Bigg] = 0 \end{aligned} \tag{2.22}$$

At this point, the choice of harmonic frequencies shows its convenience, because it

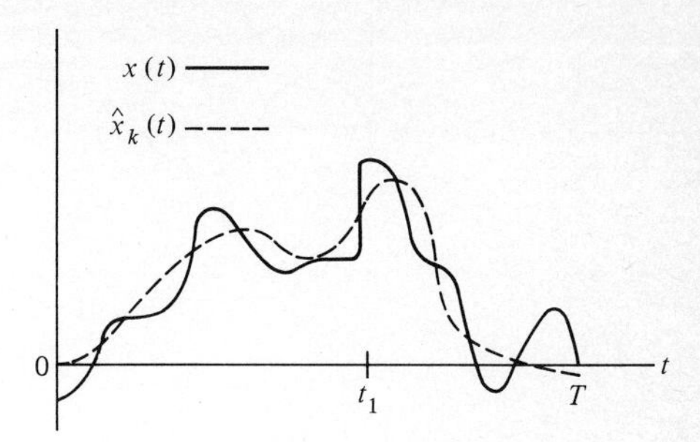

FIGURE 2.13
K-term Fourier series approximation.

makes most of the integrals in (2.22) vanish. For example, when k and m are different integers,

$$\int_0^T \cos k\omega_0 t \cos m\omega_0 t \, dt = 0 \tag{2.23}$$

by direct evaluation. [Two functions whose product integrates to zero, such as $\cos k\omega_0 t$ and $\cos m\omega_0 t$ in (2.23), are said to be *orthogonal* over the interval 0 to T. Many sets of pairwise orthogonal functions exist and are used for approximating signals.]†

Simplifications of (2.22) and all the other partial-derivative conditions lead to the minimizing coefficients

$$\hat{a}_0 = \frac{1}{T}\int_0^T x(t)\, dt$$

$$\hat{a}_k = \frac{2}{T}\int_0^T x(t) \cos k\omega_0 t \, dt \qquad \hat{b}_k = \frac{2}{T}\int_0^T x(t) \sin k\omega_0 t \, dt \qquad k = 1, 2, \ldots \tag{2.24}$$

When these coefficients are used, $\$_K$ achieves its smallest value $\hat{\$}_K$. This error measure is a collective one, reflecting effects of fitting errors at all values of t. Figure 2.13 shows a typical curve of $x(t)$ and a least-squares $\hat{x}_K(t)$ for one choice of K.

It seems reasonable that $\hat{\$}_K$ should decrease as K, the number of sinusoids, approaches infinity. The amazing property of approximation by a sum of harmonic sinusoids like $\hat{x}_K(t)$ is that

$$\lim_{K\to\infty} \hat{\$}_K = 0 \tag{2.25}$$

and that the fit gets better for large K *at each* t where $x(t)$ is continuous (excluding points like t_1 in Fig. 2.13).

For every small ε and every t between 0 and T

$$|x(t) - \hat{x}_K(t)| < \varepsilon \tag{2.26}$$

†Hamming, op. cit.

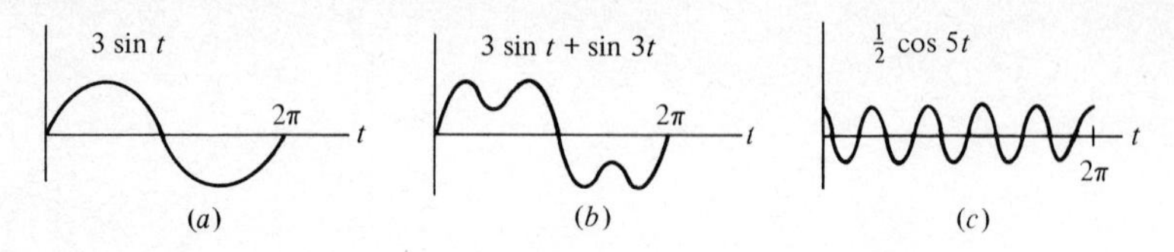

FIGURE 2.14
Low-frequency signals.

when K, the number of terms in the sum, is greater than some K_ε. This pointwise convergence (proved in standard mathematics texts†) allows us to define the infinite sum

$$\hat{x}(t) = \lim_{K \to \infty} \hat{x}_K(t) \tag{2.27}$$

in which the coefficients are defined by (2.24), and to say that $\hat{x}(t)$ is the Fourier series representation of $x(t)$ for which

$$x(t) = \hat{x}(t) \tag{2.28}$$

at all continuity points of $x(t)$. At discontinuity points, like t_1 in Fig. 2.13, $\hat{x}(t)$ equals the value halfway between the upper and lower values of $x(t)$ at the discontinuity. The Fourier series

$$\hat{x}(t) = a_0 + \sum_{1}^{\infty} (a_k \cos k\omega_0 t + b_k \sin k\omega_0 t) \qquad \omega_0 = \frac{2\pi}{T} \tag{2.29}$$

is periodic with period T; that is,

$$\hat{x}(t + T) = \hat{x}(t) \tag{2.30}$$

since the harmonic sinusoids each have this property. If the original $x(t)$ also had this kind of periodicity, equality (2.28) would hold for *all* t. Otherwise, the periodic $\hat{x}(t)$ merely represents the given $x(t)$ in the original interval of definition, $0 \leq t \leq T$.

One of the values of Fourier representations lies in their use to group together many different signals with similar properties. For example, if the only nonzero coefficients a_k and b_k are those with indices $k \leq 5$, we might say the $x(t)$ was a lower-frequency signal than a $y(t)$ with the same period T but a large a_{10}. That $x(t)$ would vary more slowly in time than the $y(t)$. Many different-looking signals satisfy this particular low-frequency constraint. A few are shown in Fig. 2.14.

The following example shows the mechanics of determining a Fourier series by evaluating the integrals in (2.24). Figure 2.15 shows a so-called half-wave-rectified sine wave formed by cutting off the negative portions of the function 3 sin πt. This function has a period of 2 s.

The coefficients can be found by direct substitution into (2.24). It should be noted, however, that the periodic nature of the integrands in those expressions allows

† For example, R. V. Churchill, "Fourier Series and Boundary Value Problems," 2d ed., McGraw-Hill, New York, 1963.

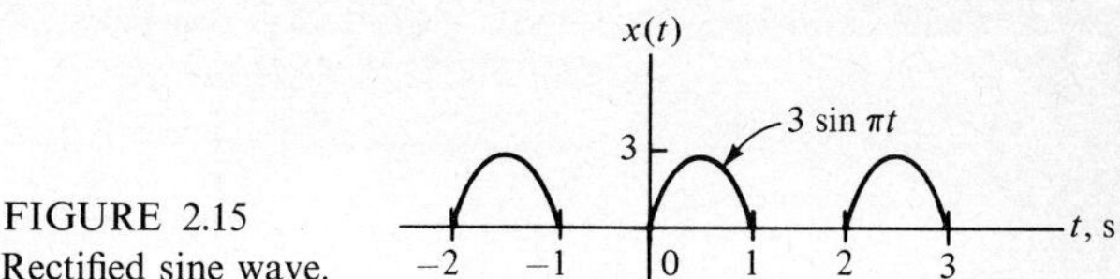

FIGURE 2.15
Rectified sine wave.

us to use *any* complete period for integration, say $-T$ to 0, or $-T/2$ to $T/2$ if it is more convenient. In this example the original 0 to T interval yields

$$a_0 = \frac{1}{2}\int_0^1 3 \sin \pi t \, dt = \frac{3}{\pi}$$

The upper limit of integration is 1 rather than 2 because $x(t) = 0$ for t between 1 and 2. The other integrals require use of some trigonometric identities

$$\begin{aligned} a_k &= \int_0^1 3 \sin \pi t \cos k\pi t \, dt \\ &= \frac{3}{2}\int_0^1 [\sin (1 + k)\pi t + \sin (1 - k)\pi t] \, dt \\ &= \begin{cases} 0 & k \text{ odd} \\ \dfrac{6}{\pi(1 - k^2)} & k \text{ even} \end{cases} \end{aligned}$$

Similarly

$$b_k = \begin{cases} 0 & k > 1 \\ \frac{3}{2} & k = 1 \end{cases}$$

Thus the half-wave-rectified sine wave, with a period of 2 s, has the Fourier series

$$x(t) = \frac{3}{\pi} + \tfrac{3}{2} \sin \pi t - \sum_{k=2,4,\ldots} \frac{6}{\pi(k^2 - 1)} \cos k\pi t$$

Although many harmonics of the fundamental frequency $\omega_0 = \pi$ rad/s are missing, the series still requires an infinite number of terms to give a perfect representation of the given function. A small number of terms will give a good approximation, however, since the high-frequency terms have small amplitudes when k is large. The $k = 10$ term is less than one-thirtieth of the a_0 and b_1 terms, for example. The main effect of adding more of the high-frequency terms is to improve the approximation to the sharp corners in the original $x(t)$.

It is convenient to represent a signal's frequency content graphically. The Fourier series provides the ingredients for such a graph, after the sine and cosine terms at each frequency have been combined into a single sinusoid through the identity

$$c_i \cos (i\omega_0 t + \varphi_i) = a_i \cos i\omega_0 t + b_i \sin i\omega_0 t \tag{2.31}$$

with

$$c_i = \sqrt{a_i^2 + b_i^2} \qquad \varphi_i = -\tan^{-1} \frac{b_i}{a_i} \tag{2.32}$$

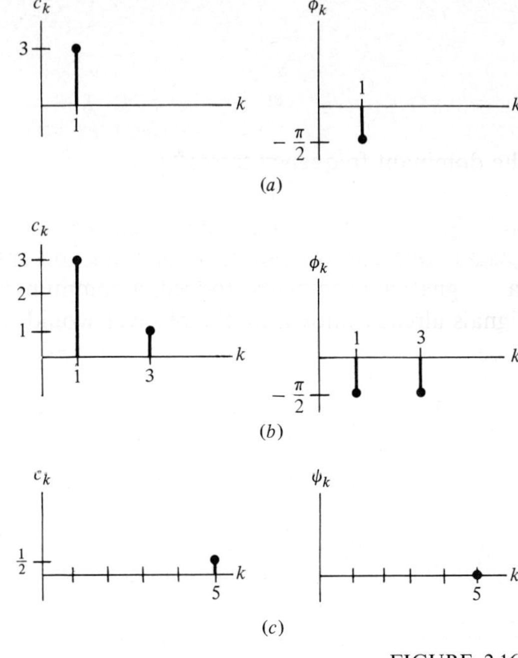

FIGURE 2.16
Signal spectra.

With this notation the Fourier series can be written

$$x(t) = a_0 + \sum_{k=1}^{\infty} c_k \cos (k\omega_0 t + \varphi_k) \tag{2.33}$$

The frequency content, or spectrum, of the analog signal is a catalog of the amplitude and phase of each harmonic sinusoid in its Fourier series (2.33). The frequency content is displayed in graphs of the magnitude c_k vs. k and of the phase angle φ_k vs. k, as in Fig. 2.16. The labels in Fig. 2.16 correspond to the similarly labeled $x(t)$ in Fig. 2.14.

It is interesting to note that the *analog* $x(t)$ defined over a time interval (or periodic repetitions of it) has a Fourier series characterized by a *discrete* set of numbers indexed by the frequency number k. The analog signal $x(t)$ is said to be in the *time domain*, and the discrete signals c_k and φ_k are in the *frequency domain*. If the functions in one domain are known, the unique corresponding functions in the other domain can be computed by an operation called a *transformation*. In particular, the frequency- to time-domain *transformation* is (2.33), and the reverse transformation can be derived from (2.32) and (2.24), in terms of weighted integrals of $x(t)$.

One possible communications application of the Fourier series representation of a function $x(t)$ in an interval $0 \le t \le T$ is to transmit the first few Fourier coefficients $c_0, c_1, c_2, \ldots, c_K, \varphi_1, \varphi_2, \ldots, \varphi_K$ instead of the signal samples $x(T_s)$, $x(2T_s)$, ..., $x[(2K + 1)T_s]$. In either case $2K + 1$ numbers must be transmitted.

Neither set is a perfect representation of the original analog signal, but both allow increased accuracy in reconstructing $x(t)$ as K gets larger. This kind of approach, using Fourier series as well as other kinds of representations, is better than sending the signal samples when the transmitter *quantizes* all signals. The advantage comes from assigning more quantization levels (less quantization error) to the coefficients of the dominant frequency terms.†

Frequency-domain representations are useful for classifying groups of signals with their nonzero c_k restricted to certain intervals, as in the low-frequency example given above. Many engineering devices and systems are designed to work in uncertain signal environments. Indeed, a communication system which only transmitted signals already known by the receiver would not be very useful. Since signals to be processed are not known precisely, system designs can be based only on signal *classes* defined, for example, by frequency-band constraints.

When a spacecraft is to photograph the surface of a planet and then scan the photograph to transmit a corresponding television picture to earth, intelligent design of the radio-transmission equipment can be based on expected general properties of the pictures. For example, a typical voltage-vs.-time curve could be hypothesized on the basis of the size of anticipated features in the photograph and the photoscanning rate. A Fourier series of such an analog signal would describe its frequency content. We shall see later that the transmission system's properties can also be described in the frequency domain, so that the system can be designed for undistorted transmission of the frequencies present in the picture (no blurring of the image).

The procedure of classifying signals according to their frequency content is called *spectral analysis*. As another example, we can think of the height of a runway surface as a disturbance signal transmitted through an aircraft-suspension system and annoying the passengers. (Automobiles on rough roads provide another example.) One approach to suspension design would be to test each candidate system on a wide variety of runways. However, it is often cheaper and faster to get general frequency properties of runway surfaces and then calculate or simulate the response of a proposed suspension for the most severe members of this class of signals. It is also necessary to classify the passenger's displacement signals in terms of those frequencies which are most uncomfortable. Spectral analysis of real data will be discussed later in considerable detail.

Before returning to the discussion of discrete-time signals and their Fourier transforms, we shall complete this analog interlude by converting the Fourier series to complex-number form. This representation uses the basic Euler identity

$$e^{j\theta} = \cos\theta + j\sin\theta \tag{2.34}$$

in which

$$j = \sqrt{-1} \tag{2.35}$$

†See H. H. Schreiber, Generalized Sampling Analysis of Basis Restricted Transformations, Ph.D. dissertation, Polytechnic Institute of Brooklyn, New York, 1972, and H. C. Andrews and W. K. Pratt. Television Bandwidth Reduction by Encoding Spatial Frequencies, *Soc. Motion Pict. Telev. Eng. J.*, vol. 77, pp. 1279–1281, December 1968.

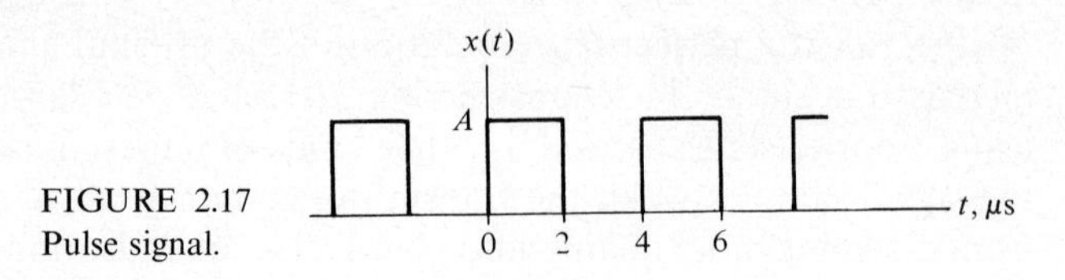

FIGURE 2.17
Pulse signal.

The *complex Fourier series* is written as a doubly infinite sum

$$x(t) = \sum_{-\infty}^{\infty} z_k \, e^{jk\omega_0 t} = \sum_{-\infty}^{\infty} |z_k| \, e^{j(k\omega_0 t + \angle z_k)} \tag{2.36}$$

in which the complex coefficients z_k are given by

$$\begin{aligned} z_k &= \frac{1}{T}\int_0^T x(t)\, e^{-j\omega_0 kt}\, dt \\ &= |z_k|\, e^{j\angle z_k} \qquad k = 0, 1, -1, 2, -2, \ldots \end{aligned} \tag{2.37}$$

As in the trigonometric case, the integration in (2.37) can be over any T-s interval, for example, $-T$ to 0 or $-T/2$ to $T/2$, since the integrand is periodic. Some choice of this interval may be more convenient than another choice for a particular $x(t)$. It is not difficult to show that this complex series is simply related to the real one in (2.33) by the relations

$$\begin{aligned} a_0 &= z_0 \\ c_k &= 2\,|z_k| \qquad k = 1, 2, \ldots \\ \varphi_k &= \angle z_k \\ z_k &= z_{-k}^* \end{aligned} \tag{2.38}$$

with the asterisk representing the complex-conjugate operation. Thus the amplitude and phase spectra, as in Fig. 2.16, can be expressed equivalently in terms of the magnitudes and angles of the z_k.

As an example, we shall find the frequency content of a voltage wave which might control the electron beam of an air-traffic-control display tube. A string of closely spaced airplanes would appear as closely spaced dots on the screen. As the electron beam is swept across the screen, it would have to be turned on and off like the square wave shown in Fig. 2.17. Knowledge of the frequency content of this waveform can be used to design the cable which must bring this signal to the display tube. The square wave shown in Fig. 2.17 is periodic with a period of $T = 4\ \mu$s. The fundamental frequency of the Fourier series is thus $1/T = \frac{1}{4}$ MHz. The complex Fourier series for this case, with t in microseconds and f in megahertz, is thus given by

$$x(t) = \sum_{-\infty}^{\infty} z_k e^{jk(2\pi/4)t}$$

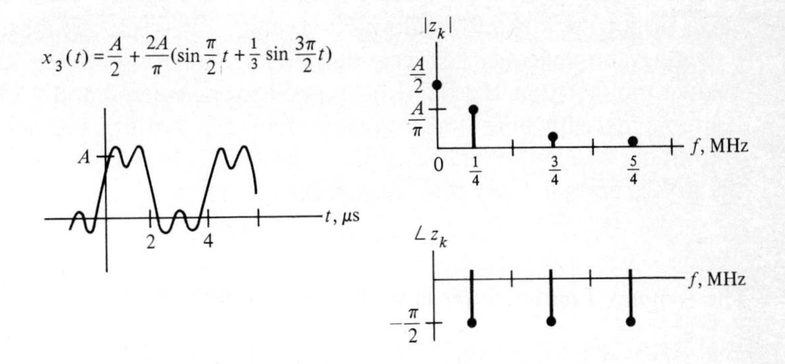

FIGURE 2.18
Three-term approximation with amplitude and phase spectra.

The complex coefficients are in turn defined by

$$z_k = \frac{1}{4}\int_0^2 Ae^{-jk(\pi/2)t}\,dt = j\frac{A}{2k\pi}(e^{-jk\pi} - 1) = \begin{cases} 0 & k \text{ even}, \neq 0 \\ -\dfrac{jA}{k\pi} & k \text{ odd} \end{cases}$$

The $k = 0$ term must be computed separately since the general expression is indeterminate for that k value. That average value z_0 is, not surprisingly, $A/2$. Thus

$$x(t) = \frac{A}{2} - j\frac{A}{\pi}\sum_{\substack{-\infty \\ k\,\text{odd}}}^{\infty} \frac{1}{k}e^{jk\pi t/2}$$

This complex series can be transformed into a real trigonometric series by collecting together the $k = \pm 1, \pm 3, \ldots$ terms, with the result

$$x(t) = \frac{A}{2} + \frac{2A}{\pi}\sum_{\substack{k\,\text{odd} \\ >0}} \frac{1}{k}\sin k\frac{\pi}{2}t$$

No cosine terms appear in this example. This will always be the case when $x(t)$ has odd symmetry about the origin after its average has been subtracted out. Figure 2.18 shows how well the first three terms in the trigonometric series approximate the original function. The amplitude and phase spectra are also shown in Fig. 2.18. Microseconds and megahertz have been explicitly shown in this figure.

In the cathode-ray-tube display example, a cable which suppressed high frequencies would produce a control signal like $x_3(t)$, making the edges of the dots extra bright, with shadows alongside the dots. It is left as an exercise to show that changing the spacings between the 2-μs pulses (or between airplanes) does not change the fact that the band of frequencies between 0 and 2.25 MHz contains all components with amplitudes greater than one-tenth of the maximum component amplitude. The pulse widths rather than their spacing determine this bandwidth if the pulses are narrower than the spaces.

2.4 DISCRETE-TIME FOURIER TRANSFORMS

The properties of Fourier series have shown us that an analog time function can often be characterized by the amplitude and phase of a small number of significant frequency components. Now we want to find similar ways of viewing the digital signals, which are our main interest. Frequency properties of analog and digital signals will be used in Sec. 2.5 to determine suitable sampling intervals for use in A/D conversion. Later, some signal processors (linear filters) will be described in terms of how they relate the frequency content of an input signal to that of an output signal. The frequency transform which we shall now develop for a discrete-time signal will also be useful for finding the essential useful properties of a long record of radar receiver noise. Knowledge of such frequency or spectral properties of the noise will be useful in designing signal processors for detecting weak signals embedded in the noise.

The discrete-time signals in Secs. 2.1 and 2.2 can also be described in the frequency domain by applying Fourier transformations similar to the Fourier series used for analog signals in the previous section. In this way, discrete-time signals can be classified, for example, as high-frequency when they are rough (change greatly between successive sample values). Digital filters can be designed using frequency ideas so that, say, a filter might pass low-frequency signals but reject or block high-frequency signals. Alternatively, a digital filter can reduce the distortion in a signal if that distortion was caused by a transmission channel which overemphasized some frequency components and underemphasized others.

The frequency-domain function $X(\omega)$ corresponding to a discrete signal x_n will be defined by the infinite sum

$$X(\omega) = \sum_{n=-\infty}^{\infty} x_n e^{-j\omega n} \tag{2.39}$$

This transformation has several interesting properties. Most of the x_n must be small if the infinite number of terms are to sum to a finite number. One acceptable case includes x_n which are zero for most n. An example of an x_n with only two nonzero values is

$$x_0 = 1 \qquad x_1 = -1 \qquad \text{and} \qquad x_n = 0 \qquad \text{all other } n \tag{2.40}$$

for which

$$X(\omega) = 1 - e^{-j\omega} = 2je^{-j\omega/2}\left(\frac{e^{j\omega/2} - e^{-j\omega/2}}{2j}\right) = e^{j(\pi/2 - \omega/2)}2\sin\frac{\omega}{2}$$

$$|X(\omega)| = 2\left|\sin\frac{\omega}{2}\right| \tag{2.41}$$

$$\angle X(\omega) = \begin{cases} \dfrac{\pi}{2} - \dfrac{\omega}{2} & 0 < \omega < \pi \\ -\dfrac{\pi}{2} - \dfrac{\omega}{2} & -\pi < \omega < 0 \end{cases}$$

This example emphasizes that $X(\omega)$ is a complex function of ω. Furthermore, this $X(\omega)$ is periodic in ω with a period of 2π, since $\sin \omega$ and $\cos \omega$ have that periodicity. One cycle each of the magnitude and angle functions for this $X(\omega)$ are shown in Fig. 2.19*a*.

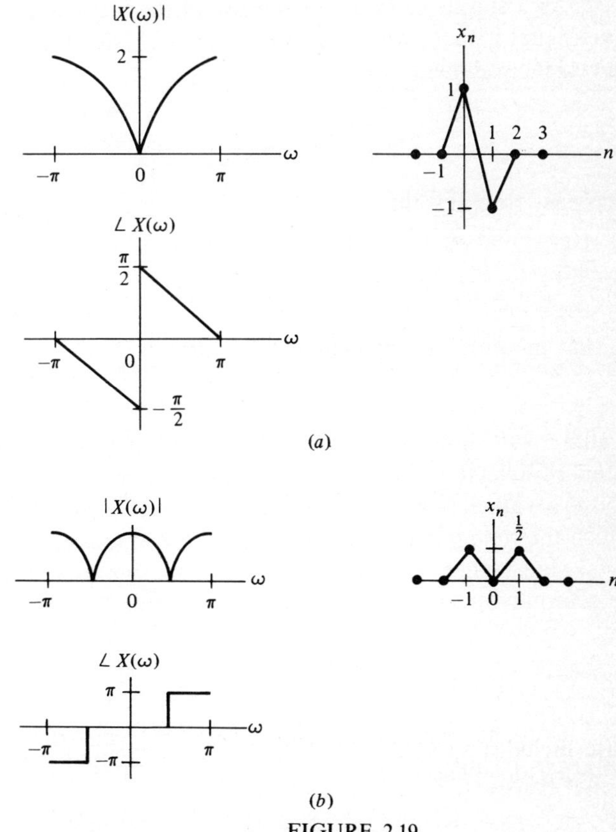

FIGURE 2.19
Discrete Fourier transform examples.

In fact, any $X(\omega)$ has period 2π, since the $e^{j\omega n}$ in the definition (2.39) has period 2π for every n. Note that the x_n sequence in the time domain is the *Fourier series* for the periodic analog function $X(\omega)$ in the frequency domain. This Fourier series differs slightly from that in the previous section in that here the analog signal is *complex* and the discrete signal is *real*, while the converse was true there.

Furthermore we have followed the conventional practice of using the negative exponential in the time-domain operation, so that the $+j$ in the Fourier series sum (2.36) is replaced here by $-j$ in the discrete Fourier transform sum (2.39). Although these two methods apply to different kinds of signals, their formal similarity means that the mathematical derivations of the usual Fourier series also justify this discrete Fourier transform and assure that it has an inverse operation

$$x_n = \frac{1}{2\pi}\int_{-\pi}^{\pi} X(\omega)e^{j\omega n}\,d\omega \qquad n = \ldots, -1, 0, 1, 2, \ldots \tag{2.42}$$

The inversion integral defines the time-sequence sample values in terms of integrals over one period of $X(\omega)$. This integral corresponds to (2.37) with the roles of time and frequency reversed.

As a simple example of the use of the inversion formula, consider a discrete-time signal with a transform $X(\omega) = \cos\omega$. Using the exponential representation for the cosine, we have

$$x_n = \frac{1}{2\pi}\int_{-\pi}^{\pi} \frac{e^{j\omega} + e^{-j\omega}}{2}\, e^{j\omega n}\,d\omega$$

It is easy to check that

$$\frac{1}{2\pi}\int_{-\pi}^{\pi} e^{j\omega k}\,d\omega = \begin{cases} 0 & k = \pm 1, \pm 2, \ldots \\ 1 & k = 0 \end{cases}$$

so that the only nonzero values of x_n are

$$x_{-1} = \tfrac{1}{2} = x_1$$

Figure 2.19*b* shows the original Fourier transform and this time function we have found by inversion. The reader should compare this time function, having two equal nonzero values (with a zero sample between them), with the previous example, in which the time function had two equal but opposite-signed adjacent samples.

Many other examples of discrete-time signals and their Fourier transforms can be computed from the geometric-sum formula for complex numbers

$$\sum_{0}^{\infty} z^n = \frac{1}{1-z} \qquad \text{if } |z| < 1 \tag{2.43}$$

(The inequality restriction ensures that $|z^n|$ gets small for large n so that the sum converges.) Thus

$$x_n = \begin{cases} \left(-\frac{1}{2}\right)^n & n \geq 0 \\ 0 & \text{otherwise} \end{cases} \tag{2.44}$$

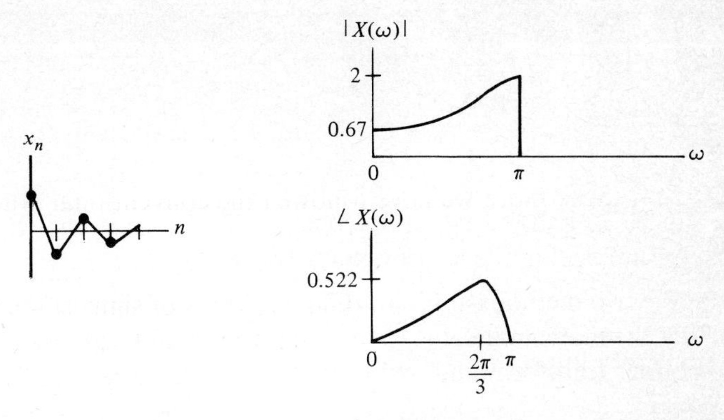

FIGURE 2.20
Discrete Fourier transform of $(-\frac{1}{2})^n$.

has the transform

$$X(\omega) = \sum_{0}^{\infty} (-\tfrac{1}{2}e^{-j\omega})^n = \frac{1}{1 + \frac{1}{2}e^{-j\omega}} \tag{2.45}$$

for which

$$|X(\omega)| = \frac{2}{\sqrt{(2 + \cos\omega)^2 + \sin^2\omega}} = \frac{1}{\sqrt{1.25 + \cos\omega}}$$

$$\angle X(\omega) = -\tan^{-1}\left(\frac{-\sin\omega}{2 + \cos\omega}\right) \tag{2.46}$$

as shown in Fig. 2.20. We can use this transform result in two ways. Mathematically we know that the original signal x_n can be recovered from it via the inversion integral (2.42). However, we are more interested in being able to relate the shapes of the transform curves to properties of the signals.

Frequency-band classification of discrete-time signals differs somewhat from that for periodic analog signals. We have seen that the periodicity of $X(\omega)$ means that all information about this spectrum is contained in the $-\pi$ to π range of ω. Figure 2.20 shows only *half* of that interval, namely 0 to π, because the symmetry suggested in Fig. 2.19 implies that the other half of the period contains no extra information.

The following steps show that $|X(\omega)|$ has even symmetry about $\omega = 0$, and $\angle X(\omega)$ has odd symmetry about that point. By definition,

$$X(\omega) = \sum_{-\infty}^{\infty} x_n e^{-jn\omega} = \sum_{-\infty}^{\infty} x_n \cos n\omega - j \sum_{-\infty}^{\infty} x_n \sin n\omega \tag{2.47}$$

Defining

$$\sum_{-\infty}^{\infty} x_n \cos n\omega = A(\omega) \qquad \sum_{-\infty}^{\infty} x_n \sin n\omega = B(\omega) \tag{2.48}$$

we can write

$$|X(\omega)| = \sqrt{A^2(\omega) + B^2(\omega)}$$

$$\angle X(\omega) = \tan^{-1}\left[\frac{-B(\omega)}{A(\omega)}\right] \tag{2.49}$$

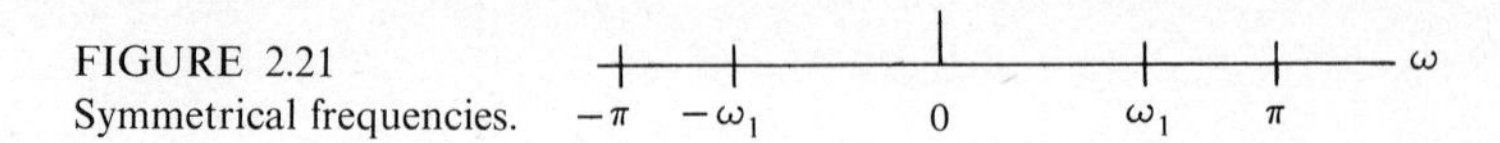

FIGURE 2.21
Symmetrical frequencies.

Figure 2.21 shows two frequencies ω_1 and $-\omega_1$ which are symmetrical about 0 rad per sample. Simple trigonometry shows

$$\cos n\omega_1 = \cos n(-\omega_1) \qquad \sin n\omega_1 = -\sin n(-\omega_1) \tag{2.50}$$

for every integer n. It follows that

$$A(\omega_1) = A(-\omega_1) \qquad B(\omega_1) = -B(-\omega_1) \tag{2.51}$$

so that

$$|X(\omega_1)| = |X(-\omega_1)| \quad \text{and} \quad \angle X(\omega_1) = -\angle X(-\omega_1) \tag{2.52}$$

Equations (2.52) are the desired symmetry properties, showing that $|X(\omega)|$ has even symmetry and $\angle X(\omega)$ has odd symmetry about 0 rad per sample.

Frequency interpretation for discrete-time signals is less direct than that for periodic analog signals. The symmetry and periodicity of the discrete Fourier transform imply that all significant frequency information here is restricted to the finite band of frequencies from 0 to π. Furthermore, there are contributions from terms at *every* frequency in that interval, rather than just at the harmonic frequencies appearing in the analog case. The inversion integral (2.42) is a mathematical description of how the frequency components combine to form the time sequence.

Although this combination of frequency components is not so easy to visualize, substantial intuitive appeal will be added later when we see how digital-signal processors are described in the frequency domain. The following comparison of the frequency content of two simple discrete-time signals also supports our interpretation that more rapidly fluctuating signals have more high-frequency content.

When the spectra of two signals are compared, the one with bigger $|X(\omega)|$ values at high frequencies will correspond to a more rapidly varying time signal and the one with bigger $|X(\omega)|$ values at low frequencies will be smoother. Figure 2.22 shows the magnitude spectrum $|Y(\omega)|$ of the signal

$$y_n = \begin{cases} (\frac{1}{2})^n & n \geq 0 \\ 0 & n < 0 \end{cases} \tag{2.53}$$

$$|Y(\omega)| = \frac{1}{\sqrt{1.25 - \cos \omega}}$$

with the spectrum of the more rapidly varying (rougher) $(-\frac{1}{2})^n$ of Fig. 2.20 superimposed.

When making this kind of interpretation, attention is restricted to the 0 to π band of frequencies, so that high frequencies are those near π and low frequencies those near zero.

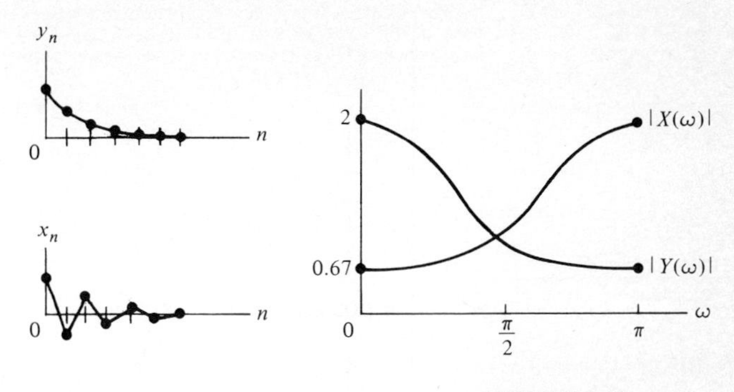

FIGURE 2.22
Comparison of spectra.

These examples of discrete Fourier transforms made use of simple exponentially decaying discrete-time signals. Problems at the end of this chapter give examples of discrete Fourier transform computations for other signals. Section 2.5 pays special attention to discrete-time sinusoids and to signals formed by sampling analog signals.

2.5 FREQUENCY CONTENT AND SAMPLING RATES

So far we have studied properties of discrete-time signals without worrying very much about where they came from. As mentioned in Sec. 2.1, discrete-time signals are often formed by sampling continuous-time signals (A/D conversion), say as the first step in a digital-signal processor. In this section we shall consider factors which influence the choice of an appropriate spacing T_s between samples. Up to this point we have assumed $T_s = 1$ s, to simplify the notation, but in practical signal processing the spacing to be used will depend on the rate at which the signal being sampled varies with time: a more rapidly varying signal will require more samples for its representation. Signals with microsecond variations will require samples to be taken microseconds apart. The samples must be chosen close enough together for the original signal to be accurately reconstructed from them, if necessary. Roughly speaking, this implies taking samples so that two or three cover the most rapidly varying portion of the original signal. As an example, consider $x(t)$ shown in Fig. 2.23*a*. Several peaks and dips appear, with three peak-valley pairs having transition regions of width 1, 1.2, and 1.5 μs, respectively. The sample spacing chosen there places three samples in a 1-μs peak-dip section. The samples are spaced 0.5 μs apart. In Fig. 2.23*b* the resultant discrete set of samples is sketched. Note that it does appear to provide a good representation of $x(t)$. Even if alternate samples were omitted, i.e., spacing samples 1 μs apart, a fairly good representation would result; i.e., the samples would follow the time variations of $x(t)$ quite well. Taking samples 2 μs apart, however, would result in a sample set that loses much of the structure of the continuous-time $x(t)$.

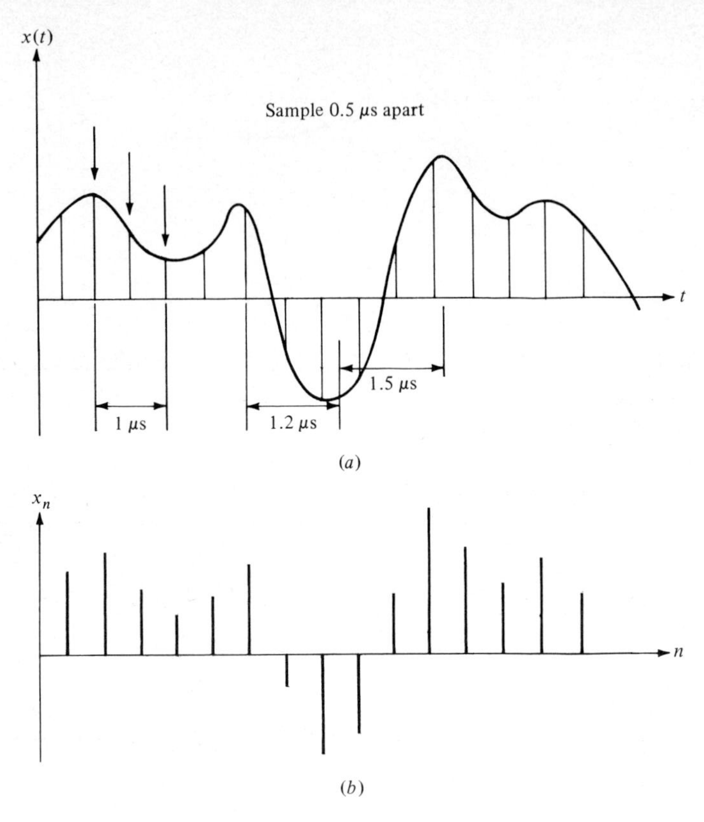

FIGURE 2.23
Discrete-time samples. (*a*) Original signal; (*b*) sampled version.

We shall quantify these remarks below by invoking the concept of the *frequency content* of a signal. We shall then show that the maximum rate of change of a signal is related to its maximum-frequency component, called the *bandwidth*. As an approximation, the peak-to-dip spacing of a signal is roughly $1/2B$, with B the bandwidth in hertz. In the example of Fig. 2.23, $B \approx 0.5$ MHz is the maximum-frequency component in $x(t)$. The *sampling theorem* says that at least one sample every $1/2B$ s (1 μs in this case), or two samples per cycle of the highest frequency, is necessary for the representation of $x(t)$. This corresponds to a *minimum* sample spacing of 1 μs in Fig. 2.23. Had the signal been varying at the rate of *minutes*, however, samples could be spaced minutes apart. We now proceed to justify these heuristic arguments more quantitatively. We first focus on the need to space samples appropriately if reconstruction of the original signal is to be possible. We then introduce the concept of the frequency content of the signal to provide a measure of the signal bandwidth.

Restrictions on the sample spacing arise if we want to be able to reconstruct the continuous signal $x(t)$ at all times t from knowledge of only its sample values $x(kT_s)$ for $k = 0, 1, 2, \ldots$. One situation we would like to avoid is that of two different

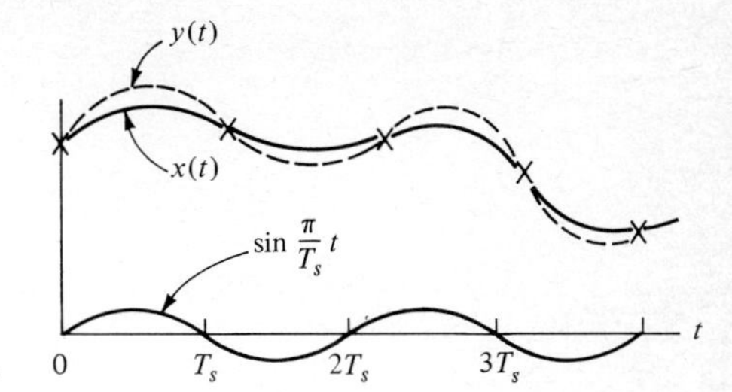

FIGURE 2.24
Ambiguous sampling: Aliasing.

continuous signals $x(t)$ and $y(t)$ having the same sample values $x(kT_s) = y(kT_s)$. In that case, a D/A converter cannot determine which continuous signal produced the samples. Figure 2.24 shows just this sort of difficulty with two signals $x(t)$ and

$$y(t) = x(t) + \sin \frac{\pi}{T_s} t$$

having the same sample values at times $t = kT_s$, where

$$\sin \frac{\pi}{T_s} kT_s = 0 \qquad \text{for } k = 0, 1, \ldots$$

Clearly, it is possible using *any* T_s to find two different signals with the same sample values. Furthermore, there is a multitude of different signals with the same sample values, not just two. [Try adding 3 sin $(4\pi t/T_s)$ to $y(t)$ in the figure.] This kind of confusion is called *aliasing error*, because one segment's samples look like another's.

Fortunately, it *is* possible to choose the unique $x(t)$ which produced a set of samples $x(kT_s)$ if we know *beforehand* that all possible signals appearing at the sampler are varying slowly enough compared to the sample spacing. Slowly varying signals cannot wiggle up and down during a T_s-s interval between samples. This would rule out the annoying sine waves in the previous example. Such restrictions are quite reasonable, because a signal processor is always designed for a *class* of possible signals with common characteristics. For example, a telephone must accurately transmit vocal sounds but not the cry of a bat or the moan of a foghorn. The rapidity of variation of the signals handled by any one processor will thus be limited. This limitation, as noted earlier, will be found to be related to the signal bandwidth.

We have suggested that signals which vary slowly enough as functions of time can be reconstructed uniquely from their equally spaced samples taken at intervals of T_s s. This class of slowly varying signals which are candidates for reconstruction can be specified by their frequency content. Frequency content will be defined in more detail in a moment, using Fourier transform arguments. The sampling theorem† states that $x(t)$ can be reconstructed precisely for all t if we know that it has no frequency content above some highest frequency B Hz which is less than $1/2T_s$, that is, $B < 1/2T_s$. Such signals are said to be bandlimited to B Hz. The conditions of this

† Schwartz, op. cit., p. 123.

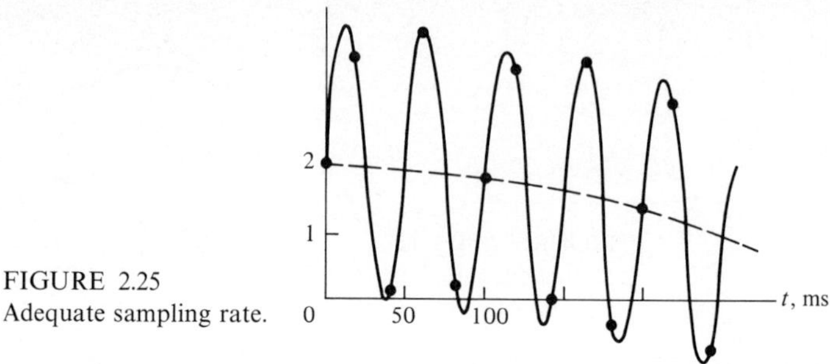

FIGURE 2.25
Adequate sampling rate.

theorem rule out the $y(t)$ in Fig. 2.24 because it contains a sine wave whose frequency equals the $1/2T_s$-Hz limit. The frequency content of that $y(t)$ is not limited to a low enough band of frequencies.

While the proof of the sampling theorem is beyond the scope of this book, we can motivate it and explain how to use it. We have already seen in the example of Fig. 2.24 that restrictions of some kind, e.g., the bandlimiting condition in the theorem, are necessary for samples $x(kT_s)$ to lead to a unique continuous $x(t)$ reconstruction. Now we shall explain how to find the bandlimit B for a given signal and how to reconstruct (or interpolate) $x(t)$.

A signal's bandlimit is defined by referring to the signal's frequency-domain representations. Two cases of continuous $x(t)$ will be considered, *periodic signals* and *pulselike signals.* The frequency representation of a periodic signal is its Fourier series. If the Fourier series has only a finite number of nonzero coefficients, it is a bandlimited signal. For example,

$$x(t) = 1 + \cos 2\pi t + 2 \sin 40\pi t \tag{2.54}$$

has a maximum-frequency component of 20 Hz. This is said to be its bandwidth. Samples of this signal spaced less than $1/2B = 0.025$ s apart should (by the sampling theorem) suffice to reconstruct $x(t)$ from the samples. As an example, say that we sample every 0.02 s. Then the samples $x[k(0.02)]$, $k = \ldots, -3, -2, -1, 0, 1, 2, \ldots$ can be used to reconstruct $x(t)$. No other periodic signal bandlimited below 25 Hz $(=1/2T_s)$ can have the same samples at these times spaced by 0.02 s $(<0.025 \text{ s})$. Figure 2.25 shows the $x(t)$ from (2.54) and the samples at this appropriate spacing.

A general periodic function with period T is bandlimited to $B = N/T$ Hz if all its coefficients (say z_n in the complex Fourier series or a_n and b_n in the trigonometric form) are zero for $n \geq N$. In the preceding example $T = 1$ s and $N = 21$ (because the highest-frequency term is the twentieth harmonic of the fundamental frequency).

The sampling theorem says that $x(t)$ can be reconstructed perfectly if the sampling rate $f_s = 1/T_s$ is greater than twice the frequency bound B for the signal. The minimal sampling rate $2B$ is called the signal's *Nyquist rate*, after the man who pointed out this condition. We must still demonstrate how to make the reconstruction which the theorem says is possible.

Some reconstruction techniques were shown in Fig. 2.4 in our introductory discussion of D/A conversion. In those cases, $x(t)$ at a time t between two sample

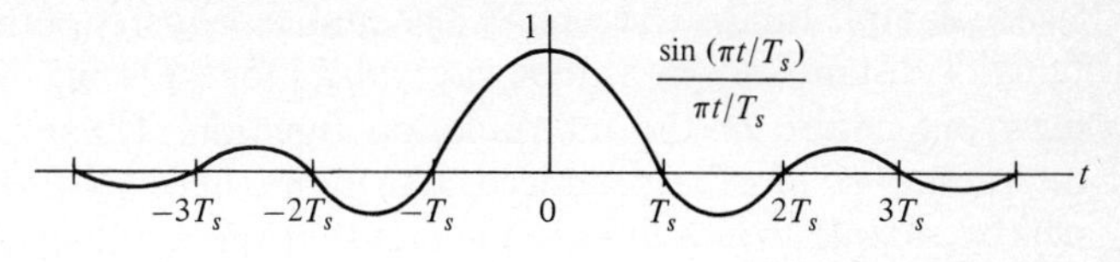

FIGURE 2.26
Interpolation function.

times, say $5T_s < t < 6T_s$, was constructed using only the two nearest sample values, $x(5T_s)$ and $x(6T_s)$. The ideal reconstruction referred to in the sampling theorem is much more complicated, since it uses all past and future sample values to compute a single $x(t)$. One reconstruction formula that can be shown to provide perfect reconstruction is defined by the infinite sum of weighted sample values

$$x(t) = \sum_{k=-\infty}^{\infty} x(kT_s) \frac{\sin [\pi(t - kT_s)/T_s]}{\pi(t - kT_s)/T_s} \tag{2.55}$$

This sum is easier to interpret if we first consider the properties of the basic interpolation function

$$s(t) = \frac{\sin (\pi t/T_s)}{\pi t/T_s}$$

shown in Fig. 2.26. This smoothly undulating time function has equally spaced zeros corresponding to the zeros of the sine-wave numerator. The function approaches zero as the t in the denominator increases away from zero (both negatively and positively). As t approaches zero, $s(t)$ appears indeterminate (0/0) but has the finite value of unity, as derivable from L'Hôpital's rule (or by recalling that $\sin \theta \approx \theta$ for small θ).

Figure 2.27 shows a few terms in the infinite sum (2.55). The kth term is a shifted version of the interpolation function $s(t)$, with its peak at $t = kT_s$. Although we shall not prove that the interpolation sum provides perfect reconstruction of $x(t)$, two observations make it plausible. First, when $t = kT_s$, the kth term equals $x(kT_s)$ and all other terms are zero. Thus the interpolation formula gives the correct value of

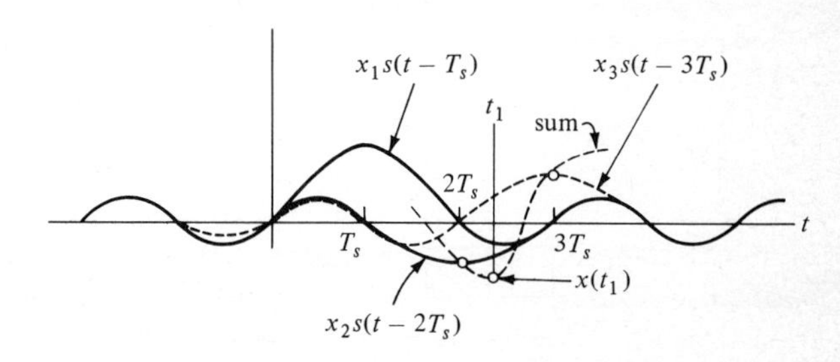

FIGURE 2.27
Several interpolation terms.

$x(t)$ at sample times. The second reasonable property of this sum is that the contributions of distant sample values, say $x(mT_s)$ for $|t - mT_s| \gg 1$, are small, due to the decaying nature of the interpolation function. Thus, as a practical matter, good interpolations can be produced at a point like t_1 in Fig. 2.27 from a finite set of nearby sample values despite the fact that the mathematical theorem calls for contributions from all past and future samples.

The sampling theorem states a lower bound on the sampling rate f_s and gives a complicated rule for perfect interpolation. It should be obvious that sampling at a much higher rate, with samples spaced more closely, will allow easier and fairly good interpolation by simple straight-line connections between sample values.

The description we have just given of the sampling theorem, Nyquist rate, bandlimits, and ideal reconstruction for periodic signals can be paralleled for pulselike signals, like those in Fig. 2.28. These signals have their significant nonzero values restricted to finite time intervals. Such a pulselike signal has its bandwidth specified in terms of its *continuous* Fourier transform $X_c(\omega)$ defined by the integral

$$X_c(\omega) = \int_{-\infty}^{\infty} x(t)e^{-j\omega t}\,dt \tag{2.56}$$

A detailed study of this new continuous transform would allow us to prove the sampling theorem. We shall avoid that path, however, since the two transforms we already know about (Fourier series and discrete Fourier transforms) are adequate for all practical purposes. In fact, each of these other transforms will now be used to approximate $X_c(\omega)$ as techniques for finding the bandwidth of a pulselike $x(t)$.

The frequency content of a single, finite-duration, continuous-time pulse, like one of those in Fig. 2.28, can be approached by thinking of that pulse as one of many in a periodic pulse train, in which the pulse spacing is so big that the other pulses are out of sight. That is, we imagine periodic repetitions of the pulse, with a large period T. The complex Fourier series for such a periodic function has coefficients z_k given by

$$Tz_k = \int_{-T/2}^{T/2} x(t)e^{-jk\omega_0 t}\,dt \qquad \omega_0 = \frac{2\pi}{T} \tag{2.57}$$

As an example, consider the periodic set of triangular pulses of Fig. 2.29*a*. It is left as a problem for the reader to show that its complex Fourier series coefficients are given by

$$z_k = \frac{A}{T}\left[\frac{\sin(\omega_k A/2)}{\omega_k A/2}\right]^2 \qquad k = \begin{cases} 0, 1, 2, \dots \\ -1, -2, \dots \end{cases} \tag{2.58}$$

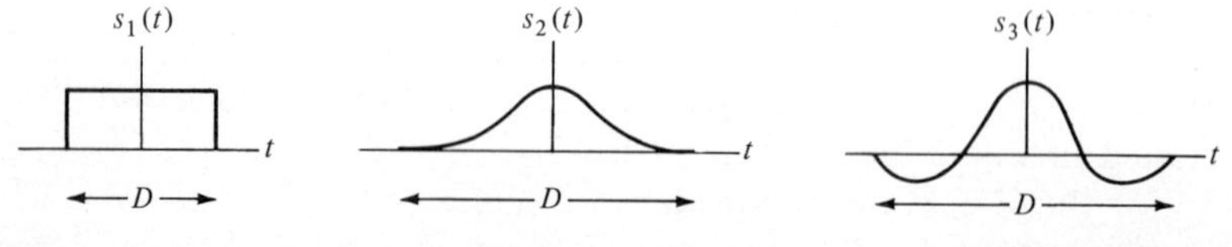

FIGURE 2.28
Pulse-type signals.

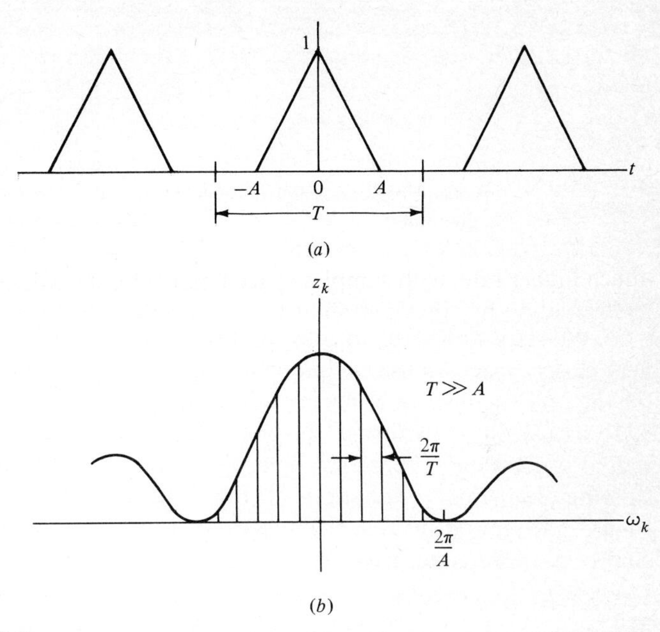

FIGURE 2.29
Spectrum of periodically repeated pulse. (*a*) Triangular pulse; (*b*) complex spectrum.

Here $2A$ is the pulse width, and $\omega_k = 2\pi k/T$. The spectrum of this set of pulses is sketched in Fig. 2.29*b*, for a repetition period T much greater than the pulse width $2A$. We can show that the shape of the spectrum, i.e., the frequencies where $|z_k|$ is big vs. the frequencies where the spectrum is small, is independent of the repetition period T.

The shape of the spectrum in Fig. 2.29*b* is given by the smooth curve through the tops of the lines, known as their *envelope*. The equation of the envelope is the continuous version of z_k in (2.58) found by replacing the discrete frequencies $k2\pi/T = \omega_k$ by the continuous variable ω. If the repetition period T were increased, the line spacing $2\pi/T$ would decrease but there would be no change in the envelope function $\{[\sin(\omega A/2)]/(\omega A/2)\}^2$. A proper limiting argument shows that as $T \to \infty$, Tz_k approaches the continuous Fourier transform $X_c(\omega)$.

For sampling-theorem purposes the bandwidth of a pulselike signal is that B for which $X_c(\omega) = 0$ when $\omega \geq 2\pi B$. When T is large, the z_k lines from the Fourier series are close enough together for discerning the shape of their envelope and thus the bandlimit B (if any) can be determined.

If $|z_k| \approx 0$ for $k \geq k_M$, a suitable bandwidth is $B = k_M/T$ Hz. This bounding-frequency approximation would be $B = 13/T$ for the pulse whose coefficients are plotted in Fig. 2.30. This use of periodic repetitions and Fourier series is fine for getting the idea of the bandwidth of a single pulse but is not practical for getting a numerical value for B from a record of a typical pulse. A practical method must be based on digital processing rather than on evaluating an integral like (2.56).

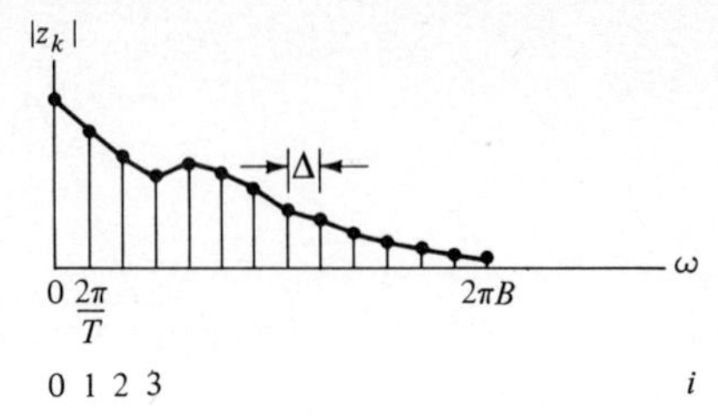

FIGURE 2.30
Spectrum of periodically repeated band-limited pulse.

We shall now explain how computation of the discrete Fourier transform of very closely spaced pulse samples will reveal the bandwidth B. We use the symbol Δ for the very small sample spacing in this argument to emphasize that this bandwidth analysis usually is performed to pick a bigger spacing $T_s > \Delta$ to be used in future digital processing of the same kind of signals. This argument is developed by considering numerical evaluation of the Fourier series coefficient integral (2.57). Figure 2.31 reminds us that the integral of a real function can be approximated by a sum of N rectangular areas

$$\int_0^D f(t)\,dt \approx \sum_{i=0}^{N-1} f(i\Delta)\Delta$$

when the width of each rectangle $\Delta = D/N$ is very small. The complex integral (2.56), when applied to one of the D-s pulses in Fig. 2.28, can be similarly approximated by the finite sum

$$\int_{-D/2}^{D/2} x(t)e^{-j\omega t}\,dt \approx \Delta \sum_{i=-N/2}^{-1+N/2} x(i\Delta)e^{-j\omega i\Delta} = X_A(\omega) \tag{2.59}$$

The summation limits were chosen to include only the nonzero samples $x(i\Delta)$ of the short pulse. The approximating sum $X_A(\omega)$ can be rearranged to look like a discrete Fourier transform of a signal with N nonzero samples.

$$X_A(\omega) = \Delta e^{-j\Delta\omega D/2N}\left\{\sum_{i=0}^{N-1} x\left[\left(i - \frac{N}{2}\right)\Delta\right]e^{-j\omega i\Delta}\right\} \tag{2.60}$$

Thus, $|X_A(\omega)|$ is simply a scaled version of the magnitude of the discrete Fourier

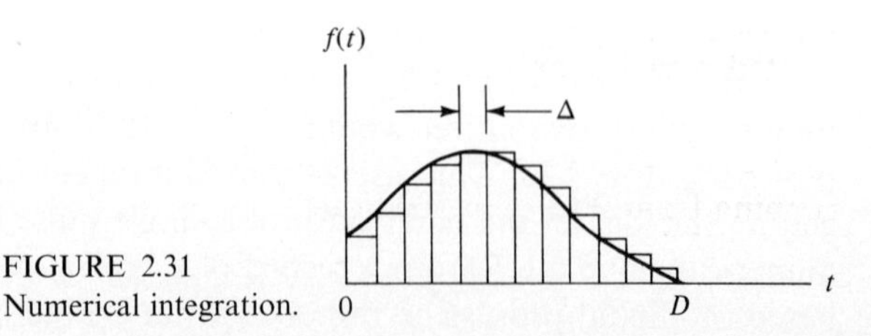

FIGURE 2.31
Numerical integration.

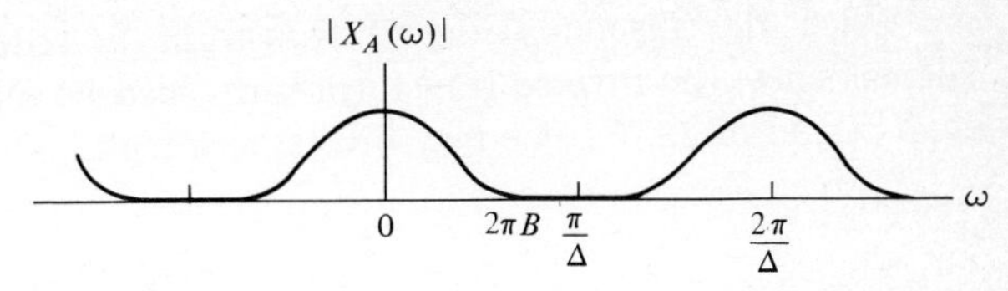

FIGURE 2.32
Discrete approximation of Fourier transform for bandwidth determination.

transform of the samples $x_i = x[(i - N/2)\Delta]$:

$$|X_A(\omega)| = \Delta \left| \sum_{i=0}^{N-1} x_i e^{-j\omega i\Delta} \right| \tag{2.61}$$

In this way the bandwidth of a finite-duration pulse can be computed by digital evaluation of the discrete Fourier transform of closely spaced samples and determination of the frequency above which $|X_A(\omega)| = 0$. This selection of B requires careful thinking, because $X_A(\omega)$ is periodic with period $2\pi/\Delta$, so that the plot under consideration will look like the one in Fig. 2.32. If the low-frequency portion (between $-\pi/\Delta$ and π/Δ) is zero for "high frequencies" $2\pi B < |\omega| < \pi/\Delta$, then that B is the bandwidth. It can be shown† that, in such cases, the *low-frequency portion* of $X_A(\omega)$ is exactly equal to the continuous $X_c(\omega)$ defined in (2.56). In the higher-frequency bands, where $|\omega| > \pi/\Delta$, $X_c(\omega)$ is zero while $X_A(\omega)$ repeats periodically. If $|X_A(\omega)|$ is not approximately zero for some band of frequencies below π/Δ, the sample spacing Δ is too big for determining the bandwidth (and aliasing errors would result if we processed samples spaced by such a large Δ).

After closely spaced samples have been transformed to get a curve like that in Fig. 2.32 for bandwidth determination, the sampling theorem tells us that future digital processing of such samples for signal detection, radar tracking, pulse communication, etc., can use a sampling period $T_s < 1/2B$ without destroying any information. When subsequent discrete Fourier transforms are computed using samples with an appropriate $T_s \neq 1$, we like to show their heritage from the original analog signal by including the sample spacing in the discrete Fourier transform; so the *transform is redefined* to be

$$X(\omega) = \sum_{n-\infty}^{\infty} x_n e^{-j\omega T_s n} \tag{2.62}$$

with a period of $2\pi/T_s$ in the ω-domain. The magnitude and angle are therefore plotted over the half period $(0, \pi/T_s)$ instead of the $(0, \pi)$ interval used earlier when $T_s = 1$. For example, if the samples used in (2.44) were taken at a spacing of $T_s = 2$ ms, the $X(\omega)$ plot corresponding to Fig. 2.20 would be on a rescaled ω axis running from 0 to $\omega = 500\pi$ rad/s.

†A. Papoulis, "The Fourier Integral and Its Applications," p. 222, McGraw-Hill, New York, 1962.

When the discrete Fourier transform is redefined as in (2.62) to show T_s-dependence, the inverse transform must also be modified to be an integral over one ω-period of $2\pi/T_s$. The new inversion integral is

$$x_n = \frac{T_s}{2\pi} \int_{-\pi/T_s}^{\pi/T_s} X(\omega) e^{j\omega n T_s} \, d\omega \tag{2.63}$$

In radar and communications work we are often interested in sampling and reconstructing pulse-type signals like those shown in Fig. 2.28. If such a signal, having a duration of T s, were to be bandlimited to B Hz, sampling at the lowest allowable rate would produce $2BT$ samples from a single pulse. The pulse should be easy to reconstruct from this set of samples, because all the other samples required for the interpolation formula (2.55) would be zero anyway. In future signal-processing discussions we shall frequently use this result that $2BT$ samples from a bandlimited T-s pulse provide all significant information about the analog signal for use by a digital processor. (Note that we are using T instead of D for pulse width now to correspond to usual radar and communications notation.)

The reduction of a pulse to $2BT$ samples is only *approximately* correct, because it can be shown that a *time-limited* pulse signal (zero except within a T-s interval) cannot be perfectly bandlimited to *any* B Hz. However, for practical purposes, we can let B be an approximate upper-frequency limit, above which $|X_c(\omega)|$ is very small. For two pulses with the same duration T, the smoother one, for example, $s_2(t)$ vs. $s_1(t)$ in Fig. 2.28, will have the lower bandlimit ($B_2 < B_1$). Generally, the sample spacing $T_s = 1/2B$ is small enough if $s(t)$ is nearly a straight line between successive samples, as we saw in Fig. 2.23.

Since a pulse signal is not truly bandlimited, any reasonably good approximation to a bandlimit B must be a high frequency. This implies that the number of significant digital samples $2BT$ will be large, i.e., much greater than 1. For a smooth pulse like $s_2(t)$ in Fig. 2.28, a good rule of thumb is $B \approx 50/T$ or $2BT \approx 100$. For a more irregular pulse like $s_3(t)$, we can use $B \approx 25/T_w$, where T_w is the width of the narrowest peak-to-valley spacing. Thus we need an increased bandwidth factor when estimating the bandwidth of a *finite-duration* pulse, in contrast to the bandwidth estimated from peaks and valleys in a *persistent* signal like the one in Fig. 2.23. This reasoning suggests that T_s must be very small if sharp corners and discontinuities in $s(t)$ are to be reconstructed well from discrete samples.

The equivalence between the discrete Fourier transform $X(\omega)$ and the continuous $X_c(\omega)$, for adequately sampled bandlimited functions, can be used to relate analog signal energy to the values of discrete samples. Signal energy is important, as we shall see later, for designing radar and communication systems. An electromagnetic wave with more energy travels farther before becoming unintelligibly weak. Of course, more energetic signals cost more to produce (energy/unit time = power). The energy dissipated in a 1-Ω resistor by voltage pulse $v(t)$ is

$$E = \int_{-\infty}^{\infty} v^2(t) \, dt \tag{2.64}$$

If $v(t) = 0$ (or approximately so) outside the 0- to T-s interval, the energy is

$$E = \int_0^T v^2(t)\, dt \tag{2.65}$$

This energy can be related to the sample values $v(kT_s)$ assuming that $v(t)$ is bandlimited and T_s is sufficiently small. We shall now show that the energy is also given by the summation of squared sample values

$$E = \frac{1}{2B} \sum_{-\infty}^{\infty} v_n^2 \tag{2.66}$$

We do so by substituting the interpolation sum (2.55) for $v(t)$ in (2.64) and interchanging the order of summation and integration to get

$$E = \sum_{m=-\infty}^{\infty} \sum_{n=-\infty}^{\infty} v_n v_m \int_{-\infty}^{\infty} \frac{\sin[2\pi(t - mT_s)/T_s]}{2\pi(t - mT_s)/T_s} \frac{\sin[2\pi(t - nT_s)/T_s]}{2\pi(t - nT_s)/T_s}\, dt \tag{2.67}$$

One factor of v in v^2 of (2.66) uses a sum with index m, while the other factor uses a sum with index n. It can be shown that the weighting functions with different indices $m \neq n$ are orthogonal, just like sines and cosines at different harmonic frequencies [see (2.23)]. This means that the integral in (2.67) is zero unless $m = n$, that is,†

$$\int_{-\infty}^{\infty} \frac{\sin[2\pi(t - mT_s)/T_s]}{2\pi(t - mT_s)/T_s} \frac{\sin[2\pi(t - nT_s)/T_s]}{2\pi(t - nT_s)/T_s}\, dt = \begin{cases} 0 & n \neq m \\ \dfrac{1}{2B} & n = m \end{cases} \tag{2.68}$$

Thus the sum of integrals in (2.67) reduces to the desired energy expression (2.66).

When the signal is nonzero for only T s, there will be $2BT$ nonzero samples and the energy is approximately

$$E \approx \frac{1}{2B} \sum_{1}^{2BT} v_n^2 \tag{2.69}$$

This expression is not a precise equality because, as mentioned earlier, a T-s pulse signal cannot be precisely bandlimited. A large B gives a good approximate bandlimit and yields a good approximation of E by the resulting sum of a large number of terms in (2.69). [Note that in (2.65) and (2.69) we have followed the radar notation, in which T represents the pulse duration, in contrast to our previous use of the same symbol for the period of a periodic signal.]

To summarize, an analog signal bandlimited to B Hz can be reconstructed perfectly using formula (2.55) to combine samples with spacings $T_s < 1/2B$. If the signal has a duration of T s, it is then specified by $2BT$ samples and the original signal's energy is related to the squares of sample values by (2.69).

The next section examines computational problems which arise when computing a discrete Fourier transform from sample values. One application of such numerical transforms is in processing closely spaced samples for determination of a

† Schwartz, op. cit., p. 598.

bandlimit B above which $X(\omega)$ is zero. Most often a precise, well-defined bandlimit does not show up. It is more common for the magnitude spectrum $|X(\omega)|$ to approach zero only asymptotically as ω approaches π/Δ. In such cases, an approximate B for the signal is defined as a frequency above which the spectral magnitude is very small. The decision of what constitutes small values must be based on experience with applications of the approximations to data reconstruction and processing problems.

2.6 NUMERICAL FOURIER TRANSFORMS

Now that frequency-domain descriptions have been defined for various kinds of signals, we can examine procedures used to compute such a frequency description from numerical signal data.

Let us recall first the several frequency models:

1. Fourier series, for an analog signal in an interval $0 \leq t \leq T$
2. Discrete transform, for a discrete signal which vanishes for large times
3. Continuous Fourier transform, for analog signals which vanish for large times

An important addition to this list, discussed in more detail later, is

4. Power spectral density, for stationary random signals (analog and discrete)

Some average properties of random signals can be related to rates of fluctuation by the power spectral density (or power spectrum). For example, the vertical motion of a passenger riding in a light private plane taxiing on a rough runway would on the average have more rapid fluctuation than the motion of a passenger riding at the same speed in a large commercial plane. The runway roughness and airplane bouncing are so complicated that it is often more convenient to describe average properties of these kinds of signals than to concentrate in detail on a single sample.

No matter which of the above frequency descriptions is wanted, the mechanics of processing numerical data will be the same. The computer takes in a sequence of numbers and generates numbers corresponding to the strength of several frequency components. For example, the discrete Fourier transform is defined as a sum of discrete numbers. Although the Fourier series coefficients are defined by integrals, a computer can only approximate these integrals by sums which look like discrete Fourier transforms. Some examples of this kind of spectral analysis will be given before examining computational details.

Figure 2.33 shows data on the number of international airline passengers per month (entering and leaving the United States) with the monthly figures connected by straight lines. Airline managers would like to be able to predict future demands based on an up-to-date version of such information.† A first reasonable step might be

† Data taken from R. G. Brown, "Smoothing, Forecasting and Prediction of Discrete Time Series," Prentice-Hall, Englewood Cliffs, N.J., 1962.

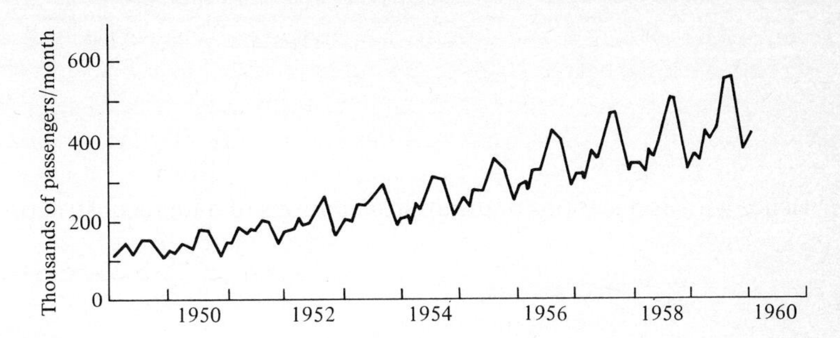

FIGURE 2.33
Airline passengers. (*From R. G. Brown, "Smoothing, Forecasting and Prediction of Discrete Time Series," Prentice-Hall, 1962, by permission.*)

to express the growing trend in these data by a least-squares fit straight line

$$\hat{x}_n^{(1)} = c_0 + c_1 n \qquad \begin{array}{l} n \text{ in months} \\ \text{January } 1949 = 0 \end{array} \tag{2.70}$$

using the methods of Sec. 2.2 for the available data (say N months). Substitution of larger values of $n > N$ would allow prediction of annual totals but would not indicate the more refined monthly variations.

A more precise analytic representation of the data might include some sinusoidal terms

$$\hat{x}_n^{(2)} = c_0 + c_1 n + a_1 \cos un + b_1 \sin un \tag{2.71}$$

Two methods come to mind for selecting a reasonable frequency u for the periodic term proposed in (2.71). The distance between the peaks in Fig. 2.33, as well as our knowledge of vacation habits, suggest $u = 2\pi/12$ rad/month. Another method, which is more useful when the data and auxiliary information are less revealing, is to examine a plot of $|\tilde{X}(\omega)|$, the magnitude of the discrete Fourier transform of $\tilde{x}_n$, the residual after the least-squares linear trend $\hat{x}_n^{(1)}$ has been subtracted from the raw data x_n:

$$\tilde{x}_n = x_n - \hat{x}_n^{(1)} \qquad n = 0, 1, \ldots, N-1 \tag{2.72}$$

The following calculation shows that the spectrum $|\tilde{X}(\omega)|$ of the N residuals $\tilde{x}_n$, defined in Sec. 2.5 by

$$\tilde{X}(\omega) = \sum_{n=0}^{N-1} \tilde{x}_n e^{-jn\omega T_s} \tag{2.73}$$

will have peaks at the dominant frequencies in $\tilde{x}_n$. In this example we have N data samples with a sampling interval T_s of 1 month. If we assume that $\tilde{x}_n$ is truly a cosine at frequency u, the spectrum at $\omega = u$ is

$$\tilde{X}(u) = \sum_{n=0}^{N-1} \frac{e^{jnu} + e^{-jnu}}{2} e^{-jnu} = \frac{N}{2} + \frac{1 - e^{-2jNu}}{1 - e^{-2ju}} \cdot \frac{1}{2} \tag{2.74}$$

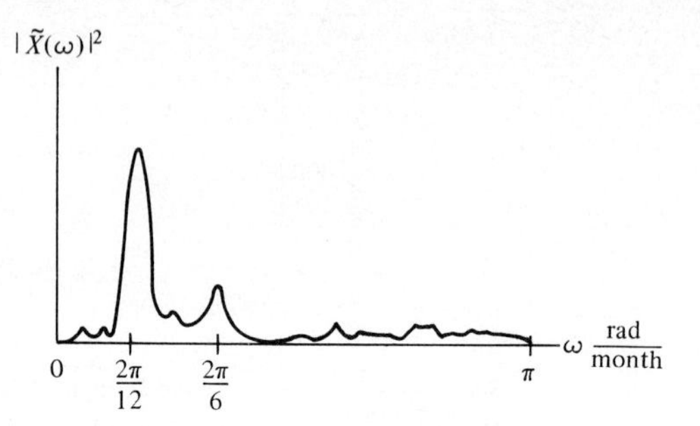

FIGURE 2.34
Amplitude spectrum for $\tilde{x}_n$.

The second term, found from the finite-geometric-sum formula, remains bounded for large N, but the first term gets very large as the number of data samples increases. For comparison, when looking at the frequency $\omega = 2u$,

$$\tilde{X}(2u) = \frac{1 - e^{-jNu}}{1 - e^{-ju}} + \frac{1 - e^{-3jNu}}{1 - e^{-3ju}} = e^{-j(N-1)u/2} \frac{\sin(Nu/2)}{\sin(u/2)}$$

$$+ e^{-j(N-1)3u/2} \frac{\sin(3Nu/2)}{\sin(3u/2)} \tag{2.75}$$

in which both terms are like the bounded second term in (2.74). Similar calculations in other frequencies yield bounded $\tilde{X}(\omega)$ values, and so the discrete Fourier transform of N data samples will have a peak at the frequency of a sinusoidal data component, for sufficiently large N. Figure 2.34 shows what the plot of $|\tilde{X}(\omega)|^2$ looks like for $\tilde{x}_n$ corresponding to the data in Fig. 2.33 after the linear trend has been removed. The largest peak corresponds to a period of 12 months, corroborating the choice of $u = 2\pi/12$ in the trend model of (2.71). The secondary peak corresponding to 6-month cycles suggests that an even better model might include an additional sinusoid at $v = 2\pi/6$ rad/month. Further refinement might allow growing amplitudes for the sinusoids, e.g.,

$$(a_1 + a_2 n) \cos \frac{n2\pi}{12}$$

in which the additional a_2 term causes no special difficulties in the least-squares approach to the estimation of the coefficients.

This example of airline-passenger counts has shown that the discrete Fourier transform of the data can reveal concomitant periodicities in the data. Once these are uncovered, amplitudes of sine and cosine terms can be fitted, and the resulting functions, as in (2.71), can be used to predict future passenger levels.

Another example of data with significant periodicities is drawn from the field of hydrology.† Figure 2.35 shows a spectrum of biochemical oxygen demand (BOD) for measurements taken at a location in the Potomac River estuary. BOD is a pollution measure which is seen to have a 12-h oscillation related to tidal action and a 24-h periodicity related to daily sewage-flow variations. If an upstream industry had natural (or imposed) variations in waste flow with a different period, appearance of a

† C. C. Kisiel, Time Series Analysis of Hydrologic Data, *Adv. Hydrosci.*, 1969.

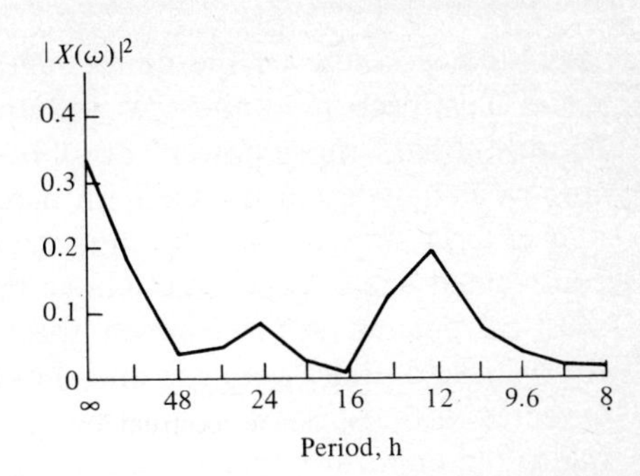

FIGURE 2.35
BOD spectrum. (*From C. C. Kisiel, Time Series Analysis of Hydrologic Data,* in *Adv. Hydrosci.*, **5**, 1969, *by permission.*)

peak at that frequency at a downstream testing point could be evidence that that industry produced significant pollution. Notice the shorthand scheme in Fig. 2.35, where frequencies are described indirectly by their periods. This saves writing all the 2π's like those in Fig. 2.34. (Why are the intervals between labeled periods different?)

Discrete Fourier transforms can also be used to establish differences between two sets of data. Figure 2.36 describes measurements of aircraft-runway elevations which can be used to decide when a runway must be resurfaced. The two undulating curves are versions of $|X(\omega)|^2$ which have been modified by methods (described

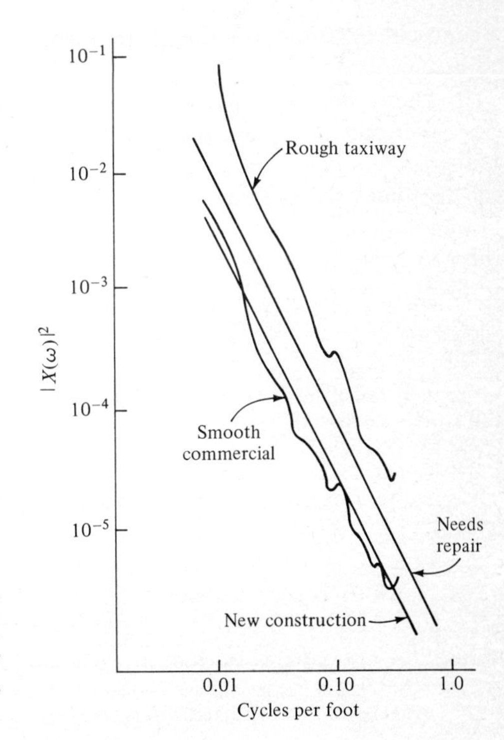

FIGURE 2.36
Runway roughness spectra. (*From G. M. Jenkins and D. G. Watts, "Spectral Analysis and Its Applications," Holden-Day, 1968, by permission.*)

later) that account for the finite length of the data record. Note that logarithmic scales have been used and that the upper curve corresponds to a rougher runway because it has "more power" at all frequencies. The choice of scales displays these spectra as having linear trends (in the frequency domain) and suggests the straight-line criteria shown there for unacceptably rough runways and acceptable newly constructed ones. A spectrum above the upper straight line means that the runway must be repaired. A new runway spectrum above the lower straight line means that the contractor did a poor job and should not be paid. What characteristic should be expected in the spectrum of a runway so worn that undulations appear corresponding to the uniformly spaced steel reinforcing rods?

Use of spectra for comparison purposes is widespread. In the medical field, attempts have been made to distinguish healthy vs. sick hearts and brains by using EKG's and EEG's. Data processing will not substitute for a trained physician, but detection of spectral peaks in routine automated checkups might help catch problems before other symptoms are evident. (See also Sec. 8 in Chap. 4.)

Another example is the use of sound spectra to identify the person whose spoken voice has been recorded. Although results are not yet precise enough for legal purposes, much work has been applied to this problem.† Figure 2.37 shows short-time spectra of human speech and a computer-controlled speaking machine. In this figure the amplitude of a spectral component is indicated by the blackness of the point.‡ Each vertical strip is the spectrum of a short segment of sound data, with frequency increasing vertically. Notice that the hissing "s" sound has considerable strength over a wide band of high frequencies, while other sounds are generally composed of three dominant frequency bands whose center frequencies are adjustable.¶

The reader may have noticed that in the foregoing examples, only the airline-passenger one included a plot of the actual data. Our reluctance to include pages of numerical data or graphs of highly irregular fluctuating curves is itself a testament to the usefulness of spectra for summarizing some data characteristics. It is worth noting that as more data become available, a new curve $|X(\omega)|^2$ still is contained in the 0 to $2\pi/T_s$ frequency range—in contrast to the increased number of pages needed to display the data. The spectrum is especially useful when the data are particularly random and have no apparent pattern. The spectrum serves to bring method out of madness.

Many practical problems and theoretical interpretations arise when numerical data are processed to get spectra like those shown above. At first glance, it seems straightforward to simply carry out the summation

$$X(\omega) = \sum_{n=0}^{N-1} x_n e^{-j\omega n} = \sum_{0}^{N-1} x_n \cos n\omega - j \sum_{0}^{N-1} x_n \sin n\omega \tag{2.76}$$

† See also A. Solzhenitsyn, "The First Circle," Harper & Row, New York, 1968, a novel about scientists forced to work on voice-recognition devices.

‡ J. L. Flanagan, The Synthesis of Speech, *Sci. Am.*, vol. 228, no. 2, p. 49, Feb. 1972.

¶ For further information, see Special Issue on Speech Communication and Processing, *IEEE Trans. Audio Electroacoust.*, vol. 21, no. 3, June 1973.

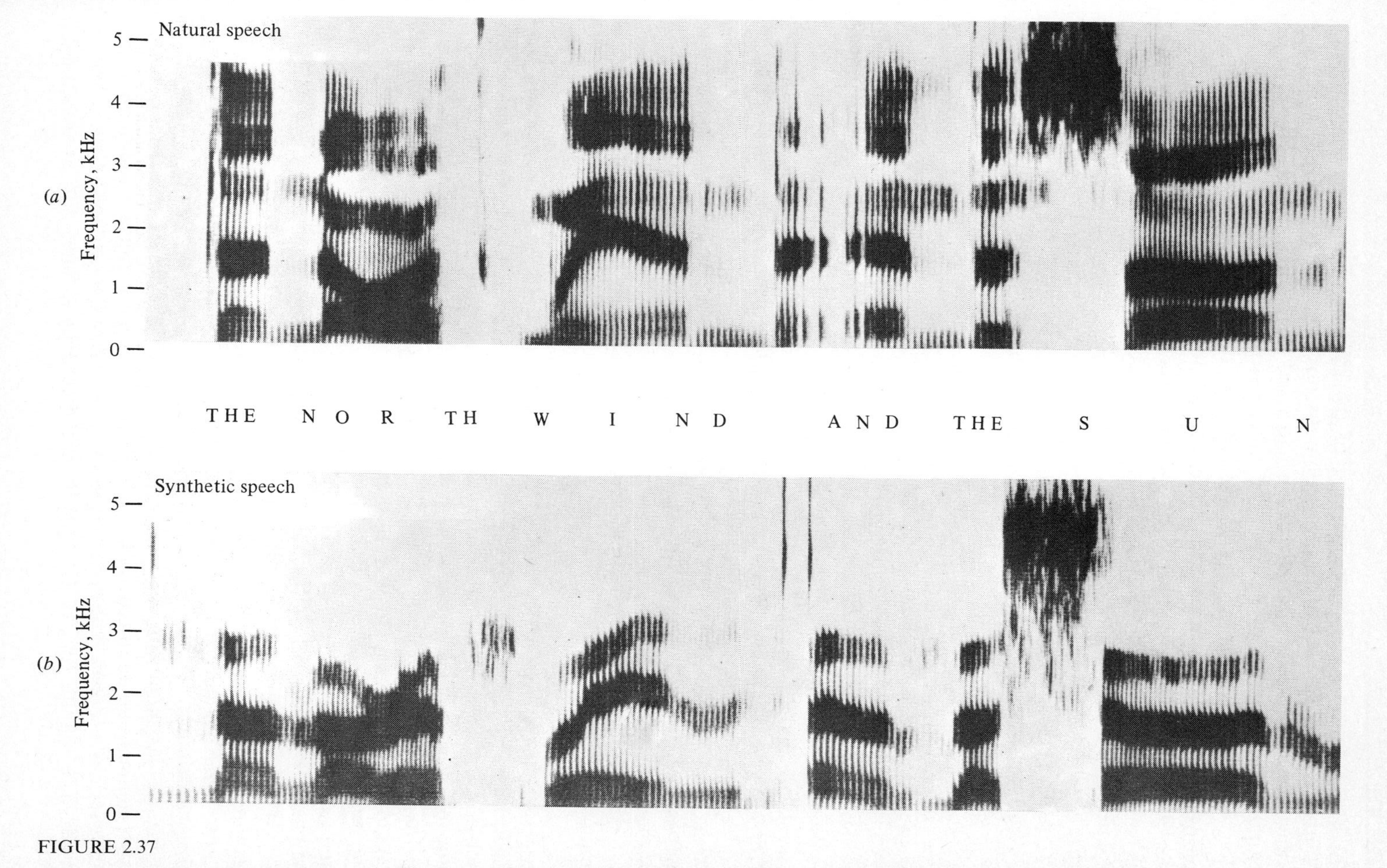

FIGURE 2.37

Spectrogram of a spoken phrase. (*From J. L. Flanagan, The Synthesis of Speech, Sci. Am.*, vol. 228, no. 2, *Feb. 1972*, p. 49, *by permission.*)

and then to form $|X(\omega)|^2$ as the sum of the squares of the sine and cosine summations in (2.76). The following sections discuss such matters as (1) the choice of frequencies at which (2.76) should be evaluated and (2) relationships between $X(\omega_i)$ computed from *long* and *short* records of data from the same source. These considerations are related to the costs of data gathering and computer time. We would prefer to be able to draw reasonable conclusions after a minimal amount of processing of a small amount of data. Sampling theorems and transfer functions will be used to study these topics. Spectral analysis of random signals will be discussed in Chap. 4.

2.7 PRACTICAL CALCULATION OF DISCRETE FOURIER TRANSFORMS

A discrete-time signal x_k consisting of samples

$$\cdots x_{-3}, x_{-2}, x_{-1}, x_0, x_1, x_2 \cdots$$

has a discrete Fourier transform $X(\omega)$ defined by the infinite summation

$$X(\omega) = \sum_{k=-\infty}^{\infty} x_k e^{-jk\omega} \tag{2.77}$$

$X(\omega)$ is a complex-valued periodic function of ω with a period of 2π. Furthermore, our restriction of the x_k to be real quantities results in additional symmetry of $X(\omega)$, and so it is completely characterized by its values in the interval $0 \leq \omega \leq \pi$ rad per sample. Note that it is immaterial whether we use k as in (2.77) or n as in (2.76) for the discrete-time index.

As mentioned earlier, discrete-signal values are often samples taken from a continuous-time function $x(t)$, that is,

$$x_k = x(kT_s) \tag{2.78}$$

in which T_s is the sampling period (time between samples). T_s must be chosen small enough to avoid missing any rapid fluctuations in $x(t)$. When $T_s \neq 1$, Eq. (2.77) is modified by replacing k with (kT_s) in the exponent. This simply makes a scale change in $X(\omega)$ so that its period becomes $2\pi/T_s$, but the general shapes of $|X(\omega)|$ and $\angle X(\omega)$ vs. ω are unchanged. Thus, although the value of T_s is an important consideration when recording data, once numbers have been recorded, the data processing can proceed assuming $T_s = 1$. The final $X(\omega)$ can always be rescaled; therefore, we shall usually assume $T_s = 1$ when examining computational problems.

It is readily apparent that there are at least two difficulties awaiting any man or computer who tries to evaluate $X(\omega)$ from (2.77) for x_k which are given as numbers (rather than by a neat analytical formula). One is that an infinite number of terms must be summed. The other is that $X(\omega)$ is desired over a continuous range of frequencies (an infinite number of ω-values). Even the fastest computer would never finish summing up an infinite number of terms. The infinite range of summation is not a serious problem, however, because any record of data will also have a finite length. The infinite sum of (2.77) will be replaced by a finite sum or truncated

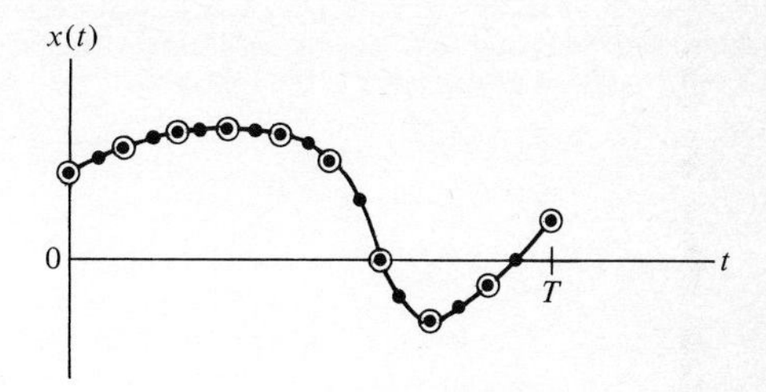

FIGURE 2.38
Comparison of sampling periods.

transform. This change will obviously introduce errors in the discrete Fourier transform $X(\omega)$. Computation time will be reduced, but errors in $X(\omega)$ will increase as fewer samples are taken from a finite-length record. There are thus trade-offs to be considered between the number of terms to be summed in (2.77) and the accuracy of the resulting spectrum (discrete Fourier transform).

The number of summation terms and amount of computer time are also affected by the choice of T_s if we are working with samples of a continuous $x(t)$. While the sampling theorem (or eyeball judgments) yield a *maximum* T_s for which $x(t)$ can be reconstructed from the samples $x(kT_s)$, interpolation will clearly be easier if T_s is much smaller (e.g., the dots vs. the open circles in Fig. 2.38). Moreover, the highest frequency in the signal is not generally a number known with great accuracy, especially before any signal processing has been carried out. For this reason it is desirable to make T_s small. The disadvantage of using a very small T_s is that the same T s of analog data will produce a much larger number $N = 1 + T/T_s$ of discrete-time samples. The best choice of T_s is a compromise between accuracy of representation of $x(t)$ and computer time required to compute the discrete Fourier transform.

As noted above, the second difficulty which arises when we attempt to compute a discrete Fourier transform using (2.77) is that it is desired for a continuum of frequencies. An entire summation is required to get $X(\omega)$ for each value of ω. It follows that in a finite amount of computing time we can evaluate $X(\omega_i)$ only for a finite set of frequencies $\omega_1, \omega_2, \ldots, \omega_m$. How shall we choose these finite frequencies? Obviously, if they are too far apart, the function $X(\omega)$ may vary radically between the frequencies chosen and we lose desired accuracy. The choice of frequencies is related to the resolution of the frequency function $X(\omega)$. We shall show below that a sort of sampling theorem in the frequency domain holds: when a finite number N of samples is summed to find $X(\omega)$ there exists a maximum-frequency spacing or resolution, proportional to $1/N$, such that the continuous $X(\omega)$ can be uniquely interpolated from values at the discrete frequency points. This frequency-domain sampling of a *time-limited* signal (only N nonzero samples) is analogous to our earlier discussion of time-domain sampling of *bandlimited* signals.

Figure 2.39 shows an example of a continuous spectrum and discrete points that might serve as the approximation to it. To demonstrate the appropriate choice of discrete frequencies, say that they are multiples $m\Omega_s$ of some frequency spacing Ω_s. If, in addition, we have only N samples of the data $x_k = x(kT_s)$ available, the discrete Fourier transform of (2.77) becomes, with the sampling period T_s introduced explicitly,

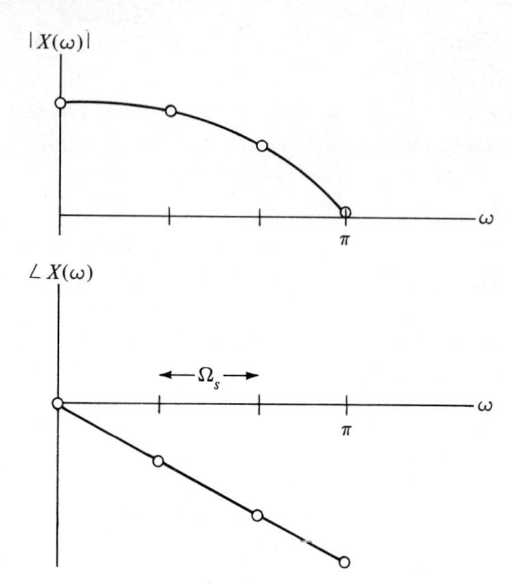

FIGURE 2.39
Discrete-frequency spectra.

$$X(m\Omega_s) = \sum_{k=0}^{N-1} x(kT_s)e^{-jkm\Omega_s T_s} \tag{2.79}$$

How shall we choose the frequency spacing Ω_s? Figure 2.40 shows a sketch of the time-limited signal $x(t)$, defined from $-T/2$ and $T/2$, from which the N samples were taken. [As noted in the figure, $T = (N-1)T_s$.] This symmetrical time interval is chosen so it is parallel with the symmetrical bandlimited spectra which enter the time-sampling theorem. It can be shown that the sampling theorem also applies to time-limited signals, in the sense that values of the continuous Fourier transform $X_c(\omega)$ of $x(t)$ can be interpolated perfectly at all ω from knowledge of values at only the discrete frequencies $m\Omega_s$ *if* the frequency spacing Ω_s is small enough. Thus, by analogy to the T_s determination in Sec. 2.5, the upper bound on the frequency spacing Ω_s can be stated as

$$\frac{1}{\Omega_s} > 2\frac{T/2}{2\pi} \tag{2.80}$$

or

$$\Omega_s < \frac{2\pi}{T} \tag{2.81}$$

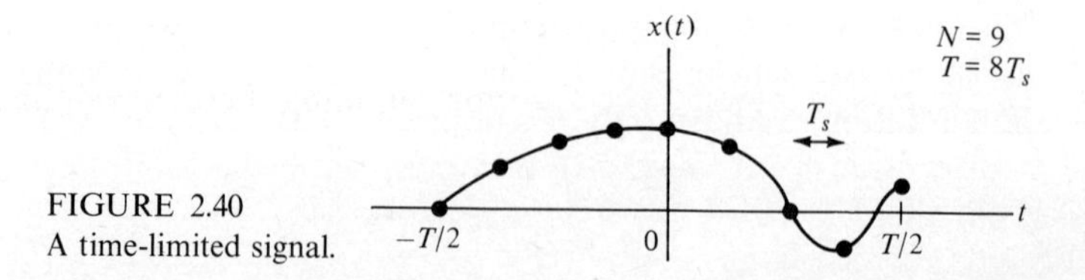

FIGURE 2.40
A time-limited signal.

Thus, if we compute the discrete Fourier transform sum at frequencies spaced by an Ω_s satisfying (2.81), a smooth curve through the resulting points should be a good representation of $X(\omega)$, the discrete Fourier transform of $x(kT_s)$. This conclusion also assumes that the samples $x(kT_s)$ are sampled at a high enough rate for $X(\omega) \approx X_c(\omega)$. A good choice for the total number M of discrete frequencies to be used will be explained below.

To summarize the foregoing arguments, a computational evaluation of the discrete Fourier transform in (2.77) will use the summations

$$X(m\Omega_s) = \sum_{k=0}^{N-1} x(kT_s)e^{-jkT_s m\Omega_s} \qquad m = 0, 1, 2, \ldots, M - 1 \tag{2.82}$$

These summations each use N data samples at spacings T_s and produce discrete Fourier transform values (complex numbers) at M frequencies with spacing Ω_s. It is useful to choose Ω_s in relation to T_s to simplify the exponents in (2.82) while still satisfying the sampling-theorem criterion in (2.81). The duration of $x(t)$ between $x(0)$ and $x(T)$ is $(N - 1)T_s$. Thus

$$\frac{2\pi}{T} = \frac{2\pi}{(N-1)T_s}$$

and an adequate frequency spacing is [see (2.81)]

$$\Omega_s = \frac{2\pi}{NT_s} \tag{2.83}$$

Substitution of (2.83) into (2.82), along with the subscript notation for the discrete functions $[X(m\Omega_s) = X_m, x(kT_s) = x_k]$, finally yields the simple form

$$X_m = \sum_{k=0}^{N-1} x_k(e^{-j2\pi/N})^{km} \qquad m = 0, 1, \ldots N - 1 \tag{2.84}$$

This summation weights each of the N data samples by an integer power of a complex exponential which has a magnitude of unity and an angle of $2\pi/N$.

When discrete-frequency values of a discrete Fourier transform are computed according to (2.84), with Ω_s and T_s related as in (2.83), we say that the resulting X_m form a *finite Fourier transform* of the data x_k. Further special properties of the finite Fourier transform will be developed later. Notice that T_s and Ω_s do not appear in these finite Fourier transform expressions. Once these parameters have been chosen appropriately, they enter the analysis only as scale factors. Comparison of (2.84) and (2.82) also shows that the highest frequency at which $X(\omega)$ is computed has been chosen to be

$$(N-1)\Omega_s = \frac{N-1}{N}\frac{2\pi}{T_s}$$

This is reasonable because the time-sampling theorem requires that

$$\omega_{\max} < \frac{2\pi}{T_s} \tag{2.85}$$

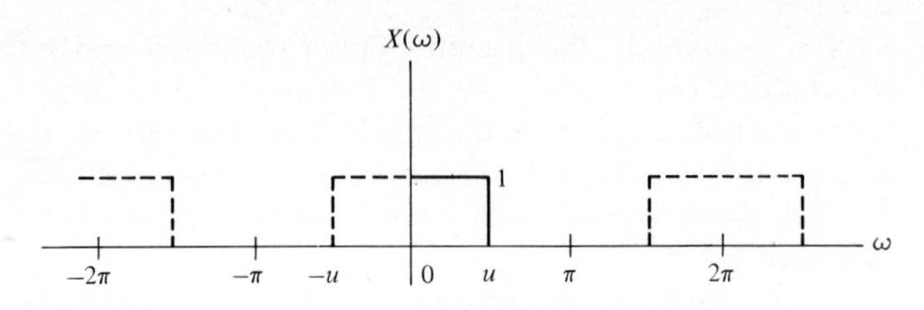

FIGURE 2.41
Simple discrete Fourier transform.

so that the discrete Fourier transform should be approximately zero at higher frequencies.

[The justification for (2.84) has effectively assumed that $x(t)$ can be time-limited to the interval $-T/2 < t < T/2$ and bandlimited so that its continuous Fourier transform is nonzero only for $-\omega_{\max} < \omega < \omega_{\max}$. Strictly speaking, no analog function can be both time-limited and bandlimited. However, a time-limited function can be effectively bandlimited if its transform is very small for frequencies higher than an $\omega_{\max}$.]

We now look at a simple example of discrete Fourier transform computation in which the finite-time and discrete-frequency sums in (2.84) are applied to a signal whose discrete Fourier transform is known analytically. Figure 2.41 shows the known discrete Fourier transform which is bandlimited between $-u$ and u ($u < \pi$). The solid portion of the curve is the part we usually graph, but the symmetrical and periodic replicas are also shown dotted because the inverse transform requires integration over one full period of $X(\omega)$.

The corresponding discrete signal is easily computed, as follows:

$$x_n = \frac{1}{2\pi}\int_{-\pi}^{\pi} X(\omega)e^{j\omega n}\, d\omega = \frac{1}{2\pi}\int_{-u}^{u} e^{j\omega n}\, d\omega$$
$$x_n = \frac{u}{\pi}\frac{\sin nu}{nu} \tag{2.86}$$

The x_n of (2.86) are shown in Fig. 2.42 for the case when the highest frequency is $u = \pi/2$. [Note that $(\sin t)/t$ approaches 1 as t approaches 0, so that $x_0 = \frac{1}{2}$, as shown in the figure.]

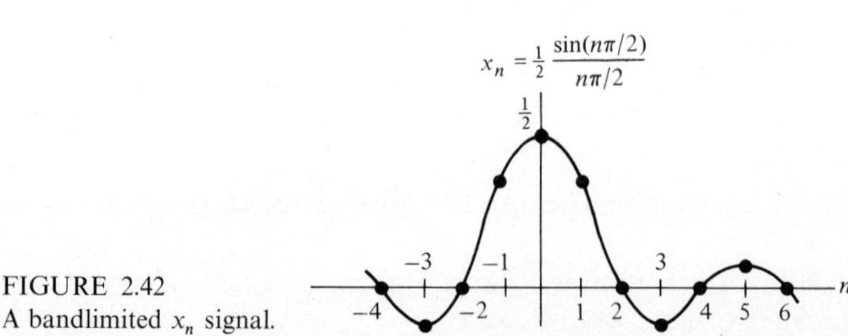

FIGURE 2.42
A bandlimited x_n signal.

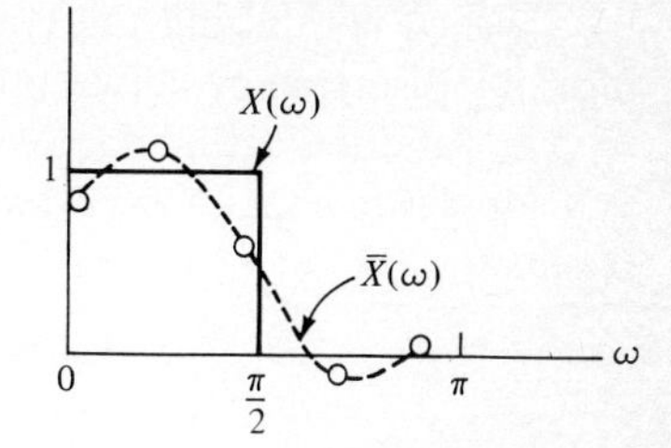

FIGURE 2.43
Frequency spreading due to truncation.

Let us now consider computer evaluation of an approximate discrete Fourier transform from a finite set of nine samples.

$$x_{-4}, x_{-3}, \ldots, x_0, x_1, \ldots, x_3, x_4$$

The discrete Fourier transform of these nine samples is

$$\bar{X}(\omega) = \sum_{-4}^{4} \frac{\sin(n\pi/2)}{n\pi} e^{-jn\omega} \tag{2.87}$$

where the overbar is used to distinguish this $\bar{X}(\omega)$ from the one plotted in Fig. 2.41 and corresponding to an infinite-sum discrete Fourier transform. Figure 2.43 compares $X(\omega)$ and $\bar{X}(\omega)$ to show the effect of truncating the discrete Fourier transform sum to a finite number of terms. The time-limited set of samples has a spectrum which is not limited to frequencies below $\pi/2$ rad/s. (Both x_n and the nine terms in the truncation are even functions of n, and so their transforms are real and no separate phase plots are necessary.)

Computer evaluation of the finite Fourier transform points $\bar{X}(m\Omega_s)$ and the $\bar{X}(\omega)$ curve could use (2.84) with

$$N = 9 \qquad T_s = 1 \qquad \Omega_s = \frac{2\pi}{9} = 0.697 \tag{2.88}$$

The five circles in Fig. 2.43 indicate the resulting $\bar{X}(m\Omega_s)$ values for $m = 0, \ldots, 4$. The other four points are not shown because they are symmetrical values located in the $-\pi < \omega < 0$ interval. Note that the discrete-frequency points do interpolate nicely to provide a good representation of the continuous $\bar{X}(\omega)$. However, $\bar{X}(\omega)$ differs considerably from $X(\omega)$ because of the truncation to nine samples. Use of a larger set of x_n samples, say 21 (for $n = -10$ to $n = 10$), would produce $\bar{\bar{X}}(m\Omega_s')$ which differed in two ways from $\bar{X}(m\Omega_s)$. First, the nonsampled discrete Fourier transform $\bar{\bar{X}}(\omega)$ would be more like the bandlimited $X(\omega)$. Second, the discrete values of the finite Fourier transform $\bar{\bar{X}}(m\Omega_s')$ will be more closely spaced points on $\bar{\bar{X}}(\omega)$ since

$$\Omega_s' = \frac{2\pi}{21} < \Omega_s = \frac{2\pi}{9}$$

Computer calculation of this 21-point discrete Fourier transform is left as a problem at the end of this chapter.

When computing the circle points in Fig. 2.43 we must be careful about the indexing discrepancy between the finite Fourier transform sum in (2.84), which starts with x_0, and the actual data, which start with x_{-4}. If we start with the $\bar{X}(\omega)$ in (2.87) and substitute $m\Omega_s$ for ω and $k = n + 4$ for n, the following finite Fourier transform results:

$$\begin{aligned}\bar{X}_m &= \sum_{k=0}^{8} \frac{\sin[(k-4)\pi/2]}{(k-4)\pi}(e^{-j2\pi/9})^{(k-4)m} \\ &= (e^{j2\pi/9})^{4m} \sum_{k=0}^{8} \frac{\sin[(k-4)\pi/2]}{(k-4)\pi}(e^{-j2\pi/9})^{mk}\end{aligned} \tag{2.89}$$

This expression differs from what would result from direct, blind substitution of the nine data values into the standard finite Fourier transform sum of (2.84). The latter would produce only the summation term in (2.89), without the linear-phase factor to the left of the sum. If a standard subroutine is available to compute (2.84), a few extra instructions can include the extra phase factor after the summations have been completed. The general phase factor is $(e^{-j2\pi/N})^{sm}$, in which s is the smallest data index (-4 in this example).

Evaluation of the finite Fourier transform summations requires many calculations of sine and cosine functions since

$$(e^{-j2\pi/9})^{mk} = \cos\left(mk\frac{2\pi}{9}\right) - j\sin\left(mk\frac{2\pi}{9}\right) \tag{2.90}$$

as well as many multiplications and additions of complex numbers. Most computer function-generator operations for $\sin\theta$ or $\cos\theta$ are fairly time-consuming because they sum several terms of power-series expansions for these trigonometric functions. Efficient finite Fourier transform routines make use of the fact that the angles needed are all integer multiples of $2\pi/N$. Once $\sin(2\pi/N)$ and $\cos(2\pi/N)$ have been computed by power series, it is faster to compute other sines and cosines by multiple-angle identities, e.g.,

$$\sin 2\theta = 2\sin\theta\cos\theta$$

$$\cos 2\theta = \cos^2\theta - \sin^2\theta$$

Further efficiency can be introduced by noting that in the overall evaluation of (2.89) for $m = 0, 1, \ldots, 8$, certain trigonometric functions occur repeatedly. For example, $\cos[4(2\pi/9)]$ occurs when $m = 1$, $k = 4$; $m = 2$, $k = 2$; and $m = 4$, $k = 1$. The so-called fast Fourier transform programs work on all N summations at the same time, so that these repeated trigonometric functions are calculated only once. The fast Fourier transform algorithms also use multiple-angle formulas and other clever organization of the calculations with the result that they can compute discrete Fourier transforms for large sets of data points very fast—and much faster than any direct program we might write. For example, with $N = 2^{13} = 8192$ points, fast Fourier transforms might take about 0.10 min (for all 2^{13} frequencies) while an unsophisticated program would take 80 min or more. The efficient organization gives a time-reduction factor of about $(\log_2 N)/N$.

Many fast Fourier transform algorithms also require the number of data points N to be a power of 2. These algorithms can be used when the actual number of points is not a power of 2 simply by adding enough additional hypothetical data points with values of zero until the augmented set has $N = 2^b$ points for some integer b. For example, the nine-point $\bar{X}(\omega)$ considered earlier could be changed to a 16-point fast Fourier transform problem by adding seven zero points at the end of the data string. The data used would then be

$$0, -\frac{1}{3\pi}, 0, \frac{1}{\pi}, \frac{1}{2}, \frac{1}{\pi}, 0, -\frac{1}{3\pi}, 0, 0, 0, 0, 0, 0, 0$$

The effects of adding zeros to the end of the data string can be described quite simply. With x_n representing the original data and x'_n the augmented data, it is clear that

$$X(\omega) = \sum_{-4}^{4} x_n e^{-jn\omega} = \sum_{-4}^{11} x'_n e^{-jn\omega} = X'(\omega) \tag{2.91}$$

That is, both sequences have the same discrete Fourier transform. However, the corresponding *finite* Fourier transforms will have different frequency spacings

$$\Omega_s = \frac{2\pi}{9} > \frac{2\pi}{16} = \Omega'_s \tag{2.92}$$

Thus, the net effect of the added zeros is to produce a finite Fourier transform with more points (nine between 0 and π rad) at a smaller frequency spacing. These extra points are actually beneficial because they lie on the same $X(\omega)$ curve and make it easier to sketch in the continuous curve. [In general, both $|X(\omega)|$ and $\angle X(\omega)$ curves will result. Special symmetry properties in previous examples gave us simpler real $X(\omega)$ values without the need for an extra $\angle X(\omega)$ plot.]

As a simple example of the effect of adding zeros, consider a discrete signal x_k whose first three samples are

$$x_0 = -\tfrac{1}{2} \qquad x_1 = 1 \qquad x_2 = -\tfrac{1}{2}$$

It is left to the reader to show that the discrete Fourier transform of this x_k is

$$X(\omega) = e^{-j\omega}(1 - \cos \omega) \tag{2.93}$$

Both x_k and $|X(\omega)|$ are shown in Fig. 2.44. Although the true x_k might have other nonzero samples, they are not available to us and computation of this $X(\omega)$ assumes that they are all zero.

A discrete transform could be computed from (2.84) with $N = 3$ and $\Omega_s = 2\pi/3$. The result must be the same as (2.93) with $\omega_m = m2\pi/3$, as indicated by circles in Fig. 2.44 for $m = 0, 1, 2$. Addition of five zeros to the end of the data

$$x_3 = x_4 = \cdots x_7 = 0$$

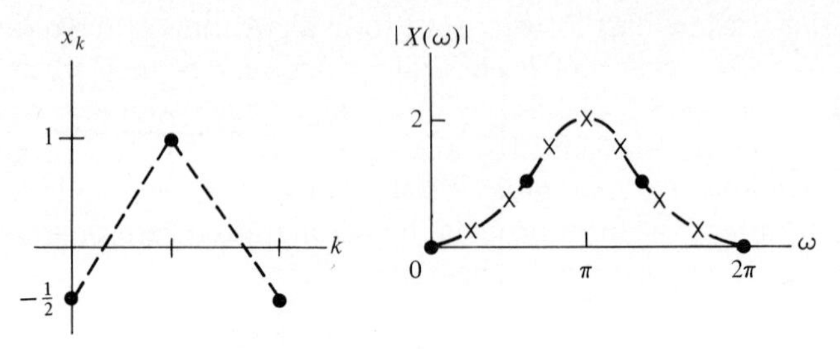

FIGURE 2.44
Signal and its Fourier transform with zeros added.

for a total of eight data samples (2^3), produces a more refined discrete Fourier transform

$$X_m = e^{-2\pi jm/8}\left(1 - \cos\frac{2\pi m}{8}\right)$$

indicated by crosses in the figure. Note that the added data zeros produce more points on the $|X(\omega)|$ curve, making interpolation simpler.

The *finite Fourier transform* sequence

$$X_m = \sum_{k=0}^{N-1} x_k(e^{-j2\pi/N})^{km} \qquad m = 0, 1, \ldots, N-1 \tag{2.94}$$

has been developed as a reasonable way to compute some discrete frequency points $X_m = X(m\Omega_s)$ on the discrete Fourier transform curve

$$X(\omega) = \sum_{0}^{N-1} x_n e^{-jn\omega} \tag{2.95}$$

The frequency spacing $\Omega_s = 2\pi/NT_s$ was justified by relating the duration of the time-limited analog $x(t)$ to the sampling theorem. Similarly the highest-frequency point $(N-1)\Omega_s$ was related to the sampling period T_s and the highest frequency in $x(t)$.

Although the finite Fourier transform was introduced in this way, it has useful properties in its own right. The complex numbers $X_0, X_1, \ldots, X_{N-1}$ truly constitute a transformation of the signal samples $x_0, x_1, \ldots, x_{N-1}$ because the x_n values can be recovered from the X_m values. This transform-pair relationship is a bit surprising because we know that the x_n can also be recovered from the continuous discrete Fourier transform $X(\omega)$ by the integrals

$$x_n = \frac{1}{2\pi}\int_0^{2\pi} X(\omega)e^{jn\omega}\,d\omega \tag{2.96}$$

Now we shall say that $X(\omega)$ is needed at only N frequencies to calculate the time sequence via the *inverse finite Fourier transform* summation

$$x_n = \frac{1}{N}\sum_{m=0}^{N-1} X_m(e^{j2\pi/N})^{nm} \qquad n = 0, 1, \ldots, N-1 \tag{2.97}$$

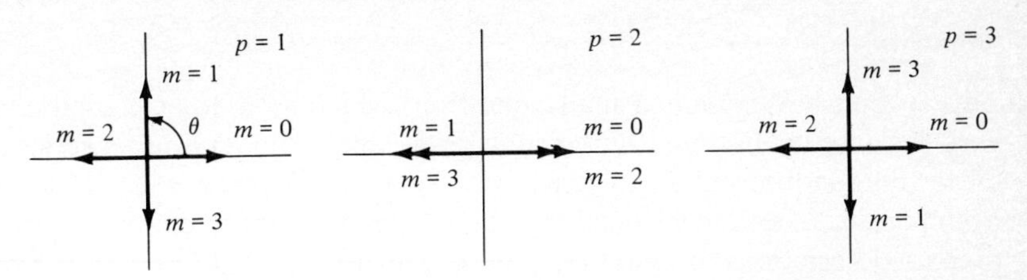

FIGURE 2.45
Cancellation of vectors.

This inverse transformation can be proved by substituting the finite Fourier transform in (2.94) into the summation of (2.97) to get

$$\frac{1}{N}\sum_{m=0}^{N-1}\sum_{k=0}^{N-1} x_k(e^{-j2\pi/N})^{km}(e^{j2\pi/N})^{nm} \tag{2.98}$$

The double sum in (2.98) reduces to x_n with the aid of the special properties

$$\sum_{m=0}^{N-1}(e^{j2\pi/N})^{(n-k)m} = \begin{cases} N & \text{if } n = k \\ 0 & \text{otherwise} \end{cases} \tag{2.99}$$

The $n = k$ case is clear, because all N summands are then unity. When $n - k = p$, a nonzero integer, the sum

$$\sum_{m=0}^{N-1}(e^{j2\pi/N})^{pm}$$

can be thought of as a sum of N equally spaced unit vectors, as shown in Fig. 2.45 for $N = 4$ and $\theta = 2\pi/N = \pi/2$. These vectors always sum to zero, as shown in the figure. The vanishing of such sums can be proved in general using the finite geometric sum formula. Thus (2.99) is true, and (2.98) equals x_n, as stated in (2.97).

The finite Fourier transform pair in (2.94) and (2.97) is the simplest Fourier transform we have encountered. In fact it is the only transform pair whose properties were actually *proved* here. We assumed that the Fourier series for periodic functions had been proved in a prior mathematics course (or text) and showed how the discrete Fourier transform pair was really a special case of a Fourier series, with the roles of time and frequency domains interchanged.

We return now to the apparent mystery concerning the information in $X(\omega)$ at frequencies other than $\omega_m = m\Omega_s$. The values at the latter finite set of frequencies suffice to compute the x_n from (2.97), while values at all frequencies are required to get the same x_n from (2.96). The secret lies in the modifying expression ($n = 0, 1, \ldots, N - 1$) in (2.97). That summation gives x_n at *only* N sample times (and repeats periodically for other values of N). On the other hand, (2.96) gives the same x_n for the original N sample times, but it also gives $x_n = 0$ for the infinity of other n-values. The discrete Fourier transform inversion integral defines a discrete-time signal of infinite extent while the finite Fourier transform inversion defines only an N-sample signal.

Summary

This section has examined many questions which arise when a digital computer is used to compute discrete Fourier transforms, approximate Fourier series coefficients, or, for the random signals to be described in Chap. 3, power spectral densities. Many examples have been mentioned to explain why we want to compute such frequency-domain descriptions of signals, e.g., finding an approximate bandlimit B on the range of frequencies over which the transform has significant amplitude, in order to sample at an appropriate spacing; reduction of complicated time signals to a few numbers representing Fourier series coefficients, for efficient transmission; and signal classification like the distinguishing between different people on the basis of the peaks in transforms of samples of their recorded speech.

We have spoken about the computational problems related to the choice of sample spacing for transformation of continuous-time signals and the analogous problem of choosing a frequency spacing separating the frequencies at which the transform of a continuous or discrete signal will be computed. The relation between the transforms of a long time signal and the transform of a truncated portion of that signal was also introduced.

These computational ideas are further developed in the problems for this chapter and in the spectral-analysis discussions of Chap. 4.

The following section concludes this chapter with an introduction to transfer functions which describe how some signal processors operate in terms of the relation between the Fourier transforms of the processor's input signal and output signal. Transfer functions are useful in many practical applications. For example, how an airplane's dynamics transmits disturbance forces (runway irregularities or wind gusts) to the body of a passenger can be summarized in a transfer function.

2.8 TRANSFER FUNCTIONS: FREQUENCY RESPONSE

We have seen how discrete-time signals can be transformed to the frequency domain and then classified according to the range of frequencies where their magnitude spectra are biggest. We shall now see how those frequency transformations can be used to describe the behavior of signal processors, or *filters*. The frequency characterization of a filter is called its *transfer function*. Transfer functions can be used to analyze both the filters used to manipulate numerical signals and the physical systems which connect physical signals, e.g., forces and velocities. An example of the latter would be the way wind-gust forces (as input signal) cause disturbance motions of a radar antenna (resulting in an error in the indicated position of a detected airplane).

Although transfer functions have numerous engineering applications, we shall use them mainly to analyze errors in the spectral-analysis processors of Chap. 4 and as building blocks for shaping signal transforms. We shall also see in Chap. 3 that the spectral descriptions of random signal inputs and outputs of a filter can be related by the filter's transfer function.

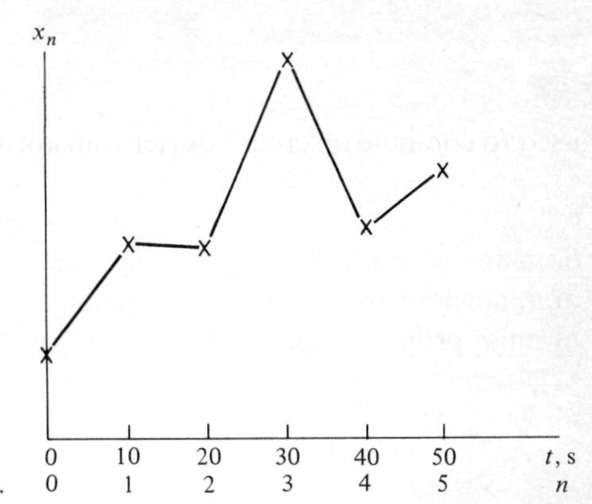

FIGURE 2.46
Smooth signal plus rough noise.

We shall use a very simplified example to introduce filters and transfer functions. Chapter 7 will describe realistic filtering problems for estimating the position and velocity of an airplane based on the sequence of range and angle measurements obtained from successive scans of a radar antenna (say, once every 10 s). A simplified version of that problem is to consider the sequence of angle measurements x_n

$$x_n = s_n + d_n \qquad n = 1, 2, 3, \ldots \tag{2.100}$$

in which the signal s_n is the true angle during scan n and d_n is the disturbance angle between the observed antenna shaft angle x_n and the actual antenna angle s_n caused by wind twisting the antenna.

We would like to determine s_n from measured values of $x_1, x_2, \ldots, x_n$. This can be achieved approximately by reducing the effect of d_n. Often the disturbance is more irregular (has more high-frequency content) than the desired signal. The inertia of an airplane and known limits on airplane accelerations will restrict the rate of actual angle changes between scans, while rapid angular changes can be caused by local wind gusts with very brief durations. Figure 2.46 shows an example of this sort of signal plus disturbance.

In order to discuss quantitative properties of filtering we shall choose examples in which the signals and disturbances have simple discrete Fourier transforms. Figure 2.47 shows a smooth desired signal

$$s_n = A[n(\tfrac{9}{10})^n] \tag{2.101}$$

and a comparatively rough disturbance signal

$$d_n = B(-\tfrac{9}{10})^n \tag{2.102}$$

In a more realistic radar-filtering problem we shall not have explicit expressions like these but only information like the relative frequency content of the desired and disturbance signals. However, choosing these explicit functions makes it possible to compute the exact response of a filter. There is still a bit of realistic uncertainty in our simplified problem because the desired-signal amplitude A and the disturbance

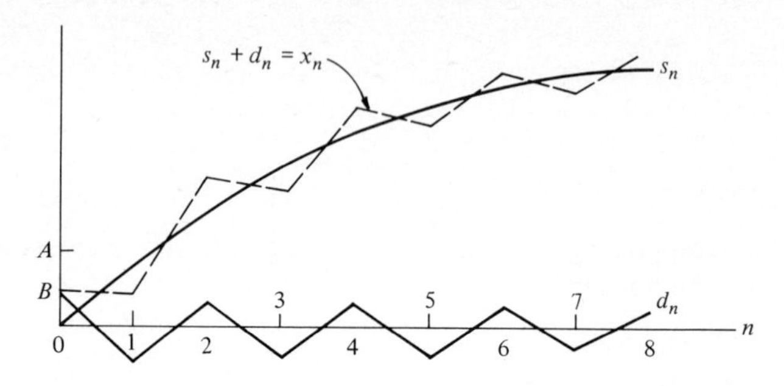

FIGURE 2.47
Signal and disturbance from (2.105) and (2.106).

amplitude B are unknown. Thus, a meaningful problem is to choose a filter which computes an output y_n from present and previous inputs $x_1, x_2, \ldots, x_n$ such that y_n is a good approximation to the desired signal s_n.

Intuitively, we might look at Fig. 2.47 and draw a smooth curve through the jagged x_n curve as an approximation to the smooth signal component. This suggests a filter which averages out the disturbance fluctuations. The simplest averaging filter is defined by the input-output relation

$$y_n = \frac{x_n + x_{n-1}}{2} \tag{2.103}$$

The output at time n is the average of the present and preceding input. We expect the filter to cancel successive noise samples (which are of opposite sign and almost equal magnitude) while causing little distortion of the desired signal because its successive values are nearly equal.

These hunches about the properties of the averager can be checked by substituting the signal functions into the filter definition (2.103), with the result

$$y_n = \frac{A}{2}[n(0.9)^n + (n-1)(0.9)^{n-1}] + \frac{B}{2}[(-0.9)^n + (-0.9)^{n-1}] \tag{2.104}$$

A little manipulation puts this into the form

$$y_n = s_n\left(1.05 - \frac{1}{1.8n}\right) - \tfrac{1}{18}d_n \tag{2.105}$$

We see that the disturbance component of the output has one-eighteenth the amplitude of the input disturbance and, except at small times n, the rest of the output is almost identical to the desired signal. Thus the averager processes the given signals as we hoped it would. Although explicit results may be harder to compute for more complicated input signals, it should be clear that the averager will be of some help in decreasing the disturbance term whenever the disturbance fluctuates rapidly and the desired signal is smoother. This *general* smoothing property will soon be verified by a frequency analysis of the averager.

We have described the averager in terms of its effects on the two input components. A filter is said to be *linear* when its output is a sum of the individual outputs

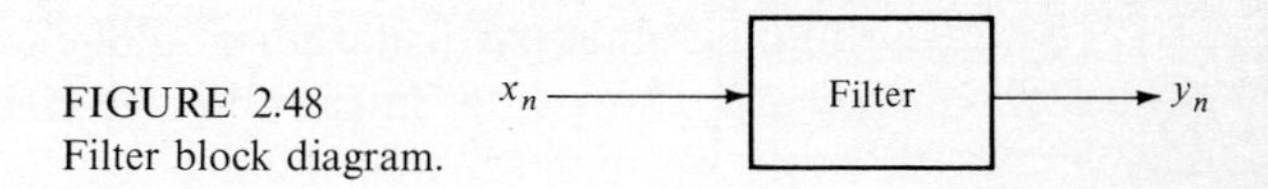

FIGURE 2.48
Filter block diagram.

corresponding to applying the inputs one at a time. Notice that the averager output can be written as

$$y_n = \frac{s_n + s_{n-1}}{2} + \frac{d_n + d_{n-1}}{2} \tag{2.106}$$

which is the sum of the output for an input of $s_1, s_2, \ldots, s_n$ plus the output for an input sequence of $d_1, d_2, \ldots, d_n$. A fairly general form for representing a linear filter with this *superposition* property is the summation

$$y_n = \sum_{k=-\infty}^{\infty} h_k x_{n-k} \tag{2.107}$$

Different filters correspond to different coefficients h_k. (An even more general form of filter with coefficients $h_{k,n}$ depending on both n and k can be defined.) Note that the averager is a special case of (2.107) when we take $h_0 = h_1 = \frac{1}{2}$ and all other $h_k = 0$. While the summation definition (2.107) of a filter is precise, we shall also often describe a filter graphically by a box with input and output, as shown in Fig. 2.48. Such diagrams are especially useful for describing complicated signal processors which may include interconnections of several different filters.

An averager is about the simplest imaginable filter of the form (2.107). Others will be examined in future examples and problems. A filter in this form with only a small finite number of nonzero coefficients, e.g.,

$$y_n = \sum_{k=1}^{10} h_k x_{n-k} \tag{2.108}$$

is often said to be a *moving-average filter*, *transversal filter*, or *nonrecursive filter*. The last name arises in contrast to another kind of filter, said to be *recursive*, and described by the relation

$$y_n = \sum_{k=1}^{\infty} g_k y_{n-k} + h_0 x_n \tag{2.109}$$

in which previous outputs (rather than inputs) are combined with the present input x_n to compute the present output y_n. The two representations (2.107) and (2.109) are both used in applications. For a particular application, one can be much simpler than the other. It is left as a problem at the end of this chapter to show that the recursive filter

$$y_n = \tfrac{1}{2} y_{n-1} + x_n$$

with a single nonzero g_k requires an infinite number of nonzero h_k coefficients when written in the moving-average form.

Before examining other particular linear filters we shall return to the frequency-transform theme and show how linear filters can be described in the frequency domain. In this way we shall see that the averager is a *low-pass* filter which passes (undistorted) low-frequency signals but which blocks or attenuates high-frequency signals.

The filter will be characterized by the way it relates the discrete Fourier transforms of its input and output. Substitution of the filter definition (2.103) into the output-transformation sum†

$$Y(\omega) = \sum_{-\infty}^{\infty} y_n e^{-j\omega n T_s} \tag{2.110}$$

yields

$$Y(\omega) = \sum_{-\infty}^{\infty} \tfrac{1}{2} x_n e^{-j\omega n T_s} + \sum_{-\infty}^{\infty} \tfrac{1}{2} x_{n-1} e^{-j\omega n T_s} \tag{2.111}$$

The second sum in (2.111) can be modified to

$$\sum_{n=-\infty}^{\infty} \tfrac{1}{2} x_{n-1} e^{-j\omega n T_s} = \frac{e^{-j\omega T_s}}{2} \sum_{m=-\infty}^{\infty} x_m e^{-j\omega m T_s} \tag{2.112}$$

by factoring out an exponential term and changing the summation variable to $m = n - 1$. Thus, each sum in (2.111) is proportional to the input transform $X(\omega)$, and we can write

$$Y(\omega) = [\tfrac{1}{2}(1 + e^{-j\omega T_s})]X(\omega) \tag{2.113}$$

The output transform is thus found from the input transform by simply multiplying by the complex function of ω in the brackets of (2.113)

$$Y(\omega) = H(\omega)X(\omega) \tag{2.114}$$

where $H(\omega)$ is the filter's transfer function,

$$H(\omega) = \tfrac{1}{2}(1 + e^{-j\omega T_s}) \tag{2.115}$$

in the case of the averager.

The beauty of the transfer-function approach is that inspection of plots of $|H(\omega)|$ and $\angle H(\omega)$ quickly shows which frequency components of the input $X(\omega)$ will be emphasized or deemphasized by the filter. Figure 2.49 shows the magnitude and phase plots for the averager, found by rearranging (2.115) into the form

$$H(\omega) = e^{-j\omega T_s/2} \cos \frac{\omega T_s}{2} \tag{2.116}$$

† We generally write discrete Fourier transforms using infinite limits when doing mathematical analysis of signals and filters. In computational applications the signals will generally have finite durations, and the limits on transform sums will then reduce to finite values like those in the previous section.

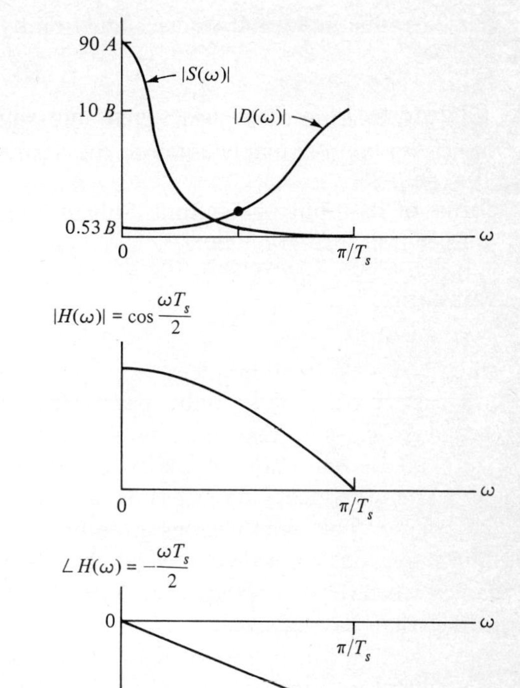

FIGURE 2.49
Averager frequency response.

for which

$$|H(\omega)| = \cos\frac{\omega T_s}{2} \qquad \angle H(\omega) = -\frac{\omega T_s}{2} \qquad 0 \le |\omega| \le \pi/T_s \quad (2.117)$$

Returning to the example under consideration, it is apparent that the transform of the averager's output depends on the frequency characteristics of the input signal and disturbance chosen in (2.101) and (2.102). It is left as an exercise to show that when $T_s = 1$, these inputs have transforms

$$S(\omega) = \frac{-0.9e^{-j\omega}A}{(1 - 0.9e^{-j\omega})^2} \qquad D(\omega) = \frac{B}{1 + 0.9e^{-j\omega}}$$

with magnitude spectra

$$|S(\omega)| = \frac{A}{2}\frac{1}{1.005 - \cos\omega} \qquad |D(\omega)| = \frac{0.746B}{\sqrt{1.005 + \cos\omega}}$$

which are also shown in Fig. 2.49. Note that s_n is an essentially low-frequency signal while the disturbance d_n has mostly high-frequency content. The basic transfer-function relation (2.114) implies that the output transform $Y(\omega)$ has a magnitude related to the input and transfer function by the relation

$$|Y(\omega)| = |H(\omega)|\,|X(\omega)| \tag{2.118}$$

Considering each input separately, the desired s_n produces an output y_n^s with transform approximately equal to the desired signal transform

$$Y^s(\omega) \approx S(\omega) \tag{2.119}$$

because, as shown in Fig. 2.49, $|H(\omega)| \approx 1$ for frequencies where $|S(\omega)|$ has substantial values. Conversely, the disturbance alone produces a small output y_n^d with transform

$$Y^d(\omega) \approx 0 \tag{2.120}$$

since $|H(\omega)| \approx 0$ at frequencies where $|D(\omega)|$ has substantial values. Thus the low-pass averager essentially passes the low frequency s_n and blocks the high frequency d_n. The reasoning used in this example is only approximate because $S(\omega)$ and $D(\omega)$ are not completely concentrated in separate frequency bands.

These frequency descriptions of the filtering action may seem less precise than the exact output computations given in (2.105). However, these arguments based on filter passbands are valuable when designing a filter to separate signals from disturbances when these two signals are only known to have general dominant frequencies [rather than having specific analytical forms like (2.101) and (2.102)].

The transform approach can also be useful for actually computing the response of a filter to a given input. If the input is given numerically (as a long string of numbers rather than a simple analytical expression), and if the filter has a large number of nonzero h_k coefficients in its moving-average representation [see (2.107)], direct evaluation of the summations like (2.107) can be very time-consuming. It can be faster to take the discrete Fourier transform of the input data, multiply the resulting values by the transfer-function values (at a finite set of frequencies), and then compute a fast Fourier transform approximation to the inverse Fourier transform

$$y_n = \frac{T_s}{2\pi} \int_{-\pi/T_s}^{\pi/T_s} Y(\omega) e^{jn\omega T_s}\, d\omega \tag{2.121}$$

The plots of $|H(\omega)|$ and $\angle H(\omega)$ shown in Fig. 2.49 are often called the *frequency response* of the filter. We have seen how the amplitude response $|H(\omega)|$ describes the frequency bands which the filter either emphasizes or attenuates. Although the phase response $\angle X(\omega)$ is not quite so easy to interpret, the simple case of a straight-line phase (as in Fig. 2.49) has a simple consequence. The negative phase slope of $-T_s/2$ corresponds to a delay of $T_s/2$ s between the input and output. This kind of delay is most easily demonstrated for an input which is a straight line

$$x_n = AnT_s \qquad n = 1, 2, \ldots \tag{2.122}$$

Admittedly this input does not have a discrete Fourier transform because it grows without bound as $n \to \infty$. However, this input is well approximated for small n by the well-behaved input

$$x_n = AnT_s(0.9999)^n$$

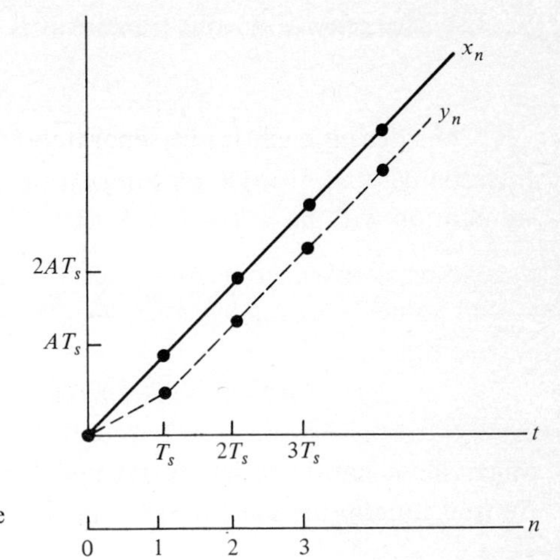

FIGURE 2.50
Delay due to averager's linear phase characteristic.

An averager with input (2.122) will have a straight-line output

$$y_n = \frac{AnT_s + A(n-1)T_s}{2} = A\left(nT_s - \frac{T_s}{2}\right) \tag{2.123}$$

Figure 2.50 shows the input and output sequences connected by straight-line segments. The output line lags the input one by $T_s/2$ s, while both have slopes of AT_s.

The delay effect of a linear phase of the form ωT_D is even easier to see if T_D is an integer. Consider w_k as a version of x_k which is delayed by three samples

$$w_k = x_{k-3} \qquad \text{all } k$$

Then the two versions of the signal have transforms

$$X(\omega) = \sum_{-\infty}^{\infty} x_k e^{-j\omega k}$$

$$W(\omega) = \sum_{-\infty}^{\infty} x_{k-3} e^{-j\omega k} = e^{-j\omega 3} X(\omega) = H(\omega)X(\omega)$$

which are related by a linear-phase factor. Clearly the phase slope of $\angle H(\omega) = -3\omega$ corresponds to the three-sample delay.

At this point we have introduced the transfer function through the frequency analysis of an averaging filter. The transfer function has been defined as the ratio

$$H(\omega) = \frac{Y(\omega)}{X(\omega)} \tag{2.124}$$

of input and output transforms. An alternate definition, useful for calculating transfer functions, can be found by using the general moving-average-filter definition (2.107) together with the change-of-summation-variable argument previously used to get

(2.112). The general output transform is

$$Y(\omega) = \sum_{n=-\infty}^{\infty} \sum_{k=-\infty}^{\infty} h_k x_{n-k} e^{-jn\omega T_s} \tag{2.125}$$

Interchanging the order of summation and replacing the n-summation by an m-summation with $m = n - k$, we get

$$Y(\omega) = \sum_{k=-\infty}^{\infty} h_k \sum_{m=-\infty}^{\infty} x_m e^{-j(m+k)\omega T_s}$$

$$= \sum_{k=-\infty}^{\infty} h_k e^{-jk\omega T_s} \sum_{m=-\infty}^{\infty} x_m e^{-jm\omega T_s} \tag{2.126}$$

Since the m-sum is clearly $X(\omega)$, comparison of (2.126) with (2.124) tells us that the transfer function is the Fourier transform of the sequence of filter coefficients

$$H(\omega) = \sum_{k=-\infty}^{\infty} h_k e^{-jk\omega T_s} \tag{2.127}$$

This relation between a filter's coefficients h_k and its transfer function $H(\omega)$ has an interesting physical interpretation. Figure 2.51 shows one way in which a nonrecursive filter could be constructed out of T_s-s delay devices, multipliers, and a summing device. The filter's transfer function is the sum of transfer functions due to each term. These are in turn $h_k e^{-jk\omega T_s}$, since a k-sample delay corresponds to a linear-phase transfer function.

The transfer function of a *differencer* filter defined by

$$y_n = K(x_n - x_{n-1}) \tag{2.128}$$

will now be derived as the transform of its coefficients. It is not difficult to imagine a situation where we would be interested in a filter like this to find the changes in a rapidly varying signal combined with a slowly varying disturbance. A person on a diet is mainly interested in the changes between daily weighings and is not as concerned about slowly varying zero-adjustment errors on his scale. Another example arises in automatic scanning of an x-ray picture in the search for a tumor. A differencer will accentuate the boundaries between two brightness levels while suppressing gradual variations in brightness which might be due to variations in the thickness of organs.

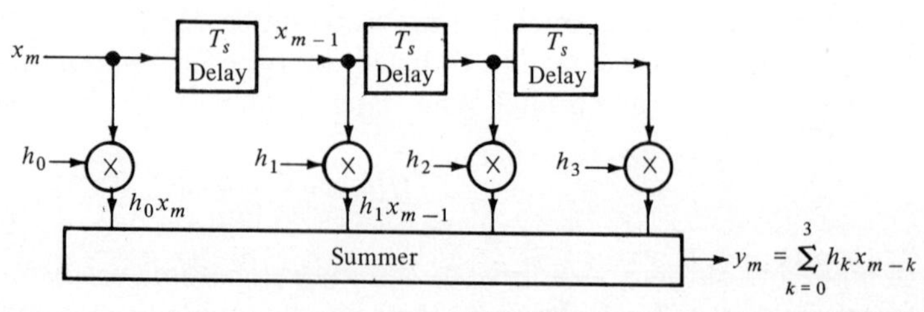

FIGURE 2.51
Construction of a nonrecursive filter.

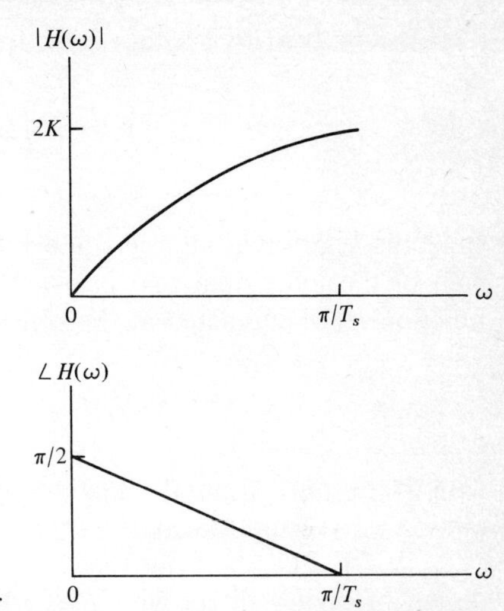

FIGURE 2.52
Differencer frequency response.

In a sense, this filter is the opposite of the averager, in that this one emphasizes high frequencies and suppresses low frequencies. Clearly if all x_n values are the same, the output of the differencer is zero. (What does the averager do to such a constant signal?)

Since the filter coefficients h_k are all zero in this case, except for $h_0 = -h_1 = K$, its transfer function is, from (2.127),

$$H(\omega) = K(1 - e^{-j\omega T_s}) = 2K \sin \frac{\omega T_s}{2} e^{j(\pi - \omega T_s)/2} \tag{2.129}$$

with the frequency response shown in Fig. 2.52. The differencer is clearly a high-pass filter. The high-frequency input signal

$$x_n = (-1)^n$$

[well approximated by the transformable $z_n = (-0.999)^n$] is represented perfectly by the output if the filter gain $K = \frac{1}{2}$. Conversely, the low-frequency signal in (2.122) as input to a differencer would produce a constant output $AT_s/2$ (the input slope or rate at which the dieter is gaining weight), totally different from the input signal.

We conclude this section by pointing out briefly that the transfer functions developed here for linear filtering of signals with discrete Fourier transforms can also be used for linear filtering of discrete-time sinusoids. This is most easily done by writing a general sinusoid at frequency u as the real part of a complex exponential function

$$x_n = C \cos(un + \varphi) = \text{Re}(Ce^{j\varphi}e^{jun}) \tag{2.130}$$

The corresponding linear-filter output will then be

$$y_n = \sum_{k=-\infty}^{\infty} h_k \text{ Re}(Ce^{j\varphi}e^{ju(n-k)}) \tag{2.131}$$

This can be rewritten as

$$y_n = \text{Re}\left(Ce^{j\varphi}e^{jun} \sum_{k=-\infty}^{\infty} h_k e^{-juk}\right) \tag{2.132}$$

since the h_k are real numbers and the real part of a sum of complex numbers is the sum of the individual real parts. We recognize the sum in (2.132) as the transfer function $H(\omega)$ evaluated at the sinusoid's radian frequency u. Thus

$$y_n = \text{Re}\,[CH(u)e^{j(un+\varphi)}] = C\,|H(u)|\,\cos\,[un + \varphi + \angle H(u)] \tag{2.133}$$

with the output amplitude equal to $|H(u)|$ times the input amplitude and the output phase shifted from that of the input by $\angle H(u)$.

We should point out that there is a large body of information† related to the design of digital filters with prescribed frequency response. If a signal-processing problem can be reduced to filtering with a suitable amplitude and phase characteristics, the filter-synthesis literature can be consulted to find techniques for assembling digital circuits or writing computer programs which achieve the desired transfer functions (or good approximations to them).

This chapter has introduced discrete-time signals which will be used in the later development of processors to detect the presence of signals of interest, e.g., radar returns from airplanes near an airport, and processors to estimate signal parameters, e.g., an airplane's position and velocity. We explained the least-squares approximation of signals by simple polynomials and sinusoids. The latter analysis led naturally to Fourier series and other Fourier transforms. In addition to providing efficient signal approximations, the frequency representation gave us insight into the choice of adequate spacing of samples in A/D conversion of continuous-time signals, a step which is necessary for presenting signals to digital processors.

The consideration of computational details related to computing Fourier transforms from real data, begun here, will be picked up again in the spectral analysis of Chap. 4. That chapter analyzes the statistical properties of digital computation of Fourier transforms using this section's transfer-function definition of linear filters.

The next chapter will review some notions from probability and use them to define discrete-time random signals. Such signals are especially appropriate as models for the disturbances in communication and control systems. Transfer functions will also be used in that context, after we define the power spectral density, which tells how a random signal's average power is distributed in the frequency domain. We shall define a standard purely random signal and then characterize any other random signal by the linear filter which produces the latter signal when its input is the purely random signal.

† For example, B. Gold and C. M. Rader, "Digital Processing of Signals," McGraw-Hill, New York, 1969.

PROBLEMS

2.1 (*a*) Which data in Fig. P2.1 can be better represented by a straight line, $x(nT_1)$ or $y(nT_2)$? Explain.
(*b*) What is $\hat{x}(3T_1)$ if $\hat{x}(nT_1)$ is the least-squares straight line fitted to the given $x(nT_1)$ data?

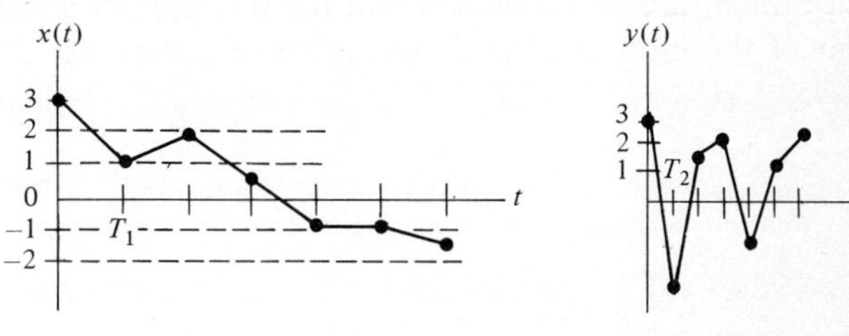

FIGURE P2.1

2.2 Fit a straight line to the data in Fig. P2.2 by eye and by least squares. Explain why the least-squares line will be the same no matter what scales are used on this graph.

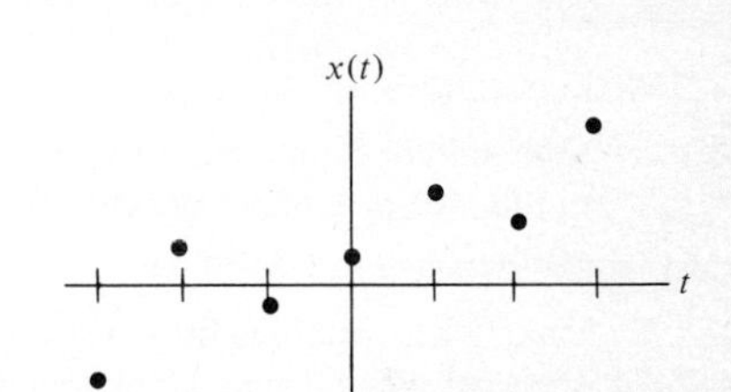

FIGURE P2.2

2.3 Get a general rule for converting

$$\hat{x}_n = \hat{c}_0 + \hat{c}_1 n \qquad n = -M, -M+1, \ldots, 0, 1, \ldots, M$$

for symmetrically oriented data, to the analog function

$$\hat{x}(t) = a + bt$$

i.e., express a and b in terms of $\hat{c}_0$ and $\hat{c}_1$, when x_{-M} is at $t = 0$ and when the time between successive samples is T_s s.

2.4 Write a FORTRAN or PL/I program to fit a straight line to an odd number of equally spaced data points. Include steps for carrying out the shifting and scaling for samples starting at time t_0 and spaced T_s s apart. Describe the necessary input data and the format in which they should appear on the input cards. Include a check on the sum-squared fitting error. Explain how the output will be presented.

2.5 Show that the least-squares fit of a parabola

$$\hat{x}_n = \hat{c}_0 + \hat{c}_1 n + \hat{c}_2 n^2$$

to symmetrical, equally spaced data $x_{-M}, \ldots, x_1, x_2, \ldots, x_M$ is defined by the coefficients satisfying

$$\hat{c}_1 = \frac{\sum_{-M}^{M} n x_n}{\sum_{-M}^{M} n^2}$$

$$\hat{c}_0 \sum_{-M}^{M} 1 + \hat{c}_2 \sum_{-M}^{M} n^2 = \sum_{-M}^{M} x_n$$

$$\hat{c}_0 \sum_{-M}^{M} n^2 + \hat{c}_2 \sum_{-M}^{M} n^4 = \sum_{-M}^{M} n^2 x_n$$

Note that the c_1 formula is the same here as in the straight-line fit but the c_0 formula is different. Formulas for sums of powers of integers are available.† For example,

$$\sum_{1}^{N} n = \frac{N(N+1)}{2} \qquad \sum_{1}^{N} n^4 = \frac{N(N+1)(2N+1)(3N^2+3N-1)}{30}$$

$$\sum_{1}^{N} n^2 = \frac{N(N+1)(2N+1)}{6}$$

2.6 The hour of sunrise is an interesting function of time which is approximately sinusoidal with a known period of 365 d. Numerical study of this approximation can be carried out by checking the sunrise time listed in daily newspapers (corrected for daylight-saving-time shifts). Sampling only every third day will reduce the data-taking tedium and the number of points per period. The average (least-squares constant polynomial) should be estimated and subtracted from the data before fitting the sinusoid.

2.7 Make a least-squares fit of a straight line to a sequence of data values such as daily closing prices of a particular stock, each year's fastest time for running the mile, etc. Use the straight line to predict the next data point. (The program written for Problem 2.4 may be used here.)

2.8 (*a*) Show that the Fourier coefficients for the periodic function of Fig. P2.8 are given by

$$z_k = \frac{a_k}{2} = \frac{2\tau}{T}\frac{1 - \cos \omega_k \tau}{(\omega_k \tau)^2} = \frac{\tau}{T}\left[\frac{\sin (\omega_k \tau/2)}{\omega_k \tau/2}\right]^2$$

(*b*) Sketch z_k vs. ω_k for $\tau/T = \frac{1}{4}$.

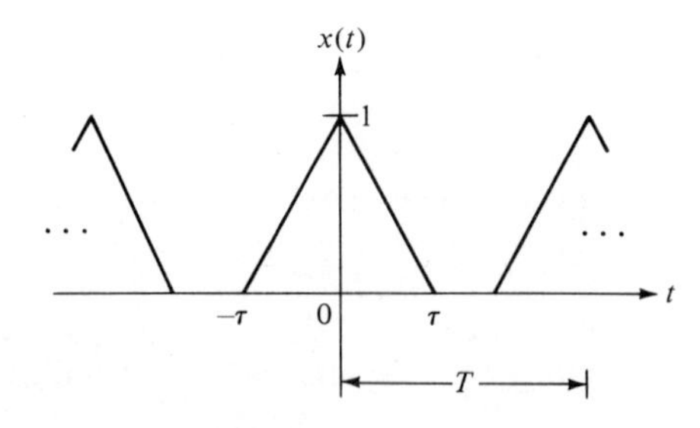

FIGURE P2.8

2.9 The periodic $x(t)$ shown in Fig. P2.9 has a complex Fourier series

$$x(t) = \sum_{-\infty}^{\infty} z_k e^{jk(\pi/2)t}$$

† For example, R. B. Blachman, "Data Smoothing and Prediction," p. 176, Addison-Wesley, Reading, Mass., 1965.

for which a portion of the amplitude spectrum is shown. Find values for A and B.

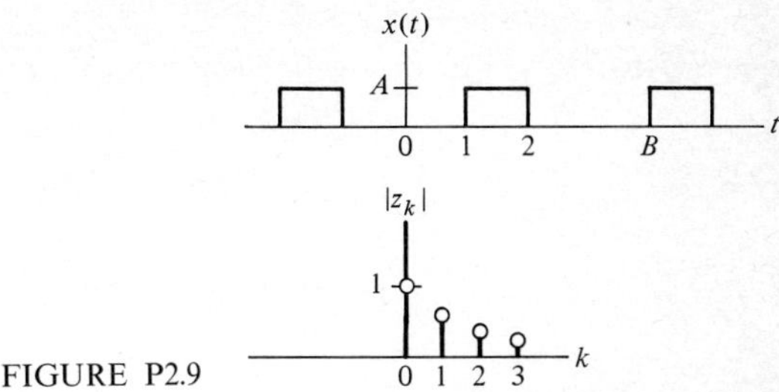

FIGURE P2.9

2.10 A pulse frequently used in communications is a *raised cosine*. The signal $x(t)$ shown in Fig. P2.10 is a periodic sequence of these pulses with spaces of equal length between them.

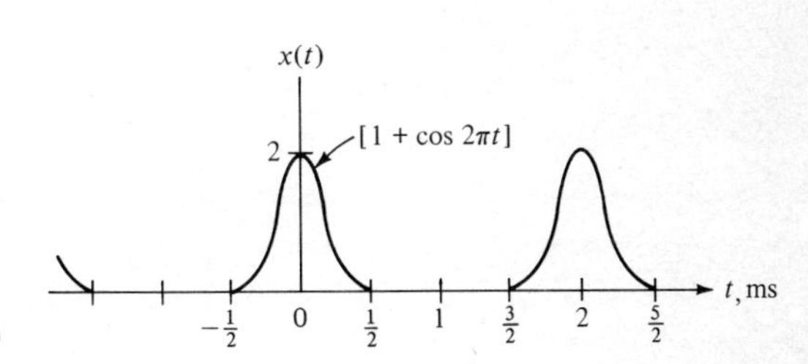

FIGURE P2.10

(*a*) From the shape of $x(t)$, what is a reasonable sample spacing T_s in milliseconds? Explain.
(*b*) Show that the first three terms (lowest frequencies) in the Fourier series for $x(t)$ are, with t in milliseconds,

$$x(t) = \frac{1}{2} + \frac{8}{3\pi}\cos \pi t + \frac{1}{2}\cos 2\pi t + \cdots$$

2.11 Given

$$x(n) = \begin{cases} (\frac{1}{3})^n & n \geq 0 \\ 0 & n < 0 \end{cases}$$

(*a*) Calculate $X(\omega)$.
(*b*) Sketch and label the magnitude $|X(\omega)|$ vs. ω and phase $\angle X(\omega)$ vs. ω.

2.12 Find the discrete Fourier transform of the signal [see (2.101)]

$$x_n = \begin{cases} n(a)^n & n = 0, 1, 2, \ldots \\ 0 & n < 0 \end{cases}$$

when $|a| < 1$, with the aid of the relations

$$\frac{d}{d\omega}\left(\frac{1}{1 - ae^{-j\omega}}\right) = \frac{-jae^{-j\omega}}{(1 - ae^{-j\omega})^2}$$

$$= \frac{d}{d\omega}\left(\sum_{n=0}^{\infty} a^n e^{-jn\omega}\right) = \sum_{0}^{\infty} -jna^n e^{-jn\omega}$$

2.13 One cycle of a discrete Fourier transform $X(\omega)$ is shown in Fig. P2.13.
(*a*) Find the corresponding x_n and sketch it for $-4 \leq n \leq 4$ when $\Omega = \pi/2$. *Note:* $e^{(3/2)\pi j} = e^{-(\pi/2)j}$.
(*b*) Is this a high- or low-frequency signal? Explain.

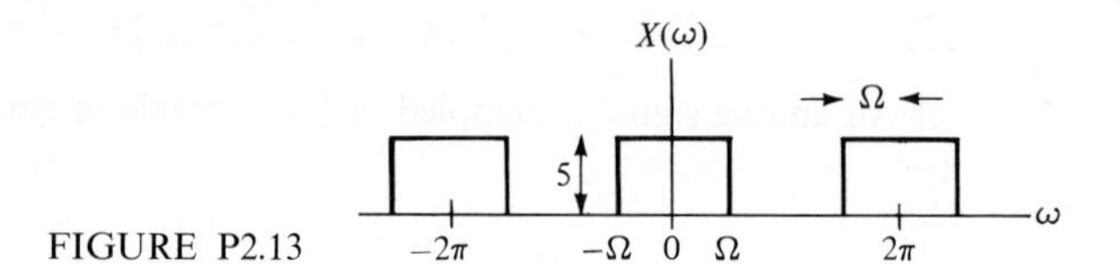

FIGURE P2.13

2.14 (*a*) Find two different higher-frequency sine waves which have the same samples at $t = 1, 2, 3, \ldots$ as the samples of $x(t) = 5\cos(t/2)$.
(*b*) Pick a reasonable sampling period for the $z(t)$ shown in Fig. P2.14 and explain your reasoning.

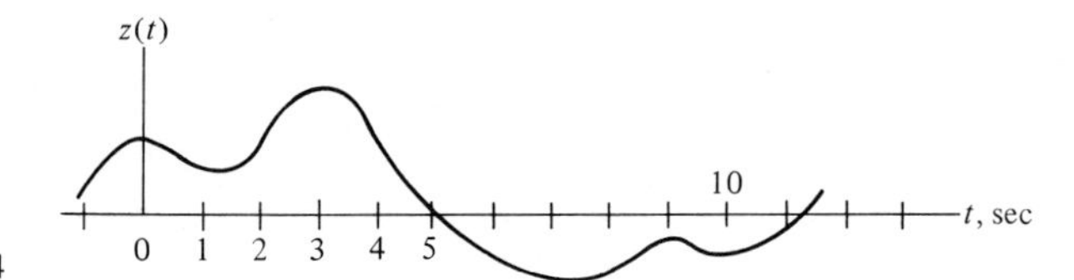

FIGURE P2.14

2.15 Sketch $x(t) = [\sin(\pi t/2)]/(\pi t/2)$ for $0 \leq t \leq 6$ and label samples at $T_s = 1$ with zeros and samples at $T_s = 3$ with ×'s. Explain why $T_s = 3$ is too big a sampling interval.

2.16 The magnitude spectrum $|X(\omega)|^2$ is desired for the following data on monthly airline passengers (in thousands) corresponding to Fig. 2.33.

	1953	1954	1955	1956	1957	1958	1959	1960
Jan.	196	204	242	284	315	340	360	417
Feb.	196	188	233	277	301	318	342	391
Mar.	236	235	267	317	356	362	406	419
Apr.	235	227	269	313	348	348	396	461
May	229	234	270	318	355	363	420	472
June	243	264	315	374	422	435	472	535
July	264	302	364	413	465	491	548	622
Aug.	272	293	347	405	467	505	559	606
Sept.	237	259	312	355	404	404	463	508
Oct.	211	229	274	306	347	359	407	461
Nov.	180	203	237	271	305	310	362	390
Dec.	201	229	278	306	336	337	405	432

First make a least-squares fit of a straight line (trend). Subtract that line and make a least-squares fit of sinusoid having a period of 12 months. Then subtract this sinusoid and compute $|X(\omega)|^2$ at frequencies

$$\omega = 0, \frac{\pi}{4T_s}, \frac{\pi}{2T_s}, \frac{3\pi}{4T_s}, \frac{\pi}{T_s}$$

2.17 Show the reduction in truncation effects on $X(\omega)$ due to increasing the number of discrete Fourier transform samples by changing the summation in (2.87) to run from $n = -10$ to $n = 10$. Both this 21-term sum and the 9-term one in (2.87) can be evaluated at several frequencies using a finite Fourier transform like (2.89). Use your own program or a standard fast Fourier transform program to evaluate both sums at 32 frequencies; i.e., let $N = 32$ and add zero x_n values as described in connection with (2.91).

2.18 (*a*) An analog signal is sampled at $\frac{1}{2}$-s intervals to get $x(0), x(\frac{1}{2}), \ldots, x(\frac{7}{2})$. What are the true radian frequencies $m\Omega_s$ of the eight finite Fourier transform components $X_0, \ldots, X_7$?
(*b*) The first four samples are used to compute a finite Fourier transform $X_0^a, \ldots, X_3^a$ and the last four samples are used to compute a finite Fourier transform $X_0^b, \ldots, X_3^b$. Show that for *even* integers m

$$X_m = X_{m/2}^a + X_{m/2}^b$$

(This is typical of tricks used to reduce the number of computations in fast Fourier transforms.)

2.19 A filter with input x_n and output y_n is defined by

$$y_n = x_n + 2x_{n-1} + x_{n-2}$$

Find the transfer function of this filter and sketch $|H(\omega)|$ and $\angle H(\omega)$ assuming $T_s = 10^{-3}$ s. *Hint:* $|H(\omega)| = (2 \cos \omega T_s/2)^2$.

2.20 (*a*) Find the discrete Fourier transform of the h_n shown in Fig. P2.20, that is, $H(\omega) = \sum_{-\infty}^{\infty} h_n e^{-jn\omega}$; all h_n not shown are zero.
(*b*) Sketch $|H(\omega)|$ and $\angle H(\omega)$.
(*c*) If this $H(\omega)$ is a transfer function and its input is $x_n = 2 \sin un$, what frequency u yields the biggest amplitude output y_n and what is that amplitude?

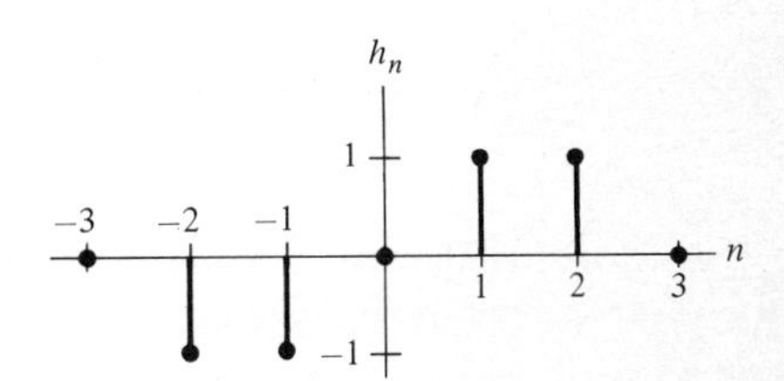

FIGURE P2.20

2.21 The filter

$$y_n = \frac{x_n + x_{n-1} + x_{n-2}}{3}$$

averages three data values.
(*a*) Show that the filter transfer function is

$$H(\omega) = \frac{Y(\omega)}{X(\omega)} = \frac{e^{-j\omega T_s}}{3} \frac{\sin (3\omega T_s/2)}{\sin (\omega T_s/2)}$$

Hint: Use the geometric-sum formula.
(*b*) Sketch $|H(\omega)|$ and $\angle H(\omega)$.

2.22 (*a*) **Find $H(\omega)$ for the infinite-term moving average filter with**

$$y_n = \sum_{-\infty}^{\infty} h_k x_{n-k} \qquad h_k = 0 \text{ for negative } k$$

$$h_0 = 1,\ h_1 = \tfrac{1}{2}, \ldots, h_k = (\tfrac{1}{2})^k, \ldots \text{ for } k \geqslant 0$$

Is this a high-pass or low-pass filter?
(*b*) Show that the recursive filter $y_n = \frac{1}{2}y_{n-1} + x_n$ has the same transfer function as the filter in part (*a*).
(*c*) Find the moving-average form for the recursive filter $y_n = \alpha y_{n-1} + x_n$ when α is any number with $|\alpha| < 1$.

2.23 (*a*) Find the transfer function $H(\omega)$ in Fig. P2.23 if an input

$$x_n = \begin{cases} (\frac{1}{2})^{-n} & n \leq 0 \\ 0 & n > 0 \end{cases}$$

produces an output whose transform is

$$Y(\omega) = \frac{2 \cos \omega}{2 - \cos \omega - j \sin \omega}$$

(*b*) Compute h_0 and h_1.

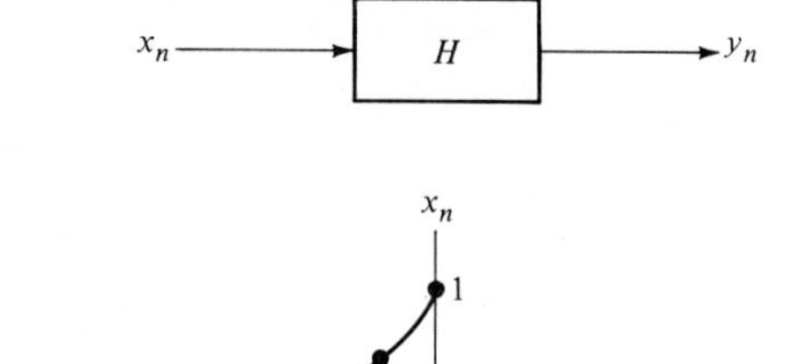

FIGURE P2.23

2.24 Two digital filters are connected in series as shown in Fig. P2.24. The difference equations describing these filters are

$$y(n) = x(n) - x(n-1) \qquad w(n) = \tfrac{2}{3}w(n-1) + \tfrac{1}{3}y(n)$$

(*a*) Show that the transfer functions $H_1(\omega)$ and $H_2(\omega)$ are given by

$$H_1(\omega) = e^{j(\pi/2 - \omega/2)} 2 \sin \frac{\omega}{2} \qquad H_2(\omega) = \frac{1}{3 - 2e^{-j\omega}}$$

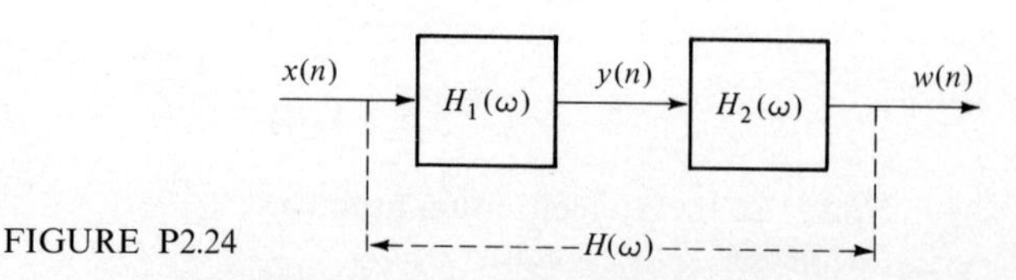

FIGURE P2.24

(*b*) Carefully sketch the magnitude characteristics for $H_1(\omega)$ and $H_2(\omega)$ over the baseband $0 \leq \omega \leq 2\pi$; then sketch the magnitude for $H(\omega)$.
(*c*) How does the overall filter behave at low frequencies?

2.25 The nonrecursive filter of Fig. P2.25 has the following inputs: $x_0 = 1$, and all other x_j's $= 0$.

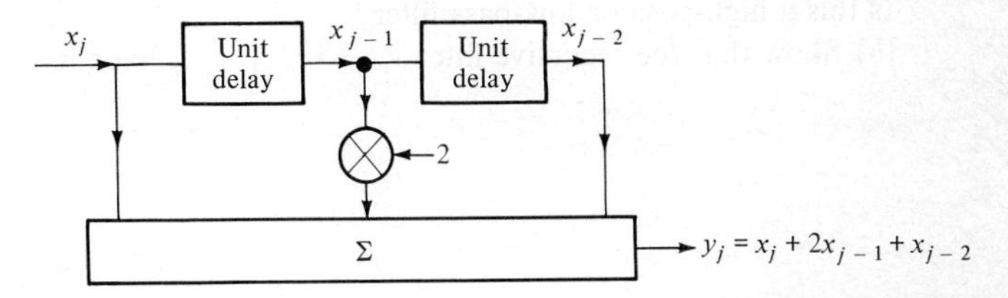

FIGURE P2.25

(*a*) Show that $y_0 = 1$, $y_1 = 2$, $y_2 = 1$, $y_3 = y_4 = \cdots = 0$.
(*b*) Find the discrete Fourier transform $Y(\omega) = \sum_{n=-\infty}^{\infty} y_n e^{-jn\omega}$ and sketch $|Y(\omega)|$.
(*c*) Calculate the finite Fourier transform $Y_m = \sum_{n=0}^{N-1} y_n e^{-j2\pi nm/N}$ of the first 12 samples $y_0, y_1, y_2, \ldots, y_{11}$. Sketch $|Y_m|$ and compare with $|Y(\omega)|$ in part (*b*), for $m = 0, 1, \ldots, 11$.

3

RANDOM DISCRETE-TIME SIGNALS

3.1 INTRODUCTION AND REVIEW OF PROBABILITY

Chapter 2 dealt primarily with the frequency analysis of known (deterministic) signals and of a set of data samples. We did not discuss the fact that in real life the signals with which we deal are generally random. A set of data samples from which we calculate the Fourier transform may not be reproducible; i.e., another set from the same source may differ considerably. What does this imply about the spectrum? Is it possible to define an average spectrum for a given data source? Questions such as these relating to the Fourier analysis of random data signals will be discussed at length in this chapter and in Chap. 4. We shall see that the concept of *power spectral density* arises naturally as a way of describing the frequency representation of random signals, and we shall consider ways of measuring it from a finite set of samples.

Random-signal models will also arise in the chapters on the detection and estimation of signals. Examples will include the radar detection of an airplane approaching an airport and means of estimating the plane's current range and bearing along with its future location. Random signals are often called *stochastic signals* or *stochastic processes.*

In these three areas of random-signal processing (spectral analysis, detection, and estimation, all of which have found significant engineering application in recent years) probability theory plays a major underlying role. We therefore review pertinent aspects of probability theory in this chapter. We shall then extend the concepts of probability theory to signals varying randomly in time. Before doing so, however, it may be helpful to introduce some examples of random signals.

Consider as one example the voltage in a radar receiver induced by energy reflected from an airplane. We say the signal is *random* if its precise value cannot be predicted in advance. Uncertainties involving the airplane, as an example, arise due to the plane's distance and attitude (head on, sideways, etc.), its pitching (due to turbulence), signal scattering by particles or rain in the air, etc. Some people argue that these nuisance factors are finite in number and could, in principle, be measured. However, whether or not such factors could be removed, the uncertainties exist in the situation being analyzed. A random model says that there is a collection of a large number of possible values or *realizations* of the signal voltage. Some specific values, for example, 7 V or 3 V, may be "more likely" than others. More generally we say that the observed value is more likely to lie in one range of values, for example, 7 ± 0.1 V, than in another range. We also speak about the average of values observed in many independent repetitions of the measurement experiment. These physically measurable notions are expressed quantitatively in terms of distribution functions (to describe the likelihood that an observation falls in an interval) and mean values (to describe average values) from probability theory.

As a second example, noise voltages in communication systems are usually unpredictable since they are caused, for example, by the erratic motion of a large number of electrons. However, knowledge of the average properties (amplitude, squared amplitude, rate of change, etc.) can be used to better process desired signals observed in the presence of such noise. Random signals and random noises can also be characterized in the frequency domain, as we shall see, by relating the average input power and output power of a filter to its transfer function. Again it is averages over many different signal realizations which are quantifiable in the frequency domain and which are useful for processing random signals.

Another example of a random signal is the waveform of a person saying the word "one." No two people will produce the same signal, but there will be enough similarities so that a typical (English-speaking) listener will be able to understand. In this case the amplitude factor affects only loudness, but time variation and frequency content are more important for intelligibility.

A random signal like the voice waveform (also called a *random function*, *time series*, or *random process*) might have a realization like the curve in Fig. 3.1. We use the symbol $\mathsf{x}(t)$ (sans serif boldface) to represent the collection of possible realizations along with a probability rule for selecting one realization for a typical observation. We do not necessarily know the selection rule nature uses, but we assume that it exists and that we can perform tests (described later) to estimate some of the average properties of the probability laws.

A discrete-time random signal

$$\mathsf{x}_j \qquad j = 1, 2, \ldots, n \tag{3.1}$$

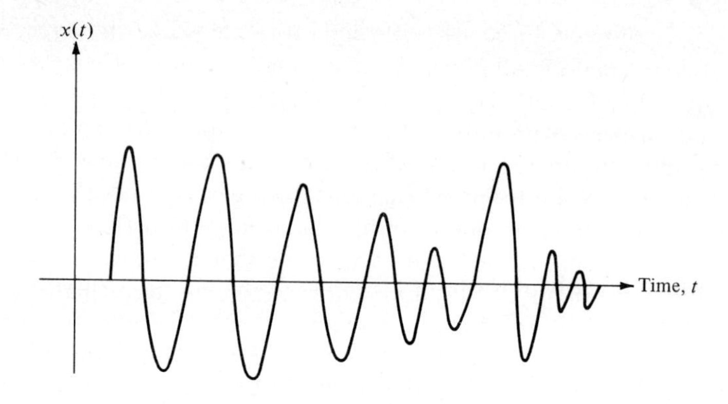

FIGURE 3.1
Example of random-signal realization.

has many realizations, a typical one of which is symbolized by

$$x_j \qquad j = 1, 2, \ldots, n \tag{3.2}$$

as shown in Fig. 3.2. At a particular time, say $j = 3$, the collection of realizations defines a random *variable* $\mathbf{x}_3$. Thus, a random signal consists of a set of random variables. We shall begin our review of probability-theory definitions and applications to signal processing by first working with properties of a single random variable and then discussing the more complicated theory of random signals.

The reader is assumed to have a background in elementary probability theory, but the basic definitions will be reviewed here to establish notation. We define an *event* by stating conditions on a random variable $\mathbf{x}$, say $\mathbf{x} < 5$. The event *occurs* if the observed realization of $\mathbf{x}$ is a number satisfying the condition. The probability of the event, say

$$P[\mathbf{x} < 3] \tag{3.3}$$

is, by convention, a number between 0 and 1, higher numbers corresponding to more likely events. Many different events might be of interest in the study of a random variable. It can be shown that all probabilities and averages of interest can be computed if we know the probabilities of a fundamental set of events. This collection of probabilities is called the *distribution function* $F_{\mathbf{x}}(x)$, defined by the probability

$$F_{\mathbf{x}}(x) = P[\mathbf{x} \le x] \tag{3.4}$$

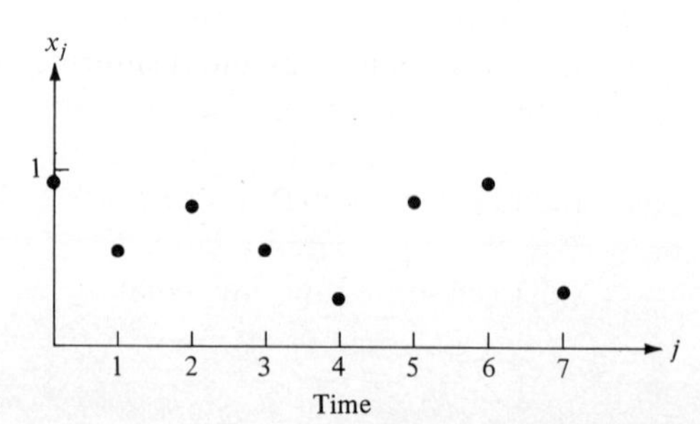

FIGURE 3.2
Discrete-time random-signal realization.

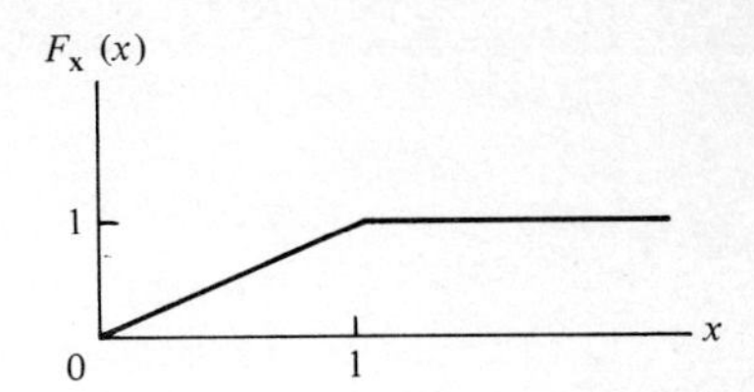

FIGURE 3.3
Uniform distribution function.

which must be known for all values of the ordinary variable x. Figure 3.3 shows the distribution function of a random variable $\mathbf{x}$ which is said to be uniformly distributed between 0 and 1. This terminology will be elaborated upon shortly. Notice that the subscript on F indicates the random variable and the argument in parentheses is an ordinary variable (nonrandom). The subscript is sometimes omitted when the argument can be expected to suggest which random variable is being described, but that simpler notation is often confusing.

For the moment we shall refer only to *continuous* random variables, i.e., with continuous distribution functions, for which *density* functions can be defined as

$$f_{\mathbf{x}}(x) = \frac{d}{dx}[F_{\mathbf{x}}(x)] \tag{3.5}$$

Figure 3.4 shows $f_{\mathbf{x}}(x)$ for the $F_{\mathbf{x}}(x)$ in Fig. 3.3.

The probability that a random variable lies in an interval can be expressed using either the distribution or the density function. Thus, for $a \leq b$,

$$P[a < \mathbf{x} \leq b] = \int_a^b f_{\mathbf{x}}(x)\,dx = F_{\mathbf{x}}(b) - F_{\mathbf{x}}(a) \tag{3.6}$$

Since probabilities are always numbers between 0 and 1 (never negative), Eqs. (3.6) require that the density be a nonnegative function with an area of 1 and that the distribution function be a nondecreasing function of x. Proper limiting arguments show that for the distribution in Fig. 3.3

$$P[\mathbf{x} = \tfrac{1}{2}] = 0 \tag{3.7}$$

and that for any continuous random variable, the probability of *any* particular value's occurring is zero. However, it is meaningful to say that a uniform random variable is equally likely to be anywhere in its range, in the sense that

$$P[0 \leq \mathbf{x} \leq 0.1] = P[0.1 \leq \mathbf{x} \leq 0.2] = \cdots = 0.1 \tag{3.8}$$

That is, the probability of a uniform random variable $\mathbf{x}$ lying in an interval of width Δ is equal to Δ (as long as the Δ interval lies entirely in the (0, 1) interval).

Since the concept of moments is extremely important in the discussion of

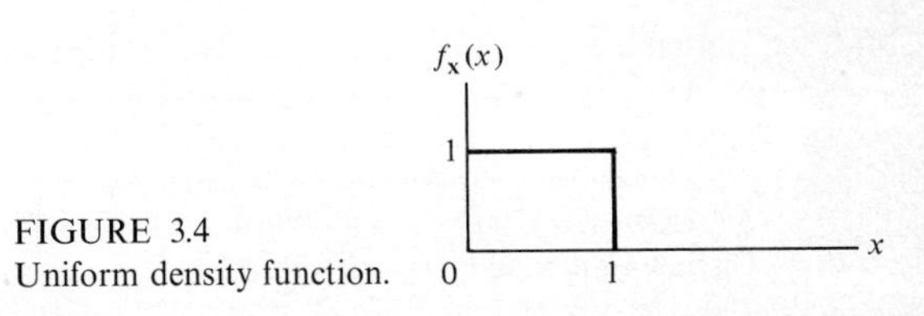

FIGURE 3.4
Uniform density function.

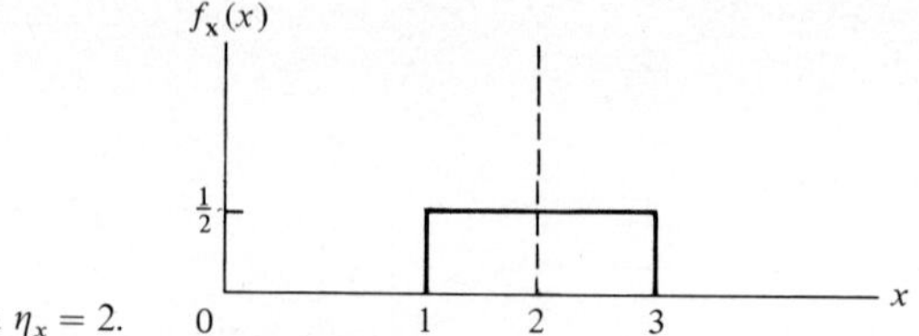

FIGURE 3.5
Density function with $\eta_x = 2$.

random variables, we review it now. Consider first the *mean value*, or expectation,

$$E[\mathbf{x}] = \eta_x = \int_{-\infty}^{\infty} x f_{\mathbf{x}}(x)\, dx \tag{3.9}$$

The mean can be interpreted as the center of gravity of the density $f_{\mathbf{x}}(x)$. For example, with a uniform density between 1 and 3 (see Fig. 3.5), the defining integral yields

$$E[\mathbf{x}] = \int_{1}^{3} x\, \tfrac{1}{2} dx = \frac{x^2}{4}\bigg|_1^3 = 2 \tag{3.10}$$

One can easily show that the mean of a random variable with a uniform density over any interval is always the center of its range.

When applying probabilistic methods to physical problems, the mean value of a random variable corresponds to the average of many independent observations of the physical quantity represented by the random variable. If the variable is an observed voltage and independent observations are randomly different due to human and instrument errors, the average of many measurements should be the mean of the assigned voltage distribution. The abscissa in Fig. 3.5 might be in units of millivolts when the true voltage is 2 mV. In another application, the density in that figure might represent the spread of azimuth angles (in milliradians of arc) of flight 201 from Boston as it is just picked up by the radar at a New York airport. Each day the plane's arrival angle may be different, but the angle is equally likely to lie in any small interval that is within the big interval between 1 and 2 mrad.

The mean tells us where a density is centered but gives no information about how spread out it is. The spread is measured by the *standard deviation* σ_x, which is the square root of another moment, the *variance* $V(\mathbf{x})$ (or var $\mathbf{x}$). These quantities are defined by

$$\sigma_x^2 = V(\mathbf{x}) = E[(\mathbf{x} - \eta_x)^2] \tag{3.11}$$

There are two ways to evaluate the expectation in (3.11). We can define a new variable

$$\mathbf{y} = (\mathbf{x} - \eta_x)^2$$

find its density $f_y(y)$, and then use the mean-value definition in (3.9) (replacing x by y).† An equivalent, and generally easier, method is to calculate

† The reader is assumed to know the techniques for computing the distribution of a new variable $\mathbf{y}$ which is a known function of an $\mathbf{x}$ whose distribution is known; see also Sec. 3.3.

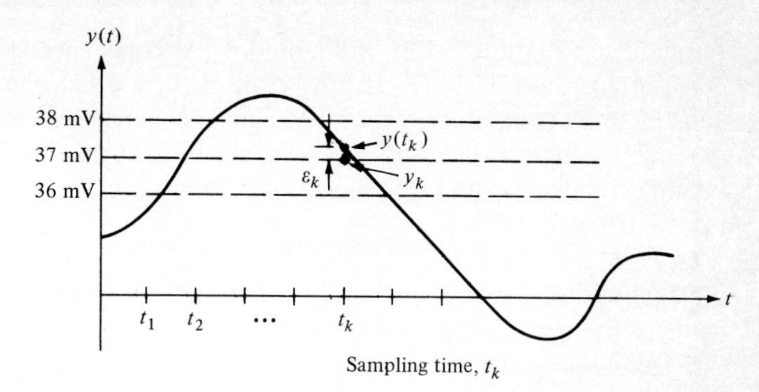

FIGURE 3.6
Quantization error.

$$\sigma_x^2 = \int_{-\infty}^{\infty} (x - \eta_x)^2 f_{\mathbf{x}}(x)\, dx \tag{3.12}$$

using the density of the original random variable **x**. For the density in Fig. 3.5 this becomes

$$\sigma_x^2 = \int_1^3 (x - 2)^2 \tfrac{1}{2} dx = \tfrac{1}{3}$$

while the density in Fig. 3.4 has the variance of $\frac{1}{12}$. One can show that the variance of a uniform density is given, in general, by

$$\sigma^2 = \frac{(\text{width})^2}{12} \tag{3.13}$$

The standard deviation σ of a uniformly distributed random variable is thus proportional to the spread of the density function. Later we shall generalize this property to other distributions.

The standard deviation, as a measure of the spread of a distribution, corresponds to the amount of uncertainty or error in a physical measurement. In the voltage-measuring example, the width of the uniform distribution corresponds to the relative inaccuracy of the meters being used.

One interesting example of uniformly distributed random variables arises in the model for quantization errors. As mentioned early in Chap. 2, inputs to digital processors must be discrete in time and in amplitude. If an analog signal with a range of ± 10 V is sampled and quantized to integral numbers of millivolts, the distance between the analog signal and its quantized value corresponds to a quantization error, as shown in Fig. 3.6. If the signal $y(t_k)$ is 37.295 mV, the quantization error will be

$$y_k - y(t_k) = 37 - 37.295 = -0.295 \text{ mV}$$

A quantized reading of 37 mV could have been caused by any analog voltage between 36.5 and 37.5 mV. In this example, for which the quantization interval of 1 mV is small compared to the 20-V signal range, it seems reasonable to assume that all values between 36.5 and 37.5 mV are equally likely. We could thus write the quantized value as the true value plus a random error ε_k

$$y_k = y(t_k) + \varepsilon_k$$

in which ε_k in this case is uniformly distributed between $\pm\frac{1}{2}$ mV. In general, the mean value of this error is zero, and its standard deviation is, as in (3.13), proportional to the size of the quantization interval. In later chapters when we discuss the estimation of signals based on noisy measurements, quantization errors will represent one possible source of the measurement noise.

A useful property of the mean-value operation is that the mean of a sum is the sum of the means. That is,

$$E[\mathbf{x}_1 + \mathbf{x}_2] = E[\mathbf{x}_1] + E[\mathbf{x}_2] \tag{3.14}$$

If c is not random, further properties are

$$E[c\mathbf{x}] = cE[\mathbf{x}] \tag{3.15}$$

and

$$E[c] = c \tag{3.16}$$

The concepts of mean and variance are of course basic to probability theory and will recur throughout our study of random signals and techniques for processing them. Specifically, we shall find them useful in studying the effects of linear filtering on a random signal.

It is appropriate at this point, however, to generalize the random-variable examples to define a simple random *signal.* Let $x_1, x_2, \ldots$ represent samples of such a signal. Each sample is the realization of the corresponding random variable $\mathbf{x}_1$, $\mathbf{x}_2, \ldots$. The sequence of random variables $\mathbf{x}_j, j = 1, 2, \ldots, n$, is the random signal $\mathbf{x}_j$.

We shall say that a signal $\mathbf{x}_j$ is *purely random* if the random variables $\mathbf{x}_1, \mathbf{x}_2, \ldots$ are statistically independent; this means that their joint distribution function

$$F_{\mathbf{x}_1, \mathbf{x}_2, \ldots}(x_1, x_2, \ldots, x_n) = P[\mathbf{x}_1 < x_1, \text{ and } \mathbf{x}_2 < x_2, \ldots, \text{and } \mathbf{x}_n < x_n] \tag{3.17}$$

factors into the product of the individual marginal distribution functions

$$F_{\mathbf{x}_1, \mathbf{x}_2, \ldots}(x_1, x_2, \ldots, x_n) = F_{\mathbf{x}_1}(x_1)F_{\mathbf{x}_2}(x_2) \ldots F_{\mathbf{x}_n}(x_n) \tag{3.18}$$

The joint density function for a purely random signal will also factor into a product of marginal densities. One of the simplest random signals is the uniform, purely random one, which has the uniform distribution of Fig. 3.3 for each of its constituent random variables. Figure 3.2 is a possible realization of this signal, all values lying between 0 and 1. This basic random signal has uniquely defined mean and variance

$$E[\mathbf{x}_i] = \tfrac{1}{2} \qquad \text{and} \qquad V(\mathbf{x}_i) = \sigma_{xi}^2 = \tfrac{1}{12} \qquad \text{for all } i \tag{3.19}$$

which are the same for the random variables at every time i.

In Chap. 2 we studied the effect of passing discrete signals through linear filters. This was basic to introducing the concept of signal processing. For random signals

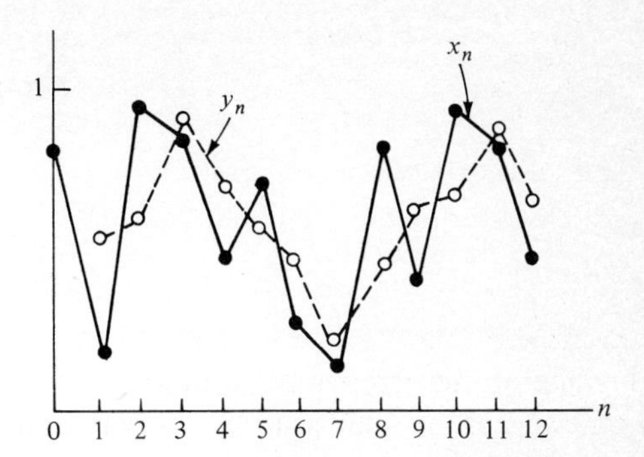

FIGURE 3.7
Output and purely random input for an averager filter.

we shall again find linear filtering appearing as one particularly simple way of processing signals. To introduce it we consider the simplest filter, the averager studied in Chap. 2. Let the $\mathbf{x}_i$ sequence just defined be the input to such an averager. What is the effect of the averager on the input sequence? How does it respond to random signal and noise sequences in general? To determine the averager's effects we focus on the corresponding output sequence. The output at time n is a new random variable $\mathbf{y}_n$

$$\mathbf{y}_n = \tfrac{1}{2}(\mathbf{x}_n + \mathbf{x}_{n-1}) \tag{3.20}$$

which is a function of two input random variables.

From the above results, the mean value of the averager output is found to be

$$E[\mathbf{y}_n] = \tfrac{1}{2}(\tfrac{1}{2} + \tfrac{1}{2}) = \tfrac{1}{2} \tag{3.21}$$

Thus the input with a constant (independent of time n) mean value produces an output signal with the same mean value.

However, the typical output realization will differ significantly from the input (see Fig. 3.7). One difference is in the amplitude of the output fluctuations, as measured by the standard deviation. In order to compute the output standard deviation σ_y we must invoke the variance property that the variance of a sum of independent random variables is the sum of the individual variances

$$V(\mathbf{w}_1 + \mathbf{w}_2) = V(\mathbf{w}_1) + V(\mathbf{w}_2) \tag{3.22}$$

We shall also use the property that multiplication of any $\mathbf{v}$ by a nonrandom c produces a new random variable with a variance increased by a c^2 factor:

$$V(c\mathbf{v}) = c^2 V(\mathbf{v}) \tag{3.23}$$

Combining these properties, which are derivable from (3.11), we have, from (3.20),

$$V(\mathbf{y}_n) = \tfrac{1}{4}V(\mathbf{x}_n + \mathbf{x}_{n-1}) = \tfrac{1}{4}(\tfrac{1}{12} + \tfrac{1}{12}) = \tfrac{1}{24}$$

Thus

$$\sigma_y = \frac{1}{\sqrt{2}}\sigma_x \tag{3.24}$$

and the output standard deviation is smaller than the input standard deviation.

The smaller output deviation means that output values are more likely than

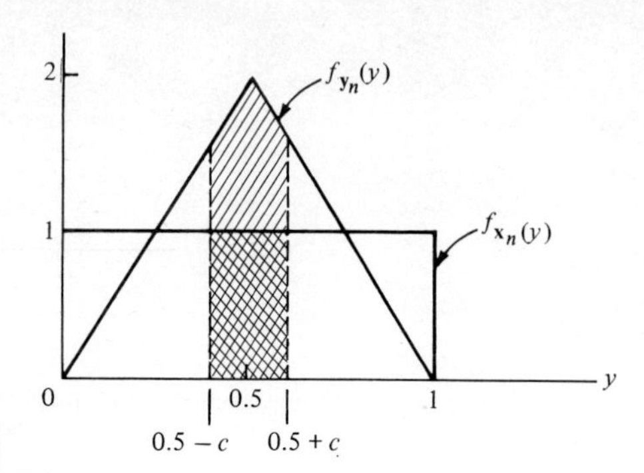

FIGURE 3.8
Input and output densities.

input values to be near the mean of $\frac{1}{2}$. The affect of averaging is thus to reduce the size of the fluctuations about the mean value of $\frac{1}{2}$. We are tempted to draw this conclusion from the fact that the width of a *uniform* density is proportional to its standard deviation. The reasoning fails here because the $\mathbf{y}_n$ do *not* have uniform densities. We omit the simple calculation which shows that $f_{\mathbf{y}_n}(y)$ is the triangular density shown in Fig. 3.8. The input density is also shown for comparison. Note the legal but novel use of y as an argument in $f_{\mathbf{x}_n}(y)$. Any variable can be used as the argument, since it appears only as a dummy variable of integration in the defining property

$$P[a \leq \mathbf{x}_n \leq b] = \int_a^b f_{\mathbf{x}_n}(y)\, dy$$

Clearly, $\mathbf{y}_n$ is more likely than $\mathbf{x}_n$ to be near the mean of $\frac{1}{2}$; that is, the probability that $\mathbf{y}_n$ is within some deviation c from $\frac{1}{2}$ is greater than the corresponding probability for $\mathbf{x}_n$:

$$P[\tfrac{1}{2} - c \leq \mathbf{y}_n \leq \tfrac{1}{2} + c] > P[\tfrac{1}{2} - c \leq \mathbf{x}_n \leq \tfrac{1}{2} + c] \tag{3.25}$$

These probabilities of being near the mean of $\frac{1}{2}$ are the areas under the respective densities and between a pair of lines like the dotted ones in Fig. 3.8. Figure 3.7 shows typical input and output realizations, with the output samples nearer to $\frac{1}{2}$ (and farther from the extremes of 0 or 1) than the input samples, on the average.

The foregoing calculations have a message-processing interpretation. The signal $\mathbf{x}_n$ can be thought of as having desired and disturbance components $\mathbf{s}_n$ and $\mathbf{d}_n$:

$$\mathbf{x}_n = \mathbf{s}_n + \mathbf{d}_n \tag{3.26}$$

with $\mathbf{s}_n = \frac{1}{2}$ and $\mathbf{d}_n$ being purely random with uniform distribution between $-\frac{1}{2}$ and $\frac{1}{2}$. The equivalence of this model can be checked by starting with (3.26) and showing that the resulting $\mathbf{x}_n$ has the same uniform distribution used in the preceding calculations. Using the distribution definition (3.4) and the signal model (3.26), we get

$$\begin{aligned} F_{\mathbf{x}_n}(x) &= P[\mathbf{x} \leq x] = P[\tfrac{1}{2} + \mathbf{d}_n \leq x] \\ &= P[\mathbf{d}_n \leq x - \tfrac{1}{2}] = F_{\mathbf{d}_n}(x - \tfrac{1}{2}) \end{aligned}$$

Figure 3.9 shows $F_{\mathbf{x}_n}(x)$ as a shifted version of $F_{\mathbf{d}_n}(x)$, as well as the corresponding

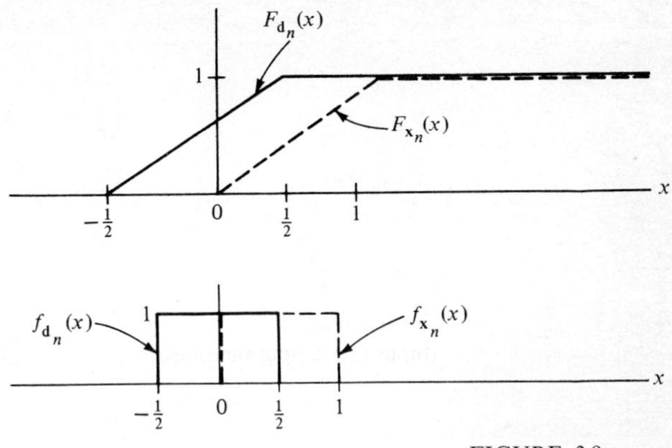

FIGURE 3.9
Shifted distributions.

densities. One can also verify that $\mathbf{x}_1, \mathbf{x}_2, \ldots$ will be statistically independent; i.e., the signal is purely random, if the $\mathbf{d}_1, \mathbf{d}_2, \ldots$ have this property.

The effect of the averaging is to produce at the output of the averaging filter a signal $\mathbf{y}_n$ with a desired component still $\frac{1}{2}$ and a disturbance component $\mathbf{d}'_n$ whose standard deviation is $1/\sqrt{2}$ that of the input-disturbance component $\mathbf{d}_n$. Clearly the output disturbance, or noise, has been reduced by the averaging filter. This agrees qualitatively with the frequency-response analysis of the averaging filter carried out in Sec. 2.8. There, although the signals considered were deterministic, the desired component was modeled as a slowly varying, or low-pass, signal, while the disturbance component was a rapidly varying, or high-pass, signal. The high-pass disturbance signal was of course attenuated by the low-pass averaging filter. Here the desired signal is a known constant value, while the disturbance is a *purely random process:* it changes independently from sample to sample and is thus rapidly varying. (The rapidity of variation of random processes will be given quantitative meaning later by introduction of the concept of autocorrelation.)

There is nothing special about the constant desired signal value of $\frac{1}{2}$. If that component has *any* constant value (which might be unknown to us), the mean of the averager's output will be that desired signal, as in (3.21), and the output will fluctuate less about that mean than the input does. For this reason, the averager can be termed an *estimator* of the desired signal. Other estimation filters will be developed in Chaps. 6 and 7 for approximating desired signals when they are observed with random disturbances.

The *uniform* distribution of the purely random input also is not crucial for the output mean and variance calculations. Another distribution of interest is the gaussian, or normal, distribution with density

$$f_{\mathbf{x}_n}(x) = \frac{1}{\sqrt{2\pi}\sigma_x} e^{-(x-\eta_x)^2/2\sigma_x{}^2} \tag{3.27}$$

an example of which is shown in Fig. 3.10. Here $\eta_x = -1$ and $\sigma_x = 2$. As the notation suggests, the η_x in the exponent of (3.27) is the mean of the gaussian random

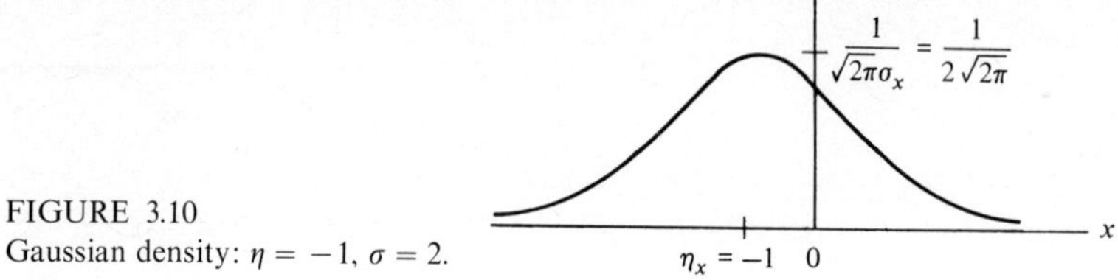

FIGURE 3.10
Gaussian density: $\eta = -1$, $\sigma = 2$.

variable, and σ_x^2 is its variance. For the averager example the input samples will have this density if all $\mathbf{s}_n = -1$, and the $\mathbf{d}_n$'s are independent zero-mean gaussian variables with variances of 4. Since the foregoing calculations of output mean and variance are independent of the form of the $\mathbf{x}_n$ distributions, we can conclude that this gaussian input produces an averager output with

$$\eta_y = -1 \qquad \sigma_y = \frac{2}{\sqrt{2}} \tag{3.28}$$

Of course, some of the output properties will be different when gaussian input samples replace uniform ones. One especially nice property of gaussian inputs is that they produce gaussian outputs, *for any linear filter*. This result follows from the linear-filter form [see (2.107)]

$$y_n = \sum_j h_j x_{n-j}$$

and the theorem† that a weighted sum of gaussian variables is also gaussian. This is a special gaussian-signal property, because input and output signals usually have different densities. (Recall that the uniform input density led to a triangular output density, for the averager.) Thus, in the present example, the $\mathbf{y}_n$ have densities of the form of (3.27) with the η and σ parameters defined in (3.28). Figure 3.11 shows the gaussian input and output densities on the same graph.

A study of the areas under those densities shows that an output sample has greater probability of being near its mean (the desired signal) than an input sample does, as described in the inequality of (3.25).

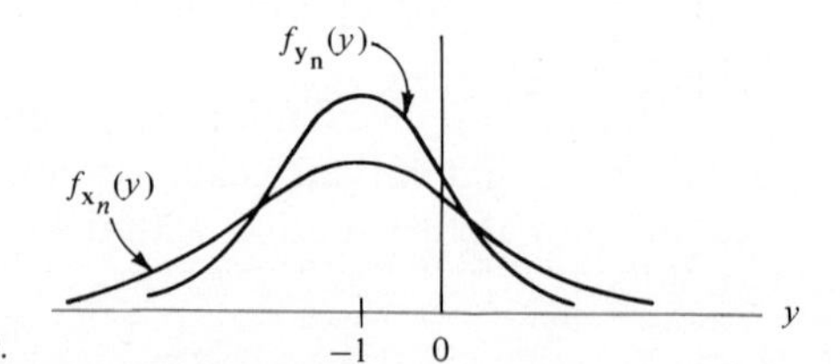

FIGURE 3.11
Input and output gaussian densities.

† E. Parzen, "Stochastic Processes," p. 90, Holden-Day, San Francisco, 1962, or A. Papoulis, "Probability, Random Variables, and Stochastic Processes," p. 253, McGraw-Hill, New York, 1965.

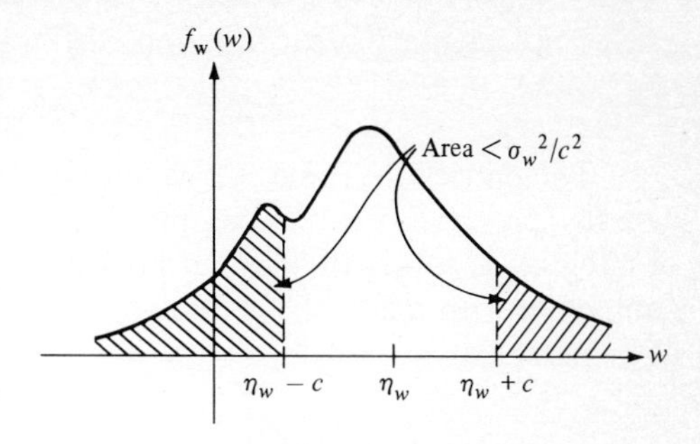

FIGURE 3.12
Tschebychev inequality example.

We again see that the averager, which was first introduced to smooth a known high-frequency disturbance observed with a known slowly varying desired signal, is also a good smoother for the class of constant desired signals and the class of zero-mean purely random disturbance signals. We have demonstrated the averager's ability to narrow the disturbance density in two special cases, i.e., when the input-disturbance samples had uniform and gaussian densities, respectively. In each case, the comparisons of input and output fluctuations followed from actual calculation of the output density.

It is possible to show that the averager has a kind of density-narrowing effect for *any* input-disturbance density. We have seen that the averager reduces the disturbance variance $\sigma_y^2 < \sigma_x^2$ and that for uniform and gaussian cases, a smaller variance means a narrower density. The Tschebychev inequality describes how the concentration of area under a density is measured by its variance. This inequality states that for *any* density $f_{\mathbf{w}}(w)$,

$$P[\,|\mathbf{w} - \eta_w| > c] \le \frac{\sigma_w^2}{c^2} \tag{3.29}$$

Here c is any constant. The probability of (3.29) is shown by the crosshatched area in Fig. 3.12. This inequality means that a small σ_w^2 implies a small probability that $\mathbf{w}$ is far from its mean (or a large probability ≈ 1 that $\mathbf{w}$ is near its mean). Thus, the variance-reducing property of the averager also reduces the probability that an output sample is far from its mean value, the desired signal, no matter what density the input disturbance has.

It is left as a problem to show that there is a density function for $\mathbf{w}$ for which (3.29) holds with the equality sign. However, uniformly distributed and gaussian-distributed variables are less likely to be far from their mean values. The probability on the left in (3.29) is easy to evaluate in the uniform case, but the gaussian case is more difficult because integrals of the density in (3.27) cannot be expressed in closed form. Table 3.1 shows the so-called error function, erf z, which is defined by the following integral of a normal, or gaussian, density with zero mean and variance of $\frac{1}{2}$:

$$\operatorname{erf} z = \frac{1}{\sqrt{\pi}}\int_{-z}^{z} e^{-\xi^2}\, d\xi = \frac{2}{\sqrt{\pi}}\int_{0}^{z} e^{-\xi^2}\, d\xi \tag{3.30}$$

$$\operatorname{erf} z = P[\,|\mathbf{z}| < z] \tag{3.31}$$

The complementary error function, also frequently used, is defined as

$$\operatorname{erfc} z = 1 - \operatorname{erf} z = P[\,|\mathbf{z}| \geq z] \tag{3.32}$$

The entries in Table 3.1 were computed by numerical integration. A general gaussian probability can be computed from this table by using changes in the variable of integration to get the desired probability into the form of (3.30). For example, a gaussian **x** with mean of -1 and variance of 4, as in Fig. 3.10, has as its probability of being negative

$$P[\mathbf{x} < 0] = \int_{-\infty}^{0} \frac{1}{\sigma\sqrt{2\pi}} e^{-(x-\eta)^2/2\sigma^2}\, dx \qquad \begin{aligned} \eta &= -1 \\ \sigma &= \;\; 2 \end{aligned} \tag{3.33}$$

When we let

$$\xi = \frac{x+1}{2\sqrt{2}} = \frac{x-\eta}{\sigma\sqrt{2}} \tag{3.34}$$

the integral becomes

$$\begin{aligned} P[\mathbf{x} < 0] &= \int_{-\infty}^{1/2\sqrt{2}} e^{-\xi^2} \frac{d\xi}{\sqrt{\pi}} = \frac{1}{2} + \frac{1}{2} \operatorname{erf} \frac{1}{2\sqrt{2}} \\ &= \frac{1}{2} + \frac{0.39}{2} = 0.695 \end{aligned} \tag{3.35}$$

The last step has also used the symmetry of the gaussian density and the fact that a gaussian variable is less than its mean with a probability of $\frac{1}{2}$. Interpolation between entries in Table 3.1 was also required.

The steps in this calculation should be thoroughly understood because many detection probabilities and error probabilities in later chapters will use these func-

Table 3.1 ERROR FUNCTION
Gaussian probabilities Eq. (3.30)

z	erf z	z	erf z	z	erf z
0.000	0.0000	0.700	0.6778	1.800	0.9891
0.050	0.0564	0.750	0.7111	1.900	0.9928
0.100	0.1125	0.800	0.7421	2.000	0.9953
0.150	0.1680	0.850	0.7707	2.100	0.9970
0.200	0.2227	0.900	0.7969	2.200	0.9981
0.250	0.2763	0.950	0.8209	2.300	0.9989
0.300	0.3286	1.000	0.8427	2.400	0.9993
0.350	0.3794	1.100	0.8802	2.500	0.99959
0.400	0.4284	1.200	0.9103	2.60	0.99976
0.450	0.4755	1.300	0.9340	2.70	0.99987
0.500	0.5205	1.400	0.9523	2.80	0.99992
0.550	0.5633	1.500	0.9661	2.90	0.99996
0.600	0.6039	1.600	0.9764	3.00	0.99998
0.650	0.6420	1.700	0.9838		

tions. For example, we shall see in Chap. 5 that the probability of making a mistake in deciding whether a message pulse is present or absent can be expressed in terms of an erf u for a suitable u. It is also convenient to understand the use of similar tables which normalize to a zero-mean, *unit*-variance gaussian variable $\mathbf{u}$ and list the area to the left of u, that is,

$$\phi(u) = P[\mathbf{u} \le u] = \int_{-\infty}^{u} e^{-u^2/2} \frac{du}{\sqrt{2\pi}} = \frac{1}{2}\left(1 + \operatorname{erf} \frac{u}{\sqrt{2}}\right) \tag{3.36}$$

A different change of integration variables will put a gaussian probability integral into this form.

Another notation we shall use frequently is that of decibel ratios. For example, the ratio of output noise variance to input noise variance σ_y^2/σ_x^2 of the averager can be expressed logarithmically. We say the ratio in decibels (dB) is

$$10 \log_{10} \frac{\sigma_y^2}{\sigma_x^2} \text{ dB} \tag{3.37}$$

This logarithmic scale is convenient because it allows a large range of ratios to be expressed by numbers lying in a small range of decibels.

Traditionally, decibels have been used to measure ratios of power (or average power). If the variable $\mathbf{x}_n$ were a zero-mean voltage across a 1-Ω resistor, the power into that resistor would be $\mathbf{x}_n^2$. The mean value of that power would then be $E[\mathbf{x}_n^2] = \sigma_x^2$. If the mean power is interpreted as an average over many realizations, we see that using a decibel measure for variance ratios is consistent with the average-power convention.

This section has introduced random signals by first reviewing random-variable theory and then considering properties of the response of an averager filter when its input is a purely random signal. The next section will discuss techniques for estimating the mean, variance, and density function of a random signal based on typical measurements of that signal. After that, we shall show how random signals can be simulated by simple digital-computer programs for the purpose, say, of testing one of the signal-processing algorithms to be developed later. The full complexity of random signals, compared with variables, will become clear when successive samples like $\mathbf{y}_1, \mathbf{y}_2, \mathbf{y}_3, \ldots$ are allowed to be *dependent*, in contrast to the samples of a purely random signal. Finally, the sampling of continuous-time random signals will be considered in much the same way as the sampling of deterministic signals in earlier chapters.

3.2 RANDOM DATA: MOMENTS AND HISTOGRAMS

There are at least two kinds of signal-processing situations in which we shall want to estimate the mean value of a random variable from actual data. Sometimes the mean value is the message-bearing part of the signal, as in the averager example in the previous section. On the other hand, the design of a signal processor might require knowledge of the mean, variance, or even more general properties of the distribution of the random noise in the observations. In such cases, large amounts of typical noise data are recorded and processed to get estimates of the desired quantities.

One general approach to the estimation of moments like the mean, variance, etc., of a random sequence is based on the intuitive connection between the mean value and the average of a large number of independent observations of a random variable. If $\mathbf{x}_1, \mathbf{x}_2, \ldots$ are statistically independent, that is, $\mathbf{x}_n$ is a purely random sequence, their common mean value can be estimated by the average of N observed values

$$E[\mathbf{x}_n] \approx \frac{1}{N} \sum_{k=1}^{N} x_k = \hat{\eta}_x \tag{3.38}$$

(A hat will often be used to designate a variable which is an estimate.)

One future example in which we shall need to know a mean value is that of tracking an airplane as it approaches an airport. We shall need to know the mean and variance of uncertainties in the plane's ground speed when it is first picked up by the airport radar. The mean speed is likely to be the nominal speed presented in the flight plan, but it might be different due to errors between the prevailing wind speed and the pilot's information about winds, or pilot psychology, or some combination of these and related factors. Careful measurements (with special precise radars, lasers, etc.) under controlled conditions could provide data for estimating the mean value, as in (3.38).

The N-sample average has been proposed in (3.38) as an estimate of the mean value. We would like to see whether this estimate is any good and if possible show that as more data become available (larger N), the estimate improves. Different realizations x_k will produce different values of $\hat{\eta}_x$. The collection of those values together with a probability distribution for them are symbolized by the random variable $\hat{\boldsymbol{\eta}}_x$. The estimator in (3.38) is said to be *unbiased* (good on the average) if

$$E[\hat{\boldsymbol{\eta}}_x] = \eta_x \tag{3.39}$$

Another good property, that the estimate improves as more data become available, is demonstrated by showing that the mean-squared estimation error decreases according to the relation

$$\lim_{N \to \infty} E[(\hat{\boldsymbol{\eta}}_x - \eta_x)^2] = 0 \tag{3.40}$$

An estimate with this property is said to be *consistent*. The mean-squared error between a consistent estimate and its mean goes to zero for large N.

Thus we are judging the quality of the mean-value estimator in (3.38) with regard to its accuracy as averaged over many applications to separate estimation situations. An unbiased estimate is correct "on the average," and if it is also consistent, its variance approaches zero for large N. When the Tschebychev inequality of (3.29) is used, these properties imply that a typical $\hat{\eta}_x$ for a large N is very likely to be near the true mean η_x.

The unbiased property of (3.39) is easily proved using a generalization of the mean-value property of (3.14), which justifies interchanging the order of the mean-value and summation operations, as follows:

$$E[\hat{\boldsymbol{\eta}}_x] = E\left[\frac{1}{N}\sum_{1}^{N}\mathbf{x}_k\right] = \frac{1}{N}\sum_{1}^{N}E[\mathbf{x}_k] = \eta_x \tag{3.41}$$

The mean-squared error is computed most easily by using the summation representations of η_x in (3.41) and $\hat{\boldsymbol{\eta}}_x$ in (3.38) to get

$$E[(\hat{\boldsymbol{\eta}}_x - \eta_x)^2] = E\left[\left[\frac{1}{N}\sum_{1}^{N}(\mathbf{x}_k - \eta_x)\right]^2\right] \tag{3.42}$$

The square of a summation is written with least confusion as a product of summations with different summation indices, and so (3.42) becomes

$$\frac{1}{N^2}E\left[\sum_{k=1}^{N}(\mathbf{x}_k - \eta_x)\sum_{r=1}^{N}(\mathbf{x}_r - \eta_x)\right] = \frac{1}{N^2}\sum_{k=1}^{N}\sum_{r=1}^{N}E[(\mathbf{x}_k - \eta_x)(\mathbf{x}_r - \eta_x)] \tag{3.43}$$

When $r = k$,

$$E[(\mathbf{x}_k - \eta_x)^2] = \sigma_x^2 \tag{3.44}$$

By definition, when $r \neq k$,

$$E[(\mathbf{x}_k - \eta_x)(\mathbf{x}_r - \eta_x)] = E[\mathbf{x}_k - \eta_x]E[\mathbf{x}_r - \eta_x] = 0 \tag{3.45}$$

since variables with different indices are assumed to be independent and

$$E[\mathbf{x}_k - \eta_x] = E[\mathbf{x}_k] - \eta_x = 0$$

{Recall from (3.18) that the joint density of independent random variables factors into marginal densities, so that the expectation of a product of functions, as in (3.45), becomes

$$\begin{aligned} E[g_1(\mathbf{x}_1)g_2(\mathbf{x}_2)] &= \int_{x_1}\int_{x_2} g_1(x_1)g_2(x_2)f_{\mathbf{x}_1\mathbf{x}_2}(x_1,x_2)\,dx_1\,dx_2 \\ &= \int_{x_1} g_1(x_1)f_{\mathbf{x}_1}(x_1)\,dx_1 \int_{x_2} g_2(x_2)f_{\mathbf{x}_2}(x_2)\,dx_2 \\ &= E[g_1(\mathbf{x}_1)]E[g_2(\mathbf{x}_2)]\} \end{aligned}$$

Thus there are exactly N nonzero terms of the form (3.44) in the double sum of (3.43), with the result that

$$E[(\hat{\boldsymbol{\eta}}_x - \eta_x)^2] = \frac{\sigma_x^2}{N} \tag{3.46}$$

which approaches zero (for consistency) as N approaches infinity.

We have just derived a more general form of (3.22), i.e., that the variance of the sum of independent random variables is the sum of the variances. The extension says that this is true for the sum of *any* number of random variables. Specifically, if we factor out the constant $(1/N)^2$ term in the right-hand side of (3.42) and note that the remaining expectation represents the variance of the sum of N-independent random variables, the mean-squared error is just $1/N^2$ times the sum of the variances. But these are all equal, so that the sum is just $N\sigma_x^2$. Equation (3.46) then results.

We shall see later that the average or sample-mean estimate defined in (3.38) has other good properties in the event that we also know that the $\mathbf{x}_k$ variables are gaussian.

The intuitive correspondence between averages and mean values similarly suggests the variance estimate

$$E[(\mathbf{x}_n - \eta_x)^2] \approx \frac{1}{N} \sum_{k=1}^{N} (x_k - \hat{\eta}_x)^2 = \hat{\sigma}_x^2 \tag{3.47}$$

A simple calculation shows that the variance $V(\hat{\boldsymbol{\sigma}}_x^2)$ of this estimate approaches zero for large N, so that the estimate *is* consistent. The mean or expected value of this estimate equals $[(N-1)/N]\sigma_x^2$, however, so that it is unbiased only for large N. [For this reason the unbiased estimate $[1/(N-1)] \sum (x_k - \hat{\eta}_x)^2$ is often used instead, but for large N there is no real distinction.]

The preceding averaging estimates are also unbiased and consistent for some cases of *dependent* samples, e.g., when the variables $\mathbf{x}_1, \mathbf{x}_2, \ldots$ are from a random signal which is not purely random.

Higher moments can also be reasonably estimated by sample averages like those in (3.38) and (3.47). Our mean and variance estimates will be further justified later by showing that they are so-called *maximum-likelihood estimates* if $\mathbf{x}_n$ is a gaussian signal. If $\mathbf{x}_n$ is known to be gaussian, its first-order probability density can be estimated by a gaussian density

$$\hat{f}_{\mathbf{x}_n}(x) = \frac{1}{\sqrt{2\pi\hat{\sigma}_x^2}} \exp\left[-\frac{(x-\hat{\eta}_x)^2}{2\hat{\sigma}_x^2}\right] \tag{3.48}$$

This uses the mean and variance estimates $\hat{\eta}_x$ and $\hat{\sigma}_x^2$, which we have just defined.

It is more difficult to estimate the density from samples of a random signal when we do not know or cannot reasonably assume that $\mathbf{x}_n$ is gaussian (or has any other particular distribution). In such cases, several points on the density-function curve can be estimated by a *histogram*. The histogram is based on the property that the area under $f_{\mathbf{x}_n}(x)$ in an interval of x-values corresponds to the probability that $\mathbf{x}_n$ will lie in that interval. This approach also uses the *law-of-large-numbers* idea that the probability of an event is approximated by the ratio of the number of occurrences of that event to the total number of independent random trials.

As an example, consider the density function $f_{\mathbf{x}_n}(x)$ shown in Fig. 3.13. Say that the random variable $\mathbf{x}_n$ is discretized into a set of values spaced 2ε units apart. Figure 3.13 shows such a cell 2ε units wide centered at $x = 3$. If ε is small enough, it is apparent that the area of that cell, an interval probability, is approximately

$$2\varepsilon f_{\mathbf{x}_n}(3) \approx P[3 - \varepsilon < \mathbf{x}_n \le 3 + \varepsilon] \tag{3.49}$$

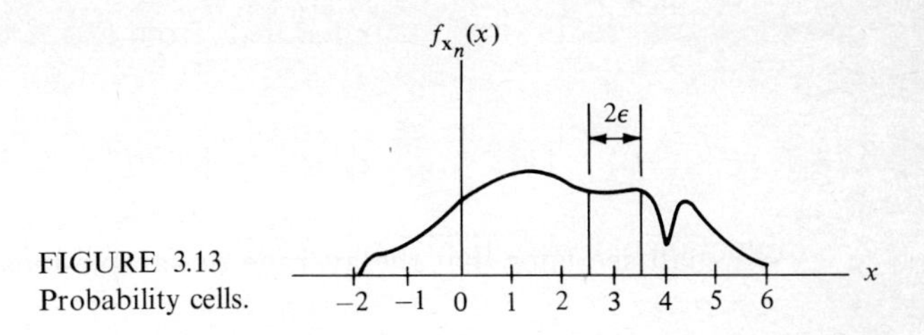

FIGURE 3.13
Probability cells.

But if $\mathbf{x}_1, \mathbf{x}_2, \ldots$ are independent and identically distributed, the probability that a typical $\mathbf{x}_n$ lies in the range $3 \pm \varepsilon$ is given approximately by

$$P[3 - \varepsilon < \mathbf{x}_n \leq 3 + \varepsilon]$$

$$\approx \frac{\text{number of } x_1, x_2, \ldots, x_N \text{ between } 3 - \varepsilon \text{ and } 3 + \varepsilon}{N} \tag{3.50}$$

When these approximations are combined, a reasonable estimate for $f_{\mathbf{x}_n}(3)$ seems to be (using $k = 1, 2, \ldots N$ as the index for *measured* values)

$$\hat{f}_{\mathbf{x}_n}(3) = \frac{\text{number of } x_k \text{ between } 3 - \varepsilon \text{ and } 3 + \varepsilon}{2N\varepsilon} \tag{3.51}$$

Obviously there is nothing special about the choice of $x = 3$ in this example. This argument will be repeated for different values of x to estimate other $f_{\mathbf{x}_n}(x)$.

The accuracy of this histogram estimate can be measured by its mean and variance. To aid in calculating these moments it is appropriate to introduce an indicator function $I_3(x_k)$. We define this to be equal to 1 if the sample x_k falls in the range $3 \pm \varepsilon$ and 0 if it falls outside the range:

$$I_3(x_k) = \begin{cases} 1 & 3 - \varepsilon < x_k \leq 3 + \varepsilon \\ 0 & \text{all other } x_k \text{ values} \end{cases} \tag{3.52}$$

In terms of this indicator function the number of values x_k falling in the range $3 \pm \varepsilon$ is given by

$$\sum_{k=1}^{N} I_3(x_k) = \text{number of } x_k \text{ between } 3 - \varepsilon \text{ and } 3 + \varepsilon \tag{3.53}$$

In this way (all subscripts on densities are omitted for simplicity), the density-function estimate of (3.51) becomes

$$\hat{\mathbf{f}}(3) = \frac{1}{2N\varepsilon} \sum_{k=1}^{N} I_3(\mathbf{x}_k) \tag{3.54}$$

when it is considered as a random variable produced from the $\mathbf{x}_k$. The mean value of this density estimate is

$$E[\hat{\mathbf{f}}(3)] = \frac{1}{2N\varepsilon} N E[I_3(\mathbf{x}_k)] \tag{3.55}$$

since the $\mathbf{x}_k$ are identically distributed. But, from (3.52), the indicator functions have mean values

$$E[I_3(\mathbf{x}_k)] = 1P[3 - \varepsilon < \mathbf{x}_k \leq 3 + \varepsilon] + 0P[\mathbf{x}_k > 3 + \varepsilon \text{ or } \mathbf{x}_k \leq 3 - \varepsilon] \quad (3.56)$$

A combination of (3.49), (3.55), and (3.56) shows that this $\hat{\mathbf{f}}$ is approximately unbiased

$$E[\hat{\mathbf{f}}(3)] \approx f(3) \quad (3.57)$$

The approximation is better the smaller ε is, as suggested in Fig. 3.13. We would like to make a more precise statement about possible bias in this estimate, but (3.57) is the best we can do in view of our complete lack of prior information about the shape of the density.

The variance of this $\hat{\mathbf{f}}$ estimate can also be approximated by similar arguments. When the deviation from the mean is written as

$$\hat{\mathbf{f}}(3) - E[\hat{\mathbf{f}}(3)] \approx \frac{1}{2N\varepsilon} \sum_{1}^{N} [I_3(\mathbf{x}_k) - 2\varepsilon f(3)] \quad (3.58)$$

the variance, which is the mean-squared value of this deviation, becomes [see (3.42)]

$$V[\hat{\mathbf{f}}(3)] \approx \frac{1}{4N^2\varepsilon^2} \sum_{k=1}^{N} \sum_{j=1}^{N} E[[I_3(\mathbf{x}_k) - 2\varepsilon f(3)][I_3(\mathbf{x}_j) - 2\varepsilon f(3)]] \quad (3.59)$$

When $k \neq j$,

$$E[I_3(\mathbf{x}_k)I_3(\mathbf{x}_j)] = E[I_3(\mathbf{x}_k)]E[I_3(\mathbf{x}_j)] \approx [2\varepsilon f(3)]^2 \quad (3.60)$$

When $k = j$, this expectation is

$$E[I_3^2(\mathbf{x}_k)] \approx 2\varepsilon f(3) \quad (3.61)$$

These last expressions combine to give a variance approximation

$$V[\hat{\mathbf{f}}(3)] \approx \frac{1}{4N^2\varepsilon^2} [N2\varepsilon f(3) + (N^2 - N)4\varepsilon^2 f^2(3) - N^2 4\varepsilon^2 f^2(3)]$$

$$\approx \frac{1}{2N\varepsilon} [f(3) - 2\varepsilon f^2(3)] \quad (3.62)$$

We see that this histogram estimate $\hat{\mathbf{f}}$ is good in the sense that its variance approaches zero as N approaches infinity.

The histogram variance in (3.62) has interesting relationships to the interval width ε. This expression has a built-in reminder of the approximate equalities used in its derivation. Those approximations are most valid for small ε, and we see in (3.62) that a meaningless *negative* variance will result if ε is too large. A conservative approach, especially desirable in view of the approximations leading to (3.62), is to ignore the negative term and say

$$\hat{\sigma}_f^2 = V[\hat{\mathbf{f}}(3)] \approx \frac{f(3)}{2N\varepsilon} \quad (3.63)$$

Possible inaccuracies in approximations like (3.49) are revealed by examining Fig. 3.13 for $f_{\mathbf{x}_n}(4)$ and seeing that the area between $4 \pm \varepsilon$ is much greater than suggested by a constant density over that interval, i.e.,

$$2\varepsilon f_{\mathbf{x}_n}(4) \ll P[4 - \varepsilon < \mathbf{x}_n < 4 + \varepsilon] = \int_{4-\varepsilon}^{4+\varepsilon} f_{\mathbf{x}_n}(x)\, dx \tag{3.64}$$

This approximation error can be reduced by using a *narrower* cell, which makes the mean of $\hat{\mathbf{f}}$ show the sharp details in $f(x)$. However, the factor to the left of the bracket in (3.62) shows that a *large* ε has the desirable property of making the variances smaller.

Thus, the choice of the width 2ε of the histogram cell must be a compromise between a desire to reduce the variance of $\hat{\mathbf{f}}$ (using a large ε) and a desire to make $\hat{\mathbf{f}}$ unbiased as in (3.57) (using a small ε). A good estimate is one in which the bias

$$B = E[\hat{\mathbf{f}}] - f$$

is small *and* the variance is small so that the estimate has a high probability of being near the true value.

Unfortunately, since it is the unknown $f(x)$ that we are trying to estimate, we do not know beforehand the width of the narrow peaks and valleys which put upper limits on the desirable cell width 2ε. In practice, results of a histogram using one set of cell widths can suggest different widths for a second histogram calculation from the same data. Such a judgment can be based on the idea that if the number

$$Q_3 = \sum_{k=1}^{N} I_3(x_k) \tag{3.65}$$

of samples in cell 3 is small compared to N, the resulting estimate

$$\hat{f}(3) = \frac{Q_3}{2N\varepsilon} \tag{3.66}$$

is not reliable because ε must have been too small. Conversely, a large Q_j suggests that the corresponding cell could be decomposed into smaller cells to improve the bias without causing too much deterioration in variance. Obviously the best results occur when ε can be made very small and N arbitrarily large (unlimited data) so that $N\varepsilon$ is large and the variance of $\hat{\mathbf{f}}$ is small.

In actually carrying out the estimation of a histogram of random data by computer the histogram program must define the location and width of several cells and then see how many data samples lie in each cell. The definition of appropriate cells is aided by first using (3.38) and (3.47) to find the mean and variance of the data. These parameters define the range of the data and hence the coverage of the cells required. Often both can be computed during one reading of each data value, rather than using all of the data to get $\hat{\eta}$ and going through the data again to get $\hat{\sigma}^2$. The details of this simultaneous calculation are described in a problem.

If the data are gaussian, most of the samples will lie within $\pm 2\sqrt{\hat{\sigma}^2}$ of the mean estimate $\hat{\eta}$. (Table 3.1 shows that 95% of these variables will lie in this interval. Use $z = 1.4$, since $\sigma^2 = 0.5$ for that table.) This assumes that $\hat{\eta}$ and $\hat{\sigma}^2$ are accurate.

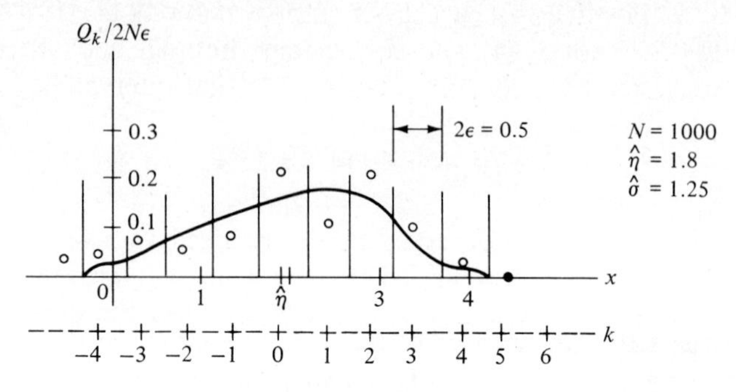

FIGURE 3.14
Histogram.

Thus it makes sense to define cells to cumulatively cover a region of width $4\sqrt{\hat{\sigma}^2}$. The cells should be wide enough for each to contain a large number of samples (say 100). In the absence of other information we might make the cells have equal widths 2ε and guess that each cell will receive an equal number of samples. In this way, the required number of cells is the *largest integer less than* $N/100$, symbolized by $[[N/100]]$. These choices of the $\pm 2\sigma$ region and the number of cells combine to define ε by

$$2\varepsilon\left[\left[\frac{N}{100}\right]\right] \approx 4\sqrt{\hat{\sigma}^2} \tag{3.67}$$

Figure 3.14 shows a typical arrangement with a possible true density superimposed. Since in this example $N = 1000$ samples were found to have $\hat{\eta} = 1.8$ and $\hat{\sigma} = 1.25$, our guidelines suggest 10 cells, 0.25 unit wide. An eleventh cell was added so the middle cell could be centered on $\hat{\eta}$. The dots in that figure represent possible histogram estimates with

$$\frac{Q_k}{2N\varepsilon} = \hat{f}(x_k)$$

and the cell centers x_k defined by

$$x_k = \hat{\eta} + 2k\varepsilon \qquad k = 0, \pm 1, \ldots \tag{3.68}$$

The two end cells are extended to run to ∞ and $-\infty$, respectively, so that even a data sample far from the mean will lie in some cell. If the true density is not concentrated near its mean, the end cells will contain large numbers of samples and a new histogram will have to be computed with more cells in these outlying regions.

The variance estimate can be used to judge the reasonableness of a fit of a density-function curve to the histogram points. For example, the $k = 1$ point in Fig. 3.14 seems far from the proposed curve. However, the distance from the curve can be better evaluated with respect to the standard deviation of the histogram estimates. If the points are based on 1000 data samples, and if the smooth curve is the true density, substitution of the height 0.18 of the curve at $k = 1$ for $f(x_1)$ in (3.63) yields

$$V[\hat{\mathbf{f}}(x_1)] \approx 0.00036 = (0.019)^2 \tag{3.69}$$

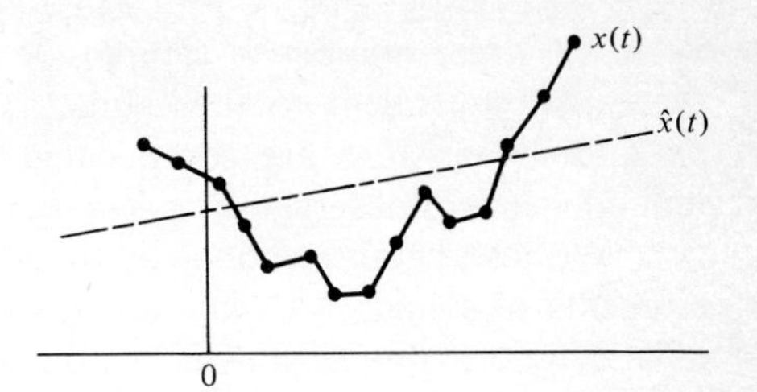

FIGURE 3.15
Noisy parabola.

This is the variance $\sigma_{\hat{f}}^2$ of the estimate assuming that the smooth curve is the true density. Thus $\hat{f}(x_1) \approx 0.12$ is more than $3\sigma_{\hat{f}} \approx 0.057$ from the curve, an event with the small probability

$$P[|\hat{\mathbf{f}} - f| > 3\sigma_{\hat{f}}] \leq \frac{\sigma_{\hat{f}}^2}{9\sigma_{\hat{f}}^2} = 0.11 \tag{3.70}$$

according to the Tschebychev inequality. This bound on the probability is actually quite generous, because each histogram point, as the sum of many independent zero or one random variables, as in (3.54), tends to have a Poisson distribution.† Furthermore, according to the central-limit theorem,† the probability that the Poisson variable $\hat{\mathbf{f}}$ is far from its mean can be approximated by the expression

$$P[|\hat{\mathbf{f}} - f| > 3\sigma_{\hat{f}}] \approx 0.003 \tag{3.71}$$

which applies to a gaussian variable with the same mean and variance.‡ Clearly, then, a proposed density curve should be closer to the $f(x_1)$ point than is the curve shown.

Another way to assess histogram accuracy is to compare the true density to the standard deviation of the histogram error. This point of view normalizes the uncertainty to a fraction or percentage of the quantity being estimated. It is left as a problem to show that the rules proposed here, namely $N/100$ cells spread over a range of $4\hat{\sigma}$, result in $\sigma_{\hat{f}} < f/10$ if the true density is uniform. While most densities being estimated are not uniform, this calculation gives an indication of the relative size of histogram errors.

One interesting use of histogram results can be a verification of the choice of trend functions which have been removed from the data. An improper trend function might show up through an unexpected histogram. If the residual errors are expected to be uniformly distributed (based on knowledge of the mechanisms causing these errors), a histogram with a sharp peak might indicate that a higher-order trend should have been used. Figure 3.15 shows data arising from noisy observations of a parabola. If the indicated straight-line trend is removed, the histogram of the residuals will have a peak to the left of its mean. (Why?)

We have developed methods for estimating the mean, variance, and density function of a purely random, stationary signal. The same methods work for random signals which are not purely random, i.e., those for which successive samples are not independent. However, in cases of signals with dependence between successive

† Papoulis, op. cit., p. 268. ‡ Papoulis, op. cit., p. 71.

values, a larger number of samples N are required to get estimates with precision (variance) equivalent to the estimates for a purely random sequence. The exact variance expressions are complicated to calculate, but it should be clear that a nonindependent observation is less useful for "averaging out" random fluctuations.†

We close this section by mentioning maximum-likelihood estimates for moments of signals with *known* density functions. If $\mathbf{x}_1$, $\mathbf{x}_2$, $\mathbf{x}_3$ are independent, identically distributed gaussian variables, their joint density is the product of their individual densities

$$f(x_1, x_2, x_3; \eta, \gamma) = \left(\frac{1}{\sqrt{2\pi\gamma}}\right)^3 e^{-[(x_1-\eta)^2+(x_2-\eta)^2+(x_3-\eta)^2]/2\gamma}$$

$$= f_{\mathbf{x}}(x_1) f_{\mathbf{x}}(x_2) f_{\mathbf{x}}(x_3) \tag{3.72}$$

The left side of (3.72) emphasizes the dependence of the density on the parameters η and γ, the mean and variance. Once specific data have been observed, e.g.,

$$x_1 = 1 \qquad x_2 = -3 \qquad x_3 = 5$$

the density can be viewed as a function of the parameters:

$$f(1, -3, 5; \eta, \gamma) \tag{3.73}$$

and we can estimate η, γ by choosing them to make the particular observed values as likely as possible. The probability of observing values near the ones *actually* observed is

$$f(1, -3, 5; \eta, \gamma)\, dx_1\, dx_2\, dx_3 \tag{3.74}$$

The maximum-likelihood estimates of η and γ are those values $\hat{\eta}$ and $\hat{\gamma}$ which maximize the probability in (3.74). (A more detailed discussion of maximum-likelihood estimation will be found in Chap. 6.)

These estimates are found easily by differentiating (3.72) with respect to η and γ and setting the resulting expressions equal to zero, with the result

$$\frac{\partial f}{\partial \eta} = [(x_1 - \eta) + (x_2 - \eta) + (x_3 - \eta)]\frac{f(x_1, x_2, x_3; \eta, \gamma)}{\gamma} = 0 \tag{3.75}$$

$$\frac{\partial f}{\partial \gamma} = \left[-\frac{3}{2\gamma} + \frac{(x_1 - \eta)^2 + (x_2 - \eta)^2 + (x_3 - \eta)^2}{2\gamma^2}\right] f(\cdot) = 0 \tag{3.76}$$

Dividing (3.75) by f/γ (which is never zero), we get the maximum-likelihood mean-value estimate

$$\hat{\eta} = \tfrac{1}{3}(x_1 + x_2 + x_3) \tag{3.77}$$

It is left as an exercise to show that N independent gaussian observations have a maximum-likelihood mean estimate

$$\hat{\eta} = \frac{1}{N}\sum_{1}^{N} x_j \tag{3.78}$$

† Further information about estimating means, variances, densities, etc., can be found in J. S. Bendat and A. G. Piersol, "Measurement and Analysis of Random Data," 2d ed., Wiley, New York, 1972.

Combination of (3.76) with (3.77) yields the maximum-likehood variance estimate

$$\hat{\gamma} = \frac{1}{3} \sum_{1}^{3} (x_i - \hat{\eta})^2 \tag{3.79}$$

Note that these are just the averaging estimates discussed earlier [see Eqs. (3.38) and 3.47)]. Thus, the mean and variance estimates we have developed are unbiased (except for small-N bias in $\hat{\sigma}^2$) and consistent for any random signals but have the additional maximum-likelihood property when the random signal is gaussian.

The best way to develop an understanding of the erratic nature of random signals and the difficulties involved in estimating their parameters is to work with real data. However, such data are cumbersome to manipulate, especially if they must be punched onto cards for computer input. In some applications this can be avoided by having the computer generate random signals. The next section describes how pseudorandom (or approximately random) signals can be generated by a computer algorithm. The resulting numbers can be used to test the techniques developed in this section, as well as other signal-processing algorithms which we shall devise later.

3.3 GENERATING AND SHAPING PSEUDORANDOM NOISE

This section, a digression from the main theme of data analysis, is concerned with the methods for artificially generating noisy data. Noise samples with prescribed statistical properties are useful for testing proposed data-processing schemes. It is possible, in principle, to store and catalog sequences of noise samples having various statistical properties. A particular sequence can then be referred to when a test is to be performed.

Storage of diverse examples of noise is impractical for two reasons. The storage space (say on magnetic tape) is expensive, and reading out the stored numbers is time-consuming. An alternative is to generate the noise as it is needed.

It is not obvious that noise with prescribed statistical properties can be generated. Moreover, the numbers which are generated by a prescribed computer algorithm can hardly be termed random. Fortunately, a computer can be programmed to generate numbers which are good approximations, in a statistical sense, to samples of random noise. These computer-generated numbers are said to be *pseudorandom.*

A wide variety of noise sequences can be generated by simple operations on a fundamental kind of random sequence consisting of numbers which are "unpredictable" and each of which lies between 0 and 1. The word unpredictable is in quotation marks because it is not well defined. Many different criteria of predictability or randomness are possible. The intuitive idea is that knowledge of the nth number should not convey any information about the $(n + 1)$st, $(n + 2)$nd, etc. Furthermore, any number in the sequence should be equally likely to have any value between 0 and 1.

An indirect but precise statement of the desired properties of a pseudorandom sequence is the following. Any statistical test on this sequence should give results which are indistinguishable from those which would result if the same test were applied to sample values of a sequence of statistically independent random variables each uniformly distributed between 0 and 1.

We shall first discuss the generation of these fundamental pseudorandom numbers, and then procedures will be given for processing these numbers (shaping) to approximate other kinds of random variables. The moment and histogram estimates of the previous section can then be used to see how well the generated numbers match their desired characteristics.

Generation

Very simple algorithms for generating good approximations to the fundamental pseudorandom sequence are available and are based on some fairly abstract results from the theory of numbers. (These results from this branch of mathematics are also significant in the construction of some codes used in communication systems.)

The algorithms of interest are based on *congruence relations*, an example of which is

$$d = bc \pmod{m} \tag{3.80}$$

These symbols mean that d is the remainder formed when bc is divided by m. For example, when $b = 2$ and $c = 5$,

$$d = 2(5) \pmod{3} = 1 \tag{3.81}$$

The recursion relation

$$k_{i+1} = ck_i \pmod{m} \tag{3.82}$$

produces a sequence of numbers which lie between 0 and $m - 1$. (Why?) Then $n_1, n_2, \ldots$, defined by

$$n_i = \frac{k_i}{m} \qquad i = 0, 1, \ldots \tag{3.83}$$

will be a sequence of numbers, each lying between 0 and 1.

Numbers generated by (3.81) and (3.82) have some of the desired pseudorandom properties, but they will not necessarily produce a long sequence of numbers which fluctuate unpredictably and uniformly between 0 and 1. For example, with $k_0 = 1$, $c = 4$, and $m = 5$, the k-sequence is

$$1, 4, 1, 4, 1, 4, \ldots$$

Theorems from number theory tell us how to choose k_0, c, and m to ensure a long string of different numbers before repetition begins. (It should be clear that if some $k_i = k_0$, then $k_{i+1} = k_1$, etc., and that this is undesired predictability.) For example, if k_0 is odd, $m = 2^b$ and $c = 8t \pm 3$ (for some integer t), then there will be 2^{b-2} different values of k_i before repetition begins.† That is, $2^b/4$ numbers are produced in the

† T. H. Naylor et al., "Computer Simulation Techniques," Chaps. 3 and 4, Wiley, New York, 1966; R. P. Chambers, Random Number Generation, *IEEE Spectrum*, vol. 4, no. 2, February 1967, p. 48.

interval 0 to $2^b - 1$. In order to get a candidate for our fundamental pseudorandom sequence, we can divide the k_i by 2^b, getting numbers between 0 and 1. However, the theorem does not rule out the undesirable possibilities that they will all be between zero and 0.5 or that they will follow a predictable pattern, e.g., being in increasing order. **(The first excess digit, in the sign place, is not simply deleted.)**

It has been experimentally observed that numbers generated by the congruence recursion (3.82), with the special parameters just described, are indeed nicely distributed over the entire interval. In addition, successive numbers seem to vary unpredictably. The problem of efficient computer evaluation of this kind of relation remains to be discussed.

It is convenient on a binary computer (like the IBM 360) to choose m to be the largest integer expressible by a single word, plus 1. This will be a large power of 2 (2^{31} on the IBM 360). This choice for m is efficient for two reasons. It produces a long nonrepeating sequence ($2^{29} = 536{,}870{,}912$), and it results in a trivial computation of the residue modulo 2^b, which is required in Eq. (3.82).

The residue calculation is a by-product of the way the computer executes the FORTRAN instruction for multiplication of integers. If the product is a number larger than can be stored in a single binary word, the excess is discarded. The binary digits which are discarded each correspond to 2^k for some k greater than or equal to the word length b; thus this discarded excess is divisible by 2^b. In addition, since the remaining number is less than 2^b, it must be the residue of the integer product, modulo 2^b. **(The first excess digit, in the sign place, is not simply deleted.)**

The IBM 360 subroutine package includes a subroutine called RANDU, with the following statements.

```
      SUBROUTINE RANDU (IX, IY, YFL)
      IY = IX*65539
      IF(IY)5, 6, 6
    5 IY = IY + 2147483647 + 1                    (3.84)
    6 YFL = IY
      YFL = YFL*0.4656613E- 9
      RETURN
      END
```

The reader should check to see that this subroutine obeys the number-theory rules. (5 corrects for overflows which change the sign digit.) YFL is the output corresponding to the number between 0 and 1. Before going from the main program to this subroutine for the first time, IX must be defined as some odd integer with nine or fewer digits. Subsequently the main program must define IX = IY before each passage to this subroutine [see (3.145)].

The histogram test of the preceding section could be used to test the distribution of the numbers coming out of this subroutine. Later we shall describe tests for checking the assumed lack of predictability of one number based on its predecessors.

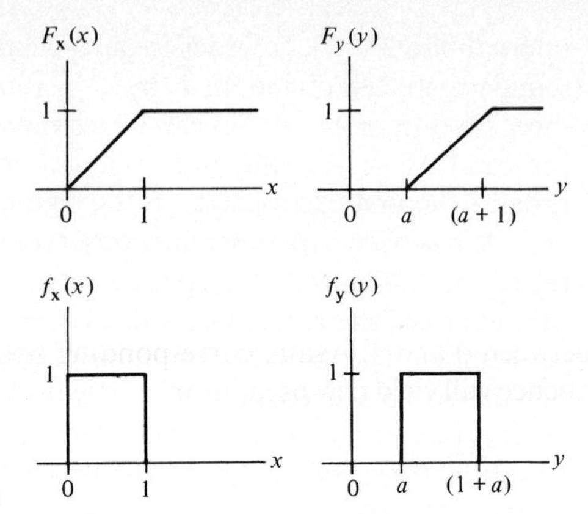

FIGURE 3.16
Shifted distribution and density functions.

Density Shaping

We have just seen how a simple algorithm can be used to generate a long sequence of numbers which "look like" sample values of independent random variables, each uniformly distributed between 0 and 1. Many other kinds of pseudorandom sequences can be produced by simple modifications of this basic sequence. The two major kinds of modification are in (1) the probability distribution of each variable and (2) the dependence between successive values (discussed in the following section).

The transformations which produce variables with more general distributions, from uniform variables, can be demonstrated by some simple examples. If $\mathbf{x}$ is uniform between 0 and 1, then $\mathbf{y} = \mathbf{x} + a$ is uniform between a and $1 + a$. This result may seem intuitively true, but intuition will not suffice for more complicated distributions. Going back to basic definitions, the probability distribution function of $\mathbf{y}$ is

$$F_{\mathbf{y}}(y) = P[\mathbf{y} \leq y] \tag{3.85}$$

Substitution of the relation between $\mathbf{y}$ and $\mathbf{x}$ yields

$$F_{\mathbf{y}}(y) = P[\mathbf{x} + a \leq y] = P[\mathbf{x} \leq y - a] = F_{\mathbf{x}}(y - a) \tag{3.86}$$

The distribution and density functions for $\mathbf{x}$ and $\mathbf{y}$, when $\mathbf{x}$ is uniformly distributed between 0 and 1, are shown in Fig. 3.16. Similarly, the distribution of the pseudorandom n_i described above could be shifted by merely adding a constant value to each n_i, for example,

$$m_i = n_i + a \tag{3.87}$$

By similar arguments, it is easy to show that random variables $\mathbf{w}$ and $\mathbf{z}$, defined by

$$\mathbf{w} = c\mathbf{x} \tag{3.88}$$

$$\mathbf{z} = c\mathbf{x} + a \tag{3.89}$$

have the uniform density functions shown in Fig. 3.17 if $\mathbf{x}$ is uniformly distributed

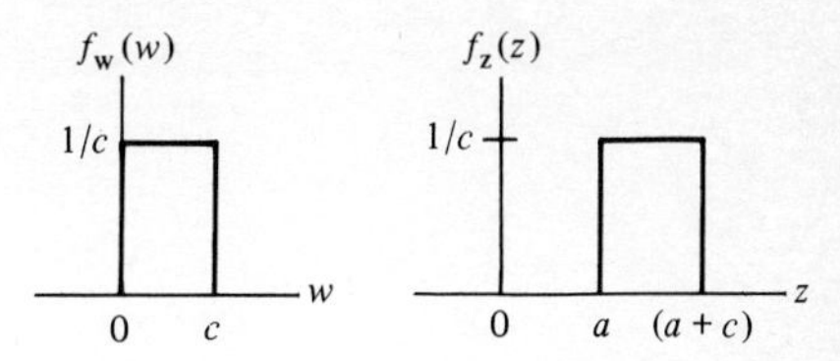

FIGURE 3.17
Scaled and shifted density functions.

between 0 and 1. Again, corresponding operations on the basic pseudorandom sequence will yield new pseudorandom sequences with distributions like those described in Fig. 3.17.

All the preceding random variables have been uniformly distributed. They have corresponded to linear transformations [that of (3.89) is the most general] on the 0-to-1 uniform random variable. Nonlincar transformations yield numbers with nonuniform distributions. A very general transformation, called the *probability transformation*, will now be described for constructing random variables with nonuniform distributions from variables with a uniform distribution between 0 and 1.

We can represent any one-to-one transformation by a monotone curve $g(y)$, as in Fig. 3.18. If the curve is not monotone [see $g_1(y)$ in Fig. 3.19], the transformation is not one-to-one since there are at least two possible input values for some single output value. The advantage of a one-to-one transformation is that it has an associated inverse function $g^{-1}(x)$. The -1 superscript does *not* indicate the usual reciprocal $1/g(x)$ but tells how to find the y from which x was computed, as shown by the example of a typical monotone function and its inverse in Fig. 3.18. Notice that $g^{-1}(x)$ can be found geometrically by rotating the graph of $g(y)$ 90° counterclockwise and then spinning it 180° about the vertical axis.

The nonlinear function $g^{-1}(x) = -\ln(1 - x)$ in Fig. 3.18 provides an example which demonstrates the *probability transformation*. If $\mathbf{x}$ is a uniformly distributed random variable between 0 and 1, then $\mathbf{y} = g^{-1}(\mathbf{x})$ will be a new random variable. We shall show that the distribution function of $\mathbf{y}$ is the $g(y)$ in Fig. 3.18:

$$F_{\mathbf{y}}(y) = \begin{cases} 1 - e^{-y} & y \geq 0 \\ 0 & y < 0 \end{cases} \tag{3.90}$$

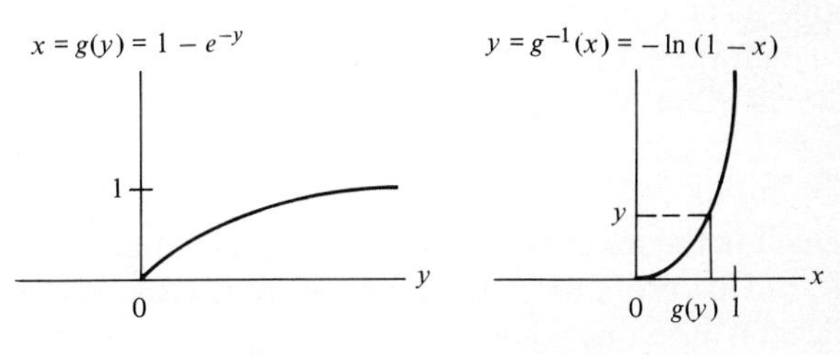

FIGURE 3.18
A monotone function.

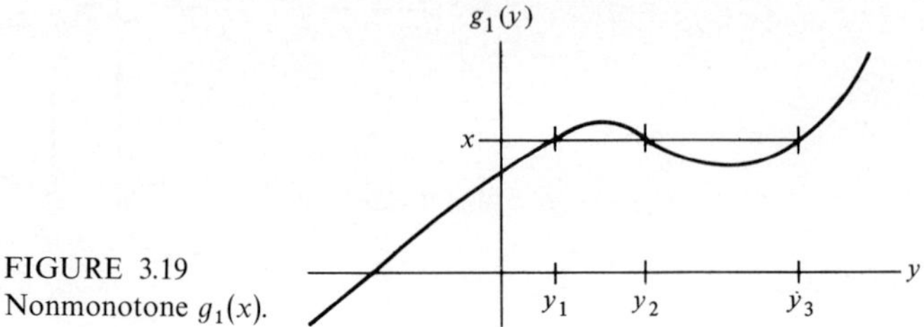

FIGURE 3.19
Nonmonotone $g_1(x)$.

with a density function

$$f_{\mathbf{y}}(y) = \begin{cases} e^{-y} & y \geq 0 \\ 0 & y < 0 \end{cases}$$

Such a **y** is an example of an *exponentially distributed* random variable obtained by nonlinearly transforming the uniformly distributed **x**. The derivation is as follows.

The distribution function of **y** is, by definition,

$$F_{\mathbf{y}}(y) = P[\mathbf{y} \leq y] = P[g^{-1}(\mathbf{x}) \leq y] \tag{3.91}$$

With reference to Fig. 3.18, it is clear that $g^{-1}(\mathbf{x}) \leq y$ if and only if $\mathbf{x} \leq g(y)$. Thus

$$F_{\mathbf{y}}(y) = P[\mathbf{x} \leq g(y)] = F_{\mathbf{x}}[g(y)] \tag{3.92}$$

Finally, reference to the 0-to-1 uniform distribution $F_{\mathbf{x}}$ in Fig. 3.16 shows that

$$F_{\mathbf{x}}[q] = q \qquad \text{for } 0 \leq q \leq 1$$

Thus

$$F_{\mathbf{y}}(y) = \begin{cases} g(y) & \text{if } 0 \leq g(y) \leq 1 \\ 1 & \text{if } g(y) > 1 \\ 0 & \text{if } g(y) < 0 \end{cases} \tag{3.93}$$

That is, the distribution function of **y** is the original monotonic function $g(y) = 1 - e^{-y}$.

Although (3.93) was derived for the transformation in Fig. 3.18, the same argument will apply for any $g(y)$ which increases monotonically from 0 to 1 as y increases from $-\infty$ to ∞. These conditions are those satisfied by every distribution function. The steps from (3.90) to (3.92) describe transformation properties of random variables. These properties also apply to pseudorandom numbers like the n_i generated to be uniform between 0 and 1 by the algorithm (3.84). Pseudorandom numbers m_i with the distribution of **y** in (3.90) can be generated by transforming the n_i according to

$$m_i = F_{\mathbf{y}}^{-1}(n_i) \tag{3.94}$$

which is the inverse function of the distribution function.

This probability transformation is often said to "shape" the uniform random variable into one whose distribution has a desired shape. This method is completely general but may be inefficient for some kinds of distributions.

An alternative approach is frequently used to generate random variables with a

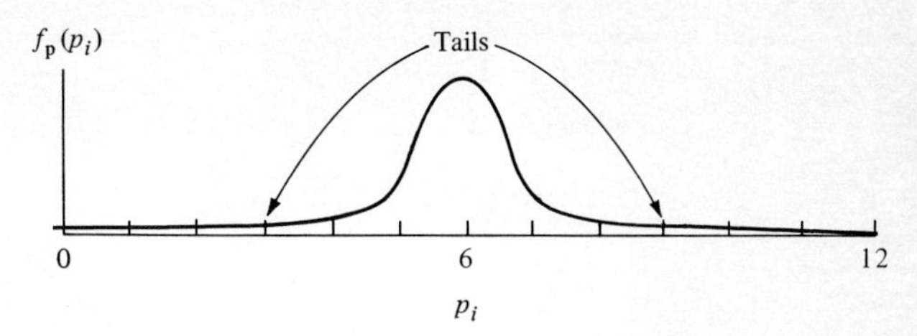

FIGURE 3.20
Density function $f_p(p_i)$.

normal (gaussian) distribution. This method is based on the *central-limit theorem* for continuous random variables,† which states that if many independent random variables are added, their sum tends to have a gaussian distribution. Approximately normal variables $p_1, p_2, \ldots$ can be generated from the n_i by the relations

$$\begin{aligned} p_1 &= n_1 + n_2 + \cdots + n_{12} \\ p_2 &= n_{13} + n_{14} + \cdots + n_{24} \end{aligned} \tag{3.95}$$

If the n_i in (3.95) were truly samples of independent random variables, the variance of their sum would be the sum of their variances. That is,

$$\operatorname{var} \mathbf{p}_j = \sum_{i=1}^{12} \operatorname{var} \mathbf{n}_i = 12 \operatorname{var} \mathbf{n}_1 = 1 \tag{3.96}$$

Equation (3.96) makes use of the easily proved result that the variance of a uniform random variable is one-twelfth of the square of the width of the distribution.

The mean value of the $\mathbf{p}_j$ is the sum of the means of the $\mathbf{n}_i$, that is,

$$E[\mathbf{p}_j] = 12(\tfrac{1}{2}) = 6 \tag{3.97}$$

The algorithm (3.95) gives a good approximation to normal random variables with unity variance and a mean of 6. The mean and variance can be easily changed by respectively adding a constant to each p_i or multiplying the p_i by a constant.

One of the limitations of (3.95) as an approximation to normal random variables is the limit on the magnitude of the p_i. It is clear from the definition of the n_i that

$$0 \leq p_i \leq 12 \tag{3.98}$$

whereas a truly gaussian random variable can take any value between $+\infty$ and $-\infty$. However, a gaussian variable with a variance of 1 [as in (3.96)] has a probability of about 10^{-9} of taking values outside of the range defined in (3.98). Thus, this limitation is usually not serious.

This central-limit-theorem approach to generating gaussian random variables yields p_i whose density function is very nearly gaussian near its mean, but the approximation is poorer for values far from the mean. The latter values are said to lie *in the tails* of the distribution (see Fig. 3.20).

† Papoulis, op. cit., p. 266.

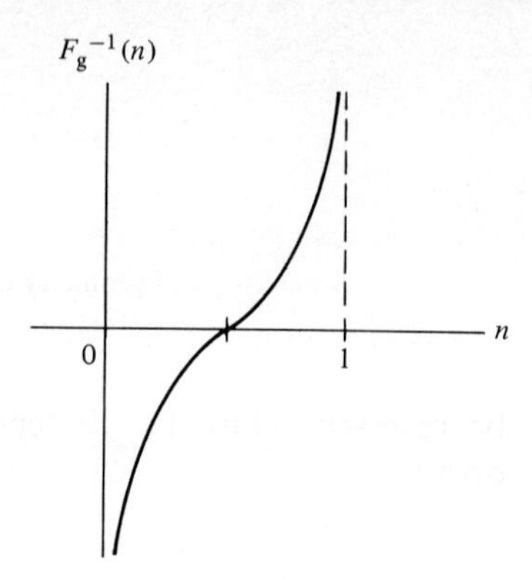

FIGURE 3.21
Gaussian inverse distribution.

Alternative methods are available for getting variables whose distribution gives a better approximation to the gaussian tails. We could use the probability transformation described above, but the required transformation $F_{\mathbf{g}}^{-1}(n)$ (shown in Fig. 3.21) for a zero-mean, unit-variance gaussian random variable is not readily available on most computers. Another equivalent approach makes use of simpler functions and a special property of independent gaussian random variables.

This second approach to generating gaussian variables, which gives a better approximation to the tails of the density, can be described in terms of errors in target shooting or dart throwing. It is well known† that if in rectangular coordinates the horizontal and vertical errors g_1 and g_2 are independent and gaussian with densities

$$f_{\mathbf{g}}(g_i) = \frac{1}{\sqrt{2\pi}\,\sigma} e^{-g_i^2/2\sigma^2} \qquad i = 1, 2 \tag{3.99}$$

then the polar coordinates r and θ of a typical point, as in Fig. 3.22, are also independent random variables with densities

$$f_{\mathbf{r}}(r) = \begin{cases} \dfrac{r}{\sigma^2} e^{-r^2/2\sigma^2} & r \geq 0 \\ 0 & \text{otherwise} \end{cases} \tag{3.100}$$

$$f_{\boldsymbol{\theta}}(\theta) = \begin{cases} \dfrac{1}{2\pi} & 0 \leq \theta < 2\pi \\ 0 & \text{otherwise} \end{cases} \tag{3.101}$$

Figure 3.22 defines the relations between these variables and shows the general shape of $f_{\mathbf{r}}$ in (3.100), which is known as a *Rayleigh density*. The angular uncertainty described by (3.101) is clearly a uniformly distributed variable between 0 and 2π rad.

The argument which leads to the properties of $\mathbf{r}$ and $\boldsymbol{\theta}$ in terms of $\mathbf{g}_1$ and $\mathbf{g}_2$ can

† M. Schwartz, "Information, Transmission, Modulation and Noise," 2d ed., p. 366, McGraw-Hill, New York, 1970.

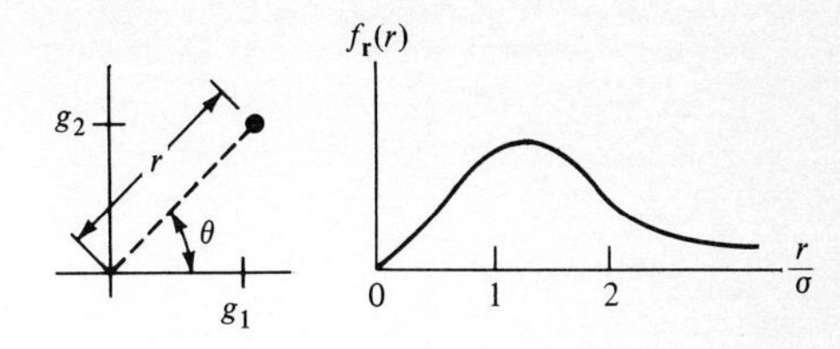

FIGURE 3.22
Density f_r and geometry of g_1, g_2, r, θ.

be reversed. Thus, two independent gaussian variables can be generated by the operations

$$g_1 = r \cos \theta \qquad g_2 = r \sin \theta \tag{3.102}$$

once r and θ values have been generated corresponding to the densities in (3.100) and (3.101). These appropriate r and θ values can easily be generated using probability transformations on two independent random variables which are uniformly distributed between 0 and 1. Details are left as a problem.

This section has described algorithms for generating numbers which approximate independent samples of a random variable with a prescribed distribution. Particular examples were uniform, gaussian, exponential, and Rayleigh distributions. A general method for getting variables with other distributions was also presented.

The next section will describe the effects that signal processors have on random signals. When the independent variables of a purely random signal are passed through a filter, the output samples will be dependent. We shall analyze the relationship between the filter parameters and the output-signal dependence, and then the signal-simulating techniques of this section can be extended to include generation of random signals with prescribed dependences between successive samples. The dependence of samples will be seen to be described by a statistical function called the autocorrelation function. Signal-processing techniques for estimating the autocorrelation function and its Fourier transform, the power spectral density, are well developed. Computer programs and laboratory instruments for carrying out such processing are in widespread use, as we shall see in Chap. 4.

3.4 FILTERED RANDOM SIGNALS: AUTOCORRELATION AND POWER SPECTRAL DENSITY

This chapter has been developing probability-theory descriptions for random signals which might be encountered in communications and radar problems, among others. The signal models have been quite simple so far, being restricted to the purely random case, in which successive samples $\mathbf{x}_1, \mathbf{x}_2, \ldots$ are statistically independent. We actually did introduce a more complicated signal that was not purely random, the output of an averager filter, but the complexity of that signal was not emphasized in Sec. 3.1.

A filter's output signal will generally have dependence among samples because

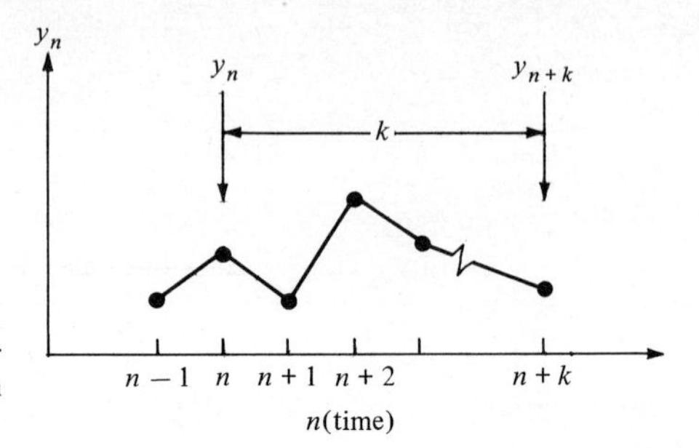

FIGURE 3.23
Random variables over which one averages in determining the autocorrelation function.

of the filter's memory. For example, two successive averager outputs are

$$y_n = \frac{x_n + x_{n-1}}{2} \qquad y_{n+1} = \frac{x_{n+1} + x_n}{2} \tag{3.103}$$

The appearance of x_n in both of these expressions indicates some kind of *functional* dependence. We might say that the filter remembers the previous input. One way of describing the *probabilistic* relationship which might exist between two random variables like $\mathbf{y}_n$ and $\mathbf{y}_{n+1}$ is termed *correlation.*

Our intuitive notion of correlation is based on colloquial use of the term, e.g., " there is a positive correlation between the amount of alcohol a man drinks before he drives and the number of automobile accidents he has per year " or " the heights of a husband and wife are correlated." These statements do not imply that every time a driver drinks a lot he will have an accident or that every tall man has a tall wife. Correlation suggests that these relationships exist when averaged over many random selections of these test events.

Similarly, the probabilistic definition of correlation between two random variables is based on an average (via the expectation operation) dependence. We shall describe the correlation between pairs of signal samples by the *autocorrelation function* defined as the expectation of the product of any two signal samples separated in time by k samples. For example, the averager output $\mathbf{y}_n$ has the autocorrelation function $R_y(k)$

$$R_y(k) = E[\mathbf{y}_n \mathbf{y}_{n+k}] \tag{3.104}$$

The time separation k, the independent variable of the autocorrelation function, is sometimes called the *lag*. Expression (3.104) is the definition of the autocorrelation function for any random signal $\mathbf{y}_n$. Figure 3.23 shows two typical signal samples whose mean product defines $R_y(k)$.

Intuitively, the autocorrelation function measures correlation in the following sense. If a signal fluctuates slowly, on the average, then two successive samples $\mathbf{y}_n$ and $\mathbf{y}_{n+1}$ will have nearly the same value and their product will be positive, on the average, making $R_y(1)$ positive. Conversely, if the signal varies rapidly, the two

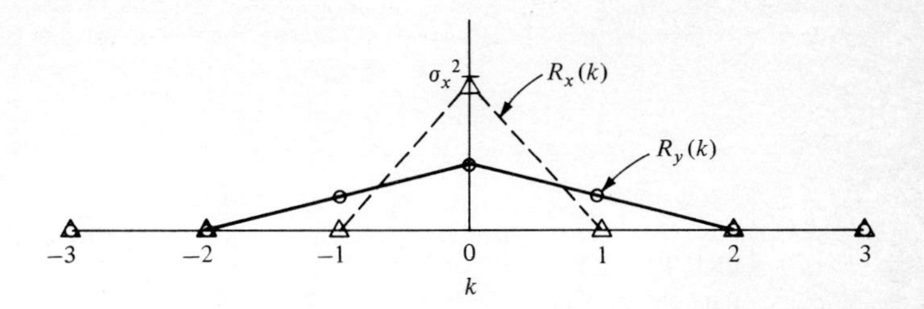

FIGURE 3.24
Averager input and output autocorrelation functions.

samples are equally likely to have the same sign or different sign. (We assume that the signal has zero average value for simplicity.) In this latter case their product is, on the average, zero. We interpret a zero autocorrelation-function value as saying that the variables are *uncorrelated*; while the larger the autocorrelation-function values, the more correlated the variables are. Since $R_y(k)$ relates signal values separated by a lag of k, we would expect the dependence of the autocorrelation function on k to be related to the signal's rate of fluctuation. Such a bandwidth property for random signals will be developed later.

We can demonstrate these intuitive ideas about autocorrelation-function properties with some examples. Consider first a zero-mean purely random signal $\mathbf{x}_n$, which could be the input to an averager filter. A little thought will indicate that its autocorrelation function is zero for all lags except $k = 0$, since

$$R_x(k) = E[\mathbf{x}_n \mathbf{x}_{n+k}] = E[\mathbf{x}_n]E[\mathbf{x}_{n+k}] = 0 \qquad k = \pm 1, \pm 2, \ldots \tag{3.105}$$

For $k = 0$, $R_x(0) = E[\mathbf{x}_n^2]$ is just the mean-squared value (or *power*) in the signal. For zero mean $\mathbf{x}_n$ this is then just the variance σ_x^2. So $R_x(k)$ is zero for *all* k in this case, except at $k = 0$, as shown in Fig. 3.24. This purely random signal displays the least possible amount of correlation and is rapidly varying.

Now assume that this purely random signal is applied at the input to the averager of (3.103). As noted in the discussion of that equation above, memory is introduced by the smoothing. Two successive output values $\mathbf{y}_n$ and $\mathbf{y}_{n+1}$ both depend on $\mathbf{x}_n$. One would thus expect $R_y(k)$ to be nonzero for $k = 1$. Values of the output spaced more than one time unit apart, however ($k = 2, 3, \ldots$), no longer have this dependence and will thus be expected to have their average equal zero. These remarks are now readily substantiated.

The autocorrelation function of the output at $k = 0$ is just the mean-squared value $R_y(0) = E(\mathbf{y}_n^2)$. For zero-mean input $\mathbf{x}_n$, $E(\mathbf{y}_n) = 0$ as well; so $R_y(0)$ in *this* case is the variance σ_y^2. We have already computed this in (3.24):

$$R_y(0) = \sigma_y^2 = \frac{\sigma_x^2}{2}$$

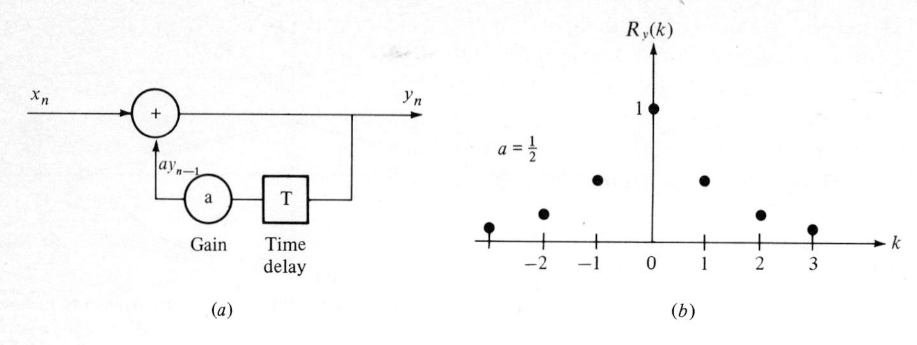

FIGURE 3.25
Recursive digital filter, purely random input.

At a lag of 1, we use (3.103) to get the output autocorrelation function

$$R_y(1) = E\left[\left(\frac{\mathbf{x}_n + \mathbf{x}_{n-1}}{2}\right)\left(\frac{\mathbf{x}_{n+1} + \mathbf{x}_n}{2}\right)\right] \tag{3.106}$$

in which all the products have zero expectations except for $\mathbf{x}_n^2/4$. Thus

$$R_y(1) = \frac{\sigma_x^2}{4}$$

For larger lags, $\mathbf{y}_n$ and $\mathbf{y}_{n+k}$ have no common samples, and so their correlation is zero.

Figure 3.24 shows the input and output autocorrelation functions. Note that they are even functions of k; that is, it is easy to check that the definition (3.104) implies $R_y(-1) = R_y(1)$. The smoother output has a nonzero $R_y(1)$, while the more rapidly fluctuating purely random input has a zero correlation for a lag $k = 1$. Both signals have samples which are uncorrelated when they are widely separated (large k).

Consider now a purely random signal $\mathbf{x}_n$ applied to the simple recursive filter shown in Fig. 3.25. Here we have

$$\mathbf{y}_n = a\mathbf{y}_{n-1} + \mathbf{x}_n \qquad |a| < 1$$

We know from previous discussions in Chap. 2 that this filter has an "infinite" memory. We would thus expect the autocorrelation function of the output to be nonzero for *all* k. *All* output samples are dependent or correlated. However, as the separation k between samples increases, one would expect $R_{\mathbf{y}}(k)$ to decrease, showing the dependence decreasing as the separation increases. (The filter memory decreases with time.) To verify this we calculate $E[\mathbf{y}_n \mathbf{y}_{n+k}]$. For $k = 0$,

$$R_y(0) = E[\mathbf{y}_n^2] = E[a\mathbf{y}_{n-1} + \mathbf{x}_n]^2 = a^2 R_y(0) + R_x(0)$$

or

$$R_y(0) = \frac{R_x(0)}{1 - a^2}$$

To get this result we assumed the filter to be in operation for a long time so that the output statistics are stationary (implying $E[\mathbf{y}_n^2] = E[\mathbf{y}_{n-1}^2]$). Furthermore, $E[\mathbf{y}_{n-1}\mathbf{x}_n] = 0$, since $\mathbf{y}_{n-1}$ depends only on $\mathbf{x}_{n-1}, \mathbf{x}_{n-2}, \mathbf{x}_{n-3}, \ldots$ [see also (3.142)] and $\mathbf{x}_n$ is uncorrelated with those earlier input samples, by the assumption of pure randomness.

Repeating these arguments for $k = 1$ gives

$$R_y(1) = E[\mathbf{y}_n \mathbf{y}_{n+1}] = E[\mathbf{y}_n(a\mathbf{y}_n + \mathbf{x}_{n+1})] = aR_y(0)$$

Similarly, it is readily shown that

$$R_y(2) = a^2 R_y(0), \qquad R_y(3) = a^3 R_y(0), \ldots$$

For k negative we get just the same sequence of values. Thus $R_y(k)$ is nonzero for *all k* and does decrease with $|k|$ increasing (although it will oscillate between positive and negative values when a is negative). This $R_y(k)$ is shown in Fig. 3.25 for $a = \frac{1}{2}$.

Justification of our autocorrelation function as a measure of the rapidity of fluctuation can be made more quantitative by examining the probability that two successive signal samples have the same sign. We shall do this for the averager of (3.103). Such a calculation requires further information about the distribution function of the signal samples. If the purely random inputs are gaussian, the output samples will also be gaussian, as mentioned earlier. It can be shown that two successive *output* samples will have the joint gaussian density

$$f(y_n, y_{n+1}) = \frac{1}{2\pi\sigma_y^2\sqrt{1-\rho^2}} \exp\left[-\frac{y_n^2 + 2\rho\sigma_y^2 y_n y_{n-1} + y_{n-1}^2}{2\sigma_y^2(1-\rho^2)}\right] \tag{3.107}$$

in which the correlation coefficient ρ is

$$\rho = \frac{R_y(1)}{R_y(0)} \tag{3.108}$$

The division by $R_y(0)$ makes ρ a number between -1 and 1; that is, the autocorrelation function is always biggest at a lag of zero.

With P_S as the probability that $\mathbf{y}_n$ and $\mathbf{y}_{n+1}$ have the same sign

$$P_S = \iint\limits_{y_n y_{n+1} > 0} f(y_n, y_{n+1})\, dy_n\, dy_{n+1} \tag{3.109}$$

This is the integral of the joint density of the two samples over the region where y_n and y_{n+1} have the same sign (or their product is positive), namely the entire first and third quadrants of the y_n, y_{n+1} plane. The integration can be carried out to show† that

$$P_S = \begin{cases} \frac{1}{2} & \text{if } \rho = 0 \\ \frac{2}{3} & \text{if } \rho = \frac{1}{2} \end{cases}$$

and, generally,

$$P_S > \tfrac{1}{2} \qquad \text{if } \rho > 0 \tag{3.110}$$

Thus, if there is a positive correlation between two successive gaussian output samples, there is a better than fifty-fifty chance that they will have the same sign. The

† Papoulis, op. cit., p. 198.

closer the correlation coefficient to unity, the higher the probability that the two signs are the same.

In this example, the uncorrelated gaussian *inputs* will also have a joint density of the form (3.107), with x's replacing y's and $\rho = 0$ because $R_x(1) = 0$. It is interesting to see that the joint density can then be written as a product of marginal densities

$$f(x_n, x_{n+1}) = \frac{1}{\sigma_x\sqrt{2\pi}} e^{-x_n^2/2\sigma_x^2} \frac{1}{\sigma_x\sqrt{2\pi}} e^{-x_{n+1}^2/2\sigma_x^2}$$
$$= f(x_n)f(x_{n+1}) \tag{3.111}$$

showing that $\mathbf{x}_n$ and $\mathbf{x}_{n+1}$ are independent random variables. Thus, uncorrelated gaussian variables are independent. In general, uncorrelated *non*gaussian variables might be dependent because correlation measures only one kind of dependence.

The correlation coefficient in (3.108) is a special case of the one usually defined in probability texts. For any two random variables $\mathbf{u}$ and $\mathbf{v}$ the correlation coefficient is defined as

$$\rho_{uv} = \frac{E[\mathbf{u} - \eta_u)(\mathbf{v} - \eta_v)]}{\sigma_u \sigma_v} \tag{3.112}$$

In our example the mean values η_y were zero, and the standard deviations were equal. Therefore

$$\sigma_{y_n} = \sigma_{y_{n+1}} = \sqrt{R_y(0)}$$

Using (3.104) and (3.112), we can easily relate our autocorrelation function to the correlation coefficient:

$$R_y(k) = R_y(0)\rho_{y_n y_{n+k}} + \eta_{y_n}\eta_{y_{n+k}} \tag{3.113}$$

Since we usually work with zero-mean random variables, this shows that $R_y(k)$ is simply a scaled (unnormalized) correlation coefficient. When a signal has a nonzero mean, we study the autocorrelation function of its *fluctuating part*, i.e., a new zero-mean signal formed by subtracting the mean from the original one. For example, $\tilde{\mathbf{y}}_n$ is the fluctuating part of $\mathbf{y}_n$ if

$$\tilde{\mathbf{y}}_n = \mathbf{y}_n - E[\mathbf{y}_n] \tag{3.114}$$

To summarize, the autocorrelation function is a collection of scaled correlation coefficients, one for each lag k. For gaussian variables, the autocorrelation function can be used to compute the probability that samples at different times have the same sign, as a measure of fluctuation rates. For nongaussian signals, such probability statements are more difficult to compute. However, the autocorrelation function does reflect the dependence of signal samples separated by a lag k. Furthermore, examples have suggested relations between a signal's autocorrelation function and its rate of fluctuation. We now quantify that relationship by introducing Fourier transforms.

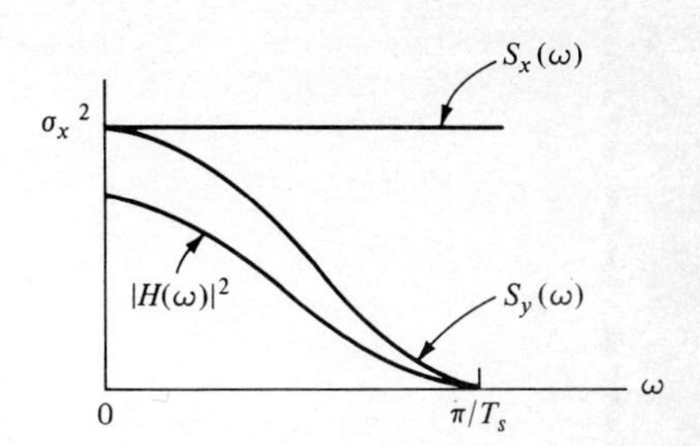

FIGURE 3.26
Input and output spectral densities.

Consider the discrete Fourier transform $S_y(\omega)$ of the autocorrelation function $R_y(k)$. We shall see that this frequency function, called the *power spectral density* (or *power spectrum*) of the signal $\mathbf{y}_n$, measures the average power vs. frequency of that random signal. We shall also see that it represents the distribution of power with respect to frequency; hence the name power spectrum. The power spectral density of a set of signal samples with autocorrelation function $R_y(k)$ is given by

$$S_y(\omega) = \sum_{k=-\infty}^{\infty} R_y(k)e^{-j\omega kT_s} \tag{3.115}$$

Consider as examples the purely random signal $\mathbf{x}_n$, discussed above, and the output $\mathbf{y}_n$ of the averager with $\mathbf{x}_n$ applied at the input. The autocorrelation functions for these two signals are sketched in Fig. 3.24. It is left to the reader to show that, using (3.115), one finds as the transforms in these two cases

$$S_x(\omega) = \sum_{k=-\infty}^{\infty} R_x(k)e^{-j\omega kT_s} = \sigma_x^2 \qquad \text{all } \omega \tag{3.116}$$

and

$$S_y(\omega) = \frac{\sigma_x^2}{2}(1 + \cos \omega T_s) \tag{3.117}$$

These input and output spectra are sketched in Fig. 3.26, along with the filter's frequency response or transfer function

$$|H(\omega)|^2 = \tfrac{1}{2}(1 + \cos \omega T_s)$$

Recall that $H(\omega)$ is the Fourier transform

$$H(\omega) = \sum_{-\infty}^{\infty} h_k e^{-j\omega kT_s}$$

of the filter weighting function h_k defined by the input-output relation

$$y_n = \sum_{-\infty}^{\infty} h_k x_{n-k}$$

As with all discrete Fourier transforms, only the ω interval from 0 to π rad per sample (or π/T_s rad/s) need be plotted. No angle spectrum appears because the evenness of $R(k)$ makes $S(\omega)$ real. These curves show that $S_y < S_x$, especially at the higher frequencies. The slower-varying signal with more correlation at a lag of 1 has a smaller high-frequency power spectral density.

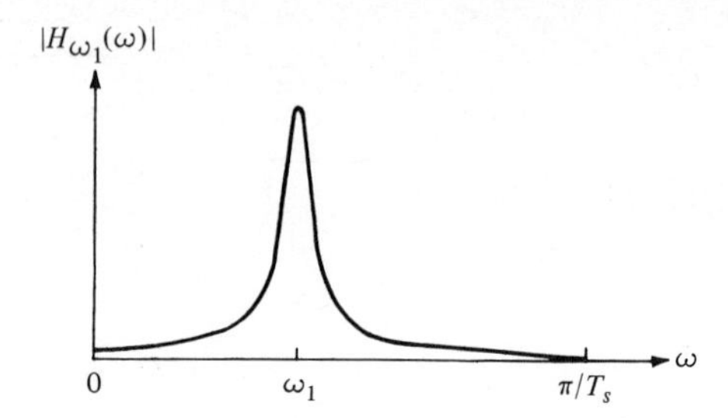

FIGURE 3.27
Narrowband filter transfer function.

Several other properties of the functions in Fig. 3.26 are noteworthy. The purely random input has a constant spectral density over all frequencies. This kind of a signal is often called *white noise*, by analogy to the property of white light containing all frequencies (which can be separated by passing the light through a prism).

The relationship between the input and output spectra and the filter transfer function is suggestive of the general result which we shall shortly prove, namely

$$S_y(\omega) = S_x(\omega)\,|\,H(\omega)\,|^2 \tag{3.118}$$

That is, the magnitude squared of the transfer function acts as a *power* transfer function. The significance of this arises when, for example, we want to compute the output-fluctuation standard deviation as a measure of the average size of output errors. The output variance σ_y^2 is identical to the zero-lag output autocorrelation function (again assuming zero-mean signals)

$$\sigma_y^2 = R_y(0) \tag{3.119}$$

which can be written in terms of spectral densities via the *inverse* Fourier transform

$$R_y(k) = \frac{T_s}{2\pi}\int_{-\pi/T_s}^{\pi/T_s} S_y(\omega)e^{+j\omega kT_s}\,d\omega \tag{3.120}$$

These last three equations combine to show the output variance is related to the input power spectral density and filter transfer function by the integral

$$\sigma_y^2 = \frac{T_s}{2\pi}\int_{-\pi/T_s}^{\pi/T_s} S_x(\omega)\,|\,H(\omega)\,|^2\,d\omega \tag{3.121}$$

The output variance will be large (and output amplitudes will be large) if the input spectrum is large in the filter passband, i.e., at frequencies where $|H(\omega)|$ is large.

Equation (3.121) is in fact the justification for the term power spectral density, because we already know the frequency properties of $H(\omega)$ from its use with deterministic signals. A narrowband filter at frequency ω_1 (as in Fig. 3.27) will pass a sine wave only if its frequency is near ω_1. If a random-signal input to the same filter produces a measurable output, we say that the input signal has average power in that band. A set of filters centered at different frequencies (or a narrowband filter with adjustable passband center frequency) can measure relative amounts of power in different frequency bands and thereby estimate $S_x(\omega)$ for given data $x_1, x_2, \ldots$. Such a device is called a *spectrum analyzer* and is shown in Fig. 3.28. A digital approach to spectrum analyzers will be discussed in Chap. 4.

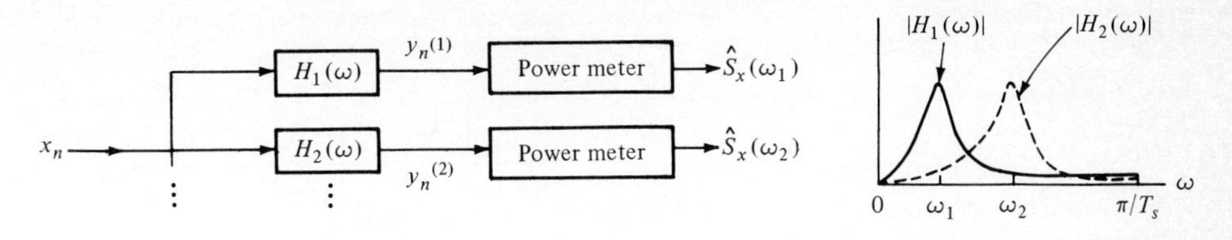

FIGURE 3.28
Spectrum-analyzer block diagram and transfer functions.

Our power-spectral-density definition has been a bit abstract, but it does satisfy intuitive notions. If a narrowband filter like the one described in Fig. 3.27 has a random-signal input, its output will look like an irregularly distorted sine wave at frequency ω_1, as in Fig. 3.29.

Our use here of the word power can be related to the usual physics definitions of average power. If $v(t)$ is a voltage across a 1-Ω resistor, the instantaneous power absorbed by the resistor is $v^2(t)$ and its average power over a time interval of T_{av} s is

$$\mathscr{P}(T_{av}) = \frac{1}{T_{av}} \int_0^{T_{av}} v^2(t)\, dt \tag{3.122}$$

If $\mathbf{v}(t)$ were a random voltage, $\mathscr{P}$ would also be random, with a different value for each T_{av}-s-long realization of $\mathbf{v}(t)$.

In the discrete-time case we could think of $\mathbf{v}_n = \mathbf{v}(n)$ as samples of a $\mathbf{v}(t)$ which is piecewise constant over 1-s intervals (see Fig. 3.30). The average power can then be written

$$\mathscr{P}(T_{av}) = \mathscr{P}(N) = \frac{1}{N} \sum_{1}^{N} \mathbf{v}_n^2 \tag{3.123}$$

When the averaging time is large ($N \to \infty$), we would expect the average in (3.123) to approach the expectation

$$\lim_{N \to \infty} \mathscr{P}(N) = E[\mathbf{v}_n^2] = \sigma_v^2 \tag{3.124}$$

if the mean value of $\mathbf{v}_n$ is zero. While it is possible to find signals which do not have this limiting property of time averages approaching expectations, most signals of interest can be assumed to have this so-called *ergodic* property. The convergence of the time average in (3.124) will be discussed further in Chap. 4, where we show that $\mathscr{P}(N)$ is an unbiased and consistent estimate of σ_v^2.

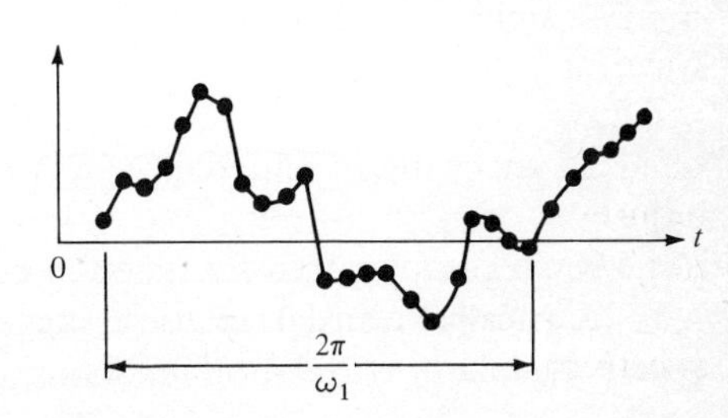

FIGURE 3.29
Narrowband random signal.

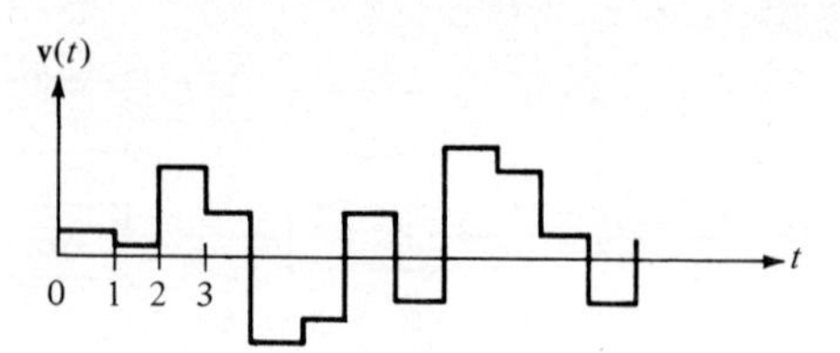

FIGURE 3.30
Piecewise constant random voltage.

To summarize, the power spectral density is defined as the Fourier transform of an autocorrelation function, as in (3.115), and this frequency function can be used to compute average powers or variances of random signals via the inverse transform, as in (3.120) and (3.121). The fact that the power spectral density indeed describes the frequency distribution of average power is demonstrated by the filter transfer relation (3.118), which we shall now derive.

If $\mathbf{x}_n$ and $\mathbf{y}_n$ are input and output, respectively, of a linear filter, the output is a weighted sum of input samples, of the form

$$y_n = \sum_{j=-\infty}^{\infty} h_{n-j} x_j \tag{3.125}$$

(In the averager, $h_i = \frac{1}{2}$ for $i = 0, 1$, and for all other weights $h_i = 0$.) $S_y(\omega)$ is related to $S_x(\omega)$ by first relating the corresponding autocorrelation functions, with the aid of (3.125).

Substituting the filter relation (3.125) into the autocorrelation-function definition, we get

$$R_y(k) = E[\mathbf{y}_n \mathbf{y}_{n+k}] = E\left[\sum_{j=-\infty}^{\infty} h_{n-j} \mathbf{x}_j \sum_{l=-\infty}^{\infty} h_{n+k-l} \mathbf{x}_l\right] \tag{3.126}$$

Interchanging the order of summation and expectation operations yields

$$R_y(k) = \sum_{j,l=-\infty}^{\infty} h_{n-j} h_{n+k-l} R_x(j-l) \tag{3.127}$$

The spectral-density relationship corresponding to (3.127) is found by using a few tedious manipulations similar to those in (2.126) after taking the discrete Fourier transform of both sides of (3.127). Details are left to the reader. The result is

$$S_y(\omega) = H(\omega)H(-\omega)S_x(\omega) = |H(\omega)|^2 S_x(\omega) \tag{3.128}$$

in which

$$H(\omega) = \sum_{-\infty}^{\infty} h_i e^{-j\omega i} \tag{3.129}$$

is the usual complex transfer function of the filter.

The power-transfer relation suggests a block-diagram relation between the time-domain autocorrelation functions, as shown in Fig. 3.31. The input and output

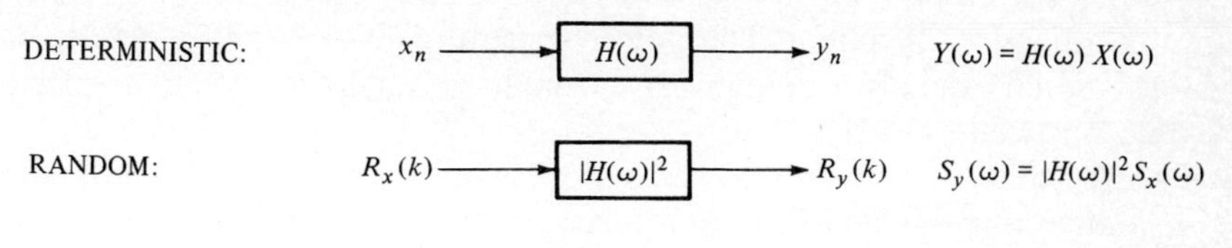

FIGURE 3.31
Signal and autocorrelation-function transfer functions.

autocorrelation functions have their transforms (power spectral densities) related by a transfer function $|H(\omega)|^2$, in a manner parallel to the transfer function relating the transforms of specific input and output signals.

The power-transfer relation we have just derived for linear filtering of random signals is useful for designing signal processors. Given an observed

$$\mathbf{x}_n = s_n + \mathbf{d}_n \tag{3.130}$$

in which the disturbance has a known spectral density $S_d(\omega)$, we could try to find a filter which is best at suppressing the noise. The signal-processing problems of detection and estimation are not as simple as this might imply, because the filter will also distort the desired signal component s_n. Methods for designing filters which are good compromises between noise suppression and signal distortion will be described in Sect. 3.5 and Chaps. 6 and 7.

The autocorrelation-function and power-spectral-density models we have been discussing strictly apply only to those random signals whose properties are independent of the reference time which we choose to call $t = 0$. For example, the noise voltage in a resistor will, on the average, be the same today as yesterday. We could say $t = 0$ is at noon today or at 3 P.M. yesterday and $v(t)$ for $0 \leq t \leq 10$ min would look about the same. Such random signals are said to be *stationary*. Although we shall usually be working with stationary random signals, it is not difficult to think of nonstationary random processes. The daily rainfall at the corner of Myrtle and Jay Streets in Brooklyn is a nonstationary sequence due to seasonal variations. The average rainfall on a day in April is much greater than for a day in January. The number of home runs hit by a ballplayer before any given day is a nonstationary sequence because it is nondecreasing. On the average, an older ballplayer will have hit a greater total number.

For a nonstationary signal, the autocorrelation function is defined as a function of both the lag k and the absolute time n:

$$E[\mathbf{y}_n \mathbf{y}_{n+k}] = R_y(k, n) \tag{3.131}$$

Even if a signal has an autocorrelation function independent of absolute time n, it could be nonstationary if its mean

$$E[\mathbf{y}_n] = \eta_y(n) \tag{3.132}$$

were a function of time n. The power spectral density is not generally defined for nonstationary signals, although a spectrum has been defined for some special cases of n-dependence in (3.131).

We may occasionally have to work with nonstationary random signals, but most often our signals will be stationary, so that the power transfer relation (3.128) will be applicable. The power spectral density is very commonly used in all types of signal-processing applications. As a matter of fact, in discussing methods of calculating $S_x(\omega)$ and $R_x(k)$ in the next chapter we shall show that $S_x(\omega)$ is approximated by appropriately averaging successive calculations of $|X(\omega)|^2$ when $X(\omega)$ is the Fourier transform of measured samples from the random signal $\mathbf{x}_n$. All our discussion in Chap. 2 and the applications of spectral analysis noted there are thus appropriate. $S_x(\omega)$ plays the role for random signals that $|X(\omega)|^2$, calculated in Chap. 2, plays for deterministic signals. Power-spectrum calculations are often made to classify signals, to look for artifacts in the frequency domain, to simplify interpretation of randomly varying waveshapes, etc. Power spectral analysis has been utilized in biomedical applications, e.g., interpretation of EEG, EKG, and other randomly varying waveshapes, in radar and sonar signal analysis, in seismic applications, in speech and video signal processing, and a host of other applications.

The power transfer relation (3.128) is useful for carrying out measurements of (unknown) transfer functions (excite a complex system with a purely random or white-noise signal and measure the power spectrum at the output), for interpretation of complex physical processes (power spectral analysis has been used in the study of turbulent atmospheric and meteorological phenomena, for example), and for the modeling of physical systems by random-signal models.

Recall, for example, that simple deterministic signals were built up as sums of powers of t (polynomials) or as sums of harmonically related sinusoids (Fourier series). A large variety of analytically tractable signals, having a variety of autocorrelation functions, can be defined by putting white noise into a variety of filters and examining the outputs. We have already done this for the averager filter, and one of the problems considers a differencer. As a final example in this section we shall look at a first-order recursive filter with a white-noise input.

The general first-order recursive filter is defined for random signals by the input-output relation

$$\mathbf{y}_n = a\mathbf{y}_{n-1} + b\mathbf{x}_n \tag{3.133}$$

(see Chap. 2). An output of this filter is a combination of the present input and the previous output. Earlier in this section we examined the input and output autocorrelation functions for such a filter with $b = 1$. We shall shortly give an alternate derivation of those results with the aid of spectral densities. Many simple physical systems can be described by a model like (3.133). For example, a very simplified model of an automobile might have the car's speed at, say, 1-s intervals, represented by y_n, with x_n representing randomly varying accelerations caused by a nervous driver. With a zero acceleration, the speed y_n is a fraction a (less than 1) of y_{n-1} due to friction and inertia. Another example, useful for our later study of aircraft tracking, uses the input x_n to represent wind gusts and y_n to represent the airplane's speed.

Air resistance and inertia relate speeds at successive sample times through the coefficient a in (3.133).

We are interested only in recursive filters for which the feedback coefficient a has magnitude less than 1, because larger values lead to nonstationary outputs. The need for this restriction is exemplified by looking at the response when $a = 2, b = 1$, $x_1 = 1$, and $x_n = 0$ for all other times n. We also assume that $y_0 = 0$, since the output of a recursive filter can be computed only after its initial conditions have been specified, just as initial capacitor voltages and inductor currents are needed to solve electric-circuit problems. These stated conditions lead to the output sequence

y_0	y_1	y_2	y_3	y_4	y_5	...
0	1	2	4	8	16	...

which shows that the given filter is unstable, in the sense that the bounded input leads to an unbounded output. If a stationary purely random disturbance were added to the simple input in this example, the output would have the sequence shown above as its time-varying (exponentially growing) mean; so the output would be nonstationary. In fact, the output variance would also be growing without bound as n increases.

Once the recursive filter in (3.133) has been restricted to being stable, a white input $\mathbf{x}_n$ applied for a long time results in an output $\mathbf{y}_n$ which is wide-sense stationary; i.e., its mean and autocorrelation function do not depend on a time reference. If the constant-input spectral density is σ_x^2, the output spectral density will be

$$S_y(\omega) = \sigma_x^2 \,|\, H(\omega)\,|^2 \tag{3.134}$$

in which the transfer function $H(\omega)$ remains to be determined. $H(\omega)$ is easily found from its property of relating the input and output discrete Fourier transforms for nonrandom signals. Thus, transforming each term in (3.133), we get for the recursive filter (see also Problem 2.22)

$$Y(\omega) = ae^{-j\omega T_s} Y(\omega) + bX(\omega) \tag{3.135}$$

and so it follows that the transfer function is

$$H(\omega) = \frac{Y(\omega)}{X(\omega)} = \frac{b}{1 - ae^{-j\omega T_s}} \tag{3.136}$$

Using (3.128), we have as the output spectral density

$$S_y(\omega) = \frac{b^2 \sigma_x^2}{(1 + a^2) - 2a \cos \omega T_s} \tag{3.137}$$

It is left as a problem for the reader to show that this $S_y(\omega)$ is the transform of the autocorrelation function

$$R_y(k) = \frac{\sigma_x^2 b^2}{1 - a^2}\, a^{|k|} \tag{3.138}$$

which is the exponential function shown at the top of Fig. 3.32 (see also Fig. 3.25).

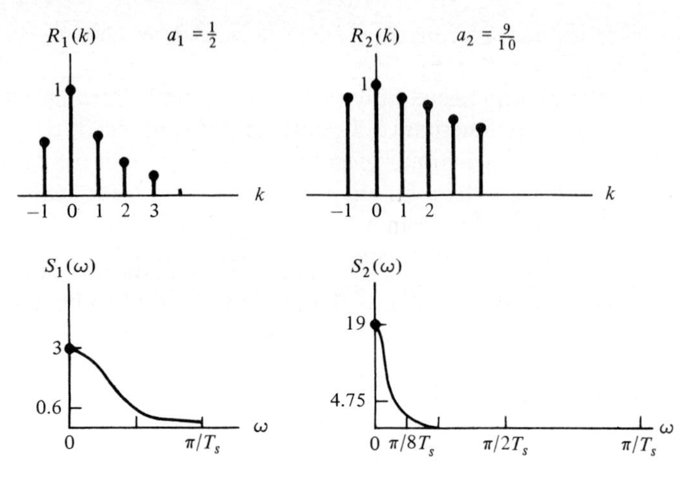

FIGURE 3.32
Recursive-filter output characteristics.

This agrees with the earlier result obtained by direct evaluation of $E[\mathbf{y}_n \mathbf{y}_{n+k}]$ after Eq. (3.106). The dependence of this autocorrelation function and spectral density on the various filter parameters shows us more about the information displayed by these descriptions of a random signal. The output variance

$$\sigma_y^2 = R_y(0) = \frac{T_s}{\pi} \int_0^{\pi/T_s} S_y(\omega)\, d\omega = \frac{\sigma_x^2 b^2}{1 - a^2} \tag{3.139}$$

is proportional to the input variance and $b^2/(1 - a^2)$. Output signals with the same variance (average power) but with different rates of fluctuation can be generated by changing the value of a but always keeping $b^2 = 1 - a^2$. Figure 3.32 shows the output autocorrelation function and spectral density for two different values of a, both with $\sigma_y = 1$.

A large value of a ($a \approx 1$) means $R(1) \approx R(0)$ and very high correlation between successive output samples. The corresponding frequency content is mostly at the low frequencies. Smaller a means more rapid fluctuations on the average and more significant high-frequency values for $S(\omega)$. Notice that the $S(\omega)$ curves are drawn on different scales and that both must have the same area of π/T_s, since $R(0) = 1$ in each case [see (3.120)].

We have seen that different choices for the parameters of the first-order recursive filter in (3.133) result in different output-signal models when the filter input is white noise. Although some choices yield signals with more power at high frequencies (say representing velocity changes for a small private plane vs. those for a heavy commercial jet), all possible power spectral densities for this model have predominant power at low frequencies. In other words, the recursive filter in this example is a *low-pass* filter. Small a corresponds to a wideband filter and large a corresponds to a narrowband filter.

Signals with predominant power at other frequencies (as in Fig. 3.28), or with several peaks in their power spectral density can be generated by using higher-order recursive filters with white inputs $\mathbf{x}_n$, for example,

$$\mathbf{y}_n = a_1 \mathbf{y}_{n-1} + a_2 \mathbf{y}_{n-2} + b_0 \mathbf{x}_n + b_1 \mathbf{x}_{n-1} \tag{3.140}$$

The same methods work for computing the output autocorrelation function and power spectral density of such filters, but the algebra is more complicated. When there is only one nonzero b term in (3.140), statisticians call the resulting $\mathbf{y}$ signal an *autoregression.* When all $a_i = 0$, the output is said to be a *moving average.*

Before leaving this discussion of signal-generating filters, we should point out that the outputs $\mathbf{y}_n$ are stationary, with the indicated power spectral densities, only in the steady state. If a filter like (3.133) is examined just as it is turned on, the output will be nonstationary (just as the response to a sinusoid is sinusoidal only after an initial transient has decayed). In particular, if $y_0 = 0$, the variances of successive outputs are unequal (the finite geometric sum is used here)

$$\begin{aligned} E[\mathbf{y}_1^2] &= b^2\sigma_x^2 \\ E[\mathbf{y}_2^2] &= a^2E[\mathbf{y}_1^2] + b^2\sigma_x^2 = b^2\sigma_x^2(1 + a^2) \\ &\cdots\cdots\cdots\cdots \\ E[\mathbf{y}_n^2] &= b^2\sigma_x^2 \sum_0^{n-1} a^{2m} = \frac{b^2\sigma_x^2}{1 - a^2}(1 - a^{2n}) \end{aligned} \tag{3.141}$$

but approach the limiting $R_y(0)$ from (3.138) as $n \to \infty$. Thus, only in the steady state is the filter output stationary. The initial transient ends when the limiting approximation in (3.141) is valid, say after a number of samples n such that $|a^{2n}| < 0.01$.

As a final note on filtering of random signals, we might consider the relation between the output probability density and that of the input in the case of the first-order recursive filter as an example. The mean values and autocorrelation functions provide only partial probabilistic information, but fortunately they suffice for answering many interesting questions. As mentioned earlier in connection with the averager, if the input signal to a linear filter has gaussian densities, the output is also gaussian. In that case the joint density of two output samples will be like (3.107) with the correlation coefficient determinable from the output autocorrelation function.

It is difficult to draw precise conclusions about the output distribution of the recursive filter when the input is nongaussian. Any conclusion in this direction must be based on the time-domain solution of (3.133)

$$\begin{aligned} \mathbf{y}_1 &= a\mathbf{y}_0 + b\mathbf{x}_1 \\ \mathbf{y}_2 &= a(a\mathbf{y}_0 + b\mathbf{x}_1) + b\mathbf{x}_2 \\ &\cdots\cdots\cdots\cdots \\ \mathbf{y}_n &= a^n\mathbf{y}_0 + b\sum_{j=0}^{n-1} a^j\mathbf{x}_{n-j} \end{aligned} \tag{3.142}$$

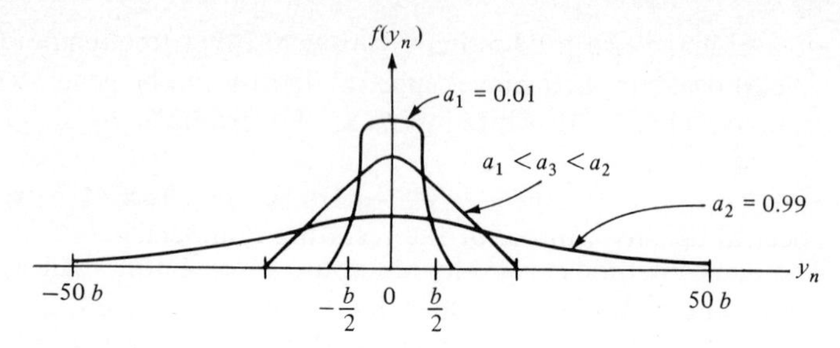

FIGURE 3.33
Output densities for low-pass filters (3-133) having different bandwidths.

For example, if the x_i are uniformly distributed between $-\frac{1}{2}$ and $\frac{1}{2}$, then a few general properties of the output distribution can be deduced as follows. The maximum value of $\mathbf{y}_n$ approaches (for large n)

$$\max |y_n| \rightarrow \frac{b}{2} \sum_{0}^{\infty} |a|^j = \frac{b}{2(1 - |a|)} \tag{3.143}$$

(recall that $|a| < 1$ for a stationary output). This limiting output magnitude corresponds to a possible but unlikely input realization with

$$x_{n-j} = \begin{cases} \frac{1}{2} & j \text{ even} \\ \frac{1}{2} (\text{sign } a) & j \text{ odd} \end{cases} \tag{3.144}$$

If $|a|$ is nearly unity, then [from (3.143)] the output can achieve very large values. In this case many terms in the convolution sum will have nearly equal weight (since $|a| \approx |a|^2$, etc.). It is possible to use a form of the central-limit theorem to show that the output of this very narrowband low-pass filter, which is a weighted sum of independent variables $\mathbf{x}_{n-j}$, is approximately gaussian.

If, on the other hand, $|a|$ is very small, only the zeroth term in the convolution sum in (3.142) will be significant and the output distribution will be approximately uniform between $-b/2$ and $b/2$. In this case the output density will look like the rectangular uniform density but with the corners rounded off.

When $|a|$ is neither very small nor very nearly equal to unity, little can be said about the output density other than that it is zero outside the range of values defined in (3.143) and is symmetrical about a single peak. Figure 3.33 shows some of the possible output densities when the input density is uniform. Similar reasoning can be used to consider other input densities which are neither gaussian nor uniform.

The recursive-filter model of (3.133) can be the basis for a computer program which simulates a random signal with a low-pass power spectral density like those in Fig. 3.32. The white input can be generated by the methods described earlier, with any density of interest, and a correlated output will be generated. A problem at the end of this chapter suggests doing this to provide data for a histogram check of the output-density curves in Fig. 3.33. Chapter 4 will discuss tests to see whether the simulated output really has the autocorrelation function and power spectral density predicted in (3.137) and (3.138).

The following set of FORTRAN instructions would generate 100 output samples corresponding to one example of the autoregression in (3.133). The reader should determine what the mean, variance, autocorrelation function, and power spectral density will be for the resulting numbers.

```
  DIMENSION Y (100)
  Y (1) = 0.
  IX = 11111
  DO 7 N = 2,101
  CALL RANDU(IX, IY, YFL)                    (3.145)
  IX = IY
  X = 2. * (YFL - 0.5)
  Y(N) = Y(N - 1) *0.6 + X
7 CONTINUE
```

Summary

This section has investigated the filtering of random signals. The autocorrelation function was introduced as a measure of dependence among samples of a stationary random process; and its Fourier transform, the power spectral density, was shown to describe the average rates of fluctuation of the random signal. Input and output spectral densities were seen to be simply related to the filter transfer function, a fact to be exploited in later signal-processing examples.

Chapters 5 to 7 will discuss detection and estimation of random signals in the presence of random noise with the aid of the signal models used here. For example, the velocity of an aircraft being tracked might have an exponential autocorrelation function like the one in Fig. 3.32, while the noise interfering with velocity measurements might be white. Section 3.5 will cover this idea of estimating the current value of a random signal based on past noisy observations.

Chapter 4 will examine the complementary problem of taking typical measurements of random signals and processing them to get the numerical power-spectral-density models, e.g., values for a and b in (3.133), used in the design of signal processors.

For typographic simplicity the use of sans serif boldface for random variables will be omitted from now on, except in situations where there might be confusion between a random variable and one of its nonrandom realization values. A probability statement like $P[\mathbf{x} \leq x]$ is one of the exceptions.

3.5 CROSS SPECTRA AND MEAN-SQUARED ERRORS

We have seen that the autocorrelation function and power spectral density are useful for describing properties of a random signal. Similar functions can be defined to characterize the interrelationship between two different random signals. These new

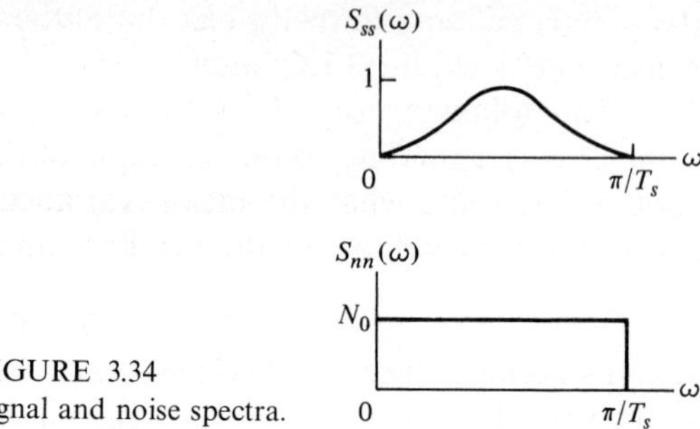

FIGURE 3.34
Signal and noise spectra.

ideas will now be demonstrated and applied to problems of filtering and system modeling.

Consider first a problem of filtering a signal from noisy observations. We have already seen demonstrations of the effectiveness of a smoothing filter in removing white noise from a *constant* signal. In such examples, the mean value of the input and output was the signal, and the output had lower-amplitude fluctuations about the signal level (lower variance) than the input (see Sec. 3.1). Here we want to examine similar filtering problems but for *random* signals. For example, we may wish to estimate $s_1, s_2, \ldots$ based on noisy observations

$$x_i = s_i + n_i \qquad i = 1, 2, \ldots \tag{3.146}$$

This kind of situation might arise when noisy radar measurements of a car's speed are used to estimate the actual speed. The noise values at successive sampling instants are likely to be independent, while the true speed will fluctuate due to road conditions, wind, and the dynamic properties of the driver and motor speed control. If we consider only deviations about the driver's desired speed, a good driver will not cause jerky, rapid changes in speed; nor will his speed error maintain the same sign for a long time.

This situation could be represented by the signal and noise spectra

$$S_{ss}(\omega) = \sin^2 \omega T_s \qquad S_{nn}(\omega) = N_0 \tag{3.147}$$

as shown in Fig. 3.34. Figure 3.35 shows typical samples of s_i and n_i indicating the

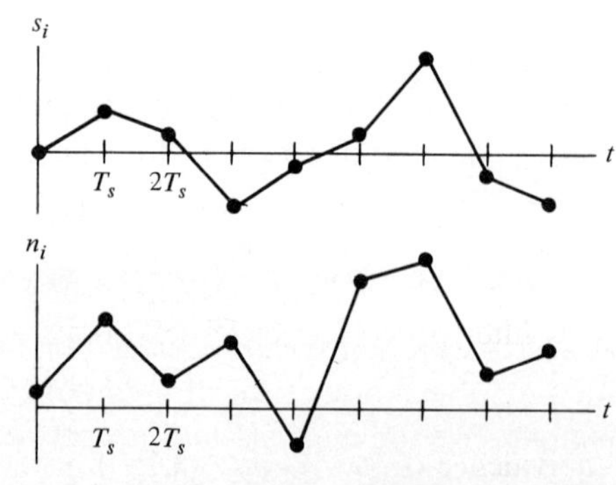

FIGURE 3.35
Signal and noise functions.

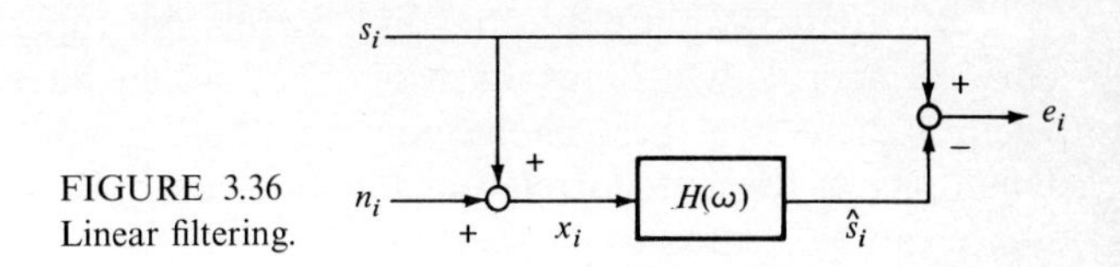

FIGURE 3.36
Linear filtering.

driver's periodic speed corrections near the frequency $\omega = \pi/2T_s$ in contrast to the more irregular noise. The signal segment has more nearly a zero average value (less low frequency) and fewer changes of slope in successive intervals (less high frequency). [The double-subscript notation $S_{ss}(\omega)$, instead of $S_s(\omega)$ used before, is introduced in anticipation of subscripts needed to describe cross spectra.] A filter can hope to remove some of the noise by exploiting the differences between the signal and noise frequency characteristics.

We want to find a linear filter to process the x_i sequence to get an estimating signal $\hat{s}_i$. Figure 3.36 describes this procedure and defines the estimation error $e_i = s_i - \hat{s}_i$. The transfer function should be chosen to minimize $\mathscr{E} = E[e_i^2]$, the mean-squared estimation error. With the usual engineering interpretation of expectations corresponding to averages, a filter chosen in this way should minimize the average squared error

$$\overline{e^2} = \frac{1}{N} \sum_{1}^{N} e_i^2 \approx E[e_i^2] \tag{3.148}$$

over a long processing interval (large N). This interpretation reminds us that power spectra describe long-time average properties. Filters designed from spectral data will have good performance on the average over a long time, as in (3.148). Chapter 6 will describe good short-time filters, which make estimates from a small number of observations.

The frequency interpretation of power spectra suggests that a filter which emphasizes the bands where the signal is strong should be effective for minimizing long-time mean-squared error. Chapter 6 will address the general problem of finding the best possible linear filter. Here we will guess a reasonable form for the filter and see how well it performs. A natural guess is

$$H(\omega) = C \sin^2 \omega T_s \tag{3.149}$$

which has a shape similar to the signal power spectral density. This bandpass transfer function is achieved by a filter equation

$$\hat{s}_i = \frac{C}{2}\left(-\tfrac{1}{2}x_{i-2} + x_i - \tfrac{1}{2}x_{i+2}\right) \tag{3.150}$$

This filter processes the observations as they arrive, and at time $i + 2$, when x_{i+2} arrives, it forms $\hat{s}_i$, an estimate of the value that the signal had at a time $2T_s$ s in the past. That is, the filter produces a delayed estimate. The reader should verify the equivalence of (3.149) and (3.150).

We shall now use spectra and correlation functions to evaluate the mean-squared error $\mathscr{E}$ which results when the chosen filter is used. We shall also see whether there is a best choice for the filter's amplification factor C. From the definitions of the autocorrelation function and power spectral density the performance can be written as

$$\mathscr{E} = E[e_i^2] = R_{ee}(0) = \frac{T_s}{2\pi} \int_{-\pi/T_s}^{\pi/T_s} S_{ee}(\omega)\, d\omega \tag{3.151}$$

It is necessary to express the error power spectral density $S_{ee}(\omega)$ in terms of the signal and noise spectra and $H(\omega)$. The calculation begins by noting that

$$R_{ee}(k) = E[e_i e_{i+k}] = E[(s_i - \hat{s}_i)(s_{i+k} - \hat{s}_{i+k})] \tag{3.152}$$

This error autocorrelation function can be expressed simply in terms of the *cross-correlation function* between the signal samples s_i and the signal estimate $\hat{s}_{i+k}$, which is defined as

$$R_{s\hat{s}}(k) = E[s_i \hat{s}_{i+k}] \tag{3.153}$$

In cross correlations the second subscript indicates the signal which is advanced by k sample times. Expanding (3.152) and separating out the resultant terms, we get

$$R_{ee}(k) = R_{ss}(k) - R_{s\hat{s}}(k) - R_{\hat{s}s}(k) + R_{\hat{s}\hat{s}}(k) \tag{3.154}$$

The cross-correlation function defined in (3.153) can be generalized to provide the cross correlation $R_{xy}(k)$ between any two variables x and y. It should be noted that the cross correlation of a signal with itself is, according to (3.153), simply the familiar autocorrelation function.

The *cross-spectral density* is simply defined as the discrete Fourier transform of the cross-correlation function, namely

$$S_{s\hat{s}}(\omega) = \sum_{-\infty}^{\infty} R_{s\hat{s}}(k) e^{-j\omega k T_s} \tag{3.155}$$

Transformation of both sides of (3.154) produces the frequency-domain relation

$$S_{ee}(\omega) = S_{ss}(\omega) - S_{s\hat{s}}(\omega) - S_{\hat{s}s}(\omega) + S_{\hat{s}\hat{s}}(\omega) \tag{3.156}$$

The desired mean-squared error will be found by integrating the terms in (3.156) according to (3.151). The first term, the signal power spectral density, is known, but further work is needed to establish the other terms.

One important relation which must be used here is a connection between cross spectra and transfer functions. Figure 3.37 shows two filters with separate inputs and outputs. It is possible to show that the cross spectrum of the output is related to the cross spectrum of inputs according to†

$$S_{yv}(\omega) = S_{xu}(\omega) H_1^*(\omega) H_2(\omega) \tag{3.157}$$

† R. J. Schwarz and B. Friedland, "Linear Systems," McGraw-Hill, New York, 1965.

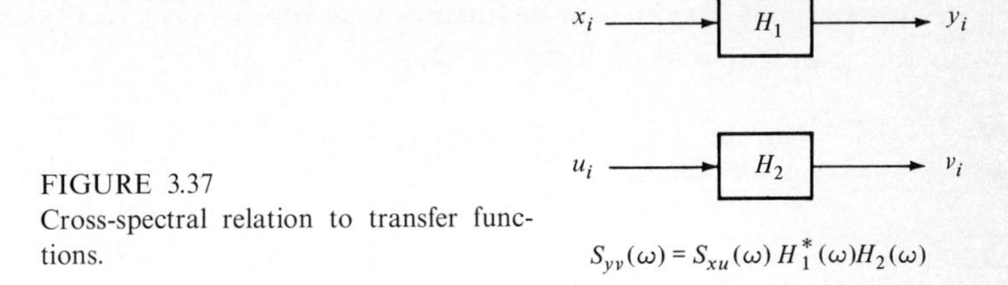

FIGURE 3.37
Cross-spectral relation to transfer functions.

Two properties of this relation are especially noteworthy. First the *order* of subscripts is important. The transfer function relating the first subscripts enters as its complex conjugate (signified by the asterisk). This ordering corresponds to the subscript order defined in (3.153), for cross-correlation functions, which associates the k shift with the second-subscript signal. The second interpretation of (3.157) is that it reduces to the familiar power-spectral-density relation (3.118) between a single filter's input and output when $v_i = y_i$ and $u_i = x_i$.

The filtering-error power spectral density in (3.156) can now be expanded in terms of the given information, with the aid of the cross-spectral relation in (3.157). Identifying the transfer functions in Figs. 3.36 and 3.37 as

$$H_1 = H \qquad H_2 = 1$$

we get

$$S_{\hat{s}s} = S_{xs}H^* \qquad S_{s\hat{s}} = S_{sx}H \qquad S_{\hat{s}\hat{s}} = S_{xx}\,|H|^2 \tag{3.158}$$

The error power spectral density now incorporates the transfer function H,

$$S_{ee}(w) = S_{ss} - S_{xs}H^* - S_{sx}H + S_{xx}\,|H|^2 \tag{3.159}$$

A little further simplification is possible when we use the given information that the signal and noise are independent. In this case we find that

$$S_{sx}(\omega) = S_{ss}(\omega) \tag{3.160}$$

for we have

$$R_{sx}(k) = E[s_i(s_{i+k} + n_{i+k})] = R_{ss}(k) + R_{sn}(k) \tag{3.161}$$

But

$$R_{sn}(k) = E[s_i n_{i+k}] = E[s_i]E[n_{i+k}] = 0 \tag{3.162}$$

due to the independence of s_i and n_{i+k} and the zero-mean value of both signal and noise. (Although we did not specifically mention these mean values at the outset, a *nonzero* mean would make $S_{ss}(0) = \infty$, contrary to the given information. This point will be reconsidered in Chap. 4.) Taking transforms of Eqs. (3.161) and (3.162), we get (3.160), the desired result. Similar arguments show that

$$S_{xs} = S_{ss} \qquad S_{xx} = S_{ss} + S_{nn} \tag{3.163}$$

An expression can now be formed from Eqs. (3.151), (3.159), (3.160), and (3.163) for the mean-squared filtering error:

$$\mathscr{E} = \frac{T_s}{2\pi}\int_{-\pi/T_s}^{\pi/T_s} S_{nn}(\omega)|H(\omega)|^2\,d\omega + \frac{T_s}{2\pi}\int_{-\pi/T_s}^{\pi/T_s} S_{ss}(\omega)\{1 - 2\,\mathrm{Re}\,[H(\omega)] + |H(\omega)|^2\}\,d\omega \tag{3.164}$$

This is the mean-squared error for the filtering problem in Fig. 3.36, for *any* H, S_{ss}, and S_{nn}. It is left as an exercise to substitute the given filter transfer function and spectral densities into (3.164). Evaluation of the resulting integrals of constants and trigonometric functions yields the mean-squared error

$$\mathscr{E} = \tfrac{1}{2} - \tfrac{3}{4}C + \frac{C^2}{16}(5 + 6N_0) \tag{3.165}$$

The optimum amplification factor C° which minimizes this error is found, by setting $\partial\mathscr{E}/\partial C = 0$, to be

$$C^\circ = \frac{1}{N_0 + \frac{5}{6}} \tag{3.166}$$

with a corresponding

$$\mathscr{E}_{\min} = \frac{1}{2} - \frac{9}{20 + 24N_0} \tag{3.167}$$

These results have several reasonable interpretations. If the noise variance N_0 gets very large, $C^\circ \to 0$ and the best estimate $\hat{s}_i = 0$ for all i. That is, when the ratio of signal-to-noise power goes to zero, the observations are useless. In this case $\mathscr{E}_{\min} = \frac{1}{2} = E[s_i^2]$ (found by integrating S_{ss}), as it should, because $\hat{s}_i = 0$ and $e_i = s_i$. On the other hand, if N_0 is very small, we would expect to be able to do a perfect job of estimating the signal, but that is not possible using a filter of the form selected in (3.149). The best filter of that form for $N_0 = 0$ has $C^\circ = \frac{6}{5}$, $\mathscr{E}_{\min} = \frac{1}{20}$. A filter of this bandpass form distorts the signal as well as removing some noise. When no noise is present, the signal distortion can be minimized for a filter of this form by using $C = \frac{6}{5}$.

Perfect filtering of a noiseless observation would result if a different kind of filter with $H = 1$ were used. However, that trivial filter will be worse than the chosen one for a filtering situation with a large noise variance N_0, since in that case $\hat{s}_i = x_i$ and $e_i = n_i$, with

$$\mathscr{E}_{H=1} = N_0$$

This will be larger than $\mathscr{E}_{\min} \approx \frac{1}{2}$ for large N_0.

To summarize, the performance of a filter used as in Fig. 3.36 to separate a random signal from additive noise can be evaluated by the integral in (3.164). That expression was derived by exploiting transfer-function relations between cross spectra, discrete Fourier transforms relating cross correlations and spectra, and the lack of cross correlation between independent zero-mean random signals. Given a filter of

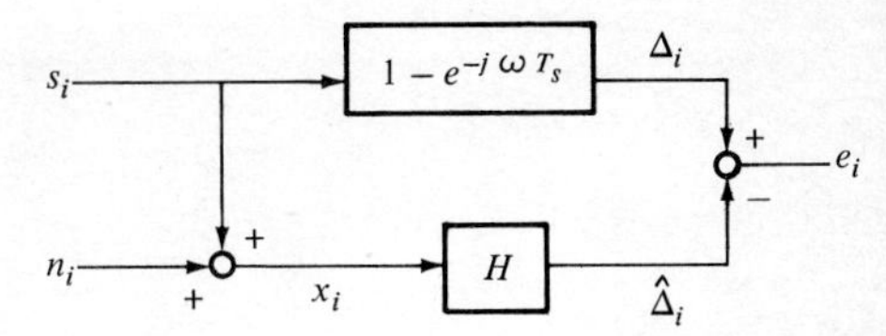

FIGURE 3.38
Linear estimation of a noisy signal's rate of change.

prescribed form, it is possible to find the best values of parameters in that filter for minimizing the mean-squared filtering error. The power-spectral-density descriptions of the signal and noise, which were *assumed* in this example, would, in practice, be derived from studies of real data according to the spectral-analysis methods which will be developed in Chap. 4. An actual filter to achieve the desired filtering could be constructed using a digital circuit or algorithm to carry out the moving average in (3.150).

This same method can be used to estimate some linear function of the signal, such as its rate of change

$$\Delta_i = s_i - s_{i-1}$$

Figure 3.38 is a block diagram, similar to Fig. 3.36, which indicates how spectral methods can be used to evaluate a candidate filter for estimating Δ_i. Details are left as a problem.

Another very interesting application area for digital filtering is in the processing of pictures and other two-dimensional images. For example, the focusing effects of lenses can be described by transfer-function properties. The mathematical analysis of digital picture processing is similar to the time-function processing described in this book, but the details are too complicated to include here. Instead of, say, a signal voltage as a function of one independent variable, *time*, a picture (black and white) is a brightness function of two orthogonal position coordinates, e.g., vertical and horizontal distances. All the ideas of transforms generalize, but two frequencies ω_1 and ω_2 are needed to characterize transfer functions.†

Figure 3.39 shows examples of two-dimensional sampling, quantization, and filtering. This kind of sampling results in squares of constant brightness corresponding to the original brightness at the center of the square (or in some cases to the average brightness over the square). Quantization means that only a few brightness levels (16 in Fig. 3.39*a*) are allowed. Sampling and quantization are useful for putting the signal into a form which can be transmitted over a noisy channel, e.g., by radio transmission from a spacecraft to the earth.

Once the crude block picture has been received, digital processing with much finer quantization levels and various transfer-function characteristics can be employed to make the picture more intelligible. The constant-brightness squares are really a very crude form of D/A conversion which can be improved by subsequent filtering. The sharp edges where brightness changes suddenly correspond to the

† A. Papoulis, "Systems and Transforms with Applications in Optics," McGraw-Hill, New York, 1968.

FIGURE 3.39
Filtered quantized pictures. (*From L. D. Harmon and B. Julesz, "Masking in Visual Recognition: Effects of Two Dimensional Filtered Noise." Science, Vol. 180, 15 June 1973, by permission. Copyright 1973 by the American Association for the Advancement of Science.*)

presence of noise errors at high frequencies, as we have seen in one-dimensional analysis of square waves. It seems reasonable to try to improve the picture by filtering out high frequencies which are more likely to be due to noise than signal. (If the picture is 2 inches and 20 blocks high, the sampling theorem tells us that there is no signal information at frequencies greater than 5 cycles/inch.)

The reader can quickly impose low-pass filtering by squinting or defocusing his eyes. A computer processor can form a low-pass filter by averaging in two dimensions so that the output at point (i, j) in rectangular coordinates is

$$y_{i,j} = \tfrac{1}{5}(x_{i+1,j} + x_{i-1,j} + x_{i,j} + x_{i,j+1} + x_{i,j-1})$$

The picture in Fig. 3.39*b* was formed from that in Fig. 3.39*a* by a more sophisticated

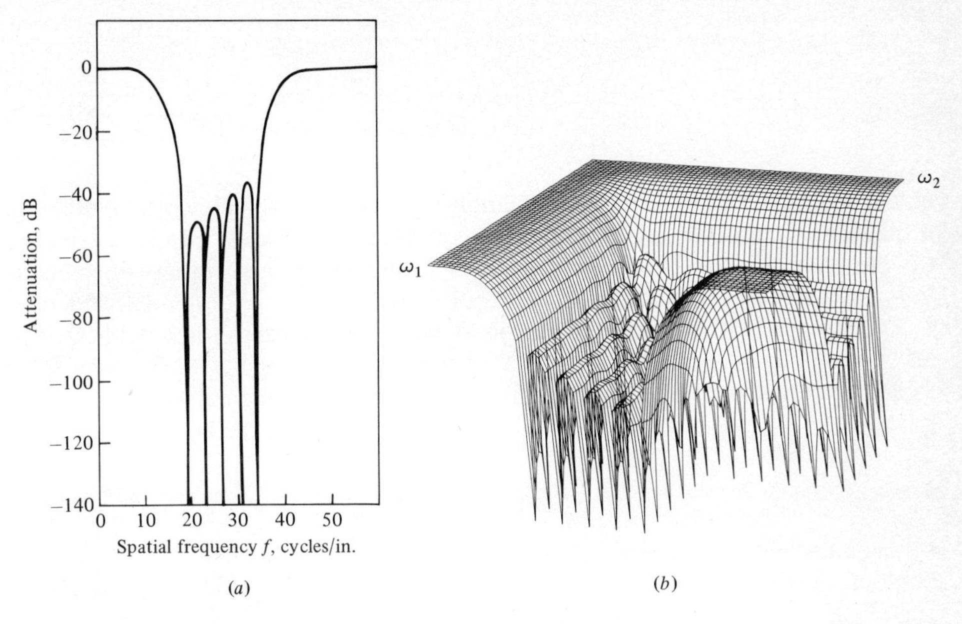

FIGURE 3.40
Band-rejection filters. (*a*) One-dimensional $|H(\omega)|$; (*b*) Two-dimensional $|H(\omega_1, \omega_2)|$ produced by horizontal and vertical one-dimensional filters. (*From L. D. Harmon and B. Julesz, "Masking in Visual Recognition: Effects of Two Dimensional Filtered Noise," Science, Vol. 180, 15 June 1973, by permission. Copyright 1973 by the American Association for the Advancement of Science.*)

low-pass filter than this simple average, having a sharp cutoff at about 6 c/in. Figure 3.39*c* was similarly formed by a 20 c/in cutoff. Filtering was achieved using fast Fourier transforms of the picture elements and multiplication by transfer functions.

The researchers who prepared these pictures were attempting to show that the observer is most sensitive to noise energy and frequencies near the highest ones in the signal and not very sensitive to much higher-frequency noise components. This is analogous to detecting the presence or absence of a tone in the presence of audio noise when the noise is at nearly the same frequency, or the noise is in a very different frequency band. Figure 3.39*d* was formed from Fig. 3.39*a* using the band-rejection filter whose transfer function is characterized in Fig. 3.40. The two-dimensional operation was achieved by two one-dimension operations, each using a filter with the characteristic in Fig. 3.40*a*. First horizontal lines of brightness were filtered, and then vertical lines of the resulting signal were subjected to the same band-rejection filtering. The reader is encouraged to establish his own ranking of the relative quality of the three picture-processing filters.

The cross-spectral density, which was introduced for evaluating mean-squared filtering errors, also has an interesting application to problems of system modeling. One example of system modeling is the process of learning the feel of driving an

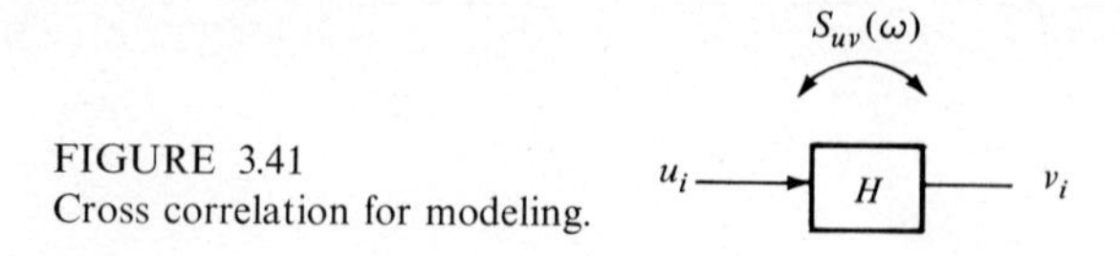

FIGURE 3.41
Cross correlation for modeling.

unfamiliar car. A driver does this by applying small signals (acceleration, braking, turning, etc.) and observing the car's response. He must correlate his inputs with the car's response. Complete modeling requires testing with a large variety of acceleration and steering commands.

Since the mathematical description of controlling a car is too complicated to show here in detail, we shall consider modeling a simple system with only one input signal u_i and one output v_i, as shown in Fig. 3.41. The fundamental transfer-function relation (3.157) can be interpreted with $x_i = u_i = y_i$ to give the input-output cross spectrum

$$S_{uv}(\omega) = S_{uu}(\omega)H(\omega) \tag{3.168}$$

In this modeling example we can select the test signal u_i. An ideal choice is a white random signal with

$$S_{uu}(\omega) = 1$$

for then the unknown transfer function equals the cross-spectral density

$$H(\omega) = S_{uv}(\omega) \tag{3.169}$$

Chapter 4 will describe methods for estimating $S_{uv}(\omega)$ from measurements of the u_i and v_i sequences. It is very common to test unknown systems with pseudorandom inputs u_i.† The resulting cross-spectrum estimate $\hat{S}_{uv}(\omega)$ can be inverted via fast Fourier transforms to estimate the corresponding weighting sequence $h_k = \hat{R}_{uv}(k)$.

Another modeling approach assumes that the unknown transfer function has a specified form but with some parameters whose values are unknown. (Recall the similar parametric approach just applied to filtering problems.) Modeling then corresponds to estimating the parameters, and in such cases this process is often called *system identification*, Chapter 6 will include further discussion of system identification from a maximum-likelihood point of view by estimating coefficients in a parametric model.

3.6 SAMPLING BANDLIMITED RANDOM SIGNALS

The preceding portions of this chapter have dealt with the description and processing of discrete-time random signals. In real signal-processing problems discrete-time signals often arise as the result of sampling continuous-time signals—if for no other reason than to take advantage of the convenience and economy of the kind of digital

† V. Rajagopal, Determination of Reactor Transfer Functions by Statistical Correlation Methods, *Nucl. Sci. Eng.*, February 1962; G. A. Korn, "Random Process Simulation and Measurements," pp. 169ff., McGraw-Hill, New York, 1966.

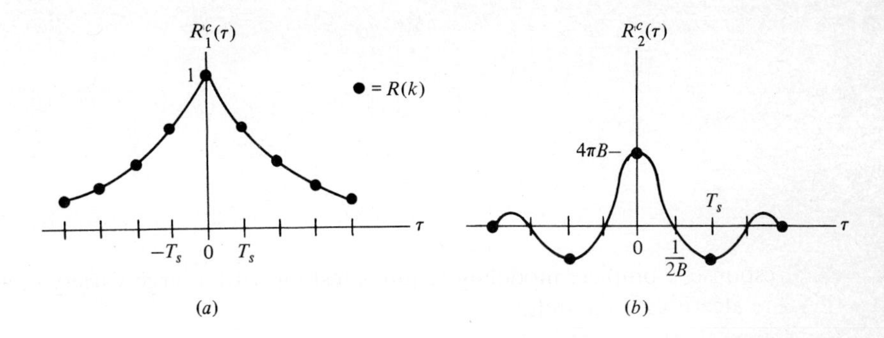

FIGURE 3.42
Continuous and discrete autocorrelation functions.

signal-processing techniques discussed in this book. We have seen in Sec. 2.5 how the sampling process can be chosen so that the discrete version of the signal contains all the essential information from the continuous signal. The sampling theorem says that if a deterministic signal $x(t)$ is bandlimited such that its Fourier transform $X_c(\omega)$ vanishes for high frequencies,

$$|X_c(\omega)| = 0 \qquad \omega \geq 2\pi B \tag{3.170}$$

then $s(t)$ can be perfectly reconstructed from samples $x(kT_s) = x_k$ if the sampling period is sufficiently small, namely

$$T_s < \frac{1}{2B} \tag{3.171}$$

Now we shall explain how essentially the same kind of sampling theorem applies to random signals. To do so, we must introduce the notions of continuous-time random signals and their spectral densities. We symbolize such a random signal by $\mathbf{x}(t)$, with the understanding that it encompasses a separate random variable for every possible value of time t. We have previously given examples of physical variables which might be represented by an $\mathbf{x}(t)$. These include the noise voltage across a resistor due to thermal motion of its electrons, the surface of a rough road as observed by a bicycle rider moving at a constant velocity, the crackling noise in a sonar caused by a hungry bed of shrimp, etc.

Our previous autocorrelation-function definition in (3.104) generalizes easily to the continuous case in the form

$$R_{xx}^c(\tau) = E[\mathbf{x}(t)\mathbf{x}(t + \tau)] \tag{3.172}$$

with a continuous-variable lag τ. The discrete autocorrelation function of samples $\mathbf{x}_k = (kT_s)$ will coincide with $R_{xx}^c(\tau)$ when τ is a multiple of T_s, that is,

$$R_{xx}^c(kT_s) = R_{xx}(k) \tag{3.173}$$

Figure 3.42 shows autocorrelation functions for continuous signals along with points on them which correspond to the discrete autocorrelation functions of the sampled

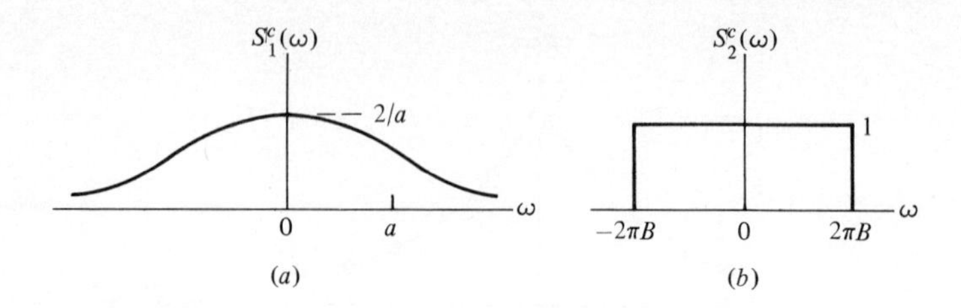

FIGURE 3.43
Continuous power spectral densities.

signals. Notice that in one case the discrete $R(k)$ values fairly well define the shape of the continuous $R^c(\tau)$ while in the other case the samples are too far apart and fail to outline the true shape of $R^c(\tau)$. These observations about sample spacing are analogous to the ones in Sec. 2.5 concerning sampling of deterministic signals. Here we relate the spacing T_s to the rapidity of variation in $R^c(\tau)$, while there the critical factor was the sharpness of peaks and valleys in $x(t)$ itself.

The frequency-domain characterization of random-signal fluctuations is contained in the continuous power spectral density

$$S^c(\omega) = \int_{-\infty}^{\infty} R^c(\tau)e^{-j\omega\tau}\,d\tau \tag{3.174}$$

which is the continuous Fourier transform (2.56) of $R^c(\tau)$. This is just the continuous version of our discrete $S(\omega)$ definition (3.115). Figure 3.43 shows the power spectral densities corresponding to the $R^c(\tau)$ functions in Fig. 3.42, namely

$$R_1^c(\tau) = e^{-a|\tau|} \qquad S_1^c(\omega) = \frac{2a}{a^2+\omega^2} \tag{3.175a}$$

$$R_2^c(\tau) = 4\pi B\,\frac{\sin 2\pi B\tau}{2\pi B\tau} \qquad \begin{cases} S_2^c(\omega) = 1 & |\omega| < 2\pi B \\ S_2^c(\omega) = 0 & |\omega| \geq 2\pi B \end{cases} \tag{3.175b}$$

By analogy with deterministic-signal ideas, we say that a random signal with the $S^c(\omega)$ in (3.175*b*) is bandlimited to frequencies less than B Hz. Continuing the analogy, we claim that a bandlimited random signal can be reconstructed perfectly from samples spaced closer than $1/2B$ s apart.

As in the deterministic case, we shall not prove the sampling theorem. Instead this section will motivate and interpret the result. Figure 3.44 depicts the sampling and reconstruction of a random signal and comparison of the result with the original signal. The interpolation in that diagram is the same as (2.55), the deterministic one used earlier, namely

$$\hat{\mathbf{x}}(t) = \sum_{k=-\infty}^{\infty} \mathbf{x}(kT_s)\frac{\sin[\pi(t-kT_s)/T_s]}{\pi(t-kT_s)/T_s} \tag{3.176}$$

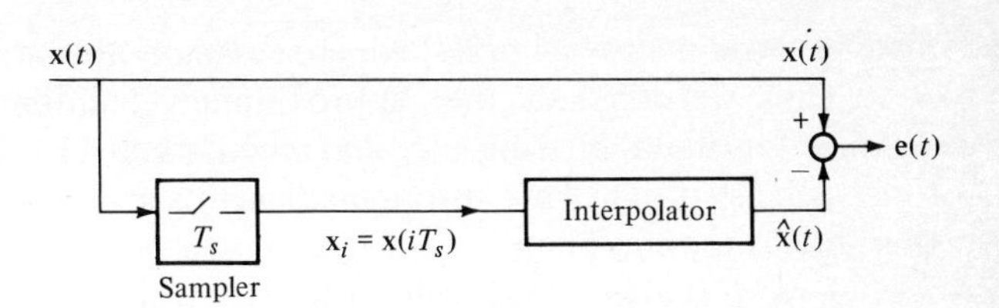

FIGURE 3.44
Sampling and reconstruction of a random signal.

It can be shown that the mean-squared reconstruction error

$$E[\mathbf{e}^2(t)] = E[[\mathbf{x}(t) - \hat{\mathbf{x}}(t)]^2]$$

is zero if $\mathbf{x}(t)$ is bandlimited to $B < 1/2T_s$ Hz.† (While Fig. 3.44 looks similar to Fig. 3.36, which was also used for a mean-squared error calculation, the present case is more difficult to analyze because it contains a mixture of discrete and continuous signals.) The T_s in Fig. 3.42*b* violates this condition for the bandlimited spectrum in Fig. 3.43*b*. It is easy to see that closer-spaced samples which do satisfy the condition will reveal all the peaks and valleys in the autocorrelation function.

We might prefer a sampling theorem which states that $\mathbf{x}(t) = \hat{\mathbf{x}}(t)$ rather than saying their mean-squared difference is zero. Actually the two statements are essentially the same because a zero $E[\mathbf{e}^2]$ implies, by a form of the Tschebychev inequality (3.29), that almost every error is zero, in the sense that

$$P[\mathbf{e}(t) = 0] = 1$$

Another interpretation of this result is that a random signal can be sampled without any loss of information if its *autocorrelation function* can be reconstructed from autocorrelation-function samples spaced at T_s s. This is true because the autocorrelation function and its transform play the roles here of the signal and its transform in the deterministic case. The points $R^c(kT_s)$ must be sufficiently close together to define the shaped $R^c(\tau)$, as indicated in Fig. 3.42.

Finally we should comment on the engineering interpretation of the idealizations inherent in this sampling theorem. First, real signals are not usually bandlimited, but small distortion is caused by picking T_s on the basis of a B above which $S^c(\omega)$ is very small. For example, in (3.175*a*) $B = 10a/2\pi$ should be adequate since

$$S^c(\omega) < \frac{S^c(0)}{100} \qquad \text{for } |\omega| > 10a$$

Second, the interpolation formula (3.176), as well as others which also exist for perfect reconstruction, requires samples from the infinite past and future. Again, as we saw in Sec. 2.5, samples at times far from the interpolation time contribute little to the reconstruction. This assumes that the distant samples are roughly of the same

† A. Papoulis, "Probability, Random Variables, and Stochastic Processes," p. 370, McGraw-Hill, New York, 1965.

magnitude as nearby samples, which is usually the case.

Thus we conclude that approximately bandlimited random signals can be sampled for digital filtering, etc., and reconstructed by simple approximate interpolators (such as straight lines through adjacent samples) without much loss of information due to the sampling.

A word is also in order about the determination of a random signal's approximate bandwidth. Chapter 4 will explain how to estimate a discrete $S_{xx}(\omega)$ based on a finite set of sample values x_k $(k = 1, 2, \ldots, N)$ from a single realization of the signal. If these samples are recorded at a very small spacing $T_s = \Delta$, the discrete and continuous power spectral densities will be equal up to a scale factor. This equivalence is demonstrated by considering approximate evaluation of the continuous Fourier transform integral in (3.174) by the following summation [see also (2.59)]:

$$S^c(\omega) \approx \sum_{k=-\infty}^{\infty} \Delta R^c(k\Delta) e^{-j\omega k\Delta} = \Delta \sum_{-\infty}^{\infty} R(k) e^{-j\omega k\Delta} = \Delta S(\omega)$$

It follows that if a reliable estimate $\hat{S}(\omega)$ can be computed by the methods developed in the next chapter, and if that power spectral density is bandlimited in the sense

$$\hat{S}(\omega) \approx 0 \qquad \text{for } 2\pi B < \omega < \frac{\pi}{\Delta}$$

then B is a suitable number to use for bandwidth and sampling considerations. If the estimated power spectral density is not bandlimited, the continuous data must be resampled at a higher rate (smaller Δ) and a new estimate computed. Once an adequate sampling rate has been determined, we can proceed to work with only the sample values when carrying out processing of random signals according to the detection and estimation techniques of Chaps. 5 to 7.

3.7 APPENDIX: RANDOM VECTORS

Signal-processing problems require simultaneous processing of several different signals. In such cases it is common to collect the several variables together and to call the collection a *vector*, to be denoted as a single quantity. For example, in a radar application in Chap. 7 we shall want to estimate range r, range rate $\dot{r}$, bearing angle θ, and bearing rate $\dot{\theta}$ at each time a radar measurement is available. These signal variables will be arranged in a column to be defined as the signal vector $\mathbf{s}_i$ at time i, with the boldface indicating the vector meaning:†

$$\mathbf{s}_i = \begin{bmatrix} r_i \\ \dot{r}_i \\ \theta_i \\ \dot{\theta}_i \end{bmatrix} \tag{3.177}$$

† Sections 3.1 to 3.6 used sans serif boldface s to represent a scalar random variable. From this point on no special designation is used for random quantities. The regular boldface $\mathbf{s}$ introduced in (3.177) will be used for both ordinary vectors and random vectors.

The top element or component of a vector, r_i in (3.177), is said to be the first element or component of $\mathbf{s}_i$, etc. When it is convenient to think of a horizontal row arrangement of components, we call that collection of variables the transpose of a column vector, e.g.,

$$\mathbf{s}_i^T = [r_i \quad \dot{r}_i \quad \theta_i \quad \dot{\theta}_i] \tag{3.178}$$

Vectors simplify notation by using a single symbol to represent several symbols. Similarly, a matrix is a single symbol representing a rectangular array of numbers, such as

$$\Phi = \begin{bmatrix} 1 & T_s & 0 & 0 \\ 0 & 1 & 0 & 0 \\ 0 & 0 & 1 & T_s \\ 0 & 0 & 0 & 1 \end{bmatrix} \tag{3.179}$$

Capital letters will generally be used to represent matrices. Furthermore, multiplication of a vector by a matrix is a shorthand notation for several simultaneous equations. For example,

$$\mathbf{s}_{i+1} = \Phi\mathbf{s}_i + \mathbf{w}_i \tag{3.180}$$

with

$$\mathbf{w}_i = \begin{bmatrix} 0 \\ u_i \\ 0 \\ v_i \end{bmatrix}$$

represents the four equations

$$\begin{aligned} r_{i+1} &= r_i + T_s\dot{r}_i \qquad & \dot{r}_{i+1} &= \dot{r}_i + u_i \\ \theta_{i+1} &= \theta_i + T_s\dot{\theta}_i \qquad & \dot{\theta}_{i+1} &= \dot{\theta}_i + v_i \end{aligned} \tag{3.181}$$

(These equations are used in Chap. 7 to describe the motion, between two measurement times, of an airplane being tracked by radar.)

The reader is assumed to be familiar with elementary matrix operations, from a mathematics course in vector analysis or linear algebra or engineering courses in electromagnetic fields, stress analysis, automatic control, or computer techniques. The main purpose of this section is to introduce random notions into vector and matrix operations.

It is often convenient to use subscripts or superscripts on a vector or matrix symbol to identify the element of a vector or matrix. For example, in Chap. 5 we shall collect m successive scalar observations $x_1, x_2, \ldots, x_m$ into a vector

$$\mathbf{x}^T = [x_1 \quad x_2 \quad \cdots \quad x_m] \tag{3.182}$$

(A scalar is a one-element vector.) In this application the time subscripts obviously indicate ordering so that x_i is the ith element of $\mathbf{x}$. However, it often happens, as in (3.177), that subscripts are already used for a different purpose; so we might use a superscript to indicate a particular component, e.g., the third element in $\mathbf{s}_i$ is

$$s_i^{(3)} = \theta_i \tag{3.183}$$

Context must tell a subscript's use (time or component). Matrix elements are identified by two indices, the first denoting row and the second column. Thus, from (3.179)

$$\varphi_{34} = T_s \tag{3.184}$$

With this notation the matrix multiplication in (3.180) can be defined by the element relation

$$s_{i+1}^{(k)} = \sum_{j=1}^{4} \varphi_{kj} s_i^{(j)} \tag{3.185}$$

If the elements x_i of a random vector $\mathbf{x}$ are random, each element has a mean variance, distribution, etc. The several components have a joint distribution and may be dependent, independent, correlated, uncorrelated, etc. The mean of a vector is defined to be the vector of mean values:

$$E[\mathbf{x}] = \begin{bmatrix} E[x_1] \\ \vdots \\ E[x_m] \end{bmatrix} \tag{3.186}$$

Generalization of the variance idea from a single variable to a vector is more complicated. Recall that the variance is the expected value of a product, e.g.,

$$\text{var } x_i = E[[x_i - E(x_i)][x_i - E(x_i)]] \tag{3.187}$$

The covariance of two different random variables has a similar definition

$$\text{cov }(x_i, x_j) = E[[x_i - E(x_i)][x_j - E(x_j)]] \tag{3.188}$$

Thus a vector of m random variables can be characterized by m variances and a covariance for each x_i, x_j combination. All the expected values of products are contained in the *covariance matrix* Q whose (i, j) element is a covariance:

$$q_{ij} = \text{cov }(x_i, x_j) \tag{3.189}$$

The calculation of covariances and mean-squared errors is aided by several other matrix-theory operations. The *inner* (or scalar) product of two vectors $\mathbf{v}$ and $\mathbf{w}$ is written as

$$\mathbf{v}^T\mathbf{w} = \sum_{i=1}^{m} v_i w_i \tag{3.190}$$

This quantity is indeed a scalar, as its name implies. In Chap. 7 we shall speak of an error vector as the difference between a signal vector $\mathbf{s}$ and its estimate $\hat{\mathbf{s}}$:

$$\mathbf{e} = \mathbf{s} - \hat{\mathbf{s}} \tag{3.191}$$

If the components of $\mathbf{s}$ are rectangular position coordinates of an airplane, the mean-squared distance error is expressible in terms of a scalar product, as follows:

$$E[d^2] = E\left[\sum_{1}^{3} (s_i - \hat{s}_i)^2\right] = E[\mathbf{e}^T\mathbf{e}] \tag{3.192}$$

Interchange of the order of multiplication of row and column vectors yields another quantity, the *outer* product $\mathbf{xy}^T$ whose (i, j) element is

$$(\mathbf{xy}^T)_{ij} = x_i y_j \tag{3.193}$$

With this notation, the covariance matrix of a zero-mean vector $\mathbf{x}$ can be written

$$Q = \text{cov}\,(\mathbf{x}, \mathbf{x}) = E[\mathbf{xx}^T] \tag{3.194}$$

It is also useful to generalize this idea to the covariance between two different zero-mean vectors, e.g.,

$$P = \text{cov}\,(\mathbf{x}, \mathbf{y}) = E[\mathbf{xy}^T] \tag{3.195}$$

One advantage of the outer-product notation for covariance matrices is that it allows easy computation of one vector's covariance in terms of the covariance of a related vector. For example, if

$$\mathbf{y} = A\mathbf{x}$$

then

$$\text{cov}\,(\mathbf{y}, \mathbf{y}) = E[(A\mathbf{x})(A\mathbf{x})^T] = E[A\mathbf{xx}^T A^T] = AE[\mathbf{xx}^T]A^T = A\,\text{cov}\,(\mathbf{x}, \mathbf{x})A^T \tag{3.196}$$

We have introduced vector and matrix shorthand for moments of random vectors. It is also convenient to use similar shorthand for the density and distribution functions of the variables contained in a random vector. We shall often write their joint density as

$$f(x_1, x_2, \ldots, x_m) = f(\mathbf{x}) \tag{3.197}$$

and expectation integrals will substitute a vector differential $d\mathbf{x}$ for the product of scalar differentials. Thus a mean-squared error might appear as

$$E[\mathbf{e}^T\mathbf{e}] = \int_{-\infty}^{\infty} \mathbf{e}^T\mathbf{e}f(\mathbf{e})\, d\mathbf{e} \tag{3.198}$$

PROBLEMS

3.1 A random variable $\mathbf{z}$ has the density shown in Fig. P3.1.

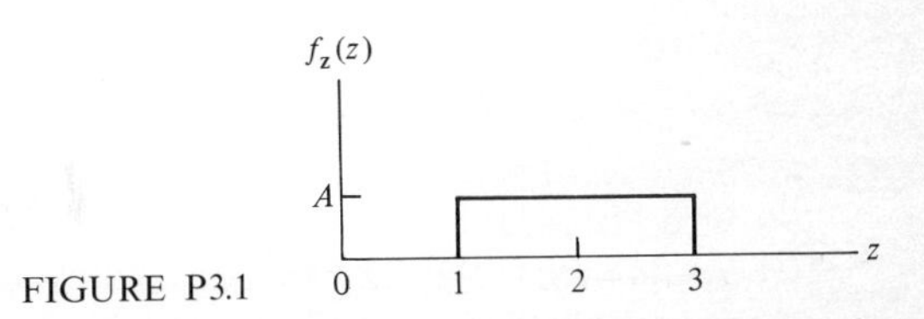

FIGURE P3.1

(*a*) Find A.
(*b*) Find $P[\mathbf{z}^2 > 4]$.
(*c*) We claim that $\mathbf{z} = a\mathbf{x} + b$, where $\mathbf{x}$ is uniformly distributed between -1 and 2. Find a and b.

3.2 A purely random input signal $\mathbf{x}_n$ has gaussian densities $f_{\mathbf{x}_n}(x)$ with means of 3 and variances of 4.
(*a*) If $\mathbf{y}_n$ is produced from $\mathbf{x}_n$ by an averager, use Table 3.1 to compute $P(0 \le \mathbf{y}_n \le 6)$.
(*b*) If $\mathbf{z}_n$ is produced from the $\mathbf{x}_n$ by a differencer

$$\mathbf{z}_n = \mathbf{x}_n - \mathbf{x}_{n-1}$$

find $E(\mathbf{z}_n)$, var $\mathbf{z}_n$, and the decibel value for the ratio σ_z^2/σ_x^2.

3.3 (*a*) Find k such that the Rayleigh density

$$f(x) = \begin{cases} kxe^{-x^2/2N} & x \ge 0 \\ 0 & x < 0 \end{cases}$$

is properly normalized. *Hint:* $f(x)$ is a perfect derivative.
(*b*) Sketch $F(x) = P(\mathbf{x} \le x)$.
(*c*) Find $E(x)$ (integrate by parts or use tables).
(*d*) What is the probability that a random variable x obeying this distribution will have values between $x = \sqrt{N}$ and $x = 2\sqrt{N}$?

3.4 In the manufacture of resistors nominally rated at 100 Ω it is found that the probability distribution of actual resistance values is very closely given by a normal (gaussian) distribution with standard deviation $\sigma = 5\ \Omega$. What percentage of the resistors manufactured lie within the range 90 to 110 Ω?
Convert to error function erf z and use Table 3.1.
Hint: Use a change of variables similar to that in the text.

3.5 A random variable x has the probability density function

$$f(x) = \begin{cases} 2e^{-2x} & x \ge 0 \\ 0 & x < 0 \end{cases}$$

sketched in Fig. P3.5.

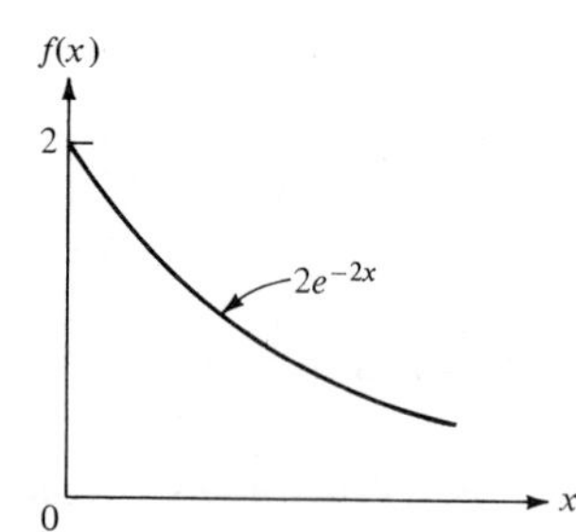

FIGURE P3.5

(*a*) Show that $E(\mathbf{x}) = \frac{1}{2}$.
(*b*) Find $P[\mathbf{x} > 2]$.
(*c*) For $\mathbf{y} = 2\mathbf{x}_1 + 4\mathbf{x}_2$, $\mathbf{x}_1$ and $\mathbf{x}_2$ *each random*, with the same density function as $\mathbf{x}$ above, find $E(\mathbf{y})$.

3.6 If $\mathbf{z}_1, \mathbf{z}_2, \ldots, \mathbf{z}_N$ are independent random variables, all with the same exponential density

$$f_{z_i}(z) = \begin{cases} \lambda e^{-\lambda z} & z > 0 \\ 0 & \text{otherwise} \end{cases}$$

and if the mean value $1/\lambda$ is unknown but $\hat{\eta}_z$, the sample mean, has been computed from a single realization of the signal, what is a reasonable density-function estimate $\hat{f}(z)$?

3.7 One realization of a purely random signal $\mathbf{x}_n$ begins with $x_1, x_2, \ldots = 1, 2, 1, 4, 3, 5, 1, 4, 2, 3 \ldots$

(*a*) Compute mean and variance estimates $\hat{\eta}$ and $\hat{\sigma}^2$ based on the observed numbers.
(*b*) Assuming that the estimates from part (*a*) are the true mean and variance, how likely is it that the next x_n, that is, x_{11}, is ≥ 8.6? *Hint:* Use the Tschebychev inequality.

3.8 (*a*) Compute the mean and variance of a random variable $\mathbf{x}$ having the density shown in Fig. P3.8.

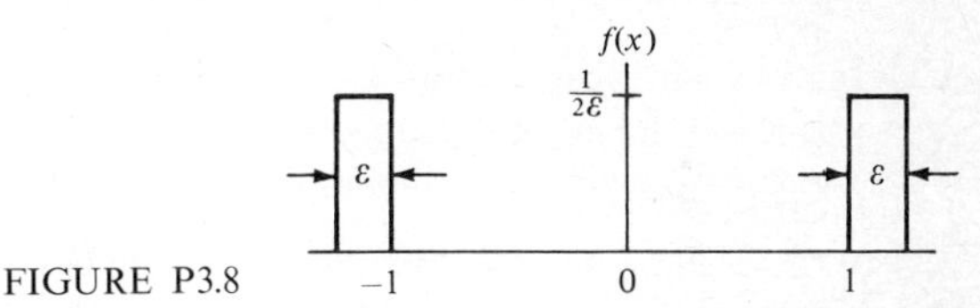

FIGURE P3.8

(*b*) When ε is very small, $\mathbf{x}$ is equally likely to have values of $+1$ or -1 and has zero probability of any other value. Show that

$$\lim_{\varepsilon \to 0} (\text{var } \mathbf{x}) = 1$$

(*c*) Show that as ε approaches zero,

$$P[|\mathbf{x}| \geq \sigma_x] \to 1$$

This demonstrates that the Tschebychev inequality is the tightest possible bound which exists for such probabilities in relation to *every* possible probability distribution.

3.9 The variance computation via Eq. (3.47) requires reading through all the data to get $\hat{\eta}$ and then reading the data again to get $\hat{\sigma}^2$. Show that $\hat{\eta}$ and $\hat{\sigma}^2$ from (3.38) and (3.47) are equivalent to the recursive estimates generated simultaneously from one pass through the data by the formulas

$$\hat{\eta}(m) = \hat{\eta}(m-1) + \frac{1}{m}[x_m - \hat{\eta}(m-1)] \qquad \hat{\eta}(0) = 0$$

$$\hat{\sigma}^2(m) = \frac{m-1}{m}\sigma^2(m-1) + \frac{m-1}{m^2}[x_m - \hat{\eta}(m-1)]^2 \qquad \hat{\sigma}^2(0) = 0$$

where $\hat{\eta}(m-1)$ is the estimate based on $x_1, x_2, \ldots, x_{m-1}$, the first $m-1$ data values, etc.

3.10 Verify the variance value given in Eq. (3.69).

3.11 (*a*) A portion of a program which processes data A(I): I = 1, 2, ..., N is shown below.

ABAR and SIGMA are previously computed estimates of the data's mean and standard deviation. Explain how this sorts the data into groups according to size. Add on steps to form a histogram estimate.

```
A(I) = (A(I) − ABAR)/(4. * SIGMA)
K = 100. * (A(I) + 1.) + .99
IF (K.LE.1) K = 1
IF (K.GE.200) K = 200
XH(K) = XH(K) + 1.
```

(*b*) For what value of N will this program give reliable histogram estimates for uniformly distributed data?
(*c*) Write a program to generate pseudorandom numbers, with your own choice of distribution, and process the numbers to get the sample mean, variance, and histogram.
(*d*) Generalize your histogram algorithm to adjust the number and width of cells to the size of N.

3.12 A reliable histogram estimate should have its standard deviation much smaller than the true value of the density being estimated. Show that if the true density is uniform, if the estimates $\hat{\eta}$ and $\hat{\sigma}$ are close to the true mean and variance, and if the histogram uses $N/100$ cells arranged over the interval $\hat{\eta} \pm 2\hat{\sigma}$, then $\sigma_{\hat{f}(x)} < f(x)/10$.

3.13 Independent observations $x_1, x_2, \ldots, x_N$ are taken of identically distributed gaussian variables (all having the same mean and variance). Compute a general expression for the maximum-likelihood estimate of their common mean value.

3.14 Write a computer program to generate independent gaussian pseudorandom numbers with means of -2 and variances of 4 according to the following steps:
(*a*) Generate a pair of uniformly distributed numbers between 0 and 1.
(*b*) Use one of the uniform numbers to get a Rayleigh number and convert the other uniform number to an interval of 0 to 2π.
(*c*) Get two zero-mean gaussian numbers via the trigonometric transformation defined in (3.102).
(*d*) Shift the mean values.

3.15 A signal $\mathbf{x}_n$ is purely random with each variable uniformly distributed between -1 and 1. Find the autocorrelation functions and power spectra of signals generated from this $\mathbf{x}_n$ according to the following filter equations:

$$\mathbf{w}_n = \mathbf{x}_n - \mathbf{x}_{n-1} \tag{i}$$

Why is $R_w(k) = 0$ for $|k| \geq 2$?

$$\mathbf{z}_n = \mathbf{x}_n + 2\mathbf{x}_{n-1} + \mathbf{x}_{n-2} \tag{ii}$$

Why is $R_z(k) = 0$ for $|k| \geq 3$?

$$\mathbf{y}_n = -\tfrac{1}{2}\mathbf{y}_{n-1} + \mathbf{x}_n \tag{iii}$$

Ans.: $R_y(k) = (-\frac{1}{2})^{|k|}R_y(0)$, $R_y(0) = \frac{4}{3}\sigma_x^2$

3.16 An ideal low-pass filter has the transfer function $H(\omega)$ shown in Fig. P3.16. (Its bandwidth is $\pi/2T_s$ rad/s, or $1/4T_s$ Hz.) The input $\mathbf{x}_n$ is a sequence of zero-mean uncorrelated random variables (white noise) with variance σ^2:

$$R_x(k) = \begin{cases} E[\mathbf{x}_n \mathbf{x}_{n+k}] = 0 & k \neq 0 \\ = \sigma^2 & k = 0 \end{cases}$$

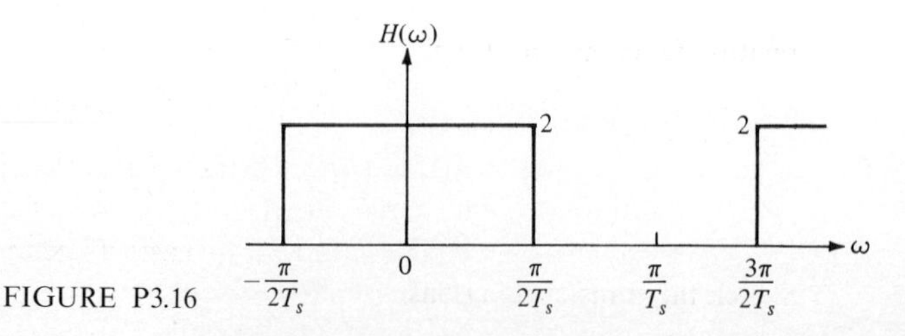

FIGURE P3.16

(*a*) Evaluate $S_x(\omega)$. Sketch $S_x(\omega)$ and $R_x(k)$.
(*b*) Determine and sketch the output spectral density $S_y(\omega)$.
(*c*) Show

$$R_y(k) = 2\sigma^2 \frac{\sin (k\pi/2)}{k\pi/2}$$

Sketch as a function of k, and compare with $R_x(k)$. What has the filter done to the uncorrelated input samples?
(*d*) Show $E(\mathbf{y}^2) = 2\sigma^2$ (power at the output of the filter) in two ways:

$$E(\mathbf{y}^2) = R_y(0) \tag{i}$$

$$E(\mathbf{y}^2) = \frac{T_s}{2\pi}\int_{-\pi/T_s}^{\pi/T_s} S_y(\omega)\, d\omega = \frac{T_s}{2\pi}\int_{0}^{2\pi/T_s} S_y(\omega)\, d\omega \tag{ii}$$

3.17 $\mathbf{x}_n$ is a purely random signal with mean $\eta_x = 0$ and variance σ_x^2. $\mathbf{x}_n$ is the input to an averager whose output $\mathbf{y}_n = (\mathbf{x}_n + \mathbf{x}_{n-1})/2$ is in turn the input to a differencer producing $\mathbf{z}_n = \mathbf{y}_n - \mathbf{y}_{n-1}$. Find (*a*) $E[\mathbf{z}]$, (*b*) $R_z(k)$ (it is nonzero for only three values of k), (*c*) $S_z(\omega)$ and sketch it, (*d*) var $\mathbf{z}$.

3.18 A signal $\mathbf{s}(t)$ is, at each time t, uniformly distributed between $-A$ and A mV. The signal is to pass through an A/D converter whose quantization interval q is to be selected. The discrete-time quantized signal is thought of as

$$\mathbf{x}_k = \mathbf{s}\left(\frac{k}{100}\right) + \mathbf{n}_k$$

the sum of the true signal plus independent quantization error $\mathbf{n}_k$. If the unquantized samples have a power spectral density

$$S_{ss}(\omega) = 10 \sin^2 \frac{\omega}{100}$$

into how many equal quanta must the $-A$ to A interval be divided in order for the mean-squared quantization error to be less than 0.05 mV^2? *Hint:* First find a numerical value for A.

3.19 (*a*) Write and execute a computer program to generate y_n with $R_y(k) = e^{-a|k|}$ from pseudo purely random numbers which are uniformly distributed between -1 and 1.
(*b*) Plot 10 successive input and output points $x_m, x_{m+1}, \ldots, y_m, y_{m+1}, \ldots$ to show the smoother behavior of the y_n for $a = 0.1$, 1, and 10.
(*c*) Generate histograms for the correlated signals and interpret the results with respect to the a-values.

3.20 Pseudorandom numbers are generated according to the programs labeled (3.84) and (3.145). Find the following properties of the random numbers which the $Y(N)$ are supposed to approximate: (*a*) mean, (*b*) variance, (*c*) maximum magnitude, (*d*) autocorrelation function, and (*e*) power spectral density.

3.21 Two filters have the same input $\mathbf{x}_n$ whose power spectral density $S_x(\omega)$ is sketched in Fig. P3.21:

$$\mathbf{y}_n = 3\mathbf{x}_n + 4\mathbf{x}_{n-1} \qquad \mathbf{z}_n = -\tfrac{2}{3}\mathbf{z}_{n-1} + \tfrac{1}{3}\mathbf{x}_n$$

(*a*) Which has the whiter output, i.e., which has a flatter power spectral density? *Hint:* Sketch the transfer functions.

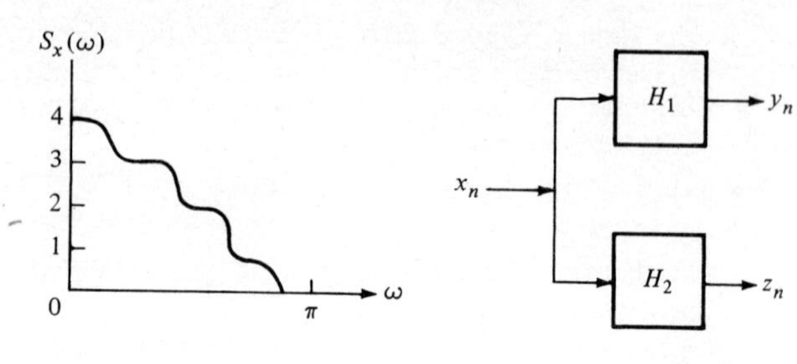

FIGURE P3.21

(*b*) Which is bigger, $E[\mathbf{y}_n^2]$ or $E[\mathbf{z}_n^2]$? Explain.

3.22 The filter shown in Fig. P3.22 is defined by

$$\mathbf{y}_n = \mathbf{x}_n - \mathbf{x}_{n-1}$$

The input is the sum of a deterministic signal, $s_n = bn$, b a known constant, and a random, zero-mean, uncorrelated sequence $\mathbf{d}_n$ with variance σ^2. Use superposition to compute the signal and noise responses separately.

(*a*) Calculate and sketch the signal component of the output.
(*b*) Calculate the noise power, i.e., the variance, in the noise component of the output.
(*c*) Evaluate and sketch the spectral density of the noise component of the output.
(*d*) What do you suppose this filter is doing? Is it effective?

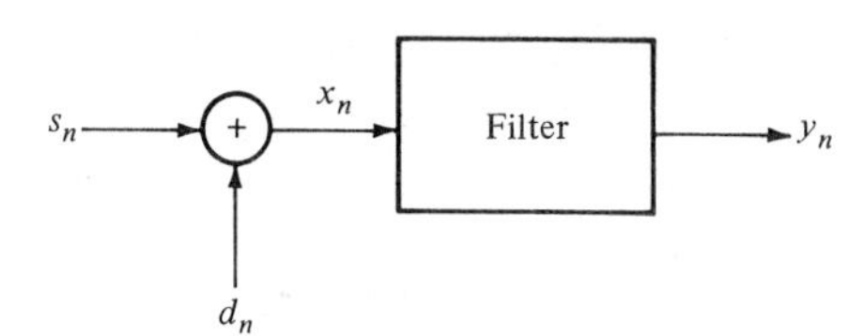

FIGURE P3.22

3.23 A first-order recursive filter with a white input

$$\mathbf{y}_n = a\mathbf{y}_{n-1} + b\mathbf{x}_n$$

$$S_x(\omega) = 1$$

has an output $\mathbf{y}_n$ with

$$R_y(k) = \frac{b^2}{1 - a^2} a^{|k|} \qquad \text{(i)}$$

$$S_y(\omega) = \frac{b^2}{1 + a^2 - 2a \cos \omega} \qquad \text{(ii)}$$

(*a*) Derive (ii) using the transfer-function relation between input and output spectral densities.
(*b*) Derive (ii) from (i) using the discrete Fourier transform relation.
(*c*) Derive (i) from the filter-equation general solution (3.142) as follows. For $|a| < 1$ and large n explain why for any realization

$$y_n \approx b \sum_1^n a^{n-i} x_i \qquad y_{n+k} \approx b \sum_1^{n+k} a^{n+k-j} x_j$$

Show that the $\mathbf{x}_i$ properties imply

$$E[\mathbf{y}_n \mathbf{y}_{n+k}] = b^2 \sum_{1}^{n} a^{2(n-i)+k} E[\mathbf{x}_j^2] = b^2 \sigma_x^2 a^k \sum_{0}^{n-1} a^{2l} \qquad l = n - i$$

and examine this last expression as $n \to \infty$ (steady-state filter response).

3.24 A linear filter is to be used to estimate a signal's rate of change

$$\mathbf{\Delta}_i = \mathbf{s}_i - \mathbf{s}_{i-1}$$

based on noisy observations

$$\mathbf{x}_i = \mathbf{s}_i + \mathbf{n}_i$$

(*a*) Compute $S_{\Delta\Delta}(\omega)$ if $S_{ss}(\omega) = \sin^2 \omega$.
(*b*) With white noise having $S_{nn} = N_0$, let the filter be of the form $H(\omega) = CS_{ss}(\omega) \times (1 - e^{-j\omega})$ and use the methods of Sec. 3.5 to compute the mean-squared estimation error. Find the best value for C.
(*c*) Find the weighting sequence h_k for the best filter.

3.25 Section 3.5 describes how to calculate the mean-squared filtering error when a random signal is being estimated from noisy observations. Fill in the missing steps in the derivation of the mean-squared error [Eq. (3.165)] and of the filter amplification which minimizes that error.

3.26 Repeat the filtering example in Sec. 3.5 but this time use a filter of the form of a low-pass averager followed by a high-pass differencer.
(*a*) Show that the combined filter is described by

$$\hat{s}_i = C(x_i - x_{i-2})$$

and that it has a bandpass form which should be effective in this filtering problem.
(*b*) Find the corresponding mean-squared estimation error and the value of C which minimizes it. Compare this filter (with the best C) to the bandpass filter in Sec. 3.5 and to a simple filter of $H = 1$ ($\hat{s}_i = x_i$). How much delay does each filter introduce? The selection of the best filter from these three will depend on the noise variance N_0. List which is best for each possible N_0.

4

SPECTRAL ANALYSIS OF RANDOM SIGNALS

4.1 INTRODUCTION

In Chap. 3 we talked about mathematical models for random signals. In future chapters we shall propose filters for processing random signals, and the design of those filters will make use of means, variances, densities, autocorrelation functions, and power spectral densities of those random signals. When a practical signal-processing problem is to be solved, the first step is often a study of records of typical data in order to establish reasonable values for these signal statistics. In Sec. 3.2 we have already discussed averaging estimates for the mean and variance as well as the histogram for estimating points on a density function.

This chapter will focus on the problem of establishing an approximate spectrum $S_x(\omega)$ [or autocorrelation function $R_x(k)$] on the basis of a record of a single realization $x_0, x_1, \ldots, x_{N-1}$ of the random signal. The procedures we develop here have much in common with the histogram. The modeling accuracy will improve if more data are available, but there is no clear-cut *best* procedure for approximating the unknown power spectral density of a random signal.

The procedures we discuss here for random signals will apply also to spectral analysis of deterministic signals as described in Chap. 2. Examination of a finite set of

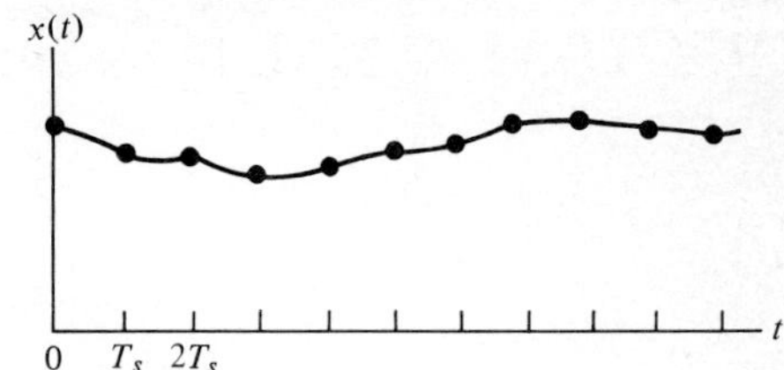

FIGURE 4.1
Random speed with a nonzero mean.

data for a strong component, on the average, in a frequency band near ω_1, that is, a large $S(\omega_1)$, is identical to looking for a deterministic sinusoid $A \cos(\omega_1 t + \varphi)$. We shall describe probabilistic properties for spectral estimates when the signals are random, but these will also be interpretable as error descriptions for estimators of deterministic spectral estimates when they must work with incomplete (truncated) data sets.

Our basic spectral-estimation study will be directed toward signals which are stationary and have zero-mean values. Raw data rarely occur in this form. Figure 4.1 shows a possible record of airplane speed which is fluctuating randomly about a nominal value due to irregular engine thrust, wind gusts, etc. If the original data are not sampled, a digital processor will first sample them. (The sample spacing T_s can be determined crudely by looking at the data or by known physical constraints on the bandwidth of the signal. In some cases the subsequent spectral analysis might indicate that the data should be resampled at a more appropriate spacing.)

The first operation on the samples $x_i = x(iT_s)$ will be to estimate the mean via

$$\hat{\eta}_x = \frac{1}{N} \sum_{0}^{N-1} x_i \tag{4.1}$$

and then to subtract this value from each data point to get the zero-mean fluctuating component $\tilde{x}_i$ of x_i:

$$\tilde{x}_i = x_i - \hat{\eta}_x \tag{4.2}$$

Another example of data needing mean removal is the runway height vs. distance used to measure the surface roughness, as in Sec. 2.6. In practice, digital spectrum analyses routinely remove an estimated mean from every data set.

More generally, the original data might even be nonstationary, like the signals in Fig. 4.2. The $z(t)$ in that figure might be the distance traveled by the plane with a velocity like that in Fig. 4.1. Even in these cases, stationary random signals can be constructed for spectral analysis by subtracting the systematic trends. A straight-line or other polynomial trend can be estimated by the least-squares method of Chap. 2. Similarly, sinusoidal components can be estimated and removed. The airline-passenger data in Fig. 2.33 will produce an approximately stationary residual after linear and sinusoidal trends have been removed.

From now on we shall assume that preprocessing has been carried out to remove the mean value and nonstationary trends from the data. The resulting N

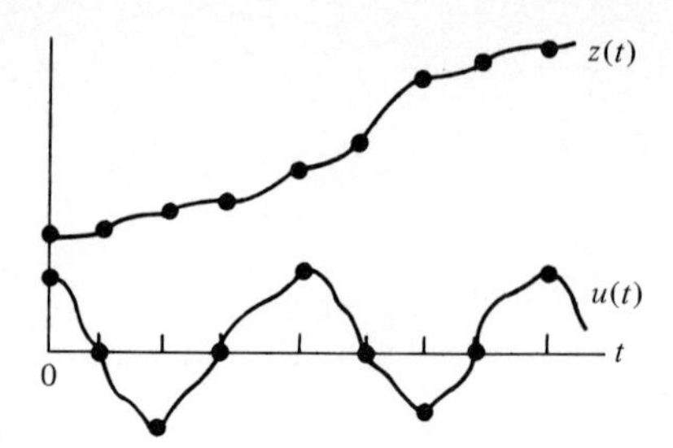

FIGURE 4.2
Nonstationary signals.

samples $x_0, x_1, \ldots, x_{N-1}$ must be processed to get estimates of $R(k)$ and $S(\omega)$. We omit the tilde over each x_i, which would indicate that trends have been removed, and also omit subscripts x on R and S since only a single signal x_i is being studied.

There are two approaches to estimating the function $R(k)$. We might try to get a numerical estimate for each of the M numbers $R(0), R(1), \ldots, R(M-1)$ for some choice of M. On the other hand, we might assume that the true autocorrelation function has some simple analytical form, e.g., the exponential form

$$R(k) = R(0)\alpha^{|k|} \qquad |\alpha| < 1 \tag{4.3}$$

as in Fig. 3.32 for the output of a first-order autoregression filter driven by white noise. This latter approach is said to be *parametric* because once the exponential form has been picked, only the numbers α and $R(0) = \sigma^2$ need be estimated. By contrast, the first method, which assumes no particular form for the autocorrelation function, is called a *nonparametric* approach.

Both parametric and nonparametric techniques are useful for different signal-processing applications. Examination of the results of the latter kind of procedure will often suggest a reasonable functional form to be used in the parametric approach. A parametric model is much more convenient for use with the signal-processor designs which will be developed in Chaps. 5 to 7.

We shall begin with nonparametric autocorrelation-function estimates. Then nonparametric power-spectral-density estimates will be studied with the aid of some special transfer functions. The final section of this chapter discusses maximum-likelihood parametric estimators.

4.2 SAMPLE AUTOCORRELATION FUNCTIONS

N samples from a realization of a stationary zero-mean random process are to be used to estimate its autocorrelation function, which has been defined as

$$R(k) = E[x_n x_{n+k}] \tag{4.4}$$

Chapter 3 has already considered the estimation of $R(0) = \sigma^2$ by means of the sample average

$$\hat{\sigma}^2 = R_N(0) = \frac{1}{N}\sum_{0}^{N-1} x_i^2 = \frac{x_0^2 + x_1^2 + \cdots + x_{N-1}^2}{N} \tag{4.5}$$

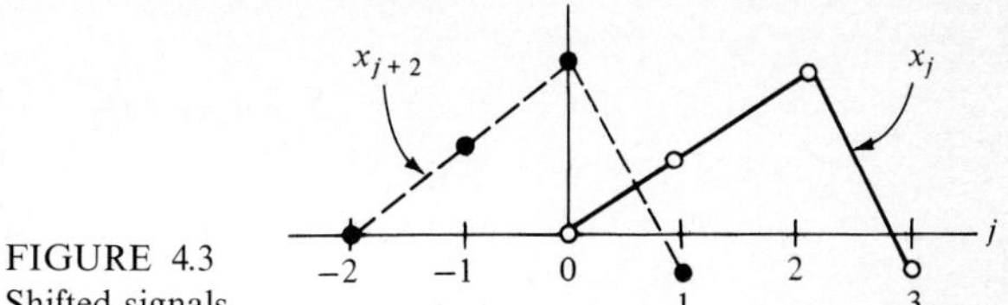

FIGURE 4.3
Shifted signals.

This average was motivated by the idea that the average of independent realizations of x_i^2 should, for large N, approach the expectation $E[x_i^2]$. This is precisely the useful connection between physically measurable quantities and abstract probability theory. In the present context, the realizations $x_0^2, x_1^2, \ldots$ need not be independent, but we still feel that if k is large, x_i and x_{i+k} will be effectively independent, so that (4.5) will still average a large number of independent terms when N is large. It should be recalled that our averaging intuition was further justified in Chap. 3 by showing that the estimate $\hat{\sigma}^2$ of (4.5) is unbiased (for large N)

$$E[\hat{\sigma}^2] = \sigma^2 \tag{4.6}$$

and consistent

$$\lim_{N \to \infty} E[(\hat{\sigma}^2 - \sigma^2)^2] = 0 \tag{4.7}$$

Thus, for large N, the σ^2 computed from a single realization has a high probability of being near the true value σ^2.

In the same spirit as the preceding estimate for $R(0)$, we propose the following estimate for the autocorrelation function at other lags:

$$\begin{aligned} R_N(k) &= \frac{x_0 x_{|k|} + x_1 x_{|k|+1} + \cdots + x_{N-1-|k|} x_{N-1}}{N} \\ &= \frac{1}{N} \sum_{i=0}^{N-|k|-1} x_i x_{i+|k|} \qquad k = 0, \pm 1, \pm 2, \ldots, \pm N - 1 \end{aligned} \tag{4.8}$$

which averages together all possible products of samples separated by a lag of k. Notice that for a large lag k, there are very few possible products (see Fig. 4.3 for the case of $k = 2$ and $N = 4$). In fact, for $k \geq N$ there are no possible pairs with this lag available, and so we arbitrarily estimate the autocorrelation function at those lags to be zero:

$$R_N(k) = 0 \qquad k \geq N \tag{4.9}$$

The autocorrelation function estimator $R_N(k)$ defined in (4.8) and (4.9) is sometimes called the *sample autocorrelation function* since it is based on an average of sample products. Its subscript N emphasizes that it uses a realization which is N samples in length. One nice property of $R_N(k)$ is that it is an even function of k, just as the true $R(k)$ is. Further properties of this estimate are found by examining its mean

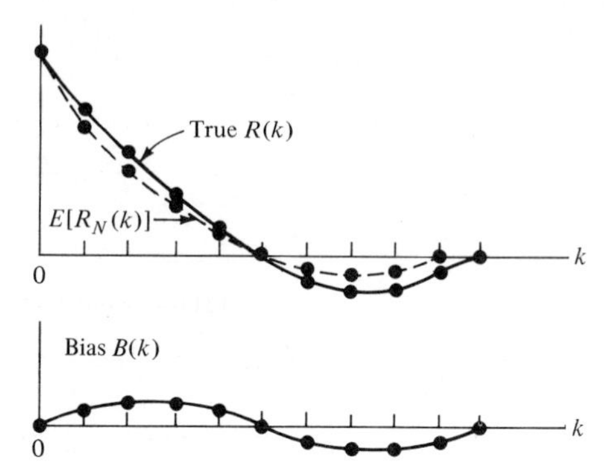

FIGURE 4.4
Bias in autocorrelation function estimate.

and variance to see how it behaves when it is used with many different N-sample realizations.

We now show that if the number of samples N is large compared to the lag k, then $R_N(k)$ is a good estimate of the true $R(k)$. It is easy to see that the mean value of the estimate is related to the true autocorrelation function as follows:

$$E[R_N(k)] = \frac{1}{N} \sum_{j=0}^{N-1-|k|} E[x_j x_{j+|k|}] = \left(1 - \frac{|k|}{N}\right) R(k) \tag{4.10}$$

Thus the sample autocorrelation function defined in (4.8) and (4.9) is *biased*, because its mean is *not* the true autocorrelation function $R(k)$ at lag k, as shown in Fig. 4.4. We can say, however, that this R_N is *asymptotically* unbiased because the $|k|/N$ term vanishes as $N \to \infty$. We could easily get unbiased estimates by using $N - |k|$ to divide the sum in (4.8), rather than N. The form used here is preferred for reasons which will be explained presently. The bias shown in (4.10) is really not important because, as we shall see, $N \gg |k|$ is needed for the variance to be small and the bias is correspondingly small in such cases. That is, unless a large number of lag products is available for an estimate at a specific lag k, the resulting $R_N(k)$ will be unreliable no matter which divisor is used.

The preceding paragraph has asserted that the variance of $R_N(k)$ decreases toward zero as N increases. Verification of this fact is fairly difficult. We can write the difference between the estimate and its mean as

$$R_N(k) - E[R_N(k)] = \frac{1}{N} \sum_{j=0}^{N-1-|k|} [x_j x_{j+|k|} - R(k)] \tag{4.11}$$

and so the variance becomes

$$\operatorname{var} R_N(k) = \frac{1}{N^2} \sum_{j=0}^{N-1-|k|} \sum_{i=0}^{N-1-|k|} E[[x_j x_{j+|k|} - R(k)][x_i x_{i+|k|} - R(k)]] \tag{4.12}$$

Evaluation of this expression requires knowledge of the *fourth-order moments*

$$E[x_j x_{j+|k|} x_i x_{i+|k|}] = \gamma_{ij}(k) \tag{4.13}$$

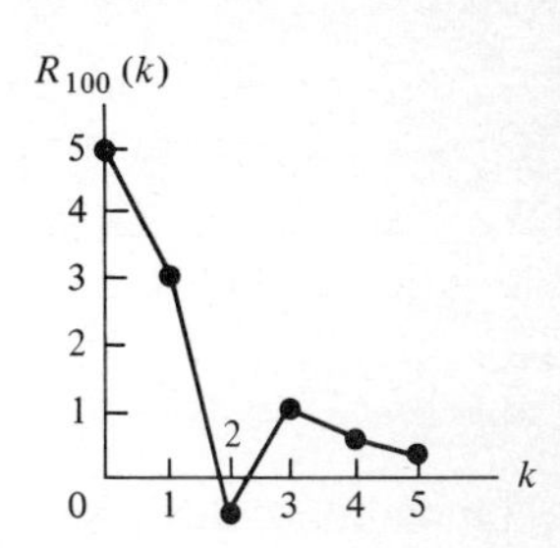

FIGURE 4.5
Typical sample autocorrelation function.

Exact or approximate evaluation of the var $R_N(k)$ thus requires some additional assumptions about fourth-order properties of the random signals whose autocorrelation function we are estimating. For most reasonable signals it can be shown that†

$$\operatorname{var} R_N(k) \approx \frac{2}{N} \sum_{r=0}^{N-1-|k|} \left[1 - \frac{|k| + r}{N}\right] \{R^2(r) + R(r+k)R(r-k)\} \tag{4.14}$$

This expression for the variance of $R_N(k)$ in terms of the true autocorrelation function $R(k)$ is exact for gaussian processes.

The variance expression in (4.14) is approximately proportional to $1/N$ for large N, by the following argument. Most true autocorrelation functions approach zero for a large lag, since samples which are widely separated in time will tend to be uncorrelated. Thus the terms in the braces will vanish for large r. For large N, the upper limit on the sum will be approximately ∞, and the term in square brackets will be approximately unity for r-values at which the term in braces is *not* very small. In this case, the sum will approach a finite value which is the sum of the terms in braces. Thus, for large N, the only dependence of var $R_N(k)$ on N is the $1/N$ factor to the left of the sum in (4.14).

To summarize, we have shown that the sample estimate $R_N(k)$ defined in (4.8) is asymptotically unbiased and has a variance which decreases as $1/N$ for a large number N of samples. Thus, when the number of samples N is much greater than the lag k, the estimate $R_N(k)$ is very likely to be near the true $R(k)$.

Figure 4.5 shows a possible result of computing $R_N(k)$ for a data sample with $N = 100$. [While (4.8) might be used to perform this calculation, we shall see that it is more common to derive $R_N(k)$ indirectly from a power-spectral-density estimate that has been based on the given data.] Note that only positive lags are shown because $R_N(k)$ is an even function, by definition. Also, only lags for which the estimates are relatively accurate, that is, $k \ll N$, are plotted. This figure might have resulted from data which we thought was white noise. The R_{100} shown here is clearly not identical to the $R(k)$ for white noise. However, a finite-length sample from a white signal *could* produce such a sample autocorrelation function. The important question is: How likely is this result if the signal is white?

A white signal has $R(k) = 0$ for $k \neq 0$, and so the variance approximation in (4.14) reduces to

† E. J. Hannan, "Time Series Analysis," Methuen, London, 1960.

$$\text{var } R_N(k) \approx \frac{2R^2(0)}{N} \qquad k \neq 0$$

The distances of $R_N(1)$, $R_N(2)$, ... from zero can then be compared to the standard deviation for those random variables, namely

$$\sigma_{R_N} \approx \frac{R(0)}{\sqrt{50}} \qquad \text{for } N = 100$$

This standard deviation is expressed in terms of the *true* $R(0)$. When $R_N(0)$ is taken as an estimate of $R(0)$, the $R_N(1)$ in Fig. 4.5 is many standard deviations from its hypothesized zero value and so it is unlikely that the given $R_N(1)$ would result if the signal were truly white. Specifically,

$$R_N(1) = 3 \gg \frac{R(0)}{\sqrt{50}} \approx \frac{5}{7}$$

This kind of confidence testing of a hypothesis is similar to arguments used in the previous chapter to relate a proposed density function to actual histogram data. Since several of the points in Fig. 4.5 are far (in terms of standard deviations) from their hypothesized values, the white-signal hypothesis must be rejected.

An obvious alternative model would be an $R(k)$ with exactly the computed values $R_N(k)$. The difficulty with such a model is that we have no way to describe $R(k)$ for larger values of k, and there is no simple analytical expression for $R(k)$ which might be handy for designing the filters and detectors in Chaps. 5 to 7 when this $R(k)$ describes the noise in the filtering problem.

Another alternative would be the simple analytical expression

$$\hat{R}(k) = \hat{R}(0)\alpha^{|k|} \tag{4.15}$$

for an exponential autocorrelation function like those in Fig. 3.32. This parametric model could be estimated with $\hat{R}(0) = R_N(0)$ and α chosen so that $\hat{R}(1) = R_N(1)$. The resulting autocorrelation estimate could be plotted over the points in Fig. 4.5 and the deviations compared to standard deviations computed from (4.14) under this hypothesis. We shall later see that these particular parameter estimates can be justified by a maximum-likelihood criterion.

Estimates for *cross-correlation functions* are also easily computed from sample values of the signals x_i and y_i. By analogy to the sample autocorrelation function in (4.8) we write the sample cross-correlation function as

$$\hat{R}_{xy}(k) = \begin{cases} \dfrac{1}{N} \displaystyle\sum_{i=0}^{N-k-1} x_i y_{i+k} & k = 0, 1, \ldots, N-1 \\ \dfrac{1}{N} \displaystyle\sum_{i=|k|}^{N-1} x_i y_{i+k} & k = -1, -2, \ldots, -N+1 \\ 0 & |k| \geq N \end{cases} \tag{4.16}$$

As with sample autocorrelation functions, the range of summations is influenced by the fact that only N samples are available for x_i and y_i. Here, however, the sample cross-correlation function is not symmetrical $[\hat{R}_{xy}(-k) \neq \hat{R}_{xy}(k)]$, in contrast to the

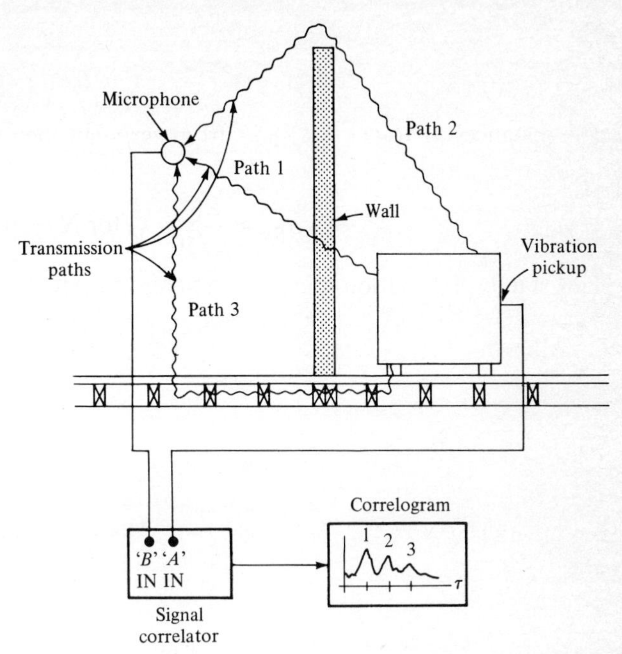

FIGURE 4.6
Noise-transmission path determination. (*From Signal Correlator and Fourier Analyzer Brochure, Princeton Applied Research Corp., Princeton, N.J., 1969, by permission.*)

familiar autocorrelation-function symmetry. A study of the mean and variance of this cross-correlation-function estimator would parallel the one we have just completed for the sample autocorrelation function. The estimate will generally get more reliable as N becomes much greater than k.

In Sec. 3.5 we spoke about using input-output cross correlations for estimating a system's impulse response. Other applications are to time-delay measurements. If the predominant effect of a system is to cause a time delay, with little distortion of the input-signal shape, then the input-out cross-correlation function will have a peak at the lag corresponding to that time delay. [The lagged products in (4.16) will all be positive, and their sum will be maximum, when the input and output waves are lined up.]

Many special-purpose analog and digital instruments are available for performing auto- and cross-correlation estimates. Figure 4.6 shows how a cross correlator can be used to determine which of several possible sound paths is most important in transmitting noise between two locations. Peaks in the cross correlation correspond to the transmission delays along different paths. The physical path corresponding to the lag of the highest peak can be determined by separately calculating and comparing transmission times along the several paths. (Recall that the speed of sound is different in air, wood, plaster, etc.) Noise-abatement actions can then be concentrated on the path which is the worst offender (highest correlation peak) in carrying the noise from one room to the next.

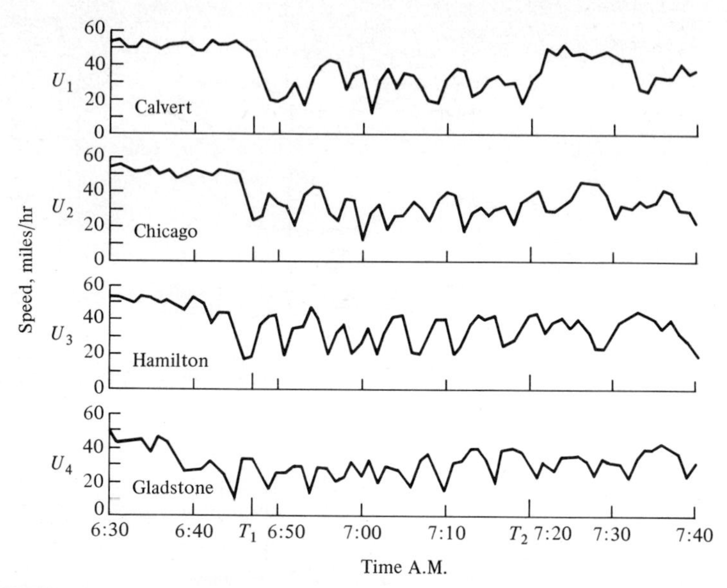

FIGURE 4.7
Records of one-minute average speeds at four locations on the John Lodge Freeway in Detroit. (*From T. Lam and R. Rothery, The Spectral Analysis of Speed Fluctuations on a Freeway, Transportation Science, vol. 4, no. 3, Aug. 1970, by permission.*)

Time-delay measurements are important in studying the detection and location of highway traffic incidents. The speed at which a wave of congestion backs up from an accident depends on geometric properties of the road as well as on the average speed and flow (vehicles per minute). The correct propagation speed for a given section of road can be determined from the peaks in cross correlations between measurements of vehicle speeds at several different locations.†

Figure 4.7 shows speed measurements taken at four locations on the John Lodge Freeway in Detroit. Sonic detectors measured instantaneous speed of nearby vehicles, and the results were averaged over 1 min. (The raw data would be very erratic due to zero-speed intervals when no vehicle was present at the detector.) Figure 4.8 shows the six possible cross-correlation functions generated from the four speed signals. Note that cross correlations do not have the even symmetry of autocorrelations. The lags corresponding to the cross-correlation-function peaks are estimates of corresponding propagation times of speed variations between the two respective sensor locations, i.e., the time delay before a slowdown at the downstream sensor is noticed at the upstream sensor. The separate time delays have been plotted against corresponding sensor-separation distances in Fig. 4.9. A compromise least-

† T. Lam and R. Rothery, The Spectral Analysis of Speed Fluctuations on a Freeway, *Transp. Sci.*, vol. 4, no. 3, pp. 293–310, August 1970.

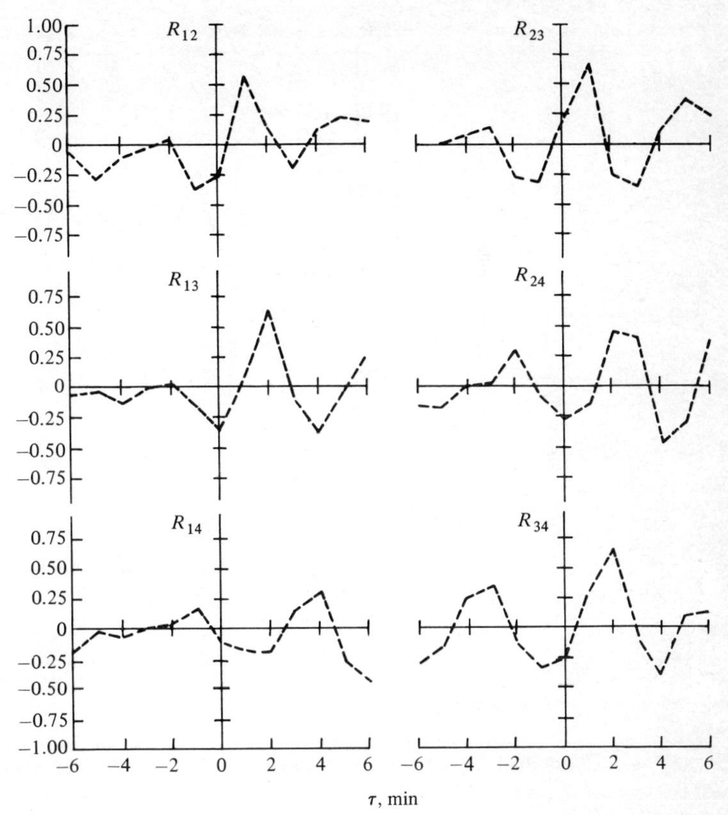

FIGURE 4.8
Cross correlations of speed signals. (*From T. Lam and R. Rothery, The Spectral Analysis of Speed Fluctuations on a Freeway, Transportation Science, vol. 4, no. 3, Aug. 1970, by permission.*)

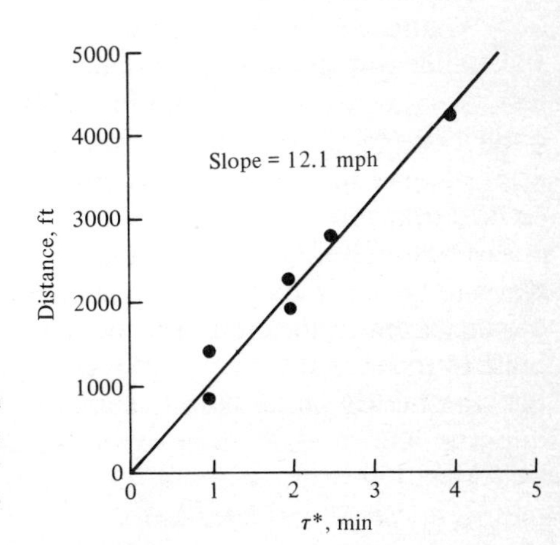

FIGURE 4.9
Peak correlation lag versus separation distance. (*From T. Lam and R. Rothery, The Spectral Analysis of Speed Fluctuations on a Freeway, Transportation Science, vol. 4, no. 3, Aug. 1970, by permission.*)

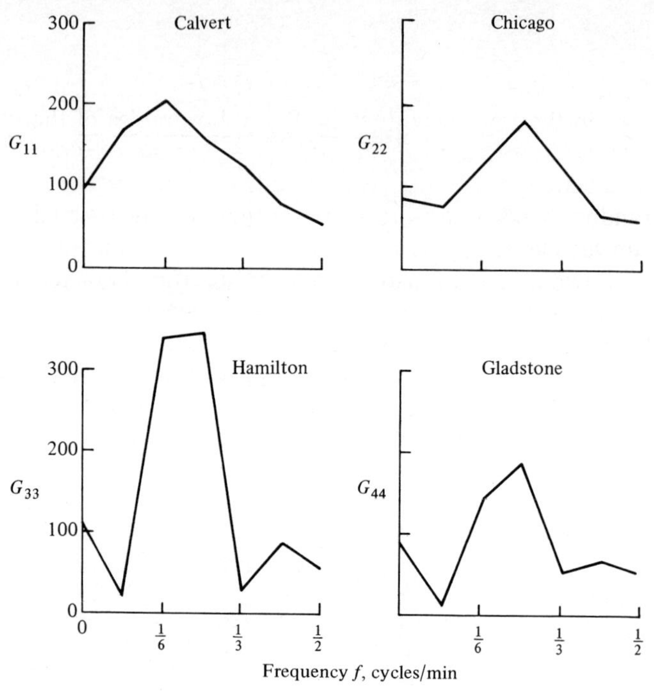

FIGURE 4.10
Spectral densities of speed signals. (*From T. Lam and R. Rothery, The Spectral Analysis of Speed Fluctuations on a Freeway, Transportation Science, vol. 4, no. 3, Aug. 1970, by permission.*)

squares straight-line fit (by the methods of Chap. 2 but modified so that $c_0 = 0$; why?) provides the speed estimate of 12.1 mi/h.

Incidentally, the vehicle speed data in Fig. 4.7 reveal the erratic speed variations characteristic of congested flow during the rush hour. Figure 4.10 shows spectral-density estimates (computed by methods to be described later in this chapter) which reveal that these speed variations are in the frequency range described by oscillation periods in the 3- to 6-min range. That characteristic behavior seems to be a property of the combined systems of vehicle dynamics and human driver speed-control behavior.

Cross correlation is also used to detect weak signals in the presence of noise, as we shall see in Chap. 5. One application of this technique is to radar testing of the distance from earth to the moon or Venus. The correct time delay for a transmitted pulse to travel to the moon and back is found by cross correlating a stored replica of the transmitted pulse with segments of the received data. If the pulse has a long duration and carefully designed irregular shape, the cross correlation with even a very weak returned pulse will be much bigger than the cross correlation with noise returns at earlier and later times.

4.3 PERIODOGRAM

Random-process models can also be fitted to experimental data by making comparisons in the frequency domain. Several examples of the utility of spectral estimates were given in Sec. 2.6. In that earlier discussion of runway roughness, river pollution as a function of time, etc., we did not describe the signals as random; however, random models are most appropriate for such signals because, for example, the runway elevation measured along two parallel but slightly displaced paths will produce two different functions, but the spectral characteristics of all such signals generated from the same runway should be similar. The measurable averages described by *expectations* in probability theory are just the kinds of quantities we might hope to use to describe such irregular signals. While random-signal ideas help us develop and understand spectral-analysis techniques, those signal-processing techniques serve to reduce masses of data and reveal hidden patterns whether the data are thought to be random or not.

We hope to find ways to compute spectral estimates which are fairly similar for separate data sets from the same signal, e.g., BOD pollution levels on different days. When the data are thought to be random, the amount of discrepancy between these evaluations from different signal realizations will be measured by the standard deviation of the estimator. Thus, probability theory is useful for describing the variability among estimates as well as the average (mean) behavior of spectral estimates from different signal realizations.

We shall see that spectral analysis of a single data set, or signal realization, will be much the same whether we view the data as being deterministic or as a realization of a random signal. That is, the magnitude spectrum $|X(\omega)|$ defined in Chap. 2 for deterministic signals can be the basis of a power-spectral-density estimate for random signals.

We shall examine several different procedures for computing power-spectral-density estimates and compare them in terms of their means and variances. Our first method defines the power-spectral-density estimate $S_N(\omega)$ as the discrete Fourier transform of the autocorrelation-function estimate $R_N(k)$:

$$S_N(\omega) = \sum_{k=-\infty}^{\infty} R_N(k)e^{-jk\omega T_s} \tag{4.17}$$

This definition relating the two estimates is motivated by the fact that the *true* power spectral density and autocorrelation function obey the similar discrete Fourier transform relation

$$S(\omega) = \sum_{-\infty}^{\infty} R(k)e^{-jk\omega T_s} \tag{4.18}$$

This spectral-density estimate of (4.17), defined in terms of the sample autocorrelation function, can be related directly to the observed data. For we can show that $S_N(\omega)$ has the following simple relationship to the discrete Fourier transform $X_N(\omega)$ of the data samples:

$$S_N(\omega) = \frac{1}{N}\left|\sum_{n=0}^{N-1} x_n e^{-jn\omega T_s}\right|^2 = \frac{1}{N}|X_N(\omega)|^2 \tag{4.19}$$

$|X_N(\omega)|$ is of course exactly the spectrum in Chap. 2, showing that the spectral estimate of a data set will be handled the same way, whether from a deterministic or random signal. Equation (4.19) as the defining relation for the spectral-density estimate has come into prominence in the past few years and is widely used now with the advent of the fast Fourier transform. This relationship is derived by substituting the R_N definition of (4.8) and (4.9) into the S_N definition of (4.17). To show this we first define a truncated signal x_n^N which equals x_n when x_n is defined and is zero for other values, namely

$$x_n^N = \begin{cases} x_n & n = 0, 1, \ldots, N-1 \\ 0 & \text{otherwise} \end{cases}$$

This is necessary in order to handle the infinite sum of (4.17) with finite data samples. In this way, (4.8) becomes

$$R_N(k) = \frac{1}{N}\sum_{n=-\infty}^{\infty} x_n^N x_{n+k}^N$$

and (4.17) is given by

$$S_N(\omega) = \frac{1}{N}\sum_{k=-\infty}^{\infty}\sum_{n=-\infty}^{\infty} x_n^N x_{n+k}^N e^{-jk\omega T_s} \tag{4.20}$$

When a factor of $1 = e^{j\omega n T_s} e^{-j\omega n T_s}$ is introduced into (4.20), this becomes

$$S_N(\omega) = \frac{1}{N}\sum_{n=-\infty}^{\infty} x_n^N e^{j\omega n T_s} \sum_{k=-\infty}^{\infty} x_{n+k}^N e^{-j\omega(n+k)T_s} \tag{4.21}$$

Changing the summation variable to $m = n + k$ in the second sum, we see that

$$S_N(\omega) = \frac{1}{N} X_N(\omega) X_N^*(\omega) = \frac{1}{N}|X_N(\omega)|^2$$

in which the transform of both the original and truncated signals is

$$X_N(\omega) = \sum_{-\infty}^{\infty} x_n^N e^{-jn\omega T_s} = \sum_{n=0}^{N-1} x_n e^{-jn\omega T_s}$$

This completes the proof that the spectral estimate $S_N(\omega)$ defined as the discrete Fourier transform of $R_N(k)$ can also be computed directly from the data, as in (4.19).

In fact, it is generally faster to compute $S_N(\omega)$ by use of (4.19) and then get $R_N(k)$ from the inverse of the discrete Fourier transform in (4.18) rather than by direct use of (4.8).

This spectral-density estimate was called a *periodogram* by statisticians who first used it to look for periodicities, e.g., seasonal trends in data. The periodogram is seen to be the magnitude squared of the discrete Fourier transform of the data divided by N. The power spectrum of a random signal is thus unrelated to the angle of the complex discrete Fourier transform $X_N(\omega)$ of a typical realization. The $1/N$ factor in (4.19) is also worth some interpretation. Ordinary deterministic signals with discrete Fourier transforms will approach zero for large $|n|$; for example, $x_n = (\frac{1}{2})^{|n|}$. However, a stationary random signal maintains the same range of values for large $|n|$ and small $|n|$. Thus, as N increases, the $|X_N(\omega)|^2$ part of $S_N(\omega)$ will keep getting bigger, on the average, so that the $1/N$ factor will help S_N converge as $N \to \infty$. [Unfortunately, we shall soon see that even with this factor, S_N does not approach the true $S(\omega)$ for large N.] In this connection we might recall Eq. (2.74); it shows that a realization which is sinusoidal will have a discrete Fourier transform whose magnitude, at the frequency of the sinusoid, grows as N. Such a realization would produce an $S_N(\omega)$ estimate which blows up as $N \to \infty$. Thus, as we shall mention again later, in Sec. 4.7, if a power-spectral-density estimate has a high narrow peak at some frequency ω_1, we are well advised to fit and remove a sinusoid of that frequency from the data in the hope that the residual signal will have a bounded power spectral density at all frequencies.

The S_N expression in (4.19) also helps justify the biased (though asymptotically unbiased) $R_N(k)$ defined in (4.8). A true $S(\omega)$ must never be negative, since otherwise we could find a narrowband filter centered at the ω_1, where $S(\omega_1) < 0$, whose output had a negative $E[y^2(t)]$ [see Eq. (3.121)]. This is clearly impossible. Since our periodogram S_N is always nonnegative, as a squared quantity in (4.19), it shares this property with the true $S(\omega)$. It can be shown that if the unbiased autocorrelation-function estimate mentioned after (4.8) were used, its discrete Fourier transform (the spectral-density estimate based on it) could have unacceptable negative values for some ω. This is the reason we avoided using the completely unbiased estimate.

Now that we have a spectral-density estimate, we must see how well it approximates the true spectral density. We shall be satisfied if the bias and variance of S_N both approach zero as N, the number of data, approaches infinity. This concern for statistical properties is important because, as we shall see in Fig. 4.21, power-spectral-density estimates are typically ragged curves requiring considerable interpretation. It might seem obvious that $S_N(\omega)$ should inherit the asymptotically unbiased and consistent properties of $R_N(k)$, of which it is a Fourier transform. However, we shall see that $S_N(\omega)$ does *not* possess all these nice statistical properties.

Referring to (4.17), we see that the mean of S_N is

$$E[S_N(\omega)] = \sum_{k=-\infty}^{\infty} E[R_N(k)]e^{-jk\omega T_s} \tag{4.22}$$

Substituting the mean of the sample autocorrelation function from (4.10), we get

$$E[S_N(\omega)] = \sum_{k=-N}^{N} R(k)\left(1 - \frac{|k|}{N}\right)e^{-jk\omega T_s} \tag{4.23}$$

With $v_k^N(k)$ defined as the triangular function in Fig. 4.11, the periodogram mean can be written as an infinite sum

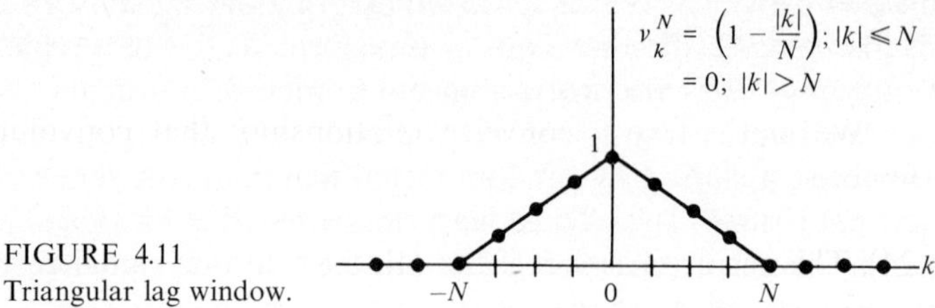

FIGURE 4.11
Triangular lag window.

$$E[S_N(\omega)] = \sum_{-\infty}^{\infty} R(k) v_k^N e^{-j\omega k T_s} \tag{4.24}$$

This sum is the discrete Fourier transform of a product of two k-functions, $R(k)$ and v_k^N. It is apparent that $S_N(\omega)$ as given by (4.17) or the more commonly used (4.19) provides a *biased* estimate of $S(\omega)$. $E[S_N(\omega)]$ in (4.24) differs from $S(\omega)$ by the presence of the v_k^N term. This term is called a *window* function and will be discussed in more detail in the sections following. It appears because of the necessity of using a finite set of data samples. We shall show that the bias effect for finite N can be changed somewhat by appropriate weighting of the data samples, resulting in other examples of window functions. We shall see that $S_N(\omega)$ is asymptotically unbiased, with the distortion effect of v_k^N in (4.24) vanishing as N approaches infinity.

In the next section we discuss window functions in more detail, emphasizing in particular their time-domain descriptions. The properties of window functions in the time and frequency domain will then allow us in Sec. 4.5 to express the periodogram mean value of (4.24) in terms of the true power spectral density $S(\omega)$ via an integral expression of the following convolution type:

$$E[S_N(\omega)] = \frac{T_s}{2\pi} \int_{-\pi/T_s}^{\pi/T_s} S(\eta) V_N(\omega - \eta)\, d\eta \tag{4.24a}$$

Here $V_N(\omega)$ is the discrete Fourier transform of v_k^N, which will be seen to influence the periodogram variance as well as its mean. The bias implied by (4.24*a*) is more troublesome than that for $R_N(k)$ in (4.10) because the spectral-density bias at frequency ω depends on the true $S(\eta)$ at frequencies other than $\eta = \omega$.

4.4 WINDOW FUNCTIONS: FREQUENCY-DOMAIN CONVOLUTION

Recall that when an input signal x_n and an output signal y_n are related by the general linear relation

$$y_n = \sum_{k=-\infty}^{\infty} h_{n-k} x_k \tag{4.25}$$

the input and output discrete Fourier transforms are related by the transfer function $H(\omega)$ according to

$$Y(\omega) = H(\omega) X(\omega) \tag{4.26}$$

$H(\omega)$ is the discrete Fourier transform of the weighting sequence h_n. The sum in (4.25) is called a *convolution* of x_n and y_n; we say that convolution in the time domain corresponds to multiplication [as in (4.26)] in the frequency domain.

We now prove a converse relationship: that convolution in the frequency domain corresponds to multiplication in the time domain. This will enable us to obtain a convolution-integral expression for the mean value of the periodogram in (4.24). This result says that if the nth term in one signal is the product of the nth terms in two other signals

$$g_n = w_n x_n \tag{4.27}$$

then the three discrete-time signals in (4.27) have discrete Fourier transforms satisfying the integral relationships

$$G(\omega) = \frac{T_s}{2\pi} \int_{-\pi/T_s}^{\pi/T_s} W(\eta) X(\omega - \eta)\, d\eta \tag{4.28}$$

or

$$G(\omega) = \frac{T_s}{2\pi} \int_{-\pi/T_s}^{\pi/T_s} W(\omega - \eta) X(\eta)\, d\eta \tag{4.29}$$

Note that these integrals look like (4.24a) and that the discrete Fourier transform of g_n in (4.27) would look like (4.24). These convolution integrals are handy for describing the truncation of finite Fourier transforms as well as discussing the bias of power-spectral-density estimates.

We shall prove that (4.28) implies (4.27) by starting with the discrete Fourier transform inversion integral

$$g_n = \frac{T_s}{2\pi} \int_{-\pi/T_s}^{\pi/T_s} G(\omega) e^{j\omega n T_s}\, d\omega \tag{4.30}$$

The substitution of (4.28) and the use of $1 = e^{jn\eta T_s} e^{-jn\eta T_s}$ yields

$$g_n = \left(\frac{T_s}{2\pi}\right)^2 \int_{\omega=-\pi/T_s}^{\pi/T_s} \int_{\eta=-\pi/T_s}^{\pi/T_s} W(\eta) e^{jn\eta T_s} X(\omega - \eta) e^{j(\omega-\eta)nT_s}\, d\omega\, d\eta$$

Interchanging the orders of integration and changing to $z = \omega - \eta$ for ω, we have

$$g_n = \left(\frac{T_s}{2\pi}\right)^2 \int_{\eta=-\pi/T_s}^{\pi/T_s} W(\eta) e^{jn\eta T_s} \int_{z=-\eta-\pi/T_s}^{\pi/T_s - \eta} X(z) e^{jnzT_s}\, dz\, d\eta$$

When $\eta = 0$, the z-integral equals x_n. However, the integrand in that integral has a period of 2π, and the range of integration is one full period, no matter what the value of η. Therefore, the value of that integral is independent of η and equals x_n for all η. Thus,

$$g_n = \frac{x_n T_s}{2\pi} \int_{\eta=-\pi/T_s}^{\pi/T_s} W(\eta) e^{jn\eta T_s}\, d\eta = x_n w_n$$

and so g_n is the product of x_n and the weight w_n, as asserted earlier.

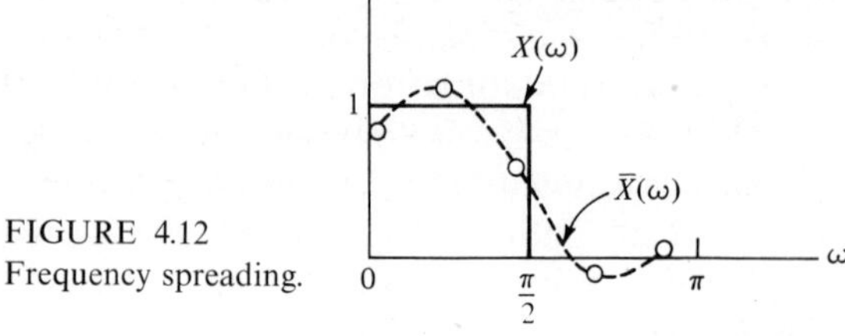

FIGURE 4.12
Frequency spreading.

This result on time-domain multiplication and frequency-domain convolution can be used to explain the effect of truncating a deterministic discrete Fourier transform to a sum of a finite number of terms. The window-function concept noted at the end of the last section arises naturally from this discussion. It also allows generalization to truncation and *weighting* of data samples to obtain various types of window functions. These results will be used in the next section to compute the mean value of a periodogram, discuss the bias effect, and extend the periodogram analysis to include weighted data samples.

The effect of truncating a deterministic discrete Fourier transform was introduced in Sec. 2.6, where, as an example, we approximated the bandlimited $X(\omega)$ of an infinite-duration time function $x_n = [\sin (n\pi/2)]/n\pi$ by the truncated transform $\bar{X}(\omega)$ involving a finite sum. Thus we had (with $T_s = 1$)

$$X(\omega) = \sum_{n=-\infty}^{\infty} \frac{\sin (n\pi/2)}{n\pi} e^{-jn\omega} \tag{4.31}$$

while the truncated transform $\bar{X}(\omega)$ for nine samples was written as the finite sum

$$\bar{X}(\omega) = \sum_{n=-4}^{4} \frac{\sin (n\pi/2)}{n\pi} e^{-jn\omega} \tag{4.32}$$

These frequency functions were sketched in Fig. 2.43, repeated here as Fig. 4.12. The idea of convolution is introduced by writing the truncated sum as an infinite sum for a suitable truncated signal $\bar{x}_n$:

$$\bar{X}(\omega) = \sum_{n=-\infty}^{\infty} \bar{x}_n e^{-jn\omega} \tag{4.33}$$

It is apparent that the truncated signal $\bar{x}_n$ appearing here may be written as the product of x_n with a weighting function w_n:

$$\bar{x}_n = w_n x_n$$

where the weights w_n are obviously defined by

$$w_n = \begin{cases} 1 & |n| \leq 4 \\ 0 & \text{otherwise} \end{cases} \tag{4.34}$$

The sequence w_n, an example of a *window* function, is sketched in Fig. 4.13. The idea

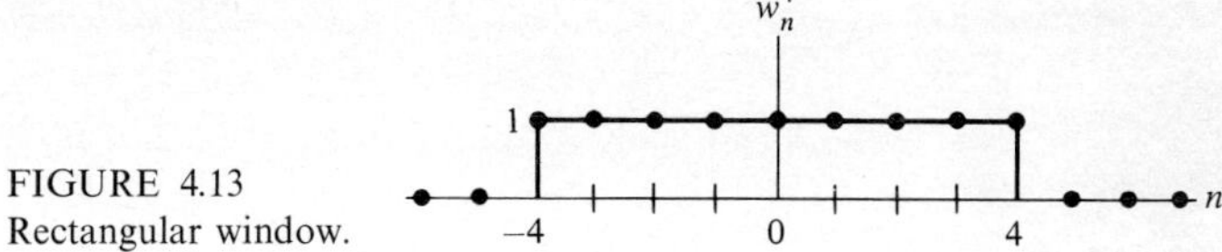

FIGURE 4.13
Rectangular window.

of a window function is that $\bar{x}_n$ is the portion of x_n which can be seen through the window. Equation (4.28) tells us that $\bar{X}(\omega)$ is the convolution of $X(\omega)$ with $W(\omega)$, the discrete Fourier transform of w_n. The spectral window $W(\omega)$ can in turn be found using the geometric-summation formula. Thus, in the case of $T_s = 1$,

$$W(\omega) = \sum_{n=-4}^{4} e^{-jn\omega} = e^{j4\omega} \sum_{k=0}^{8} e^{-jk\omega} = e^{j4\omega} \frac{1 - e^{-j9\omega}}{1 - e^{-j\omega}}$$

The last equality follows from the finite-geometric-sum formula. A little further algebra yields

$$W(\omega) = \frac{e^{j4\omega} e^{-j(9/2)\omega} \left(e^{j(9/2)\omega} - e^{-j(9/2)\omega}\right)}{e^{-j\omega/2} \left(e^{j\omega/2} - e^{-j\omega/2}\right)} = \frac{\sin\left(\frac{9}{2}\omega\right)}{\sin(\omega/2)}$$

In general, the rectangular window w_n will have a width depending on the number of data samples used. Say that it exists between $n = -p$ and p; then there are $2p + 1$ data samples used. (In the example above there were nine samples, and $p = 4$.) We denote this general rectangular window by w_n^p. This will have the transform $W_p(\omega)$ given by

$$W_p(\omega) = \frac{\sin\left(p + \frac{1}{2}\right)\omega}{\sin(\omega/2)} \tag{4.35}$$

As an example, the spectral window discussed above, $W_4(\omega)$, is sketched in Fig. 4.14. Notice that this function has a period of 2π in the ω-domain and the figure shows two complete periods. A little thought shows that $W_p(\omega)$ will cross through zero p times between 0 and π, since the numerator has that many zeros and the denominator

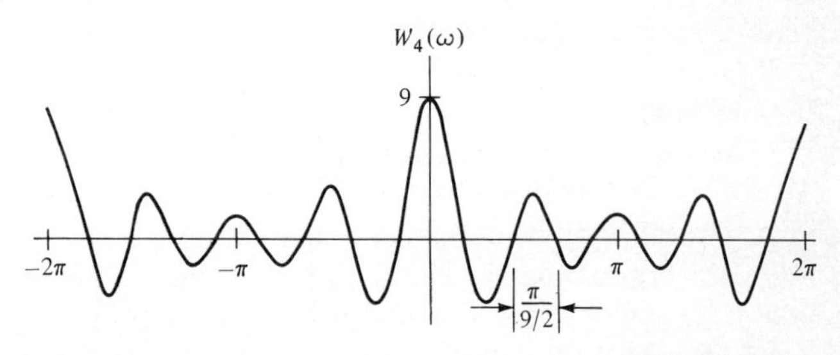

FIGURE 4.14
Spectral window.

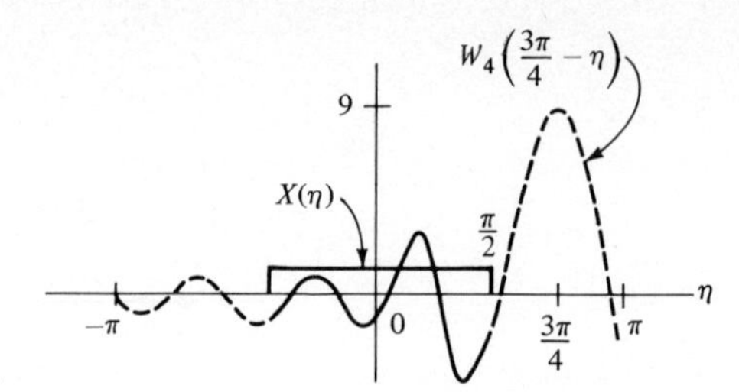

FIGURE 4.15
Convolution integrand.

sin $(\omega/2)$ is an increasing function over that interval. The small-argument property of $\sin\theta \approx \theta$ shows that the highest peak at $W_p(0)$ equals $2p + 1$; and the inverse discrete Fourier transform integral tells us that

$$w_0^p = \frac{1}{2\pi}\int_{-\pi}^{\pi} W_p(\omega)\, d\omega = 1$$

so that the area under one cycle of $W_p(\omega)$ equals 2π.

The reason for introducing the window or weighting function w_n in describing the truncated transform $\bar{X}(\omega)$ is now apparent. Note from (4.33) that $\bar{X}(\omega)$ is the transform of the product of two signal terms x_n and w_n. From our previous discussion of multiplication in the time domain leading to convolution in the frequency domain, i.e., Eq. (4.27) leading to (4.29), we have

$$\bar{X}(\omega) = \frac{T_s}{2\pi}\int_{-\pi/T_s}^{\pi/T_s} W(\omega - \eta)X(\eta)\, d\eta$$

with $W(\omega)$ the spectral window discussed above (the transform of w_n) and $X(\omega)$ the transform of x_n. The effect of truncation is thus described by convolving the infinite-sample transform $X(\omega)$ with a spectral window $W(\omega)$ which depends on the total number of data samples. The effect of convolution is made more explicit in Fig. 4.15, in which, as an example, $X(\eta)$ of Fig. 4.12 and $W(\omega - \eta)$ in the convolution integral are sketched for $\omega = 3\pi/4$, $T_s = 1$, and $p = 4$ (nine data samples used). Since $X(\eta)$ is either 0 or 1, the integrand is either zero or the solid portion of the W_4 curve in that figure. The area under those solid portions is approximately -0.10, giving the value $\bar{X}(3\pi/4) = -0.10$ indicated in Fig. 4.12. The numerical value is more easily found directly from the summation in (4.32). However, the convolution integral and window-function interpretation allows us to see how opening the window, i.e., using a larger p for less truncation, leads to a discrete Fourier transform closer to the discrete Fourier transform of the infinite sequence.

As p is increased (the number of samples $2p + 1$ increased), the oscillations of W_p occur over smaller ω intervals, so that, except for the ever-bigger main peak, adjacent peaks will be of opposite sign but almost equal amplitude. The peak amplitudes are approximately $[\sin(\omega/2)]^{-1}$ evaluated at the appropriate ω. Thus, the large-p version of Fig. 4.13 would yield an $\bar{X}(3\pi/4)$ even closer to the zero value of the true $X(3\pi/4)$. Furthermore, the same reasoning shows that most of the 2π units of area under W_p must become concentrated under the large narrow main peak. This indi-

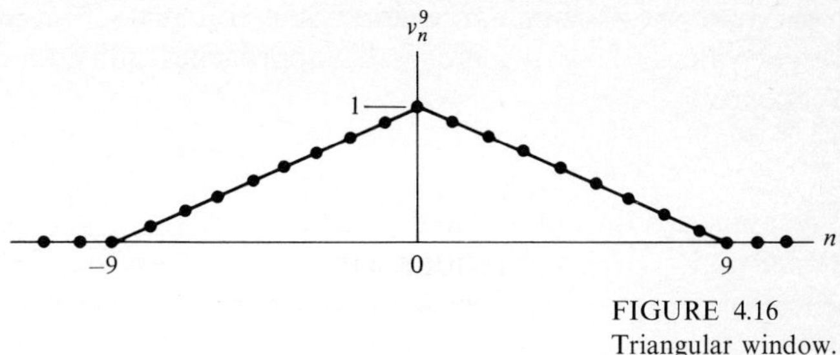

FIGURE 4.16
Triangular window.

cates that when a large-p spectral window $W_p(\omega - \eta)$ overlaps a nonzero portion of $X(\eta)$, the convolution integral is approximately

$$\bar{X}(\omega) \approx \frac{1}{2\pi} X(\omega) \times \text{area under peak} \approx X(\omega)$$

In summary, the spectral window corresponding to a rectangular time window yields truncated transforms which are blurred or smoothed versions of the infinite-sequence transforms. That is, sudden changes in $X(\omega)$ produce gradual changes in $\bar{X}(\omega)$. As the time window widens and the spectral window sharpens, the convolution operation "transmits" $X(\omega)$ into $\bar{X}(\omega)$ with less blurring.

As one further trick with convolutions, we can derive another time-and-frequency window pair which will be used in the next section to compute the mean values of the periodogram and other power-spectral-density estimates. A little thought shows that the triangular window in Fig. 4.16 results from a convolution of w_n^4 with itself (and division by 9) to obtain the window shown with a normalized peak of unity. More generally, assume that we have a convolution of rectangular windows for any value of p,

$$v_n^{2p+1} = \frac{\sum_{k=-\infty}^{\infty} w_k^p w_{n-k}^p}{2p+1} \tag{4.36}$$

It follows from Sec. 2.8 that the corresponding spectral window is the product

$$V_{2p+1}(\omega) = \frac{[W_p(\omega)]^2}{2p+1} = \frac{1}{2p+1} \frac{\sin^2 (p+\frac{1}{2})\omega}{\sin^2 (\omega/2)} \tag{4.37}$$

shown in Fig. 4.17 for $p = 4$.

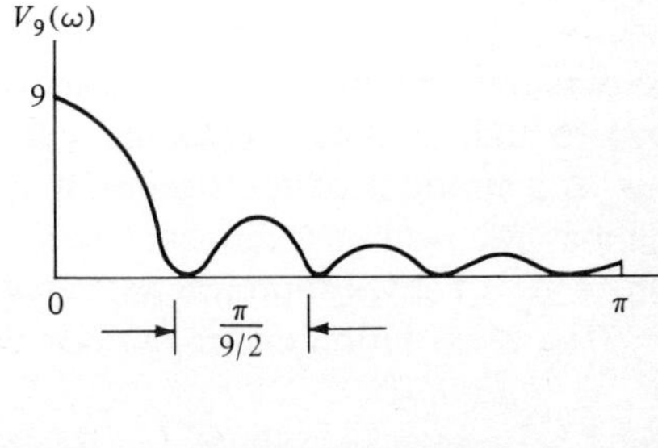

$V_9(\omega)$

9

0 $\quad \frac{\pi}{9/2} \quad \pi \quad \omega$

FIGURE 4.17
Discrete Fourier transform of triangular window.

Like the rectangular window, the triangle v_n^{2p+1}, as it opens up (larger and larger values of $2p + 1$ are used), approaches unity for all n and a "windowed" sequence $\bar{x}_n$

$$\bar{x}_n = v_n^{2p+1} x_n$$

will approach the untruncated x_n. The convolution smoothing effect of the triangular windows $V_{2p+1}(\omega)$ in the frequency domain can be described just like the analogous effect of $W_p(\omega)$ for the rectangular window. Note that the multiplication of x_n by v_n^{2p+1} corresponds to a truncation and *weighting* of the data samples. Window functions other than the rectangular and triangular ones described here can obviously be introduced as well. By the property of convolution in the frequency domain the effect of any one of these on the infinite-sample transform $X(\omega)$ can be determined. Various kinds of window functions are used in spectral estimation or periodogram analysis. In the next section we use window functions to evaluate the bias and variance of the periodogram. Then in Sec. 4.6 we shall consider the advantages and disadvantages of several different window functions for decreasing the bias and variance of spectral-density estimates.

4.5 PERIODOGRAM REVISITED AND MODIFIED

We now return to the question left at the end of Sec. 4.3. Recall that we had introduced (4.19) as a possible (and very simple) estimate of the spectral density $S(\omega)$. We then showed in (4.24) that the expected value or mean of this estimate appeared as the transform of the autocorrelation function *multiplied* by a window function v_k^N. It is now apparent from the discussion of the previous section that this must correspond to the convolution of $S(\omega)$ with a spectral window given by the transform of v_k^N. This spectral estimate is thus biased. We explore the consequences of that bias in this section.

Specifically, we shall now use convolution integrals to examine the bias of a periodogram estimate. This bias at a particular frequency, $B_N(\omega_a)$, is the difference between the true power spectral density and the mean value of its estimate

$$B_N(\omega_a) = S(\omega_a) - E[S_N(\omega_a)] \tag{4.38}$$

The equation for the periodogram mean [see (4.24)] is

$$E[S_N(\omega_a)] = \sum_{k=-\infty}^{\infty} R(k) v_k^N e^{-j\omega_a T_s}$$

The window function v_k^N is the same kind of triangular time function as the one in Fig. 4.16 and Eq. (4.36). Thus, the periodogram mean is the discrete Fourier transform of a product of the *true* autocorrelation function and a triangular window function. The resulting frequency function can be expressed entirely in the frequency domain by a convolution integral, as we saw in the previous section.

The convolution expression for the mean power spectral density is

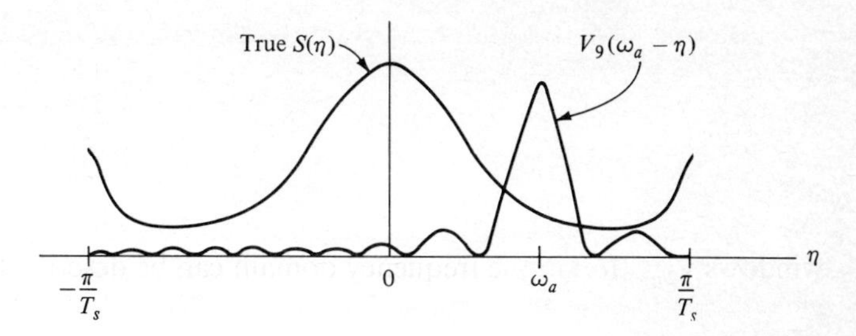

FIGURE 4.18
Convolution integrand.

$$E[S_N(\omega_a)] = \frac{T_s}{2\pi} \int_{-\pi/T_s}^{\pi/T_s} S(\eta) V_N(\omega_a - \eta)\, d\eta \tag{4.39}$$

with the integrand functions displayed in Fig. 4.18 for $N = 9$ (in order to use Fig. 4.17). The general expression for V_N is [see also Eq. (4.37)]

$$V_N = \frac{1}{N} \left[\frac{\sin\,(N/2)\omega}{\sin\,(\omega/2)} \right]^2$$

The main contributions to the integral of (4.39) will occur in frequency bands where one or both integrand factors are large. For large N, the spectral window $V_N(\omega)$ will have a high narrow main peak with area of 2π, and its side lobes will be very narrow. In such cases, $S(\eta)$ will be approximately constant at $S(\omega_a)$ over the narrow width of the main lobe of V_N, and the integrand will be almost zero elsewhere; so for large N and any ω

$$E[S_N(\omega)] \approx \frac{T_s}{2\pi} \int_{\omega - 2\pi/NT_s}^{\omega + 2\pi/NT_s} S(\omega) V_N(\omega - \eta)\, d\eta = S(\omega) \tag{4.40}$$

Thus we see that our estimate $S_N(\omega)$ is asymptotically unbiased since $B_N(\omega_a)$ in (4.38) will vanish for large N. Furthermore, we can see that N must be larger to achieve an equivalent bias at a frequency where the true power spectral density has a narrow peak, for example, $\omega = \pi/T_s$ in Fig. 4.18, compared with the estimate at a frequency where the true $S(\omega)$ is smooth, for example, $\omega = \pi/2T_s$ in Fig. 4.18. This follows because small bias requires $S(\omega)$ to be constant over a width of the main lobe of V_N. To achieve this small bias, N must be large enough for the main lobe of V_N to be narrower than the narrowest peak in $S(\omega)$.

Figure 4.19 summarizes the three window functions we have been discussing. The finite-data restriction corresponds to the rectangular data window w_n^N which results in the triangular autocorrelation-function window v_k^N. The corresponding power-spectral-density bias is found using the spectral convolution window $V_N(\omega)$. Various examples of the window effect on bias will be discussed in connection with Fig. 4.22 after we first develop analogous properties for the power-spectral-density variance and its relation to the number N of data samples.

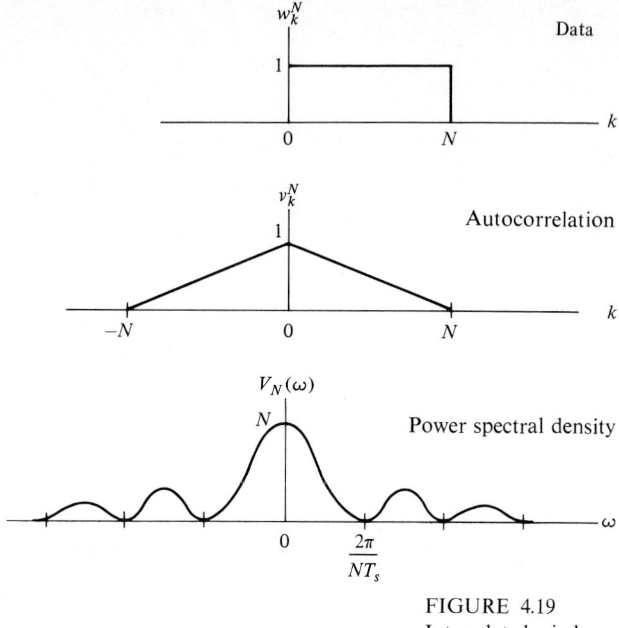

FIGURE 4.19
Interrelated windows.

Variance of the Periodogram

We have shown that the experimental spectral density $S_N(\omega)$, or periodogram, defined in (4.17) and (4.19) can have a small bias if the number of data samples N is sufficiently large. This means that the average of many N-sample periodograms will be near the true spectrum $S(\omega)$, at all ω. We would prefer to be able to say that *each* N-sample periodogram is a good estimate of $S(\omega)$, a conclusion which would follow if $S_N(\omega)$ had the additional property that its variance became vanishingly small as N is increased.

Unfortunately, var $S_N(\omega)$ is generally not small, even for large N. As an example, if the data x_n come from a gaussian process, it can be shown† that

$$\lim_{N \to \infty} \text{var } S_N(\omega) = S^2(\omega) \tag{4.41}$$

i.e., the variance of $S_N(\omega)$ approaches the square of the true spectrum at each frequency ω. Using the ratio of mean to standard deviation as a kind of signal-to-noise ratio

$$\frac{E[S_N(\omega)]}{\sqrt{\text{var } S_N(\omega)}} \approx \frac{S(\omega)}{S(\omega)} = 1 \tag{4.42}$$

† G. M. Jenkins and D. G. Watts, "Spectral Analysis and Its Applications," Holden-Day, San Francisco, 1968.

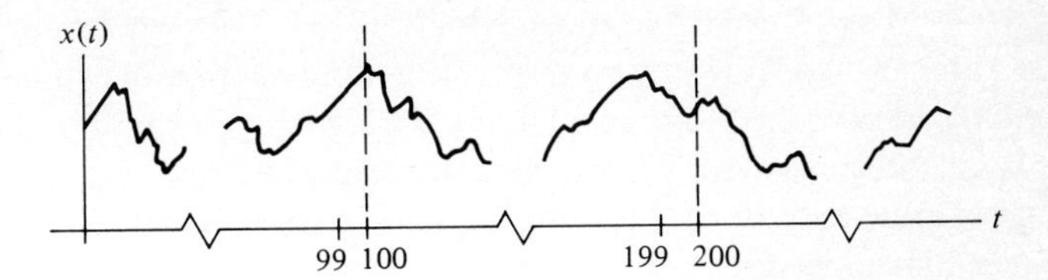

FIGURE 4.20
Segmenting of data.

we see that the signal (true spectrum) is only as big as the noise [uncertainty in $S_N(\omega)$], for large N. The variance expression in (4.41), which also applies approximately for nongaussian data, shows that calculations using different sets of N samples from the same x_n process will yield vastly different values for $S_N(\omega)$ even when N is large.

This annoying property of the periodogram, that its variance does not vanish for large N, can be circumvented. We certainly should be able to get good spectral-density estimates when a large number of data are available. One way to do this is to average together several periodograms. Given 1000 data samples, we could compute 10 separate periodograms of length $M = 100$. The mth periodogram, for example, would then be given by

$$S_{100,m}(\omega) = \frac{1}{100}\left|\sum_{k=100(m-1)}^{100m-1} x_k e^{-jk\omega}\right|^2 \qquad m = 1, 2, \ldots, 10 \tag{4.43}$$

This is of course done by cutting the original data into 10 equal-length segments, as shown in Fig. 4.20. An *averaged periodogram* $\hat{S}_M^K$ can then be computed from K individual periodograms of length M, like (4.43). For example, when $K = 10$,

$$\hat{S}_{100}^{10}(\omega) = \frac{1}{10}\sum_{m=1}^{10} S_{100,m}(\omega) \tag{4.44}$$

Since the 10 subsidiary estimates are identically distributed periodograms, the averaged spectral estimate will have the same mean value [see (4.39) and (4.40)] as any of the subsidiary estimates.

However, the averaged estimate will have a smaller variance. We know that the variance of the average of K identical independent random variables is $1/K$ of the individual variances [see (3.46)]. If the K periodograms were statistically independent, the variance of the averaged estimate would be

$$\text{var } \hat{S}_M^K(\omega) = \frac{1}{K}\text{var } S_M(\omega) \approx \frac{1}{K}[S(\omega)]^2 \tag{4.45}$$

Thus, the averaging of 10 periodograms as in (4.44) would result in approximately a factor-of-10 reduction in power-spectral-density estimation variance.

This K-fold variance reduction [or $\sqrt{K}$ improvement in the signal-to-noise ratio of (4.42)] is not quite possible because there is some dependence between the subsidiary periodograms. Unless the signal being analyzed is white, adjacent samples will be correlated. This means, for example, that x_{99} and x_{100}, which are used in separate subsidiary periodograms, are correlated and dependent. It follows that there is some dependence between $S_{100,\,1}$ and $S_{100,\,2}$. Such dependence will be small when there are many samples per periodogram, so that the reduced variance in (4.45) is a good approximation in such cases.

To summarize, we have developed a reasonable approach to estimating the power spectrum $S(\omega)$ of a random signal x_n. K periodograms are computed separately for segments of the data, and these are averaged to get an estimate $\hat{S}_N^K(\omega)$ with a smaller variance. Both the bias and variance of this averaged periodogram, unfortunately, are clearly related to the unknown spectral density we are trying to determine. This circular situation is much like that concerning properties of the histogram described in Chap. 3.

Just as the histogram cell width had to be chosen by a trial-and-error sequence, as a compromise between variance and resolution of narrow peaks in the true density, we must choose a suitable length for subsidiary periodograms. In the previous 1000-point example we could also have constructed 100 segments, each with 10 data values. The resulting $\hat{S}_{10}^{100}(\omega)$ would have had a much smaller variance than the previously described $\hat{S}_{100}^{10}(\omega)$. However, these shorter segments correspond to wider spectral windows: $V_{10}(\omega - \eta)$ vs. the previous V_{100} (see Fig. 4.18), so that there would be a loss in resolution of narrow peaks in the true spectral density. The $\hat{S}_{10}^{100}(\omega)$ estimates have less variability, but they will be close to the *wrong* values at frequencies where the true spectrum has narrow peaks or valleys. The wide side lobes in V_{10} can also cause mean-value errors at frequencies where the true spectrum is smooth, as we shall see later.

The interrelationships between bias, variance, the number of samples M per periodogram, and the number K of periodograms averaged together are best demonstrated with an example. For this purpose periodograms were computed from 96 samples $x_0, x_1, \ldots, x_{95}$ which had been generated using an autoregression equation

$$x_{i+1} = -0.9x_i + n_i \tag{4.46}$$

with a white input signal n_i. The white-signal samples were generated using the computer program RANDU with zero mean and unit variance, discussed in Chap. 3. From Eqs. (3.137) and (3.138) we can compute the true autocorrelation function and power spectral density for this random signal

$$R(k) = 5.26(-0.9)^{|k|} \tag{4.47}$$

$$S(\omega) = \frac{0.555}{1.005 + \cos \omega} \tag{4.48}$$

(We have taken $T_s = 1$ since any other value merely introduces a scale-factor change.)

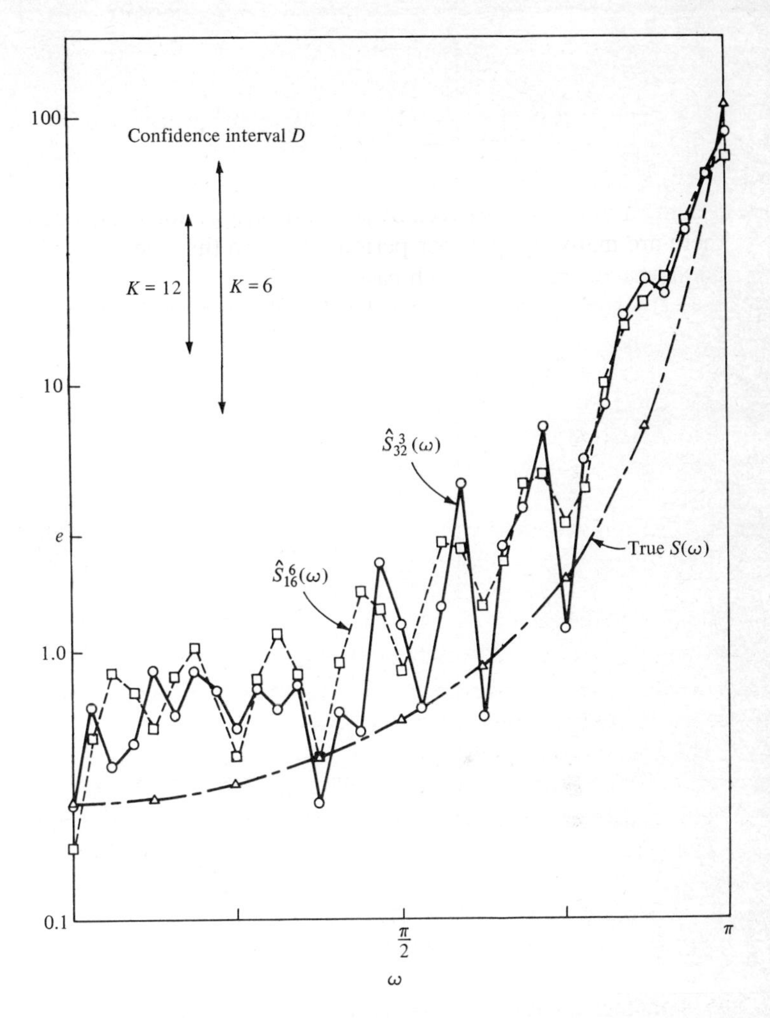

FIGURE 4.21
True and estimated power spectral densities.

The true $S(\omega)$ for this random signal is shown in Fig. 4.21 along with two averaged periodogram estimates. Notice that the vertical scale is logarithmic and that curves have been drawn through the discrete points where estimates have been computed. $\hat{S}_{16}^{6}(\omega)$ is the average of six periodograms, each computed from 16 samples; $\hat{S}_{32}^{3}(\omega)$ is the average of three periodograms, each computed from 32 samples. In calculating the individual periodograms, Eq. (4.19) was used. The 33 frequency points shown in Fig. 4.21 were obtained by using a fast Fourier transform algorithm [to compute the discrete Fourier transform in (4.19)] with $N = 64$, and 32 or 48 data zeros added, respectively, when 32 or 16 actual data points were used to get the periodogram. The other 31 frequency points produced by the algorithm are uninformative because of the even symmetry of $|X(\omega)|$.

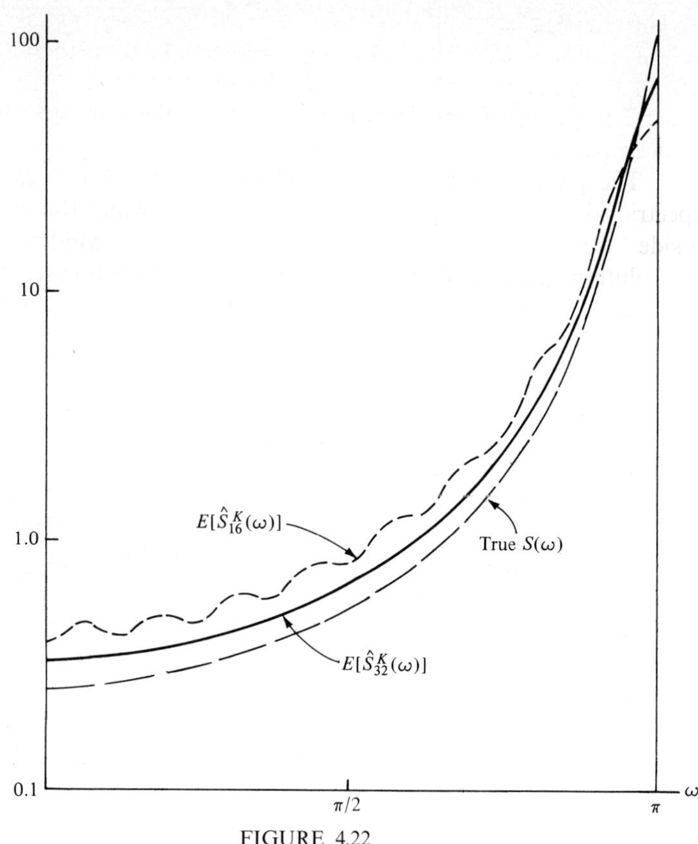

FIGURE 4.22
Mean values of averaged spectral-density estimates.

Neither averaged periodogram in Fig. 4.21 is a very precise representation of the true power spectral density, although either might be sufficient if we were only looking to see whether the signal was white or had more power at high or low frequencies. In addition, the general shape of either estimate might suggest that the signal is from a first-order autoregression whose parameters could be estimated by methods we shall discuss later for parametric power-spectral-density estimation.

Note that the estimate $\hat{S}_{32}^{3}$ using larger data segments does a better job of representing the sharp peak of $S(\omega)$ at $\omega = \pi$ but has many random peaks due to large variances. The other estimate is smoother (smaller variance), but it flattens out the sharp peak and is generally farther from the true curve. The errors in this smoother curve are predominantly bias errors.

The bias effects of these two estimators are revealed more clearly in Fig. 4.22, which shows the *mean* values of the spectral estimators. Note that the estimate using 32 samples has a smaller bias [its mean is closer to the true $S(\omega)$] than the one using 16 samples. Those mean values can be calculated in this artificial example only because the true spectrum is known. In a real spectral-estimation problem such mean values cannot be calculated. Although numerical evaluation of the convolution integral (4.39) could have been used in determining these mean values, it was faster to use a fast Fourier transform algorithm to carry out the truncated discrete Fourier transform in (4.23).

The peaks in the mean spectral estimate for the shorter-sample case (wider spectral window) correspond to frequencies ω_i for which the spectral window V_{16} has a side-lobe peak at the true spectrum peak when the window is centered at ω_i. The convolution integral (4.39) gives a large value in such cases. Comparison of Figs. 4.21 and 4.22 shows a strong bias effect for the estimate from this choice of data segments; i.e., the estimation errors in Fig. 4.21 are shaped like the bias errors in Fig. 4.22.

Similar peaks and valleys are imperceptible in the other mean-value function because the side-lobe peaks in $V_{32}(\omega)$ are narrower and smaller than the true peak. If the number K of periodograms being averaged were very large, the power-spectral-density estimates $\hat{S}_{16}^K$ and $\hat{S}_{32}^K$ would look like their mean values in Fig. 4.22.

When designing a spectral-estimation calculation, it is helpful to have some idea of the general shape of the unknown spectrum. For example, if the true spectrum were thought to be somewhat like (4.48), the bias calculation shown in Fig. 4.22 might indicate that an averaged estimate of the 32-point form S_{32}^K has an acceptably small bias. The next step would be to choose the number K of periodograms to be averaged together. One rule of thumb is to make the standard deviation and bias error about equal. Since both these quantities are functions of frequency, this criterion would have to be checked at several different frequencies. As a typical calculation, consider $\omega = \pi/2$. From (4.48)

$$S\left(\frac{\pi}{2}\right) = 0.552$$

while Fig. 4.22 shows that

$$\left|B\left(\frac{\pi}{2}\right)\right| = |0.55 - 0.67| = 0.12$$

(logarithmic interpolation must be used). The variance will equal the square of the bias if the number of periodograms being averaged is

$$K = \left(\frac{0.552}{0.12}\right)^2 \approx 21$$

It is useful at this point to realize that the systematic side-lobe-bias errors apparent in $\hat{S}_{16}^6$ are real results whether the data are considered to be random or deterministic. As we saw in the previous section, the relationship between the transform of a long deterministic-data record and the transform of a shorter, truncated data set is determined by a window convolution, just like the bias error in random-signal analysis.

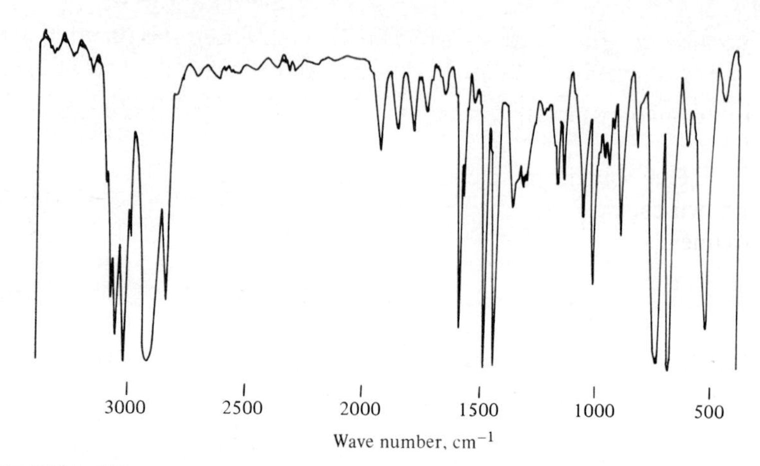

FIGURE 4.23
Polystyrene spectra. (*From E. D. Becker and T. C. Farrar, "Fourier Transform Spectroscopy," Science, vol. 178, October 27, 1972, p. 361, by permission. Copyright 1972 by the American Association for the Advancement of Science.*)

One interesting example of these signal-processing considerations arises in infrared spectroscopy, where truncation effects for nonrandom signals must be dealt with in the design of optical interferometers.† The article referred to relates data truncation length to the physical dimensions of an optical system of lenses and mirrors. The article discusses the effectiveness of such measurement equipment. Figure 4.23 shows two infrared absorption spectra of a polystyrene sample, each computed in less than 1 s by fast Fourier transforms of an optical-interference pattern. The speed and reproducibility (it is difficult to see where the two curves differ) of these results indicate that these spectral tests can be applied to streams of gas effluents to check for the presence of various components, e.g., air pollutants. The wave number k used in the figure is proportional to frequency; $k = 1/\lambda = c\omega/2\pi$, where λ is the wavelength and c the speed of light.

The true power spectral density will not be known in a real spectral-estimation situation. Faced with curves like the wiggly ones in Fig. 4.21, we find it difficult to make precise statements about the true power spectral density. We should certainly reject any spectral estimate in which bias effects due to the spectral-window side lobes are identifiable. (We always *do* know what the spectral window looks like.) No matter how many data are available, averaged periodograms without obvious bias problems will usually be very irregularly shaped. Of course, the amplitude of the irregularities should decrease as the total N and number of periodograms K increase if the true spectrum is a relatively smooth function of ω. One usually tries to approximate the power spectral density by curves corresponding to simple analytical functions. For example, one might assume that the true power spectral density is a function like

† E. D. Becker and T. C. Farrar, Fourier Transform Spectroscopy, *Science*, vol. 178, p. 361, Oct. 27, 1972.

$$\tilde{S}(\omega) = \frac{a}{b + c \cos \omega} \tag{4.49}$$

[motivated by (4.48)] in which the a, b, and c parameters could be chosen for a good fit.

One way, in principle, of determining a good fit would be to choose the unknown coefficients, such as a, b, c, to minimize a sum of squared errors at, say, Q frequencies

$$\sum_{i=1}^{Q} [\hat{S}(\omega_i) - \tilde{S}(\omega_i)]^2 \tag{4.50}$$

However, the equations for the best coefficients would generally be nonlinear. This is apparent in (4.49), where b and c do not enter as the simple additive proportionality coefficients we have used in other least-squares problems.

Either least-squares or eyeball fitting of simple curves to averaged periodogram curves is aided by the logarithmic ordinate scale used in Fig. 4.21. The reason is that the standard deviation of the estimation error is proportional to the true power spectral density [see (4.41)]. If $S(\omega_a)$ is big at a frequency ω_a, then the error $\hat{S}(\omega_a) - S(\omega_a)$ has a relatively large standard deviation. Conversely, a small $S(\omega_b)$ makes $\hat{S}(\omega_b) - S(\omega_b)$ have a small standard deviation. Because of the logarithmic scale, one unit of physical vertical distance from $S(\omega_a)$ corresponds to a proportionally bigger estimation error than an equal distance away from $S(\omega_b)$. However, these distances are equally significant, since they correspond to equal multiples of the respective standard deviations. Thus it makes sense to weight fitting-distance errors equally at all frequencies when the curves are plotted on a logarithmic scale.

To be more quantitative about this property of the logarithmic scale, note that from (4.45)

$$\sqrt{\text{var } \hat{S}_M^K(\omega)} \approx \epsilon S(\omega)$$

where $\epsilon = 1/\sqrt{K}$ is generally a small number. The points located $\pm$ two standard deviations about the true power spectral density therefore have the values

$$\log [(1 + 2\epsilon)S(\omega)] = \log [S(\omega)] + \log (1 + 2\epsilon)$$
$$\log [(1 - 2\epsilon)S(\omega)] = \log [S(\omega)] + \log (1 - 2\epsilon)$$

The distance between these points, which describes the limits of 95 percent of the estimation errors,† is

$$D = \log (1 + 2\epsilon) - \log (1 - 2\epsilon)$$

independent of the true $S(\omega)$. A vertical line segment of this length is often drawn on the page of a power-spectral-density estimate, as in Fig. 4.21, to indicate the inherent uncertainty. A $K = 3$ segment is not included on the figure because the correspond-

† This argument uses the facts that a gaussian variable has a 0.95 probability of being within $\pm 2\sigma$ of its means and that for large M and K, $\hat{S}_M^K(\omega)$ is approximately gaussian, by the central-limit theorem.

ing ϵ is too big for the gaussian-error assumption, as indicated by the logarithm of a negative number appearing in the D expression. [When ϵ is very small, the approximation $\ln(1 + 2\epsilon) \approx 2\epsilon$ reduces D to

$$D \approx 4\epsilon \log e$$

where $\log e$ is the distance from 1 to $2.718 \cdots$ on the vertical logarithmic scale.]

Once it has been decided to fit a simple analytical model to an $S(\omega)$ curve, another approach is to return to a representation like that of (4.46), e.g.,

$$x_{i+1} = \alpha x_i + \beta n_i \tag{4.51}$$

This represents a difference equation which could have generated x_i from white noise. It is then possible to make maximum-likelihood estimations of the parameters α and β. This approach will be developed later.

Up to this point our spectral-estimation discussion has omitted the actual details of digital-computer algorithms. In fact, there is little new to be said regarding these points. The definition of $S_N(\omega)$ in (4.19) in terms of the discrete Fourier transforms of the data allows us simply to use the methods introduced before in Chap. 2, including the fast Fourier transform. The considerations of sample spacing, frequency spacing, adding zero data points, etc., apply here also.

We should mention other approaches to variance reduction and resolution improvement which are frequently used in actual data processing. There is an advantage to using *overlapping* data segments, so that each of the K segments can have more than N/K samples. The resulting subsidiary periodograms are more statistically dependent, so that there is less than a $1/K$ variance reduction, but the reduced bias may yield a net improvement in the estimates. Alternately, by overlapping data samples one may obtain more than K segments, each with N/K samples. Although the subsidiary periodograms are more statistically dependent, the increased number of segments averaged reduces the variance and if properly done can more than compensate for the increased dependence, reducing the variance substantially.† (As an example, if 100 data samples were originally broken into 10 segments of 10 samples each, overlapping by 50 percent would produce 20 segments of 10 samples each. K has thus increased from 10 to 20. Although the variance reduction is not the usual 2 to 1 because of increased statistical dependence, there will be a significant improvement.)

A different and historically earlier approach to variance reduction in spectral estimates will be discussed in the next section.

4.6 SMOOTHED SPECTRAL ESTIMATES AND WINDOWS

We have seen that the variance of a power-spectral-density estimate based on N data samples can be reduced by chopping the data into shorter segments and then *averaging* periodograms computed separately from the segments. Here we describe an

† P. D. Welch, The Use of Fast Fourier Transforms, The Estimation of Power Spectra: A Method Based on Time Averaging over Short, Modified Periodograms, *IEEE Trans. Audio Electroacoust.* vol. AU-15, pp. 70–73, June 1967.

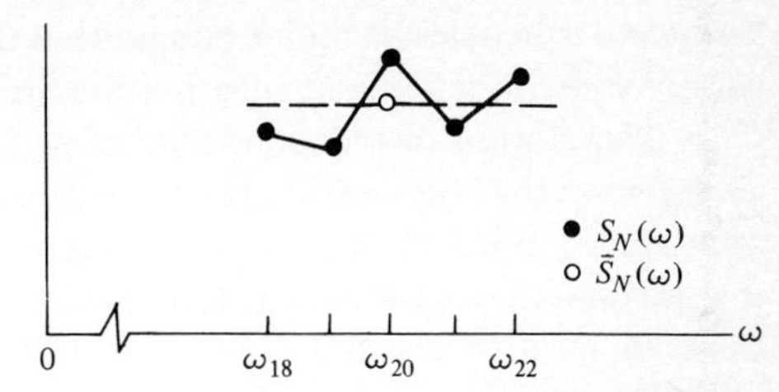

FIGURE 4.24
Rectangularly smoothed periodogram ($L = 2$).

alternate method for computing a reduced-variance estimate through a *smoothing* operation on the single periodogram which can be computed from all the data. Several different kinds of periodogram-smoothing operations will be considered, along with their bias and variance properties. Computational aspects of these smoothed estimates will also be examined.

The simplest *smoothed* periodogram can be defined as the average of several adjacent periodogram values. For example, we shall define $\bar{S}(\omega_i)$ as the average

$$\bar{S}(\omega_i) = \frac{1}{2L + 1} \sum_{j=i-L}^{i+L} S_N(\omega_j) \tag{4.52}$$

of $2L + 1$ periodogram values near the frequency ω_i. Figure 4.24 demonstrates this smoothing for $L = 2$ and $i = 20$, with $\bar{S}(\omega_{20})$ as the average of $S_N(\omega_j)$ values at frequencies ω_{18} to ω_{22}. This averaging of S_N at discrete frequencies is a frequency-domain analogy of the averager filtering of discrete-time signals discussed in earlier chapters. Recalling that an averager filter's output has a smaller variance than its input, we expect the smoothed $\bar{S}(\omega_{20})$ to have a smaller variance than the raw periodogram $S_N(\omega_{20})$. This conclusion can be justified and quantified as follows.

We would like to be able to argue that the $2L + 1$ periodogram values being averaged in (4.52) are independent and identically distributed random variables. If so, those properties would make $\bar{S}(\omega_i)$ an unbiased estimate of the true $S(\omega_i)$, with a variance less than the variance of a single periodogram value $S_N(\omega_i)$ by a factor of $(2L + 1)^{-1}$. Such independence of errors in adjacent periodogram values seems reasonable when we recall the frequency-domain sampling theorem from Sec. 2.7, which said that a Fourier transform $X_N(\omega)$ can be reconstructed at all frequencies from its values at discrete frequencies if those frequencies are separated by no more than $\Omega_s = 2\pi/NT_s$. The same kind of reconstruction applies to $S_N(\omega)$ since it is directly related to $X_N(\omega)$ via (4.19). Figure 4.25 shows a typical $S_N(\omega)$ and suitable

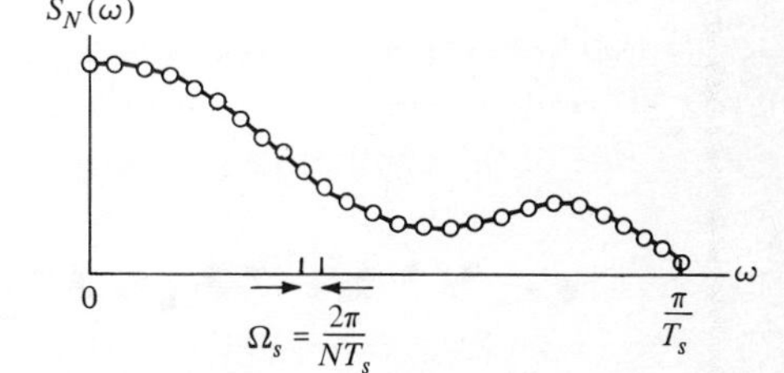

FIGURE 4.25
Spacing of discrete frequencies for spectral smoothing.

discrete frequencies. It might be said that each of the $S_N(m\Omega_s)$ values at $m = 1, 2, 3, \ldots$ carries some independent information since its presence is necessary.

This notion of independence of periodogram errors can be made more precise by showing that these discrete values $S_N(m\Omega_s)$ become statistically independent as N approaches infinity.† It is just this independence of S_N values at adjacent frequencies that gives periodograms their typical irregular shapes. Even if the *true* spectrum is the same at frequencies $m\Omega_s$ and $(m+1)\Omega_s$, the independent errors in periodogram estimates at those adjacent frequencies can be very different.

Periodogram smoothing for variance reduction is similar to the histogram problem. A wider rectangular smoothing window (larger L) will yield a greater variance reduction, since the variance of the average is inversely proportional to the number of terms in the average:

$$\text{var}\ \bar{S}(\omega_i) = \frac{1}{2L+1}\ \text{var}\ S_N(\omega_i) \tag{4.53}$$

However, if L is too large, the $S_N(\omega_j)$ being averaged will not all have the same mean value. The *true* spectrum must be constant over the frequency band $2L\Omega_s$ to avoid bias problems, as we now explain in more detail.

The bias problem in smoothed periodograms is easier to interpret if we view (4.52) as an approximation to a *continuously smoothed* periodogram $\tilde{S}(\omega)$ defined in terms of the periodogram $S(\eta)$ as a function of a continuous frequency variable

$$\tilde{S}(\omega_i) = \frac{1}{2\lambda}\int_{\omega_i-\lambda}^{\omega_i+\lambda} S_N(\eta)\,d\eta \approx \bar{S}(\omega_i) \tag{4.54}$$

The discrete smoothing in (4.52) and the continuous smoothing in (4.54) are approximately equivalent at $\omega = \omega_i$ because the continuous $S_N(\omega)$ will be smooth between the discrete frequencies $\omega_j = j\Omega_s$, according to our previous sampling-theory argument. Thus the sum in (4.52) should be a reasonable approximation to the integral in (4.54) (and vice versa).

The advantage of the integral-smoothing form in (4.54) is that the integral there can be written as a convolution of the raw periodogram $S_N(\omega)$ with a rectangular frequency window $Q(\omega)$. Thus,

$$\tilde{S}(\omega) = \int_{-\pi}^{\pi} S_N(\eta)Q_\lambda(\omega-\eta)\,d\eta \tag{4.55}$$

where

$$Q_\lambda(\omega-\eta) = \begin{cases} \dfrac{1}{2\lambda} & |\omega-\eta| < \lambda = 2L\Omega_s \\ 0 & \text{otherwise} \end{cases} \tag{4.56}$$

as pictured in Fig. 4.26. This convolution form allows the bias of $\tilde{S}(\omega)$ to be studied in terms of window properties, just like the bias of the raw periodogram $S_N(\omega)$. The mean value of the continuously smoothed estimate is

$$E[\tilde{S}(\omega)] = \int_{-\pi}^{\pi} E[S_N(\eta)]Q_\lambda(\omega-\eta)\,d\eta \tag{4.57}$$

† Jenkins and Watts, op. cit.

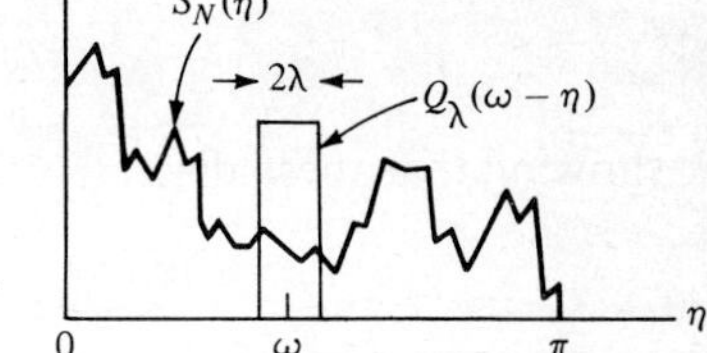

FIGURE 4.26
Integrand factors for continuously smoothed periodogram.

When N is large, the periodogram $S_N(\omega)$ using all the data is unbiased, as we saw in (4.40). Thus, for large N, the present smoothed estimate has a mean of

$$E[\tilde{S}(\omega)] \approx \int_{-\pi}^{\pi} S(\omega)Q_\lambda(\omega - \eta)\, d\eta \qquad \text{large } N \tag{4.58}$$

We see that the continuously smoothed estimate $\tilde{S}(\omega)$ has a mean-value expression (4.58) which is of the same form as the mean of the periodogram $S_N(\omega)$ given in (4.39) except for the use of a different window [$Q_\lambda(\omega - \eta)$ instead of $V_N(\omega - \eta)$]. Thus, the bias arguments here for the window Q_λ which we have *imposed* will be parallel to those for the short-sample periodogram with its *inherent* window caused by the finite number N of samples. Here we assume that N is large, so that the approximate equality in (4.58) is valid. [We should recall that even though $S_N(\omega)$ has no bias for large N, it has an unacceptably large variance which we are improving upon here by smoothing.] A pictorial representation of the relation between bias and window width λ is shown in Fig. 4.27. The circle at ω_1 represents the mean value for the estimate $\tilde{S}(\omega_1)$ [average of $S(\omega)$ within the window band], which is significantly different from the true power-spectral-density value. At other frequencies where $S(\omega)$ is smoother the bias will be smaller.

The bias and variance of the discretely and continuously smoothed periodograms $\bar{S}$ and $\tilde{S}$ can be easily summarized by considering these estimates as approximations of each other. The discretely smoothed one clearly indicates the variance reduction due to averaging of $2L + 1$ effectively independent simple periodogram values within the band of $2\lambda = 2L\Omega_s$. The continuously smoothed one describes the bias in terms of the width of the window relative to the widths of peaks in the true $S(\omega)$.

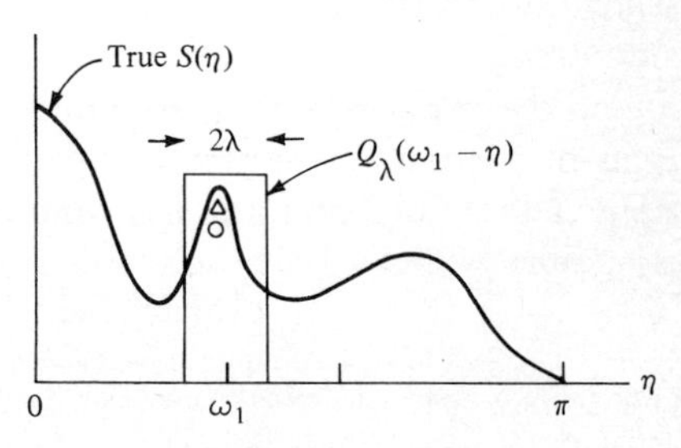

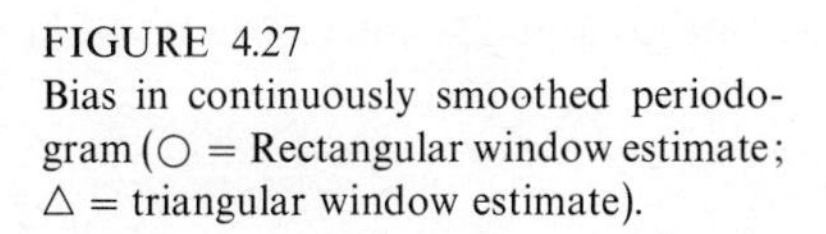
FIGURE 4.27
Bias in continuously smoothed periodogram (○ = Rectangular window estimate; △ = triangular window estimate).

Triangular Windows

Windows other than the rectangular Q_λ have been used for both kinds of smoothed periodograms. A discrete triangular window would form an estimate like

$$\bar{S}^t(\omega_i) = \frac{1}{L} \sum_{j=i-L+1}^{i+L-1} \left(1 - \frac{|j-i|}{L}\right) S_N(\omega_j) \tag{4.59}$$

giving less weight to $S_N(\omega_j)$ at frequencies farther from ω_i. The triangular weighting $(1/L)(1 - |j|/L)$ is like the time-domain triangle in Fig. 4.16 but normalized so that if all the periodogram values $S_N(\omega_j)$ happened to be equal, $\bar{S}^t(\omega_i)$ would equal $S_N(\omega_i)$. This triangular window results in less bias than the rectangular window because it gives more weight to raw periodogram values near the center frequency. The continuously smoothed version of a triangular frequency window

$$\tilde{S}^t(\omega) = \frac{1}{\lambda} \int_{-\infty}^{\infty} \left(1 - \frac{|\omega - \eta|}{\lambda}\right) S_N(\eta)\, d\eta \tag{4.60}$$

would produce an estimate whose mean value at ω_1 in Fig. 4.27 is at the point marked with a triangle, with less bias error than the circle point, which was produced from the rectangular window.

This bias reduction associated with a triangular window is obtained at the price of a greater variance than the rectangular window. Assuming that the $S_N(\omega_j)$ in (4.59) are independent and have identical variances of $S^2(\omega_i)$ (which are good assumptions when N is large and the triangular window has a small width L), the variance of that weighted sum is proportional to the sum of the *squares* of the weights†

$$\text{var}\ \tilde{S}^t(\omega_i) \approx S^2(\omega_i) \sum_{j=i-L+1}^{i+L-1} \frac{(L - |j-i|)^2}{L^4} \tag{4.61}$$

A little algebra and use of the summation formula

$$\sum_{n=1}^{N} n^2 = \frac{N(N+1)(2N+1)}{6}$$

allow the variance of this triangularly smoothed estimate to be written as

$$\text{var}\ \tilde{S}^t(\omega_i) \approx S^2(\omega_i) \frac{1}{2L+1} \left(\frac{4}{3} + \frac{2L^2 + 2L + 1}{3L^3}\right) \tag{4.62}$$

which is clearly greater than the variance given in (4.53) for the rectangular smoothing.

We have seen that smoothed estimates using triangular and rectangular frequency windows with the same "bandwidth" $2L\Omega_s$, that is, both combining $2L + 1$ periodogram values, have different means (and biases) and variances. The triangular window leads to smaller bias but greater variance. Many other windows

† This variance property can be derived using arguments like those associated with (3.42) for an *equally* weighted sum of independent random variables.

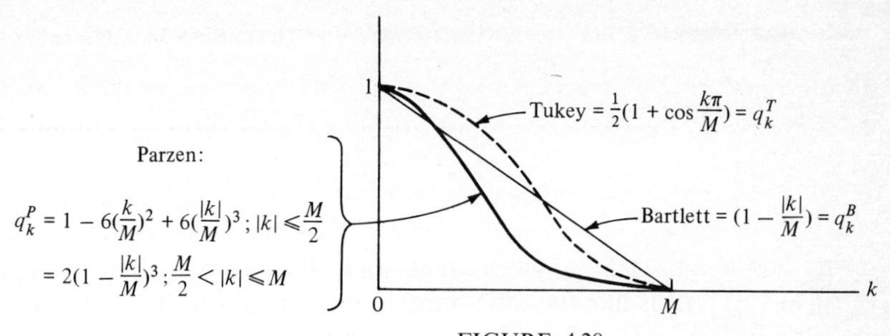

FIGURE 4.28
Weighting functions for spectral windows.

have been studied for their trade-offs between these two kinds of error.† In discussing other window functions it is convenient to study both frequency-domain and time-domain descriptions.

The rectangular and triangular frequency windows introduced above have inverse discrete Fourier transforms

$$q_k = \frac{\sin \lambda k}{\pi k} \qquad k = 0, \pm 1, \pm 2, \dots \tag{4.63}$$

and

$$q_k^t = \left[\frac{\sin (k\lambda/2)}{k\pi\lambda}\right]^2 \qquad k = 0, \pm 1, \dots \tag{4.64}$$

respectively. The convolution integrals in (4.55) and (4.60) are equivalent to discrete Fourier transforms of the lagged products of the time windows with the sample autocorrelation function

$$\tilde{R}(k) = R_N(k)q_k \tag{4.65}$$

and

$$\tilde{R}^t(k) = R_N(k)q_k^t \tag{4.66}$$

Thus, an alternate way to compute $\tilde{S}(\omega)$ would be to find it as the discrete Fourier transform of a weighted version of the sample autocorrelation function, as in (4.65). When this route is followed, it is more common to use lag windows which have only a finite number of nonzero values, in contrast to those in (4.63) and (4.64). Some of these common finite-length lag windows, with $2M - 1$ nonzero values, are shown in Fig. 4.28, where they are identified by the names of the statisticians who suggested them. Each window has a different trade-off between bias and variance error, and one might be best for a true power spectral density with a particular shape to its peaks (which is unknown, of course, when a spectral analysis is being carried out).

Computational Procedure for Periodogram Smoothing

The procedure for processing N data samples to get a smoothed (e.g., Bartlett window) estimate $\tilde{S}(m\Omega_s)$ is outlined in Table 4.1. This operation requires three fast Fourier

† Jenkins and Watts, op. cit.; A. Papoulis, Minimum Bias Windows for High Resolution Spectral Estimates, *IEEE Trans. Inf. Theory*, vol. 19, no. 1, pp. 9–12, 1973.

Table 4.1 PROCEDURE FOR SMOOTHING SPECTRAL-DENSITY ESTIMATES

1. Get $X_N(m\Omega_s)$, $m = 0, 1, \ldots, N-1$ as fast Fourier transform of the data $x_0, x_1, \ldots, x_{N-1}$
2. Use Eq. (4.19) to get $S_N(m\Omega_s)$ from $X_N(m\Omega_s)$
3. Get $R_N(k)$, $k = 0, 1, \ldots, N-1$, as inverse fast Fourier transform of $S_N(m\Omega_s)$
4. Multiply $R_N(k)$ by q_k^B (or some other lag-window function in Fig. 4.28) with specified truncation point M to get $\tilde{R}(k)$
5. Compute $\tilde{S}(m\Omega_s)$ as the fast Fourier transform of $\tilde{R}(k)$

transforms (steps 1, 3, and 5), $2N$ multiplications (squaring real and imaginary parts in step 2), and M multiplications in step 4. This procedure may seem overcomplicated, especially when we notice that most of the weighted autocorrelation-function values are zero,

$$\tilde{R}(k) = 0 \qquad k \geq M \tag{4.67}$$

due to the zero value of q_k^B for large lags. Thus it seems that steps 1 to 3 in the table could be replaced by a single step of using (4.8) directly to compute the sample autocorrelation function $R_N(k)$ for lags $k = 1, 2, \ldots, M$. This direct procedure was indeed the usual one before the advent of the fast Fourier transforms. Now, the steps in Table 4.1 are more commonly used and are much faster to execute, as a result of the fantastic speed of fast Fourier transform algorithms. Moreover, even when one desires only an *autocorrelation-function estimate*, like the one in Fig. 4.5, for example, the fastest way to get it from random-signal samples is via the power spectral density, stopping at step 3 or 4 in the procedure.

When one of the finite-length lag windows in Fig. 4.28 is used in the procedure of Table 4.1, the resulting bias and variance properties can be examined in terms of the corresponding frequency windows. Such an analysis is more complicated than our earlier discussion of rectangular and triangular *frequency* windows since these lag windows have more complicated frequency windows. For example, q_k^B with $M = 9$ is the same as the triangle function v_k^9 in Fig. 4.16. Thus, the related frequency window is the same as $V_9(\omega)$ shown in Fig. 4.17. The nonzero values of $V_9(\omega)$ are not restricted to a finite band of frequencies. We see then that the computationally efficient lag-window procedure of Table 4.1 has the side effect of introducing the side-lobe type of bias errors encountered in the averaged periodograms of the previous section.

It is interesting to compare the bias and variance characteristics of averaged and smoothed power-spectral-density estimates based on the same N data samples. Consider once again the 96 data points used to get the averaged estimates in Fig. 4.2.1. When $\hat{S}_{16}^6(\omega)$ was computed from 6 sets of 16-point data segments, a variance reduction of $\frac{1}{6}$ resulted. The bias curve in Fig. 4.22 resulted from convolution of the true $S(\omega)$ with $V_{16}(\omega)$ similar to the curve in Fig. 4.17 but with a main-lobe width of $4\pi/16$ instead of $4\pi/9$ rad/s. A smoothed $\tilde{S}(\omega)$ with the same bias window would result from the Table 4.1 procedure using a Bartlett window with length $M = 16$. This smoothed estimate will then have a mean value, analogous to (4.57), of

$$E[\tilde{S}(\omega)] = \int_{-\pi}^{\pi} E[S_N(\eta)]V_{16}(\omega - \eta)\frac{d\omega}{2\pi} \tag{4.68}$$

or

$$E[\tilde{S}(\omega)] \approx \int_{-\pi}^{\pi} S(\eta)V_{16}(\omega - \eta)\frac{d\omega}{2\pi} \tag{4.69}$$

which is the same as the mean of the averaged estimate given in Eq. (4.39).

The two kinds of estimates also yield very similar variance-reduction factors. As mentioned earlier, the averaged estimate $\bar{S}$ has a variance reduction of $\frac{1}{6}$. To find the variance of the smoothed estimate we can try to parallel the derivation of (4.62). The number of approximately independent periodogram values within the main lobe of $V_{16}(\omega)$ is found by dividing the width of that lobe $(4\pi/16)$ by the "independence spacing" $2\pi/NT_s = 2\pi/96$ (in this case $T_s = 1$ for simplicity). The frequency window is thus smoothing (with unequal weights) about 12 independent periodogram values. The corresponding variance reduction can be approximated by (4.62) with $L = 6$. The net result is an approximate variance-reduction factor of $\frac{1}{8}$. While this seems a little better than the $\frac{1}{6}$ factor for the averaged estimate having the same bias error, the small difference is not really significant. The important fact is that the order of magnitude of variance reduction is the same. (Moreover, since the assumption of independence of periodogram-value space by $2\pi/NT_s$ is less accurate than the independence of periodograms from different data segments, the $\frac{1}{8}$ factor is overoptimistic. Recall also that the variancc-reduction factor for averaged estimates can be improved by *overlapping* the segments to get $K > N/M$ segments.)

In summary, computing an averaged periodogram $\hat{S}_M^K$ by averaging K periodograms of length $M = N/K$ gives estimates of about the same quality (bias and variance) as a smoothed periodogram $\tilde{S}$ using a Bartlett window of length M with the periodogram from the entire N samples. Experimentation with different smoothing-window widths to achieve good balance between bias and variance errors is achieved by changing the truncation point M of the lag window. A third equivalent alternative to improve estimation accuracy uses both devices, by averaging together separately smoothed spectra from separate data segments.

Bandpass-Filter Interpretation

Smoothed spectral estimates can also be interpreted as results from spectrum analyzers like the one in Fig. 3.28, which passes the input through parallel bandpass filters and computes the average output power in each band. If an input signal x_i produces an output y_i from a filter with transfer function $H(\omega)$, the average output power may be defined as

$$\hat{\sigma}_y^2 = \frac{1}{N}\sum_{0}^{N-1} y_i^2 \tag{4.70}$$

(We assume zero-mean signals for simplicity.) The spectrum-analyzer point of view assumes that this $\hat{\sigma}_y^2$ is a good estimate of the mean-squared output

$$\sigma_y^2 = \int_{-\pi}^{\pi} S_x(\omega)|H(\omega)|^2\frac{d\omega}{2\pi} = E[y_i^2] \tag{4.71}$$

as expressed in terms of the true input power spectral density $S_x(\omega)$ and the transfer function. This is desirable because the mean-squared output σ_y^2 will be proportional to the unknown input power spectral density at frequency ω_1, $S_x(\omega_1)$

$$\hat{\sigma}_y^2 = \epsilon S_x(\omega_1)$$

if $H(\omega)$ is chosen to have a narrow passband of width ϵ near $\omega = \omega_1$.

It is not difficult to show, as we shall see in a moment, that the time-average output power from (4.70) can be expressed as

$$\hat{\sigma}_y^2 = \int_{-\pi}^{\pi} S_{N_x}(\omega)|H(\omega)|^2 \frac{d\omega}{2\pi} \tag{4.72}$$

in terms of the input periodogram. Thus, $\hat{\sigma}_y^2$ is just a smoothed power-spectral-density estimate of the form of $\tilde{S}(\omega)$ given in (4.55), with $|H(\omega)|^2$ playing the role of the spectral window. Our previous discussion of such estimates shows that if $|H(\omega)|$ has a narrow passband, then $\hat{\sigma}_y^2$ will have a small bias as an estimate of σ_y^2. Furthermore, as the number of data N increases, the number of independent values being averaged in (4.72) increases in proportion to N. Thus σ_y^2 is a consistent power estimate because its variance approaches zero as N approaches infinity.

This bandpass-filter interpretation of smoothed power-spectral-density estimates can be completed by verifying that the periodogram and power estimate $\hat{\sigma}_y^2$ play the roles in (4.72) which are analogous to the true variance and power spectral density in (4.71). Derivation of (4.72) uses truncated input and output signals x_i^N, as defined in (4.20), in order to attain the simplicity of infinite summation limits which permit easier accounting of interchanges in the order of summations. Thus

$$\hat{\sigma}_y^2 = \frac{1}{N} \sum_{-\infty}^{\infty} (y_i^N)^2$$

and substitution of the linear-filter convolution sums produces

$$\hat{\sigma}_y^2 = \frac{1}{N} \sum_{i=-\infty}^{\infty} \left(\sum_{k=-\infty}^{\infty} h_k x_{i-k}^N \right) \left(\sum_{m=-\infty}^{\infty} h_m x_{i-m}^N \right)$$

$$= \sum_{k=-\infty}^{\infty} \sum_{m=-\infty}^{\infty} h_k h_m \left(\frac{1}{N} \sum_{i=-\infty}^{\infty} x_{i-k}^N x_{i-m}^N \right)$$

The sum over i is recognized as the sample autocorrelation function $R_{N_x}(m-k)$, which can be expressed in terms of the periodogram by using the inverse Fourier transform of (4.17), with the result

$$\hat{\sigma}_y^2 = \sum_{-\infty}^{\infty} \sum h_k h_m \int_{-\pi}^{\pi} S_{N_x}(\omega) e^{j\omega(m-k)} \frac{d\omega}{2\pi}$$

$$= \int_{-\pi}^{\pi} S_{N_x}(\omega) \sum_{-\infty}^{\infty} h_k e^{-j\omega k} \sum_{-\infty}^{\infty} h_m e^{j\omega m} \frac{d\omega}{2\pi} \tag{4.73}$$

The last expression is recognized as being equivalent to the desired result in (4.72).

Thus the time-averaged power $\hat{\sigma}_y^2$ has the form of a smoothed periodogram, and is a good estimate of the power spectral density at the narrowband filter's center frequency.

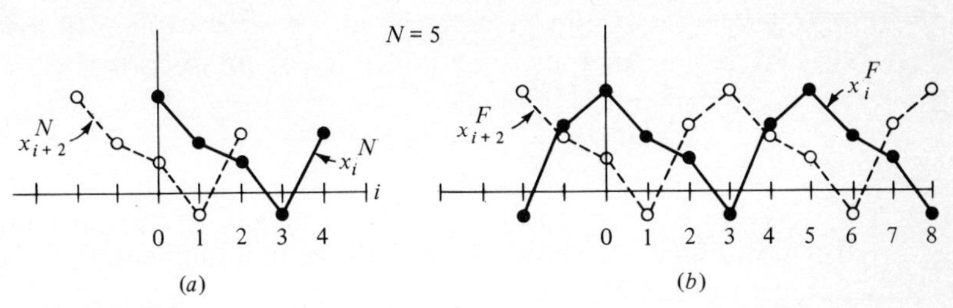

FIGURE 4.29
Truncated signal and its periodic repetition. (*a*) Truncated signal; (*b*) finite inverse of truncated transform.

Padding with Zeros

There is one digital-processing consideration which arises when computing smoothed spectral estimates but which does not apply to averaged estimates. When the inversion integral

$$R_N(k) = \int_0^{2\pi} S_N(\omega) e^{jk\omega} \frac{d\omega}{2\pi} \tag{4.74}$$

is approximated by a discrete sum in step 3 of Table 4.1, errors are introduced. This might seem surprising since we showed in Sec. 2.7 that if $X_N(\omega)$ is the finite Fourier transform of a truncated x_i^N, the inverse finite transform of $X_N(n\Omega_s)$ reproduces the original x_i^N. The difficulty here is that we are inverting $S_N(\omega) = |X_N(\omega)|^2/N$ rather than $X_N(\omega)$. If we define the signal x_i^F as the finite inverse transform of $X_N(n\Omega_s)$ *for all i*

$$x_i^F = \frac{1}{N} \sum_0^{N-1} X_N(n\Omega_s) e^{jni\Omega_s} \tag{4.75}$$

then that earlier discussion tells us that x_i^F coincides with the truncated x_i^N

$$x_i^F = x_i^N \qquad i = 0, 1, \ldots, N-1$$

in the original time interval. However, *outside* that interval, e.g., for $i \geq N$, $x_i^N = 0$, while x_i^F in (4.75) clearly repeats periodically ($e^{jni\Omega}$ is periodic as a function of i).

Figure 4.29*a* shows a truncated signal and a shifted version of it; Fig. 4.29*b* shows the corresponding periodic x_i^F and a shifted version of it. Errors can arise in a finite approximation to (4.74) because the two signals in that figure have different sample autocorrelation functions

$$\begin{aligned} R_N(k) &= \frac{1}{N} \sum_{i=0}^{N-1} x_i^N x_{i+k}^N \\ R_N^F(k) &= \frac{1}{N} \sum_{i=0}^{N-1} x_i^F x_{i+k}^F \end{aligned} \tag{4.76}$$

The first three terms in each sum are the same, but others are not. For example, $x_3^N x_{3+2}^N = 0$ while $x_2^F x_{3+2}^F \neq 0$. The resulting difference between these two autocorrelation functions is significant because, as we shall explain in a moment, finite inversion of $S_N(\omega)$, as an approximation to (4.74), produces

$$R_N^F(k) = \frac{1}{N} \sum_0^{N-1} S_N(n\Omega_s) e^{jkn2\pi/N} \qquad \Omega_s = \frac{2\pi}{NT_s} \tag{4.77}$$

when we really want $R_N(k)$, which is the sample autocorrelation function of the given signal.

The error caused by the extra overlapping in Fig. 4.29*b* is easily removed, at the expense of extra storage in the fast Fourier transform computation. This is achieved by adding N zero data values at the end of the given nonzero data sequence (a process sometimes called *padding* the data). The reader should convince himself that if five zero points $x_5 = x_6 = \cdots = x_9 = 0$ are added to the signal in Fig. 4.29*a*, the periodic version in Fig. 4.29*b* will not have extra overlaps for the range of lags $k = 0, 1, \ldots, 4$ of interest. Thus, after padding

$$R_N(k) = R_N^F(k) \qquad \text{for } k = 0, 1, \ldots, 4$$

and the finite inversion in (4.77) gives correct sample autocorrelation-function values for window weighting, as in step 4 of Table 4.1.

This padding with zeros has no deleterious effects other than requiring more data-storage locations. As we saw in Chap. 2, the effect on $X(n\Omega_s)$ and $S_N(n\Omega_s)$ is merely to produce more closely spaced points; i.e., smaller Ω_s makes it easier to interpolate between the discrete-frequency points. Recall that padding is also used to bring the total number of data points up to a power of 2, as required for many fast Fourier transform subroutines.

The padding discussion will be completed by explaining why the finite inversion in (4.77) produces the erroneous autocorrelation function $R_N^F(k)$. That summation can be rewritten as

$$\frac{1}{N^2} \sum_0^{N-1} X(n\Omega_s) X^*(n\Omega_s) e^{jkn2\pi/N} \tag{4.78}$$

This sum is the finite inverse transform of a product of two finite transforms. Recall that in Secs. 2.8 and 4.4 we showed how a product of functions in one domain corresponds to convolution of their transforms. Those observations can be extended to the present case of finite transforms, so that the inverse of the product in (4.78) equals the convolution of the inverses of the factors $X(n\Omega_s)$ and $X^*(n\Omega_s)$. The finite inverse of $X(n\Omega)$ is x_i^F, while the inverse of $X^*(n\Omega_s)$ is

$$x_i^{F*} = x_{-i}^F$$

[It is easy to see from the finite-transform definition in (2.94) that taking the conjugate by changing the sign in front of j is equivalent to computing the transform with the sign of i changed.] Thus (4.78) equals the convolution

$$\frac{1}{N} \sum_{r=0}^{N-1} x_{k-r}^F x_{-r}^F = R_N^F(k)$$

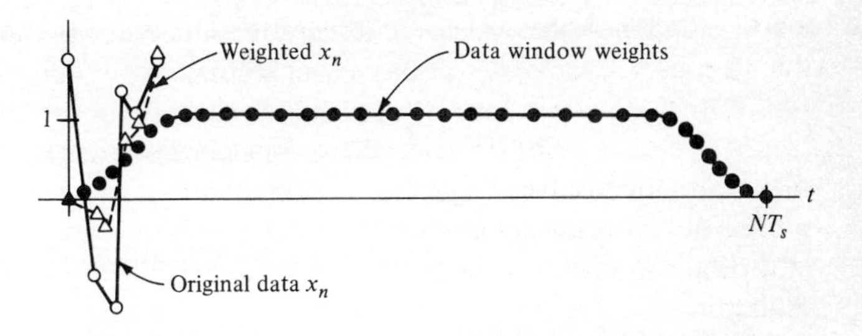

FIGURE 4.30
Data window weighting.

This sum is seen to be the same as that in (4.76) by noting that the summand is periodic so that the same value would result by summing over $r = 0, -1 \ldots, -N-1$. Thus, the finite inversion of $S_N(n\Omega_s)$ as in (4.77) produces $R_N^F(k)$ instead of $R_N(k)$.

Data Windows and Tapering

Before concluding our discussion of window functions we should also mention *data windows.* These windows result from applying weights directly to the data samples before computing a periodogram or sample autocorrelation function. Figure 4.30 shows a possible data-window weighting function which smoothes or tapers the ends of the data sample. In this case the first five data values are reduced to $\bar{x}_n$ values according to the relation

$$\bar{x}_n = x_n \frac{[1 - \cos(n\pi/6)]}{2} \qquad n = 0, 1, \ldots, 5 \tag{4.79}$$

The final five points are symmetrically reduced.

There are two ways to assess the effects of such weighting. We can think of this operation as a slight signal distortion to smooth out the turning-on and turning-off discontinuities which are aspects of the data-gathering procedure, rather than true jumps in the signal. Those discontinuities in the original data make the signal appear to have extra-high-frequency content, which, among other things, will cause a peak at π/T_s in the periodogram, with associated bias errors propagating through spectral-window side lobes.

An alternate viewpoint is to follow through the steps paralleling the mean-value expression in (4.23) to see that the weighting in (4.79) effectively changes the lag window, caused by finite data, from the triangular function in Fig. 4.11 to a similar one with a smoother transition to zero for large lag values. The triangular lag window is a convolution of a rectangular data weighting, i.e., *no* weighting, with itself; while convolution of the weighting in Fig. 4.30 with itself produces a lag window decreasing linearly for small k, with a smoother, slower decrease as k approaches N. This smoother time-domain transition produces a frequency window whose side lobes decay faster and thereby reduces the bias effects at other frequencies due to a peak at π/T_s (or at any other frequency).

Other data windows which change more than just the end values of the data record are also used. The net effect of such windows can be found by convolving them with themselves to get the corresponding lag window which multiplies the sample autocorrelation function. That lag window can then be used to get the effects of the data window on the mean and variance of the ultimate power-spectral-density estimate.

This section has discussed variance reduction by smoothing periodograms with windows which are usually applied by weighting $R_N(k)$ in the time domain. The smoothed estimates $\tilde{S}_N(\omega)$ described here are comparable to the averaged estimates $\hat{S}_M^K(\omega)$ described in the previous section. The data windows introduced to reduce end effects on finite samples can be applied in either case. It is left as a problem to show that the averaging approach is more efficient if a triangular weighting (Bartlett window) is to be used for the smoothing. The smoothing method from Table 4.1 requires three *long* fast Fourier transforms of length $N = MK$, as well as $2M$ multiplications in step 4. The averaging method requires more (K) fast Fourier transforms, but they are shorter (length $= M$), so that the net number of computer operations is much less.

Some other helpful operations for use in conjunction with these basic power-spectral-density estimation procedures will be described in the next section.

4.7 IMPROVEMENT OF POWER-SPECTRAL-DENSITY ESTIMATION THROUGH PREPROCESSING

We have said several times that spectral estimation is a statistical procedure, and we may wish to submit a single data set to repeated cycles of power-spectral-density estimation. Examination of the results from one estimation, for example, $\hat{S}_{16}^6(\omega)$, may suggest reprocessing, e.g., with longer segments, to get $\hat{S}_{32}^3(\omega)$. A large bias effect due to window side lobes in $\hat{S}_{16}^6$ indicates that periodograms should be computed from longer segments. The roughness of $\hat{S}_{32}^3(\omega)$ may suggest that more than 96 samples are needed to get an acceptable power-spectral-density estimate.

The bias problem in our experimental example of Eqs. (4.46) to (4.48) was due to the sharp peak at $\omega = \pi$ in the true power spectral density. By contrast, if the data were generated from an autoregression with less correlation between samples, e.g.,

$$x_{i+1} = -0.5x_i + n_i \tag{4.80}$$

the peak of the true spectrum would be broader with the resulting $\hat{S}_{16}^K$ bias shown in Fig. 4.31. Thus, a sample length of 16 could be adequate if the unknown true spectral density had no sharp peaks.

The transfer-function properties of power spectral densities can be used to advantage when the true power spectral density has (or is believed to have) a sharp peak. The original signal can be passed through a digital filter which passes relatively little power in the frequency band where the original power spectral density has a

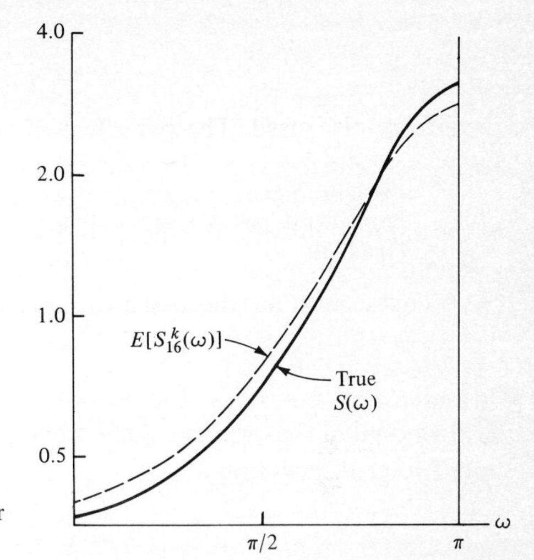

FIGURE 4.31
Bias in 16 sample estimate of power spectral density with broad peak.

sharp peak. The output spectrum $S_y(\omega) = |H(\omega)|^2 S_x(\omega)$ will thus be flatter and easier to estimate since more periodograms (from shorter segments of the total data set) can be averaged without getting large bias errors. An estimate $\hat{S}_x(\omega)$ for the original power spectral density can then be formed by using the *known* filter-transfer function to modify the computed output spectral estimate $\hat{S}_y(\omega)$

$$\hat{S}_x(\omega) = \frac{\hat{S}_y(\omega)}{|H(\omega)|^2} \tag{4.81}$$

$\hat{S}_y(\omega)$ is used here to symbolize either a smoothed or averaged power-spectral-density estimate.

This process of prefiltering to get a flatter spectrum for spectral analysis is called *prewhitening* because it attempts to produce the flat spectrum of white noise. In the example introduced in (4.48) and Fig. 4.21, with a peak at the high frequencies ($\omega = \pi$), a low-pass prefilter is required. We might try using an averager filter which generates as its output at time i, $y_i = \frac{1}{2}(x_i + x_{i-1})$. As shown in Chap. 2, this has the transfer function $|H(\omega)| = |\cos(\omega/2)|$ (see Fig. 2.49). By cascading the autoregressive process or recursive filter of (4.47) with this averager filter the resulting output will have a power spectral density given by

$$S_y(\omega) = \frac{0.555}{1.005 + \cos\omega}\left(\cos\frac{\omega}{2}\right)^2 \tag{4.82}$$

(Of course, in a real data-processing problem the true spectral density would be unknown, and we would simply justify low-pass prefiltering on the basis of results like those in Fig. 4.21 produced without prefiltering.) Figure 4.32 shows the resulting flatter $S_y(\omega)$. The approximate whiteness of the y_i sequence can also be seen in this idealized test case by combining the averager and autoregression equations to get

$$y_i = \tfrac{1}{2}(x_i + x_{i-1}) = \tfrac{1}{2}(-0.9x_{i-1} + n_i + x_{i-1}) = \tfrac{1}{2}(0.1x_{i-1} + n_i) \tag{4.83}$$

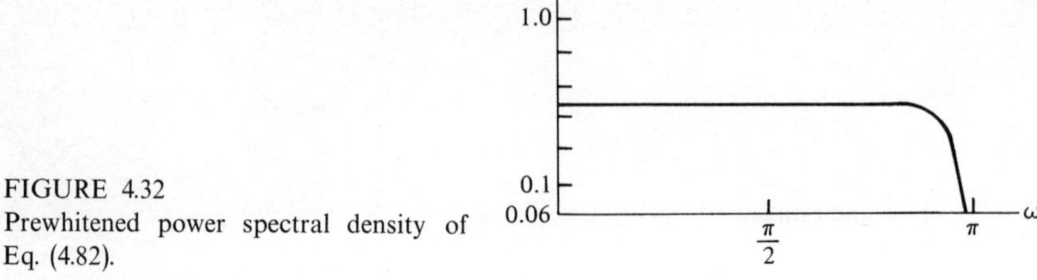

FIGURE 4.32
Prewhitened power spectral density of Eq. (4.82).

This equation shows that the output y_i is predominantly the white noise n_i plus a small amount of the original signal. Ideal prewhitening could be achieved if we used a similar filter to generate

$$z_t = x_t + 0.9x_{t-1} = n_t \tag{4.84}$$

This filter exactly removes the correlation that the autoregression produced. In an actual spectral-analysis situation such perfect results would require a very lucky guess for the prewhitening filter.

We have addressed most of our remarks to the single example of the signal with a sharp peak in its power spectral density at $\omega = \pi$ (as in Figs. 4.21 and 4.22). It should be clear that similar bias errors would result if a true power spectral density has a comparable peak at some other frequency. For example, the autoregression

$$x_i = +0.9x_{i-1} + n_i \tag{4.85}$$

with high *positive* correlation between successive samples will have an $S_x(\omega)$ which is simply the reflection about $\omega = \pi/2$ of the $S(\omega)$ shown in Fig. 4.21 (for the process coefficient $a = -0.9$). In this case the sharp *low*-frequency peak will cause bias-error problems, which could be reduced by prewhitening with a *differencer* filter.

Signals generated from white noise by higher-order (bandpass) filters can have sharp peaks at intermediate frequencies between zero and π. In such cases prewhitening can be achieved with a *notch filter*,† which has low transmission at an intermediate band of frequencies only.

It is instructive at this point to recall the admonition at the beginning of this chapter to the effect that systematic trends should be estimated and removed from the data before attempting spectral analysis. It is easy to show that a nonzero *average* value in the data will result in a sharp peak in the power spectral density at $\omega = 0$, just as Eq. (2.74) shows that a sinusoidal component introduces a sharp peak at its frequency. Such true power-spectral-density peaks would cause bias-error problems just like the ones we have been discussing. Thus, some experimentally observed bias problems might be due to systematic trends. Using the frequency of the observed

† B. Gold and C. M. Rader, "Digital Processing of Signals," McGraw-Hill, New York, 1969.

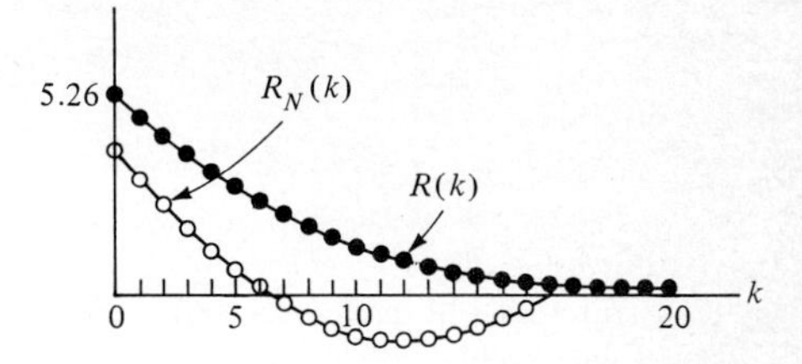

FIGURE 4.33
True and estimated autocorrelation functions.

peak as a "known" frequency, the method of Sec. 2.2 could be used to fit and remove a sinusoidal component as a kind of prewhitening operation.

In prewhitening we get a power-spectral-density estimate by using a known transfer-function relation between the random signal of interest and a signal whose averaged or smoothed periodogram can be computed more easily. The same kind of transfer-function relationship can be exploited when a measuring device distorts, in a known way, the signal being measured. For example, the roughness of a road might be studied from a recording (say a movie) of the vertical motion of a car being driven at a constant velocity over the road. Development of a transfer-function model connecting the road surface through springy tires, shock absorbers, etc., to the car's mass is left as an exercise. The power-spectral-density estimate of the car's motion could be modified by the magnitude squared of that transfer function to get a power-spectral-density estimate for the road surface.

Another interesting consequence of preprocessing to remove the mean value should be mentioned. If the autocorrelation function of a signal generated by (4.85) were to be estimated, we would be hoping to get the true monotonically decreasing exponential function $R(k)$ shown in Fig. 4.33. A typical experimental autocorrelation function will look more like the other curve $R_N(k)$ in that figure, with two striking properties. In contrast to the ragged power-spectral-density estimates in Fig. 4.21, autocorrelation-function estimates are very smooth functions. This does *not* mean that they are more reliable. It merely shows that the autocorrelation-function estimation errors are highly correlated at nearby lag values k, while the periodogram estimation errors tend to be independent at nearby frequencies. The other interesting observation in Fig. 4.33 is that $R_N(k)$ overshoots and goes negative as k increases. At first glance it is difficult to see how this negative correlation is introduced. The answer is that the subtraction of the estimated mean value from each data point introduces this kind of negative dip in the mean $E[R_N(k)]$ of the autocorrelation-function estimate.† Thus, bias problems are also inherent in autocorrelation-function estimation. If the mean were not removed, the periodograms would have added errors in the shape of the window side lobes. Correspondingly, $R_N(k)$ would be raised by a triangle function (the k-domain window), removing the negative dip shown in Fig. 4.33 but inserting a different kind of distortion in the autocorrelation-function estimate.

† Jenkins and Watts, op. cit.

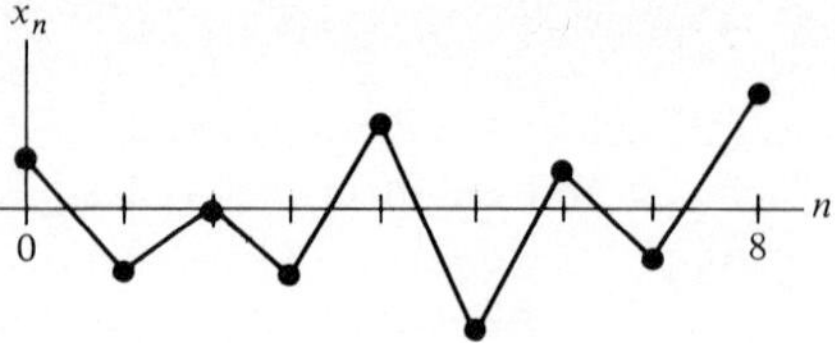

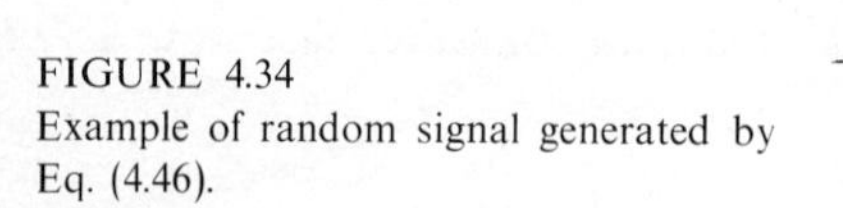
FIGURE 4.34
Example of random signal generated by Eq. (4.46).

In addition to bias errors, the sampling rate is another important consideration in spectral analysis. While the predominant high frequencies in the spectra of Fig. 4.21 are quite possible for a signal which is fundamentally defined at discrete times, it is highly suspicious for the samples of a continuous-time signal. Recall from Sec. 2.5 that we want to be sampling at such a rate that the signal has no power at frequencies above or near π/T_s. Examination of a sequence of typical data samples, as in Fig. 4.34 for the signal studied in Fig. 4.21, would also suggest that the sampling was below or precariously near the Nyquist rate. The large negative correlation between successive samples makes them very likely to have opposite signs, with exactly two samples per cycle of oscillation.

Thus, if the experimental curves in Fig. 4.21 had resulted from data consisting of samples from a continuous-time signal, we should resample the original signal at a much higher rate. In fact, most continuous-time signals are not exactly bandlimited. They just have very small power spectral densities at very high frequencies. The best we can hope for is power-spectral-density estimates which are much smaller at frequencies near π/T_s than at some lower frequencies (rather than being exactly 0 at $\omega = \pi/T_s$).

This chapter gives only an introduction to the study of spectral analysis. More advanced texts† and journal articles‡ should be consulted for additional statistical and computational details.

The discussion presented here has assumed that the nonparametric spectral analysis is to be carried out on a general-purpose digital computer. The wide applicability of this signal-analysis technique has led to the development of a wide variety of analog and digital instruments for spectral analysis. Some have banks of narrow passband filters or a single narrow filter whose center frequency can be automatically moved across a band of frequencies. Others do fast Fourier transforms directly, or they may first compute the autocorrelation-function estimate $R_N(k)$ and then transform it.

The next section provides several examples in which the spectral-analysis methods of this chapter have been applied using either general-purpose-computer algorithms or specialized instruments.

† Jenkins and Watts, op. cit., and R. K. Otnes and L. Enochson, "Digital Time Series Analysis," Wiley-Interscience, New York. 1972.

‡ L. R. Rabiner and C. M. Rader (eds.), "Digital Signal Processing," IEEE Press, New York, 1972.

4.8 SPECTRAL-ESTIMATION APPLICATIONS

Spectral analysis is a tool whose signal-analysis applications appear in all branches of science and engineering. The speed of digital-computer fast Fourier transform algorithms and the convenience of special-purpose solid-state instruments for correlation and spectral estimation have led to greatly increased use of these techniques in recent years. Chapter 2 gave several examples of spectral studies in the areas of speech analysis and synthesis, runway-roughness tests, pollutant concentrations, and passenger-traffic trends. This section presents a few examples in which spectral estimates may yield answers directly or the spectra are intermediate results which are themselves subjects of subsequent processing.

Presence or absence of energy in specific frequency bands can often be very informative. Astrophysicists determine the presence of specific molecules in distant regions of space by the absorption or resonance spectra of those regions. Presence or absence of a particular frequency component has similarly been used to study the development of brain dominance in man. While it is known that the two sides of the brain behave differently and that the right side generally dominates for visual activities, much remains to be learned about how these behavior patterns develop. The tests we refer to here exposed newborn babies to lights flashing at 3 Hz while EEGs were recorded separately from the two sides of the brain. The EEG recordings are irregular waves, in which the effects, if any, of switching on the light flashes are difficult to determine. Power spectral densities calculated from the EEGs, however, can reveal increases in the region of 3 Hz and its harmonics. Figure 4.35 shows that the right side in a newborn child's brain responds to the light while the left side seems not to respond. An adult brain produces responses that are similar on both sides. This test indicates that the cooperative interaction of the two halves has not yet developed and that the right side may dominate in adult visual activity because it develops that capability before the left side does.

The next example is conceptually more complicated because it relates spectral properties of two devices rather than merely looking for the presence of energy in a certain frequency band. It is well known that mechanical vibrations can be very destructive, but quantitative analysis of potential damage requires a study of the relationship between the spectrum of applied vibrations and the frequency response of the endangered part. It is possible for vibrations from the differential gears at the rear of an automobile to destroy a water pump in the front of the car.† The spectrum of vibrations from the differential (and other systems) can be computed from accelerometer recordings. In addition, the vibration transfer function of the pump can be determined by transforming its motional response (sensed by an accelerometer) to a mechanical impulse produced by a hammer (see Fig. 4.36). If the differential will supply energy in the 10-Hz range, redesign of the pump case or its mountings to reduce the sharp resonance may prevent case cracking or loosened connections.

The next two examples use points on estimated power spectral densities as numbers which are themselves viewed as signals for future processing. First we shall consider a refinement of the road- or runway-surface study described in Chap. 2.

†H. P. Fourier Analyzer Brochure, Hewlett Packard, Palo Alto, Calif., 1973.

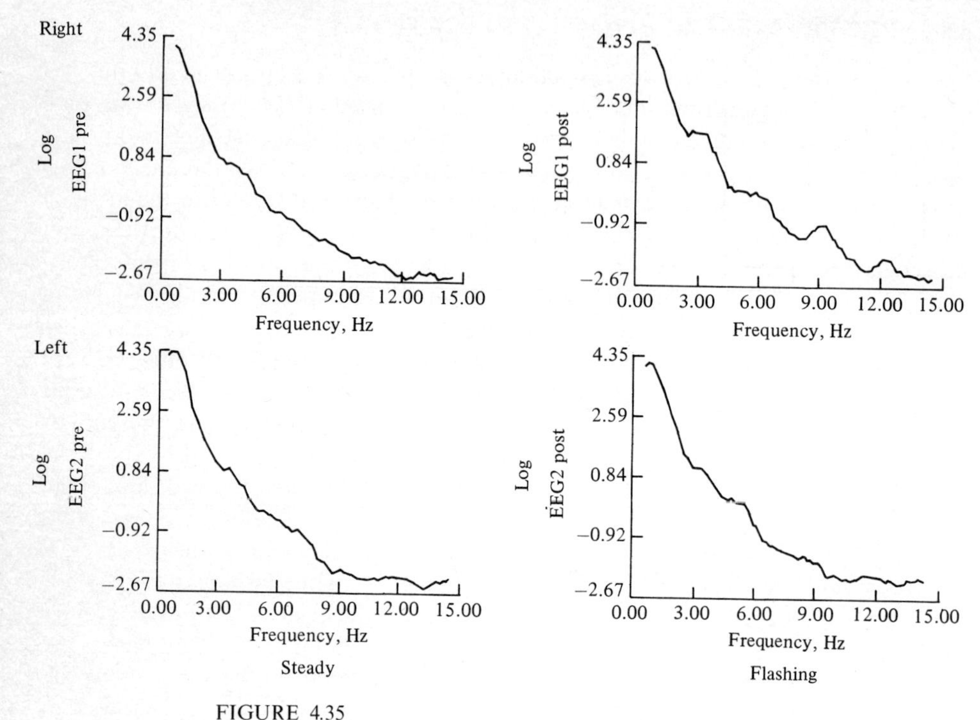

FIGURE 4.35
Newborn EEGs. (*From D. H. Crowell et al., "Unilateral Cortical Activity in Newborn Humans: An Early Index of Cerebral Dominance?" Science, vol. 180, April 13, 1973 by permission. Copyright 1973 by the American Association for the Advancement of Science.*)

Figure 4.37 shows a system for measuring road-surface roughness.† A potentiometer arm is moved by a lever connected to a wheel which rolls on the road. The sensed motion is relative to the potentiometer, which is mounted in a bouncing truck. It is necessary to eliminate the effects of the truck's motion on its suspension and springy tires in order to have a signal which represents only the variation in road surface. This correction is achieved by separately computing the truck's vertical motion as

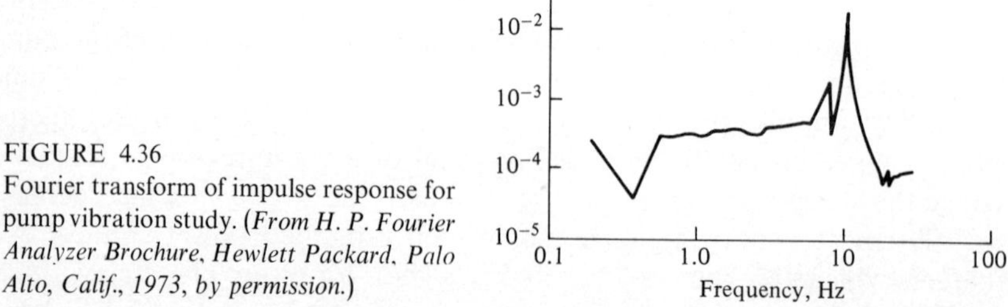

FIGURE 4.36
Fourier transform of impulse response for pump vibration study. (*From H. P. Fourier Analyzer Brochure, Hewlett Packard, Palo Alto, Calif., 1973, by permission.*)

† R. S. Walker et al., Modeling Riding Quality with Spectral Analysis Methods, *IEEE Decis. Control Conf.*, New Orleans, December 1972.

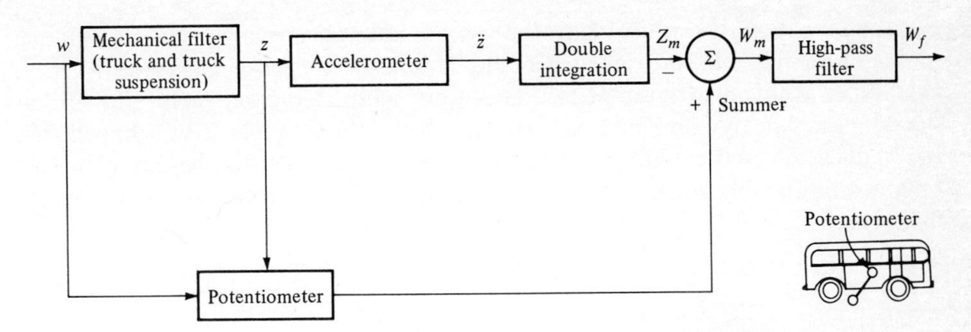

FIGURE 4.37
Road-surface measuring system.

the double integral of its sensed vertical acceleration. The input W on the figure represents the road-surface variations, and the output W_f is the electrical version of the road surface after the potentiometer output has been corrected for truck bouncing.

The purpose of the study referred to here was to establish an algorithm for computing a road-surface figure of merit based on surface measurements. This figure of merit, called the serviceability index (SI), had to be consistent with the present serviceability rating (PSR) as determined by a panel of test riders. A perfect road has a PSR of 5, while a zero rating means that the road is impassable at high speeds. Thus, the SI is a combination of physical road conditions and subjective rider reactions. While it is clear that the straight-line criteria in Fig. 2.36 will be adequate to define very rough and very smooth surfaces, the frequency dependence of a rider's discomfort suggests that a more complicated criterion will be needed to make finer distinctions in road quality. A road with a higher power spectral density at all frequencies is worse than one with a lower power spectral density, but how do we compare them if one is rougher at low frequencies and the other is rougher at higher frequencies? The proposed SI form was a linear combination of the amplitudes of several frequency components (plus a term for road-surface flexibility). Logarithmic amplitudes were used to equalize the uncertainties in the spectral estimates, as described in Sec. 4.5. Numerical experiments were performed by choosing a set of frequencies whose amplitudes would be included and then getting the coefficients α_i in an expression like

$$\text{SI} = \alpha_0 + \alpha_1 S_N(\omega_1) + \alpha_2 S_N(\omega_2) + \alpha_3 S_N(\omega_1) S_N(\omega_2)$$

to minimize the sum-squared error

$$\sum_{j=1}^{M} (\text{SI}_j - \text{PSR}_j)^2$$

over M different road samples. The technique used for finding the best α_i is a simple generalization of the least-squares fitting methods we developed in Chap. 2. The difficult part of this problem is to decide which frequency terms to include in the SI function. [Actually, the results reported here used smoothed estimates of Fourier amplitudes $|X_N(\omega)|$ rather than the power spectral density. Cross-amplitude estimates relating left and right wheel measurements were also included.] Average values, i.e., averages over the M test sites, were subtracted from each term so that the quantities being combined had zero average values. While it was hoped that the frequencies used could be justified on physical grounds, the best results were obtained by the following expression, which was not completely amenable to such interpretation:

$$\begin{aligned} \mathrm{SI} = 3.24 - 1.47X_1 - 0.133X_2 - 0.54X_3 + 1.08XC_1 - 0.25XC_2 \\ + 0.08X_2X_3 - 0.91X_3X_4 + 0.67X_6X_{10} + 0.49T \end{aligned}$$

where
$$\begin{aligned} X_1 &= \log A_{0.023} + 2.881 \\ X_2 &= \log A_{0.046} + 4.065 \\ X_3 &= \log A_{0.069} + 4.544 \\ X_4 &= \log A_{0.093} + 4.811 \\ X_6 &= \log A_{0.139} + 5.113 \\ X_{10} &= \log A_{0.231} + 5.467 \\ XC_1 &= \log C_{0.023} + 3.053 \\ XC_5 &= \log C_{0.116} + 5.659 \end{aligned}$$

A_i = average right- and left-wavelength amplitude, in
C_i = cross amplitude, in
i = frequency component, c/f
T = 1 for rigid pavements and 0 for flexible pavements

Although terms in this expression are not easy to explain, the large negative coefficient (-1.47) on the lowest-frequency term shows that it causes the most discomfort. However, the significant positive coefficient on the cross-spectral term (1.08) at that frequency suggests that if both sides rise and fall together at that rate it is less objectionable than if only one side is bouncing at that rate.

This road-surface example combined quantities related to spectral-density estimates at several frequencies to get a measure of road quality. A similar approach has been considered for detecting the presence of dyslexia, a learning disability, by processing a child's EEG waves.† Dyslexia is a symptom characterized by an individual's inability to learn to read and spell at a level corresponding to his intelligence. Many children who are very good at abstract thinking have severe difficulty learning to read. They confuse mirror-image letters *b* and *d*, *p* and *q*, and upside-down pairs such as *W* and *M*. They may substitute "saw" for "was" or "funny" for "laugh." It would be desirable to have a fast definitive test for this condition because the children and their parents suffer greatly until other causes for the learning difficulty (emotional, motivational, etc.) have been ruled out.

EEGs from normal and dyslexic children do not show any significant differences, but power spectral densities of those waves have revealed differences. Figure 4.38 compares spectra (averages from many children) of normal and dyslexic children, both at rest and while reading. The differences between these curves, especially at rest, suggested that a test function much like the SI in the previous

† B. Sklar et al., A Computer Analysis of EEG Spectral Signatures from Normal and Dyslexic Children, *IEEE Trans. Biomed. Eng.*, vol. 10, no. 1, pp. 20–26, January 1973.

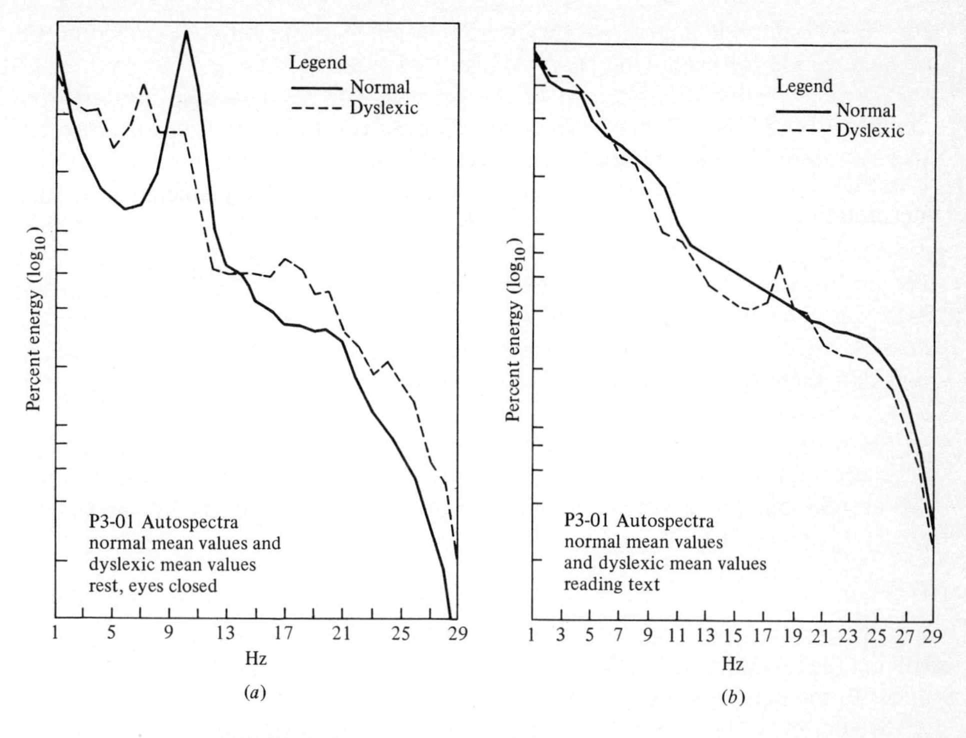

FIGURE 4.38
Normal and dyslexic EEG spectra. (*a*) Autospectral averages—rest, eyes closed; (*b*) autospectral averages—reading text. (*From B. Sklar et al., A Computer Analysis of EEG Spectral Signatures from Normal and Dyslexic Children, IEEE Trans. on Biomedical Engineering, vol. 10, no. 1, Jan. 1973, pp. 20-26, by permission.*)

example could be developed from the smoothed power-spectral-density heights at several frequencies (using separate power spectral densities and cross-spectral densities from various locations on the head). It seems that a good test function of this form can be found such that a child is likely to be dyslexic if his EEGs produce a sufficiently large test value.

The last two examples have mentioned the use of estimates of cross-spectral properties. The next section will discuss some of the computational and statistical aspects of such estimates.

4.9 CROSS-SPECTRAL ESTIMATES

The definition and some applications of cross-correlation functions and their transforms, the cross-spectral-density functions, appeared in Sec. 3.5. Further discussion of the estimation and use of cross correlations for time-delay measurements was given in Sec. 4.2. This section will outline some of the steps to be followed when estimating the cross-spectral density $S_{xy}(\omega)$ of signals x_i and y_i. Such estimates can be used to determine the transfer function of a filter relating x_i and y_i, as mentioned in

Sec. 3.5, or to provide ingredients for discrimination tests like those described in the previous section for classifying road surfaces or EEGs.

In view of our development of the periodogram for power-spectral-density estimates it is natural to define a cross-spectral-density estimate $S_{xy}(\omega)$ as the discrete Fourier transform of the sample cross correlation of (4.16). Thus,

$$\hat{S}_{xy}(\omega) = \sum_{k=-\infty}^{\infty} \hat{R}_{xy}(k)e^{-jk\omega T_s} \tag{4.86}$$

where we have chosen to indicate these estimates by hats rather than the subscript N, which was used in the estimates of autocorrelation functions and power spectral densities. The xy subscript is used to indicate the "cross" property, and it is understood that these estimates are based on N samples from each signal: x_i and y_i for $i = 0, 1, \ldots, N - 1$.

As with the periodogram, it is possible to relate the $\hat{S}_{xy}(\omega)$ estimate directly to the Fourier transforms of the data instead of the indirect definitions of (4.86) via the cross correlation. An argument similar to the one beginning with (4.20) shows that the direct expression for $\hat{S}_{xy}(\omega)$ is

$$\hat{S}_{xy}(\omega) = \frac{1}{N} X_N^*(\omega) Y_N(\omega) \tag{4.87}$$

involving the transforms $X_N(\omega)$ and $Y_N(\omega)$ of the data sequences. Notice that (4.87) reduces to the periodogram in (4.19) if $x_i = y_i$.

Studies of the bias, variance, and computational properties of $\hat{S}_{xy}$ parallel those we have given for the periodogram,† but they are more complicated because this estimate (as well as the true cross-spectral density) is a complex function of frequency. Longer data sequences will reduce the bias, but variance reduction requires either averaging of several separate $\hat{S}_{xy}$ estimates or smoothing them with spectral windows.

The complex-valued cross spectrum has some interesting properties and uses which have no counterparts in power-spectral-density analysis. When the magnitude and angle of the Fourier transforms are shown explicitly, $\hat{S}_{xy}$ becomes

$$\hat{S}_{xy}(\omega) = \frac{1}{N} |X_N(\omega)| \, |Y_N(\omega)| \exp\{j[\angle Y_N(\omega) - \angle X_N(\omega)]\} \tag{4.88}$$

If, for example, y_i is simply the x_i signal delayed by p steps, then

$$y_i = x_{i-p}$$

and, as discussed in Sec. 2.8, Y_n is the same as X_n except for an additional phase angle proportional to ω (linear phase)

$$Y_N(\omega) = X_N(\omega)e^{-jp\omega T_s}$$

Thus when y_i is a delayed version of x_i, the cross-spectral estimate will be

$$\hat{S}_{xy}(\omega) = \frac{|X_N(\omega)|^2}{N} e^{-jp\omega T_s} \tag{4.89}$$

† Jenkins and Watts, op. cit.

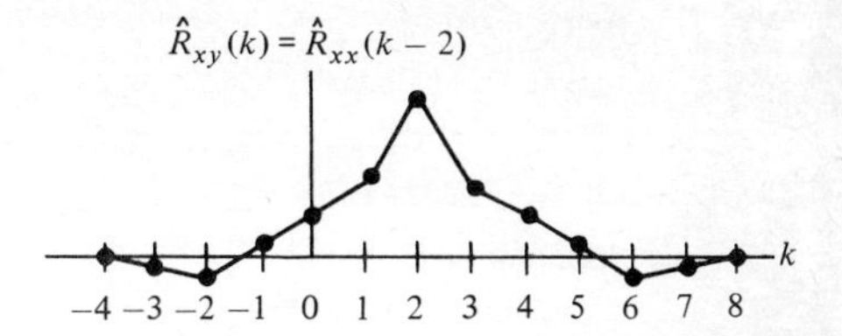

FIGURE 4.39
Cross correlation of x_i and $y_i = x_{i-2}$.

The cross-correlation estimate will be a shifted version of the sample autocorrelation function, with its peak at lag $k = p$ corresponding to the phase factor in (4.89), as in Fig. 4.39.

To reverse the argument of the previous paragraph, if a cross-spectrum estimate appears to have a linear phase or a phase which is a linear term in ω plus some other variation, part of the relation between x_i and y_i can be considered to be a delay. For example, if x_i is the setting of a valve controlling flow in a chemical plant and y_i is the concentration of a chemical in the plant output, there will be a delay corresponding to the travel time of the input fluid from the valve to the reactor, where it enters a transfer-function relation with the output y_i. We saw in Sec. 3.5 that the x-y transfer function can be estimated by

$$H(\omega) \approx \frac{\hat{S}_{xy}(\omega)}{\hat{S}_{xx}(\omega)} \tag{4.90}$$

It can be shown† that when the transfer function includes a pure delay (sometimes called *transportation delay*), the rest of the transfer function is estimated more accurately by decomposing the two effects into the product

$$\hat{H}(\omega) = \hat{H}'(\omega)e^{-jp\omega}$$

with

$$\hat{H}'(\omega) = \frac{1}{\hat{S}_{xx}(\omega)} \overline{\frac{X_N^*(\omega)Y'_N(\omega)}{N}} \tag{4.91}$$

where $Y'_N(\omega)$ is the transform of the shifted output data $y_p, y_{p+1}, \ldots, y_{p+N-1}$, and the overbar indicates smoothing or averaging of the revised cross spectrum. This alignment of the output data with the input data results in a smoothed cross-spectral estimate with smaller bias. The alignment technique for improving cross-spectrum estimates is conceptually similar to the power-spectral-density technique of removing an indicated trend or spectral component and recomputing a power-spectral-density estimation.

It is generally easier to detect an apparent delay relation between x_i and y_i by examining their sample cross correlation for a shifted peak. If the cross correlation is calculated for this purpose, by a finite transform of the cross-spectral estimate

$$\hat{R}_{xy}(k) = \frac{1}{N} \sum_{n=0}^{N-1} \hat{S}_{xy}(n\Omega_s)^{jnk2\pi/N} \tag{4.92}$$

† Ibid.

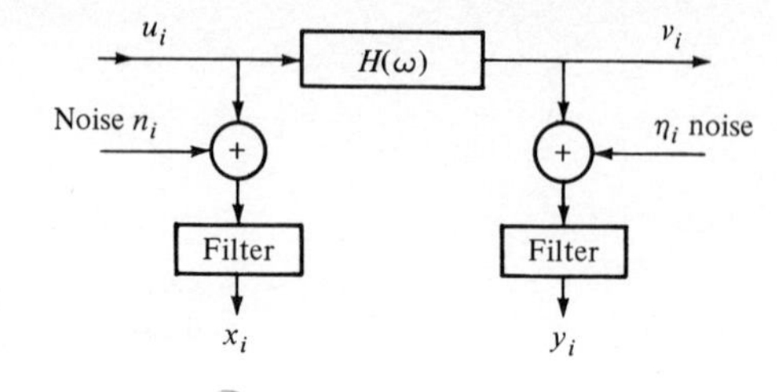

FIGURE 4.40
Noisy input-output measurements for transfer-function estimation.

as an approximation to the inverse transform

$$\hat{R}_{xy}(k) = \int_0^{2\pi} \hat{S}_{xy}(\omega) e^{jk\omega} \frac{d\omega}{2\pi} \tag{4.93}$$

then we must again heed the admonition given in Sec. 4.6 concerning padding the signal data with zeros. The sum in (4.92) will coincide with the integral in (4.93) for $|k| < N$ if N zeros are added to the x_i and to the y_i data.

Figure 4.40 shows a typical measurement situation by which data are recorded and a cross-spectral estimate made in order to estimate an unknown transfer function $H(\omega)$. The only available input and output measurements are those corrupted by noise (errors in the recording devices, for example). If the measurements are sampled at a high rate, the errors will generally be uncorrelated from sample to sample, while the desired input u_i and output v_i will be slowly varying. The effects of the noise can thus be reduced by low-pass filtering. (Recall that large power-spectral-density values at high frequencies, which would result in the absence of this filtering, can cause bias errors in the power-spectral-density estimates at the low frequencies of interest, due to the inevitable presence of window side lobes associated with finite data sets.)

We would like to compute the transfer-function estimate

$$\hat{H} = \frac{\hat{S}_{uv}(\omega)}{\hat{S}_{uu}(\omega)} \tag{4.94}$$

but u_i and v_i are not directly available. However, if we assume that the filters have identical transfer functions $F(\omega)$ and that they remove all the noise (this cannot be completely true), then, using (3.157) gives

$$\frac{S_{xy}}{S_{xx}} = \frac{S_{uv}|F(\omega)|^2}{S_{uu}|F(\omega)|^2} = H(\omega) \tag{4.95}$$

In this way $\hat{S}_{xy}/\hat{S}_{xx}$ is seen to be a reasonable estimate of the unknown transfer function. Figure 4.41 shows the magnitude of such an estimate based on data relating the power demand on a turboalternator to the resulting line voltage.† The portion of

† Ibid., and K. N. Stanton, Measurement of Turbo Alternator Transfer Functions Using Normal Operating Data, *Proc. IEEE*, vol. 110, no. 11, p. 2001, 1963.

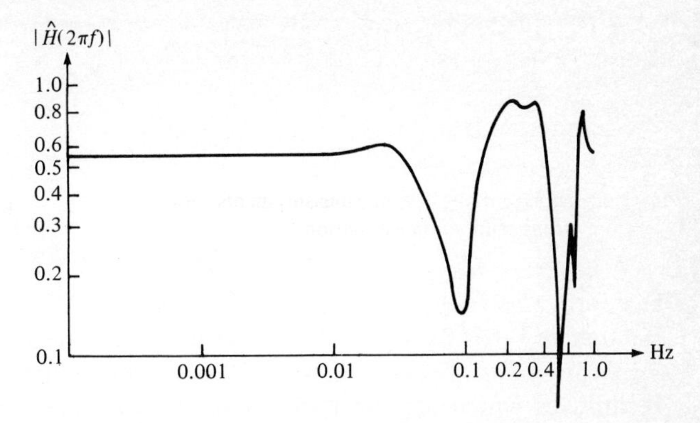

FIGURE 4.41
Estimate of power demand-to-voltage transfer function. (*From G. M. Jenkins and D. G. Watts, "Spectral Analysis and Its Applications," Holden-Day, 1968, by permission.*)

the curve above 0.1 Hz is essentially meaningless because the low-pass filters had sharp cutoffs at that frequency, so that

$$|\hat{H}(\omega)| = \frac{|\hat{S}_{xy}(\omega)|}{|\hat{S}_{xx}(\omega)|}$$

has the unpredictable 0/0 form in that frequency range.

Figure 4.42 shows the corresponding $\hat{R}_{xy}(k)$ estimate which has a peak at zero indicating no pure delay factor. (If u_i and v_i had any pure delay factor, the identical low-pass filters would induce the same delay relation between x_i and y_i.) The negative peak shows that increased demand causes decreased voltage. That negative relation does not show up in the magnitude plot of Fig. 4.41 but would appear in the transfer-function phase estimate.

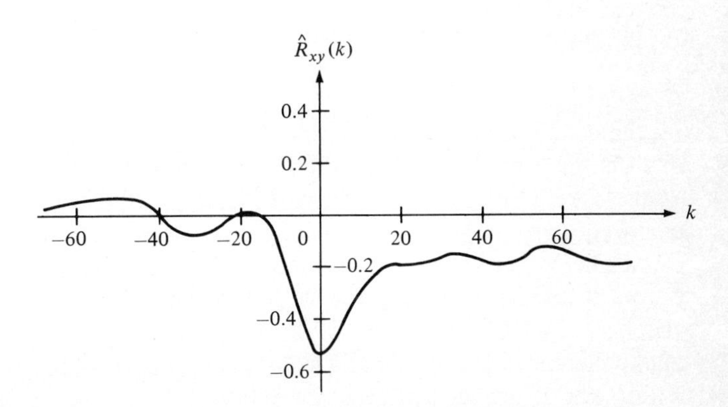

FIGURE 4.42
Estimate of power demand-to-voltage cross correlation. (*From G. M. Jenkins and D. G. Watts, "Spectral Analysis and Its Applications," Holden-Day, 1968, by permission.*)

The turboalternator study referred to here proceeded from Fig. 4.41 to a transfer-function estimate of the form

$$|H(\omega)| = \frac{c^2}{(1 + a^2 + b^2) - 2a(1 - b)\cos\omega - 2b\cos 2\omega}$$

corresponding to a filter described by

$$y_i = ay_{i-1} + by_{i-2} + cx_i$$

This form was suggested since it is the simplest way of achieving a magnitude curve like that of Fig. 4.41 for $f < 0.1$ Hz, with a slight peak at $f \approx 0.03$ Hz and a linearly decreasing slope of about a factor of 100 for an increase in frequency by a factor of 10. Techniques which are the basis for such judgments are beyond the scope of this chapter. It will suffice here to say that a transfer function can be chosen to approximate a curve like $|\hat{H}(\omega)|$, and it can be further adjusted so that it comes close to $\angle\hat{H}(\omega)$. It is left as a problem to compare an $|H(\omega)|$ of the given form to the curve in Fig. 4.41.

While the above modeling procedure is effective for simple systems, it is now more common to use cross-spectral analysis to find pure delay terms and the general nature (high-pass, low-pass, high-frequency decay rate, etc.) of a transfer function and then compute the parameters of an appropriate model directly from the given input-output data.† This latter procedure for identifying transfer-function model parameters is closely related to the parametric power-spectral-density estimation ideas to be presented in the next section.

One more cross-spectral property should be mentioned here because of its widespread use. That is the *coherence* function, whose square is defined as

$$K_{xy}^2(\omega) = \frac{|S_{xy}(\omega)|^2}{S_{xx}(\omega)S_{yy}(\omega)} \tag{4.96}$$

This is a frequency function whose value at a particular ω is a measure of similarity of the strength of components in x_i and y_i at that frequency. If x_i and y_i are the input and output of *any* transfer function, then

$$S_{xy} = S_{xx}H \qquad S_{yy} = S_{xx}|H|^2 \tag{4.97}$$

and the coherence function is unity ($K_{xy} = 1$). If y_i is a sum of x_i plus some other n_i which is uncorrelated with y_i, then the coherence will be less than 1. Thus, $K_{xy}(\omega)$ behaves like a correlation coefficient [see (3.112)] for x_i and y_i components at the same frequency. High coherence suggests a cause-and-effect relationship.

Coherence estimates were used in the EEG studies described in the previous section. In the light-flashing test with newborn babies there was small coherence at 3 Hz between EEGs from the left and right sides of the brain. In the dyslexia study, small coherence between signals from different parts of the brain was another indication of lack of proper learning interaction.

† G. P. E. Box and G. M. Jenkins, "Time Series Analysis: Forecasting and Control," Holden-Day, San Francisco, 1970; D. Graupe, "Identification of Control Systems," Van Nostrand Reinhold, New York, 1972; Special Issue on System Identification and Time Series, *IEEE Trans. Autom. Control*, December 1974.

We now move from the study of nonparametric spectral estimation, in which power spectral densities and cross spectra were not assumed to have any preconceived shape or functional form, to the study of parametric estimates. Nonparametric estimates are often computed in order to pick likely parametric models whose unknown parameters are then estimated according to the techniques of the next section.

4.10 PARAMETRIC SPECTRAL ESTIMATES

In this section we show how to find autocorrelation-function and power-spectral-density estimates from a record of data by assuming that the data are typical of a random signal whose autocorrelation function has a known form, for example, $R(k) = Aa^{|k|}$, with a small number of unknown parameters (A and a in this example). The nonparametric power-spectral-density estimates of the previous section, e.g., averaged or smoothed periodograms $S_M^K(\omega)$ or $\tilde{S}(\omega)$, will reveal general dominant frequencies, suggest better sampling rates, and generally reveal information not apparent in the raw data. However, other signal-processing applications require simple analytical expressions for $S(\omega)$, rather than numerical curves of power-spectral-density estimates. As an example, a digital processor for detecting the presence or absence of a radar pulse can be most efficiently constructed if the noise in the measurements is represented by a power spectral density with a simple analytical form. Similarly, a tracking filter which estimates the plane's future position based on noisy determination of the plane's present and past locations must have convenient models for the fluctuations in past errors as well as the uncertainties in the plane's future motion. Examples like these will be encountered in Chaps. 5 to 7.

This parametric approach contrasts with the nonparametric methods of Sec. 4.6, where no assumptions were made about the form of $R(k)$ or $S(\omega)$ and where a large number of quantities, for example, $S(\omega_i)$ for 32 values of ω_i, were estimated. In a typical signal-processing situation, a nonparametric averaged periodogram (along with engineering experience) will suggest a suitable analytical form $R(k)$ or $S(\omega)$. While one could try to fit, by adjusting parameters, a curve of that assumed form to $\hat{S}_M^K(\omega)$ or $R_N(k)$, it is usually easier and more effective to return to the original data to get estimates of the unknown parameters. We now discuss a maximum-likelihood procedure for getting these estimates.

Just as smoothed periodogram estimates are analogous to histograms (with bias problems and windows), parametric spectral analysis is much like parametric probability density analysis. Recall from Chap. 3 how we got maximum-likelihood estimates of mean and variance by assuming a gaussian form for the unknown density of the data.

The simplest example of a parametric spectral analysis results when we assume that the data have been generated by a first-order autoregression

$$x_i = ax_{i-1} + bu_i \tag{4.98}$$

in which the u_i are samples from a purely random zero-mean, unit-variance gaussian signal. We shall show later that the results found here are sensible even if the u_i are not gaussian. The assumption that the signal x_i is generated in this way is equivalent to assuming that it is a gaussian signal with the following autocorrelation function and power spectral density [see (3.138) and (3.137)]:

$$R(k) = \frac{b^2}{1 - a^2} a^{|k|} \tag{4.99}$$

$$S(\omega) = \frac{b^2}{1 + a^2 - 2a \cos \omega} \tag{4.100}$$

Thus we can think of the estimation of a and b in (4.98) as parametric estimation of either an autocorrelation function or a power spectral density.

The maximum-likelihood choice of parameters is based on the idea that the underlying system should be the one that makes the observed data more likely than any other data. As we saw in getting maximum-likelihood estimates for the mean and variance in Chap. 3, the first step in this argument is to get the joint density of the observations in terms of the unknown parameters. In this case we want to find the joint density

$$f(x_0, x_1, \ldots, x_{N-1}; a, b)$$

of the N observations. This density is computed using the assumed autoregression model and the fundamental conditional density relation†

$$f(x_0, x_1, \ldots, x_{N-1}) = f(x_0)f(x_1 \mid x_0)f(x_2 \mid x_0, x_1) \cdots f(x_{N-1} \mid x_0, \ldots, x_{N-2}) \tag{4.101}$$

This relation is a generalization of the more familiar two-variable joint density relation

$$f(x_0, x_1) = f(x_1 \mid x_0)f(x_0) \tag{4.102}$$

As explained earlier, the zero-mean gaussian input will produce a zero-mean gaussian output whose variance is the zero-lag value of $R(k)$ in (4.99). Thus, the first observation has the density

$$f(x_0) = \frac{1}{\sigma_x \sqrt{2\pi}} e^{-x_0^2/2\sigma_x^2} \qquad \text{where } \sigma_x^2 = \frac{b^2}{1 - a^2} \tag{4.103}$$

The conditional density $f(x_1 \mid x_0)$ of x_1 given x_0 in (4.102) is easily found from the autoregression equation because, with x_0 given or fixed, the ax_0 on the right side of

$$x_1 = ax_0 + bu_1 \tag{4.104}$$

is not random and merely shifts the density of the gaussian variable bu_1. We shall define that input term bu_1 as a new variable m_1. That input term is a zero-mean gaussian variable with variance b^2 (recall that u_1 has unit variance), and so its density is

$$f_{\mathbf{m}_1}(m_1) = \frac{1}{b\sqrt{2\pi}} e^{-m_1^2/2b^2} \tag{4.105}$$

† A. Papoulis, "Probability, Random Variables, and Stochastic Processes," p. 236, McGraw-Hill, New York, 1965.

Shifting the density by the known amount ax_0

$$f(x_1 \mid x_0) = f_{\mathbf{m}_1}(x_1 - ax_0)$$

yields

$$f(x_1 \mid x_0) = \frac{1}{b\sqrt{2\pi}} e^{-(x_1 - ax_0)^2/2b^2} \tag{4.106}$$

The same argument can be used to get any conditional density in (4.101) as

$$f(x_i \mid x_0, \ldots, x_{i-1}) = \frac{1}{b\sqrt{2\pi}} e^{-(x_i - ax_{i-1})^2/2b^2} \tag{4.107}$$

since all the preceding samples are useful only in telling us the value of ax_{i-1} in (4.98). They convey no information about u_i because (even though x_{i-1} is a function of u_{i-1}) u_i is independent of all previous samples of the purely random input. Thus, given the conditioning information of all previous sample values, we know that x_i has a mean of ax_{i-1} and is a gaussian variable with the same variance as bu_i.

Substitution of (4.106) and (4.107) into (4.101) results in the compact expression

$$f(x_0, x_1, \ldots, x_{N-1}; a, b) = \frac{f(x_0)}{(b\sqrt{2\pi})^{N-1}} \exp\left[-\frac{1}{2b^2}\sum_{i=1}^{N-1}(x_i - ax_{i-1})^2\right] \tag{4.108}$$

in which the product of exponentials has been written as a single term whose exponent is the sum of the individual exponents.

In the maximum-likelihood approach we substitute observed values, for example, $x_0 = 17.5$, $x_1 = -32.7, \ldots,$ into (4.108) and then see which values of a and b maximize the resulting expression. That maximization can be achieved by setting partial derivatives equal to zero and solving the resulting equations. When we let f_N represent the joint density in (4.108) and use (4.103) for $f(x_0)$, the differentiations lead to

$$\frac{\partial f_N}{\partial a} = f_N\left[\frac{ax_0^2}{b^2} - \frac{a}{\sqrt{1 - a^2}} + \frac{1}{b^2}\sum_{i=1}^{N-1}(x_i - ax_{i-1})x_{i-1}\right] = 0 \tag{4.109}$$

$$\frac{\partial f_N}{\partial b} = f_N\left\{\frac{1}{b^3}\left[x_0^2(1 - a^2) + \sum_{1}^{N-1}(x_i - ax_{i-1})^2\right] - \frac{N}{b}\right\} = 0 \tag{4.110}$$

The maximum-likelihood estimates of a and b are found by solving (4.109) and (4.110) simultaneously, but some simplifications are possible for large N (long data records). For example, the summation term in (4.109) will get very large and dominate the other terms when N is large, to give the simpler condition

$$\sum_{i=1}^{N-1}(x_i - \hat{a}x_{i-1})x_{i-1} = 0 \tag{4.111}$$

This can be solved for the maximum-likelihood estimate $\hat{a}$ to give

$$\hat{a} = \frac{\sum_{1}^{N-1} (x_i x_{i-1})}{\sum_{1}^{N-1} x_{i-1}^2} \tag{4.112}$$

Similarly, since the x_0 term in (4.110) becomes unimportant compared with the other terms for large N, that equation simplifies and leads to

$$\hat{b} = \pm \sqrt{\frac{1}{N} \sum_{1}^{N-1} (x_i - \hat{a} x_{i-1})^2} \tag{4.113}$$

The sign of b cannot be determined from the data in this model since changing its sign and the sign of all u_i would produce identical data; the gaussian input numbers are equally likely to have values $u_1 = 3, u_2 = -1, \ldots$ as $u_1 = -3, u_2 = 1, \ldots$. Furthermore, the parametric models in (4.99) and (4.100) do not require knowledge of a sign for $\hat{b}$. The estimate $\hat{a}$ is used in (4.113) since $\hat{b}$ results from simultaneous solution of both optimization conditions.

The calculations leading up to (4.112) and (4.113) and the appearance of these estimates might seem formidable at first, but they are really quite reasonable when related to earlier estimation results. For example, (4.112) can be written as

$$\hat{a} = \frac{NR_N(1)}{NR_N(0) - x_N^2}$$

or

$$\hat{a} \approx \frac{R_N(1)}{R_N(0)} \tag{4.114}$$

This ratio of autocorrelation-function estimates is quite a reasonable estimate since the true autocorrelation function in (4.99) obeys a similar relation to the true parameter a

$$a = \frac{R(1)}{R(0)} \tag{4.115}$$

We have already commented that the sample autocorrelation-function estimates $R_N(k)$ are unbiased and consistent, and so there is a high probability that the numerators and denominators in (4.114) and (4.115) are, respectively, nearly equal. It would be additionally satisfying if we could compute the mean and variance of the $\hat{a}$ in (4.112), but its very complicated relationship to $x_0, x_1, \ldots$ precludes any hope for simple results along those lines.

The other estimate (4.113) bears a similar interpretation. If a were known perfectly, we could use (3.47) to get a consistent unbiased estimate of the variance b^2 of the scaled input $m_k = bu_k$. Equation (4.98) could be solved exactly for every input

$$m_k = x_k - ax_{k-1} \tag{4.116}$$

and the variance estimate would be

$$\hat{\hat{b}}^2 = \frac{1}{N-1} \sum_{1}^{N-1} (x_i - ax_{i-1})^2 \tag{4.117}$$

This is the same as the square of (4.113) except that the estimate $\hat{a}$ is used there in the absence of perfect knowledge of that parameter. Estimation of b^2 instead of b is adequate, since the sign of b is irrelevant to the resulting power spectral density or autocorrelation function.

It is important to notice that these reasonableness comments about the estimates in (4.112) and (4.113) would apply to *any* random process generated from (4.98) by uncorrelated u_i inputs, i.e., a white input signal. These estimates are optimal in the maximum-likelihood sense when the input is gaussian, but they are quite reasonable even if we have no reason to believe that the data are gaussian.

The statistical nature of these estimates should also be reemphasized. We could blindly plug any data into these estimation formulas and get numerical answers. The resulting autocorrelation function and power spectral density of the forms (4.99) and (4.100) might, however, be very poor representations of the data. There are several ways to check this accuracy. The simplest test is to form estimates of the input function

$$\hat{m}_i = x_i - \hat{a}x_{i-1} \tag{4.118}$$

and submit the resulting signal $\hat{m}_i$ to a nonparametric spectral analysis to see if it is white. (See discussion of Fig. 4.5.) If that test is negative or inconclusive, a more complicated parametric model of the form

$$x_i = a_1 x_{i-1} + a_2 x_{i-2} + b_0 u_i + b_1 u_{i-1} \tag{4.119}$$

might be considered. The transfer function relating u_i to x_i is, in this case,

$$H(\omega) = \frac{b_0 + b_1 e^{-j\omega}}{1 - a_1 e^{-j\omega} - a_2 e^{-2j\omega}} \tag{4.120}$$

so that the output power spectral density is

$$\begin{aligned} S_x(\omega) &= H(\omega)H^*(\omega) \\ &= \frac{b_0^2 + b_1^2 + 2b_0 b_1 \cos \omega}{1 + a_1^2 + a_2^2 + 2a_1(a_2 - 1)\cos \omega - 2a_2 \cos 2\omega} \end{aligned} \tag{4.121}$$

Maximum-likelihood conditions can be computed for this kind of model, too, although the resulting zero-derivative equations may be nonlinear with nonunique solutions. Computer algorithms have been developed to search automatically for the parameters $a_1, a_2, b_0, b_1, \ldots$ which maximize the likelihood function corresponding to (4.108).†

Again, these higher-order results could be checked for reasonableness by using the $\hat{a}_1$ and $\hat{a}_2$ estimates and the original data to get an estimate $\hat{m}_i$ of the effective input function

† S. A. Tretter and K. Steiglitz, Power Spectrum Identification in Terms of Rational Models, *IEEE Trans. Autom. Control*, vol. AC-12, pp. 185–188, April 1967.

$$\hat{m}_i = x_i - \hat{a}_1 x_{i-1} - \hat{a}_2 x_{i-1} \tag{4.122}$$

The true input

$$m_i = b_0 n_i + b_1 n_{i-1} \tag{4.123}$$

has a triangular autocorrelation function (much like a simple averager's output) with

$$\begin{aligned} R_m(0) &= b_0^2 + b_1^2 \\ R_m(1) &= b_1^2 \\ R_m(k) &= 0 \qquad k \geq 2 \end{aligned} \tag{4.124}$$

and so the estimated input $\hat{m}_i$ can be tested nonparametrically to see whether its autocorrelation function and/or power spectral density correspond to (4.124) with the estimated values of the $\hat{b}_i$ substituted. This procedure can be repeated with higher-order models (more a_i and b_j terms) until the fit seems adequate.

The maximum-likelihood estimates of system parameters to be discussed in Sec. 6.5 are similar to those just described. In that case we shall see that the system is assumed to be a moving-average or nonrecursive filter only [this corresponds to all the a_i terms in (4.119) set to zero] and, more importantly, the inputs u_k are observable along with the x_k. More general maximum-likelihood parametric procedures for estimating system models from input-output measurements are highly developed and widely applied, as mentioned in the preceding section on transfer-function estimation.

4.11 SUMMARY

This chapter has discussed techniques for processing typical samples of random signals to determine their average frequency content and correlation properties, as embodied in power spectra and correlation functions. Typical applications of these functions as data-reduction techniques were time-delay measurements, based on cross correlations, and EEG classification based on spectral peaks. Later chapters will present other kinds of applications in which spectral models developed by the techniques of this chapter will be used as ingredients for the design of signal detectors and estimators.

Estimation of correlation functions and spectra were seen to be statistical problems, the quality of estimates being described in terms of their mean values and variances. Methods were described for improving estimates when more data became available, subject to the familiar trade-off limitations between bias and variance errors.

Computational problems inherent in processing large but finite amounts of data were also considered. Time- and frequency-domain properties of convolutions and products of functions were exploited to take advantage of computationally efficient computer algorithms.

PROBLEMS

4.1 An algorithm for computing $R_N(k)$ using Eq. (4.8) might be debugged by using simple data sequences. With $N = 10$, compute $R_N(k)$ for the two cases:

$$x_n = 1 \qquad n = 0, 1, \ldots, 9 \tag{i}$$

$$x_n = (-1)^n \qquad n = 0, 1, \ldots, 9 \tag{ii}$$

4.2 A realization $y_0, y_1, \ldots, y_{999}$ was used to compute the sample autocorrelation function shown in Fig. P4.2.

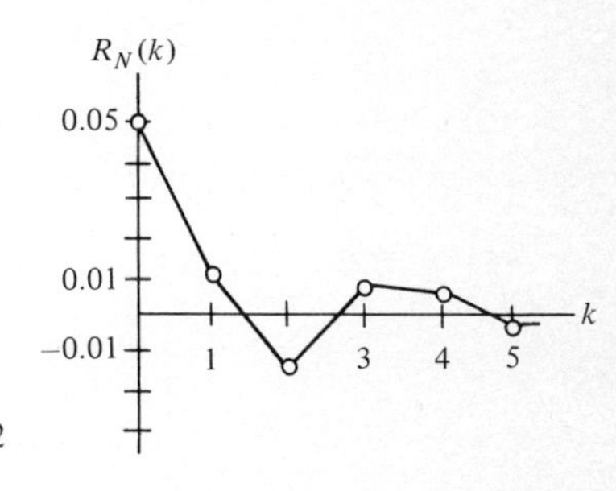

FIGURE P4.2

(*a*) If the data were from a white process, which k has the least likely $R_N(k)$ value? Use the Tschebychev inequality to bound the probability of getting a value so far from the expected value.
(*b*) Get a tighter bound by assuming that $R_N(k)$ is gaussian. (This assumption can be justified by a central-limit-theorem argument.)
(*c*) Show that a first-order autoregression $y_n = ay_{n-1} + bx_n$ with white-noise input has $R_y(1) = aR_y(0)$. Use $R_N(0)$ and $R_N(1)$ to estimate a and b for this nonwhite y-model.

4.3 (*a*) Find the maximum-likelihood estimate of the exponent α in the autocorrelation-function model of Eq. (4.15) based on the data in Fig. 4.5.
(*b*) Using the exponential model from part (*a*) and Eq. (4.14), compute the standard deviations of $R_N(k)$ for $k = 0, 1, \ldots, 5$ by digital computer.
(*c*) Combine the previous results to determine which point in Fig. 4.5 is farthest from the maximum-likelihood estimated exponential curve, in terms of units of corresponding standard deviation. Get a bound on the probability that such an $R_N(k)$ could occur if the answer to part (*a*) is the true autocorrelation function.

4.4 A record of N samples $x_0, x_1, \ldots, x_{N-1}$ is to be broken into K segments, each of length $M = N/K$ samples. The periodograms from the segments are averaged to get an estimate of $S_{xx}(\omega)$. We know that the true spectrum has a single peak as shown in Fig. P4.4, but we do not know where this peak is centered.

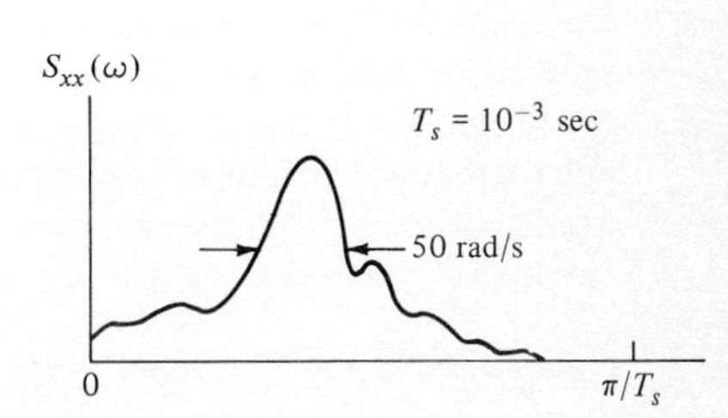

FIGURE P4.4

(*a*) Assuming that N is very large, how big should M be to make the spectral window narrower than the peak?
(*b*) Use a sentence to explain why it is best to make M no larger than the value found in part (*a*).

4.5 Samples $x_0, x_1, \ldots, x_{N-1}$ are used to estimate $S_x(\omega)$ by averaging together 10 periodograms, each computed from $N/10$ different samples.
(*a*) What is the variance of the resulting $\hat{S}^{10}_{N/10}(\omega)$ at $\omega = \pi/2T_s$ if the true power spectral density is $S_x(\omega) = 1 - \cos 2T_S\,\omega$, as shown in Fig. P4.5?

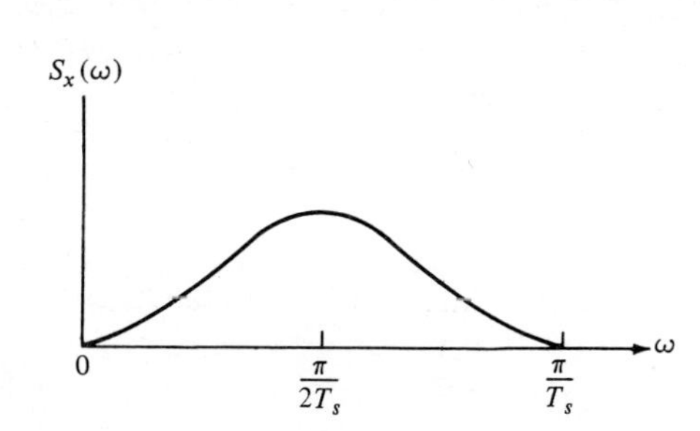

FIGURE P4.5

(*b*) Using the answer to part (*a*), find an upper bound on the probability

$$P\left[\left|\hat{S}^{10}_{N/10}\left(\frac{\pi}{2T_S}\right) - E\left[\hat{S}^{10}_{N/10}\left(\frac{\pi}{2T_S}\right)\right]\right| > 1.26\right]$$

that the estimate will be more than 1.26 units from its mean.
(*c*) How big must N be for this spectral estimate to be essentially unbiased? That is, for

$$E\left[\hat{S}^{10}_{N/10}\left(\frac{\pi}{2T_S}\right)\right] \approx S_x\left(\frac{\pi}{2T_S}\right)$$

(*d*) If the variance σ_x^2 were estimated by

$$\hat{\sigma}_x^2 = \frac{1}{N}\sum_{0}^{N-1} x_k^2$$

what value would you expect to get?

4.6 Try to reproduce the results in Fig. 4.21 by generating appropriate pseudorandom data and evaluating fast Fourier transforms. Different sets of data will not give exactly the same results, but the bias and variance errors should be similar in magnitude and shape. Verify that the confidence intervals shown in that figure are correct.

4.7 Compare the bias of rectangular and triangular windows for continuously smoothed power-spectral-density estimates, as follows:
(*a*) Sketch a true power spectral density $S(\omega) = (\sin\omega)^4$ and superimpose on it a rectangular window of width $\pi/4$ rad/s centered at $\omega = \pi/2$. Compute the bias of the resulting $S(\pi/2)$ according to (4.58). Remember that the window must have an area of unity.
(*b*) Repeat part (*a*) using a triangular window whose base has a width of $\pi/4$.

4.8 Derive the triangular-window variance expression given in (4.62).

4.9 Equation (4.53) for the variance of a smoothed periodogram is wrong when $\omega_i = 0$. Explain why the correct expression in that case is

$$\text{var } \bar{S}(0) = \frac{2}{2L+1} \text{ var } S_N(0)$$

because, for example, $S_N(-\Omega_S) \equiv S_N(\Omega_S)$, so several of the periodogram values being averaged are not independent.

4.10 Verify the lengths of the confidence-interval segments shown in Fig. 4.21 for $K = 6$ and 12.

4.11 A signal bandlimited to 10 kHz is to have its power spectral density estimated using a fast Fourier transform algorithm which can handle a maximum of 1024 data points. The variance of the estimate is to be reduced by using a rectangular window of width 500 Hz.
(*a*) Assuming a sampling interval of 40 μs (is that sufficiently small?), how many discrete-frequency periodogram values $S_N(n\Omega)$ will be averaged within the window?
(*b*) It is alleged that if we are really only interested in the 0- to 5-kHz portion of the spectrum, it pays to pass the signal through a low-pass filter cutting off at 5 kHz and resampling at half the rate (80 $\mu s = T_s$). Explain why 1024 wider-spaced time samples produce an $\tilde{S}(\omega)$ with smaller variance when a rectangular window with the same 500-Hz width is used.

4.12 As indicated in Sec. 2.7 the fast Fourier transform of N data points requires about $N \log_2 N$ complex-valued multiplication-addition operations to get the N-transformed values. (This is in contrast to the N^2 operations required for unsophisticated evaluation of the transform sums.) Use this information along with an accounting of lag-window multiplication-time requirements to explain why averaging of eight periodograms of 128 points each takes less than $\frac{4}{15}$ the time needed to compute an equivalent smoothed spectrum estimate from a 1024-point periodogram. What is the equivalent lag window which must be used in the latter case?

4.13 (*a*) Show that padding with N zeros has the desired effect by sketching the resulting x_i^F and x_{i+2}^F corresponding to Fig. 4.29*b*.
(*b*) Compute the true $R_N(2)$ and the erroneous one which would result from the original Fig. 4.29*b*.

4.14 (*a*) Find and sketch the power spectral density of x_n produced from zero-mean unit-variance white n_n noise by

$$x_n = 0.95x_{n-1} + 2n_n$$

(*b*) If the power spectral density of x_n is to be estimated, explain why the differencer filter (with input x_n)

$$y_n = x_n - x_{n-1}$$

will produce a y_n signal with a flatter power spectral density.
(*c*) Explain why it is easier to estimate $S_y(\omega)$ than $S_x(\omega)$ and how an estimate of the latter can be computed from an estimate of the former.

4.15 Reconsider the airline passenger data from Prob. 2.16. Remove a straight-line trend and add zeros to make a total of 128 data points. Compute smoothed power-spectral-density estimates using Bartlett windows with truncation lengths of $M = 20$, 40, and 60. (Note that since the fast Fourier inverse transform operates over the frequency range 0 to 2π, the q_k^B weights must be applied to samples 0, 1, ..., M and symmetrically to $N - 1 - M$, ..., $N - 1$. Compare your results with Fig. 2.34.

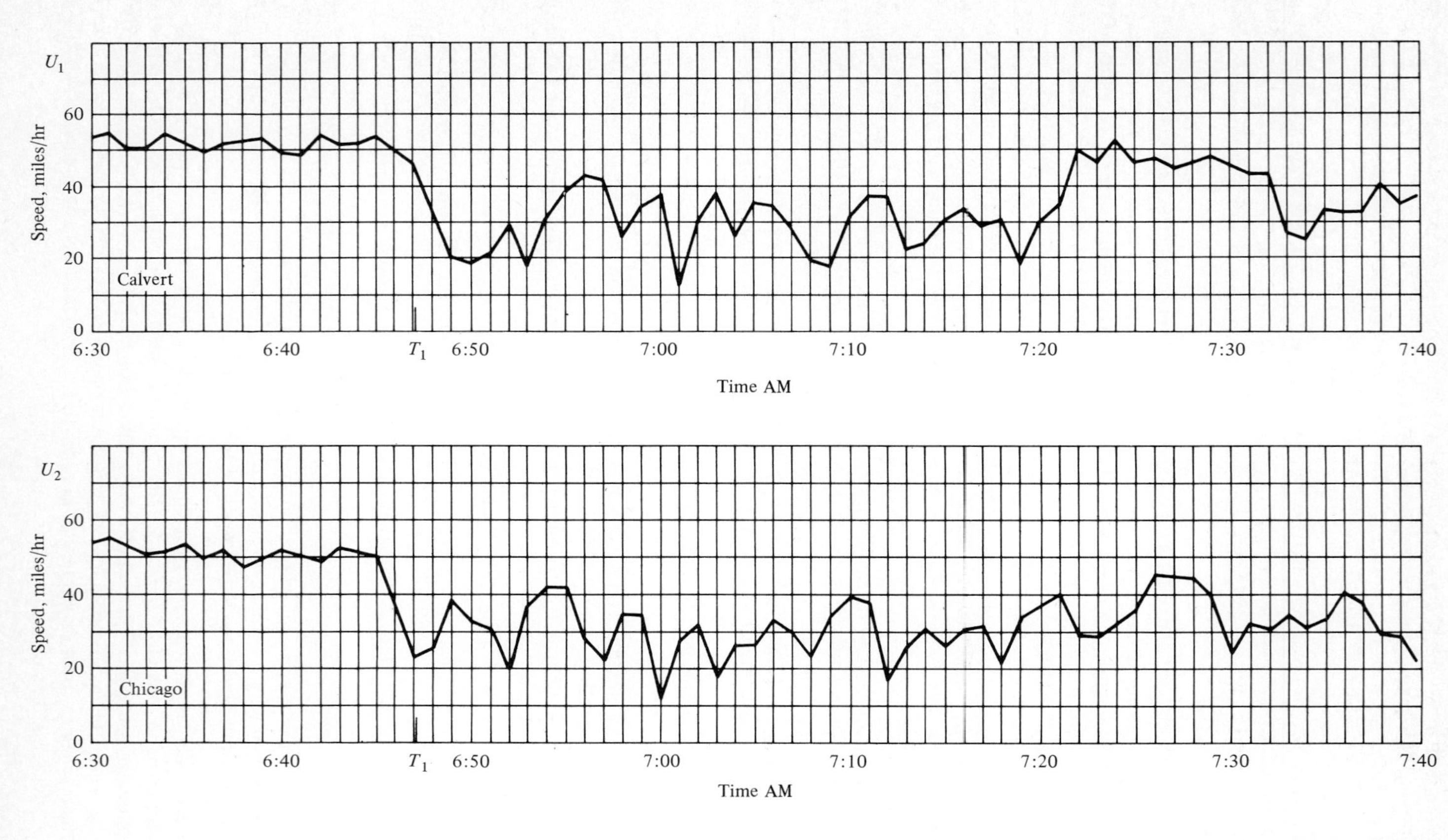

U_1
Speed, miles/hr
60
40
20
0
Calvert
6:30
6:40
T_1
6:50
7:00
7:10
7:20
7:30
7:40
Time AM
U_2
Speed, miles/hr
60
40
20
0
Chicago
6:30
6:40
T_1
6:50
7:00
7:10
7:20
7:30
7:40
Time AM

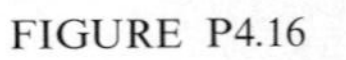

FIGURE P4.16

4.16 Use the methods of this chapter to get the cross correlation R_{12} of Fig. 4.8 and the power spectra G_{11} and G_{22} of Fig. 4.10 from U_1 and U_2 in Fig. 4.7. The latter curves have been reproduced here as Fig. P4.16 on a scale more suitable for reading off numerical values. Compare G_{11} curves computed using different values of T_s.

4.17 An unknown transfer function is to be estimated by cross-spectral analysis from noiseless input and output measurements x_i and y_i. It is claimed that if a filter F can be found which will convert x_i into a white signal u_i [$S_{uu}(\omega) = 1$], then the u_i and v_i shown in Fig. P4.17

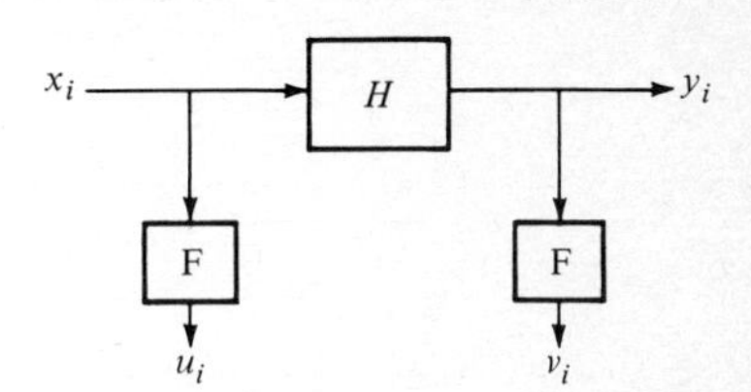

FIGURE P4.17

can be used to get the simple estimate

$$\hat{H}(\omega) = \hat{S}_{uv}(\omega)$$

Explain this.

4.18 x_i and n_i in Fig. P4.18 are uncorrelated zero-mean white signals with $S_{xx}(\omega) = 1$, $S_{nn}(\omega) = N_0$. Evaluate S_{yy}, S_{zz}, S_{xy}, S_{xz}, and the coherence functions K_{xy} and K_{xz}. Discuss the variation in K_{xz} as N_0 changes from zero to infinity.

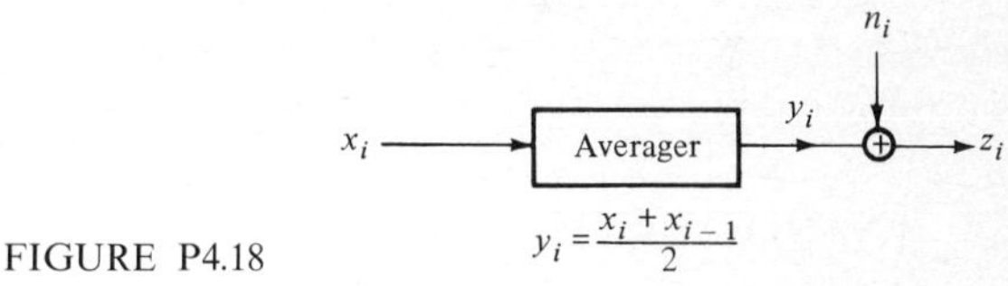

FIGURE P4.18

4.19 Show that a filter of the form

$$y_i = ay_{i-1} + by_{i-2} + cx_i$$

has a transfer function like the one in Fig. 4.41 (below 0.1 Hz) when $T_s = 0.5$ s.

Ans.: $a = 0.183$, $b = -0.865$, $c = 0.0192$

5

DETECTION OF SIGNALS IN NOISE

5.1 INTRODUCTION

In Chap. 1 we discussed rather briefly the radar systems associated with the air-traffic-control system in the United States. We described the characteristics of the radars to provide simple examples of systems in which signal-processing functions are carried out. The two specific functions of the radars noted there were those of aircraft detection and tracking: the radars continually search the sky "looking" for aircraft. Once the presence of an aircraft is detected, tracking begins: the aircraft location, direction of flight, and speed are estimated and updated regularly to enable the traffic-control function to be carried out.

We indicated in that chapter that the two tasks of aircraft *detection* and *estimation* of its characteristics, e.g., velocity and location for tracking purposes, involved signal-processing problems because of the unavoidable presence of noise, as well as other possible deleterious effects. In this chapter and the one following we focus on these two signal-processing tasks, detection and estimation, in detail. For simplicity, we shall discuss the detection problem first, considering in this chapter both ad hoc and more systematic approaches to developing appropriate signal processors for carrying out the detection functions. In the next chapter we move on to the more complex problem of signal estimation.

Although the radar systems discussed earlier provide realistic and useful examples for both the detection and estimation problems, many other applications of course exist. The problem of detecting the presence or absence of signal in the continuous presence of noise is often encountered in the broad area of signal processing. As an example, visualize a radio antenna pointing into the sky. It continually picks up thermal energy radiated from the earth's atmosphere, from the sun if pointed in that direction, from far-off stars, etc. Say the antenna is being used in an experiment to detect the presence of hydrogen molecules in a particular star cluster in space. It would thus normally be tuned to a resonant frequency characteristic of the rotation spectrum for hydrogen molecules. But the background thermal energy appears at this range of frequencies as well. (This background radiation is often modeled as *white noise*, with its spectral components present at all frequencies in the radio spectrum.) The antenna and receiving system, as well, generate their own thermal noise. If the source of the signal to be detected is extremely far away, very little signal energy may appear at the antenna: how does one detect the signal, if present, in the ever-present background noise field?

A similar problem arises in optical spectroscopy. Sonar systems, like the radar systems already mentioned, depend on schemes for determining the presence or absence of signals in additive noise. Digital-communication systems and computers tied together over digital communication channels also require receivers that detect signals in the presence of noise.

All these examples are characterized by the common problem of detecting the presence of a particular type or class of signals in the presence of noise. This is the simplest form of detection problem and the one we shall stress in this book. (In some detection problems occurring in practice several types of signals may be present simultaneously. The problem then is to see whether the signal of interest is present among this group. The other signals are then labeled *interfering signals* and may be lumped in with the ever-present noise. In other detection problems signal-induced noise may appear in addition to the noise, independent of signal, under discussion here. An example would include reverberation noise encountered in signal propagation through water; random reflections of the propagating signal from wavelets, from the usually turbulent surface, from a possibly rough bottom—all these contribute to the reverberation noise. Although the stress in this book will be on the simplest form of noise model, in which noise independent of the signal *adds* to the signal, we shall have occasion to consider other examples as well.)

Specifically, assume that, whatever the application, the signal $s(t)$ whose presence we are interested in detecting is a pulse of constant amplitude A known to last T s when it appears. (The varying-amplitude signal and further extension to *random* signals will be considered later.) Present at *all* times is a noise term $n(t)$, independent of the signal. The composite received waveshape that we must process to determine the presence of $s(t)$ is then

$$x(t) = s(t) + n(t) = A + n(t)$$

or

$$x(t) = n(t) \tag{5.1}$$

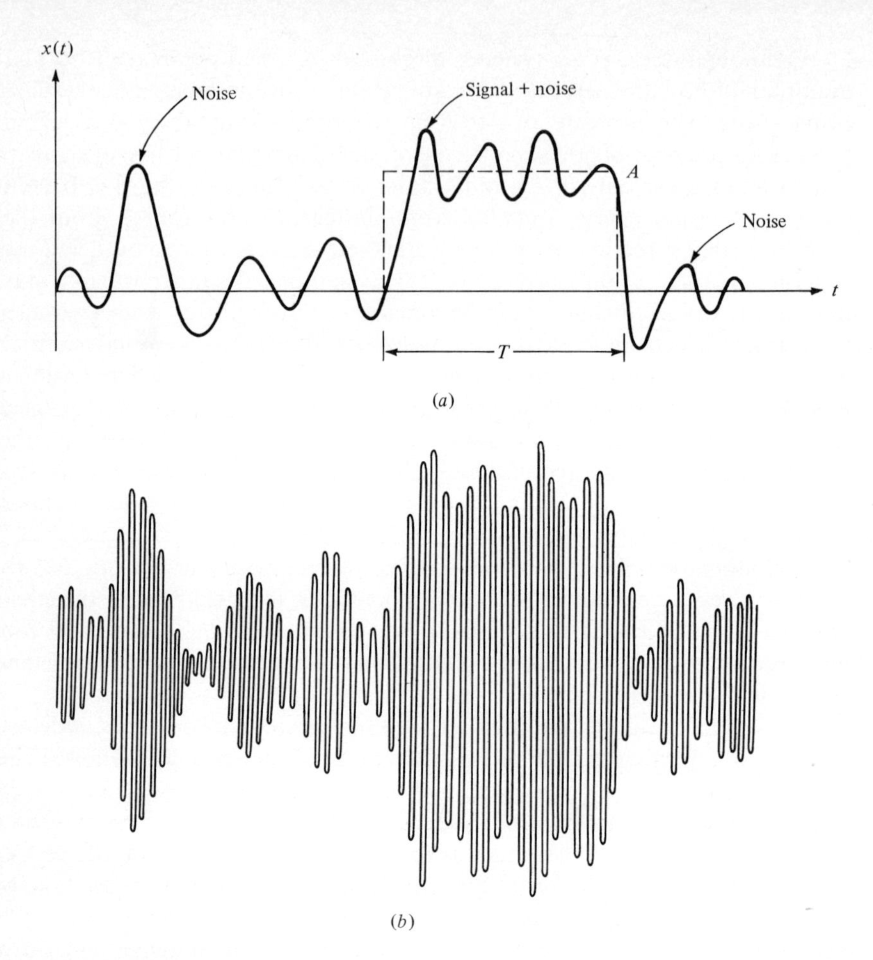

FIGURE 5.1
(*a*) Received analog waveshape (baseband). (*b*) Modulated version of curve in Fig. 5.1(*a*).

The signal when it does appear thus results in a change in the dc level or bias of $x(t)$. An example is shown in Fig. 5.1*a*. The signal is shown raising the level of $x(t)$ by A units during the T s it is present. (Note that although the original received signal and noise commonly appear as high-frequency terms, as in Fig. 5.1*b*, we are assuming here for simplicity that the system heterodynes them both down to so-call baseband frequencies. This is common practice in radar receivers and most radio receivers, whether used at home or for commercial purposes. The so-called carrier frequency may be raised or lowered at will by mixing with a locally generated frequency.)†

We are interested in high-speed computational techniques that carry out the

† M. Schwartz, "Information, Transmission, Modulation and Noise," 2d ed., chap. 4, McGraw-Hill, New York, 1970.

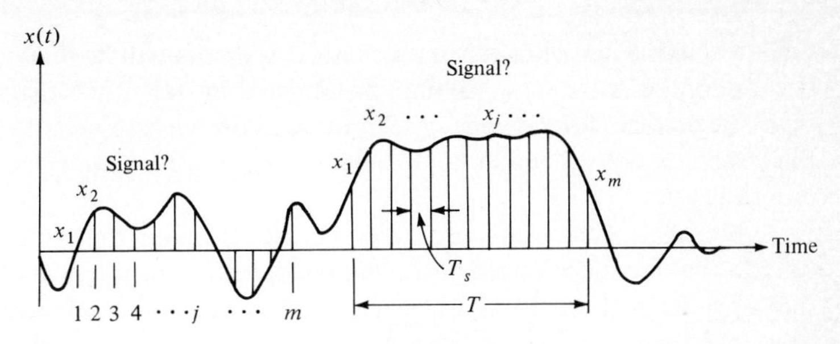

FIGURE 5.2
Taking samples to carry out signal detection.

decision-making process automatically. Assume for this purpose that we continually sample $x(t)$ m times in T-s intervals (see Fig. 5.2). We are then interested in processing the m samples $x_j, j = 1, 2, \ldots, m$, in any T-s interval to determine the presence or absence of $s(t)$. How does one determine the sample spacing T_s in this case?

The usual model for problems of this type (and one that is often quite realistic) is to assume that the noise before entering the system is white or at least that its bandwidth is much higher than that of the signal $s(t)$ with which we have to deal. We note also that there is some bandwidth B sufficient to pass the signal $s(t)$ essentially undistorted. [In the case of the constant-amplitude pulse $s(t)$ under consideration here it is of the order of several times $1/T$. The exact value of B is not critical. As we saw in Sec. 2.5, B need only be large enough to ensure that $2BT >> 1$. This will be discussed further later on.] In most detection systems $x(t)$ is filtered to bandwidth B *prior* to the sampling process to reduce the effect of the noise. [Since B is sufficient to pass the signal undistorted, this filtering does not change the signal component of $x(t)$. For white noise, however, the mean noise power is proportional to the bandwidth, and so restricting the noise bandwidth as much as possible can only help in the detection process. All systems have a certain amount of innate filtering anyway due to the inertia of the physical elements of which they are composed.] We shall assume that $n(t)$ is thus bandlimited white noise of bandwidth B.

An interesting theorem that can be invoked says that bandlimited noise samples taken $1/2B$ s apart are uncorrelated.† As is intuitively apparent (we shall demonstrate it later), the larger the number of uncorrelated samples of noise or signal-plus-noise we can collect, the better our chance of detecting a signal when it appears. (We have more data with which to smooth the noise.) We shall thus select as our sampling interval T_s exactly $1/2B$, with a corresponding maximum of $m = 2BT$ samples in any T-s interval.

How do we now go about processing the m samples x_j, $j = 1, 2, \ldots, m$, to determine the presence or absence of $s(t)$? Obviously there are many different ways of processing the m samples. We can add them, multiply them, check to see how many are positive or negative, whether a specified number lie above a certain level, etc. Here the designer must play a key role in choosing the appropriate processing

† Ibid., p. 597; Sec. 3.6, this book.

scheme. No matter what scheme he chooses, errors will be made: signals may be lost because of the noise, or noise may be detected mistakenly for signal. He must decide on a reasonable performance criterion based on his assessment of the significance of the errors possibly made and then design a processing scheme satisfying this criterion.

We shall give some examples in the material that follows of performance criteria that have been used in solving the detection problem and processing schemes that arise from them. First, however, we shall consider some simple ad hoc schemes that appear reasonable for the detection of signals in noise. We shall then compare the performance of these schemes with those developed more directly on the basis of satisfying specified performance criteria. Obviously, the ad hoc schemes will be sub-optimum with respect to these criteria; yet in some situations they may nevertheless be desirable: they may be simpler to implement, requiring less memory, fewer computations, less hardware, etc.

As the first example of an ad hoc approach, consider the simple addition of all m samples with division by m (motivated by the idea that the uncorrelated noise samples will average out):

$$y(m) = \frac{1}{m} \sum_{j=1}^{m} x_j \tag{5.2}$$

The random variable $y(m)$ is called the *sample mean*. This type of filtering to reduce disturbances was discussed in Chap. 3 and is essentially an extension of the averaging filter discussed in previous chapters. What are its properties? We assume, as is usually the case in practice, that the noise has zero mean. We have also stipulated that the noise samples be chosen to be uncorrelated with variance σ_n^2. Thus

$$E[n_j] = 0 \qquad E[n_i n_j] = \begin{cases} 0 & i \neq j \\ \sigma_n^2 & i = j \end{cases} \tag{5.3}$$

Then we have as the expected value (average or mean value) of y

$$E[y] = E[x_j] = \begin{cases} A & \text{if } s \text{ is present} \\ 0 & \text{if } s \text{ is absent} \end{cases} \tag{5.4}$$

The variance of $y(m)$ is similarly given by

$$\sigma_y^2(m) = E[y(m) - E(y)]^2 = \frac{1}{m^2} \sum_{i,j=1}^{m} \sum E[n_i n_j] = \frac{\sigma_n^2}{m} \tag{5.5}$$

(The details of the calculation are left to the reader; see Chap. 3.)

This result leads to the important observation that as the number of samples m increases, the variance of $y(m)$ decreases as $1/m$, approaching zero for large m. This is exactly the point made earlier in specifying that m be as large as possible. This means that for large m, $y(m)$ has a probability density closely concentrated about its mean. From the Tschebychev inequality (Chap. 3) the probability that $y(m)$ is close to its mean value is almost 1 for large m. The sample mean would thus appear to be a good decision quantity, since it is likely to be near A or 0, according as the signal is present or absent. In particular, we may decide that the signal is present if $y(m) > A/2$ and absent if $y(m) < A/2$. It is apparent that as m increases, the chance of errors occurring becomes smaller. We shall return to this particular detection processor after discussing performance criteria and the processing algorithms developed on the basis of them.

That this processing technique is appropriate is apparent: the presence of a fixed-amplitude signal shifts the bias or mean level of $x(t)$ by that amount. One then looks for a procedure to detect this change. If the signal amplitude A is very small compared with the noise (a common difficulty in detection problems), the signal contribution to any individual observation sample may be very small; we indicate this by saying that $A \ll \sigma_n$. The random variation due to noise masks the average change in the received-signal level due to the appearance of A. By taking enough samples and averaging them, however, we effectively increase the change due to signal relative to the noise variations.

Another processing technique that appears intuitively to have promise is that of checking the level of each individual sample x_j. One would expect that if a signal were present, more samples would exceed a specified positive level than if the signal were absent. For example, we could count the number of x_j's that exceed $A/2$ (or some other predetermined positive level). One would expect this number to be a larger portion of the m samples if A is present, and we might well decide that a signal is present if more than a fixed number of samples exceed the test level. This scheme is particularly useful since its implementation is readily carried out with counts based on simple logic circuitry. It is thus attractive in terms of implementation. We shall see later that it also performs fairly well under appropriate performance criteria.

5.2 OPTIMUM DETECTION ALGORITHMS: MINIMUM PROBABILITY OF ERROR

We have mentioned several times in passing the desirability of setting up appropriate performance criteria for the detection problem. This serves two purposes: (1) It forces us to focus on the properties we would like our detector to have; i.e., it forces us to set up quantitative measures on which to base a design, and (2) it often gives rise to specific detector structures that can be implemented or with which other suboptimum schemes can be compared. We now consider in detail this question of detection processors satisfying specified performance criteria.

The radar problem serves as a useful example of what we have in mind. It is apparent that no matter what we do, we shall sometimes mistake noise for signal when the signal is absent and sometimes miss the signal when it does appear. An ideal radar signal detector would be one which maximizes the probability of detecting the signal when it appears and minimizes the probability of mistaking noise for signal when it is absent. Unfortunately, as we shall see later in discussing the radar problem in detail, both probabilities cannot be optimized simultaneously. The best we can do is to keep the noise probability (or *false-alarm probability*, as it is often

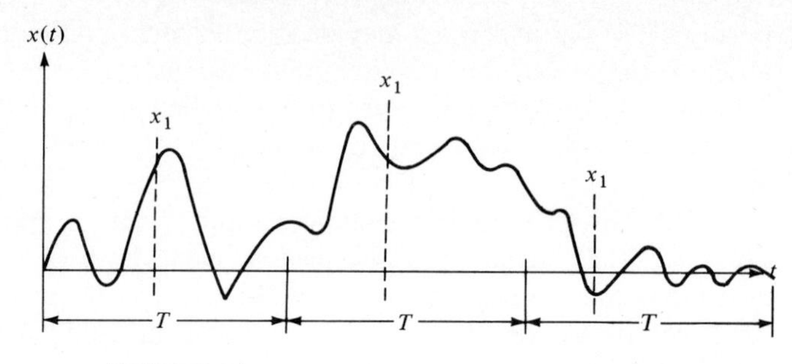

FIGURE 5.3
Independent decision in each time interval—single-sample case.

called) at some tolerable level and maximize the corresponding signal probability. This is in fact the performance criterion usually adopted for radar systems.

The radar case is an example of a detection problem in which the signal appears relatively infrequently, with no predetermined (a priori) statistics. The noise is always present, however. Instead of beginning with this example, it is simpler and equally instructive to consider first the detection problem in cases where signals appear with some regularity—say in a binary (digital) communication system in which the presence of a signal denotes the transmission of a 1 while the absence of a signal denotes the transmission of a 0. Here the presence of the signal is a more regular occurrence than in radar, and we shall in fact assume prior knowledge of the probability of occurrence of a signal. The discussion of this type of detection problem is then readily extended to radar problems.

In this communication type of detection problem the performance criterion usually chosen as the basis for signal-processing design is that of minimizing the probability of error of the system. This guarantees the smallest number of mistakes over the long run in system operation and is obviously the kind of performance one would normally strive for in a digital-communications system.

How do we now develop the appropriate detection algorithms that correspond to this minimum-error-probability criterion? It is simpler to begin with a one-sample case and then extend it to the use of m samples of the received signal $x(t)$ to determine the presence or absence of a signal. Assume then that the communication pulses are T s long. The received signal is then broken into strips T s long, and one sample x_1 of the received signal $x(t)$ is taken in each T-s-long time interval. Independent decisions must be made in *each* time interval (Fig. 5.3). From the sample x_1 in each time interval, we wish to determine with as small a probability of error as possible whether the signal is present (the 1 case) or absent (the 0 case). Two types of error may occur: the signal may be missed when it is present, or the noise may be mistaken for signal when the signal is not present (often called a false alarm). The overall probability of error is due to both. Call the probability of the first type of error P_{e1} and the second P_{e0}. These are both examples of *conditional* probabilities since they are conditioned, respectively, on the signal's being present and absent. Since these errors are due to mutually exclusive events (the presence of a signal obviously precludes its absence),

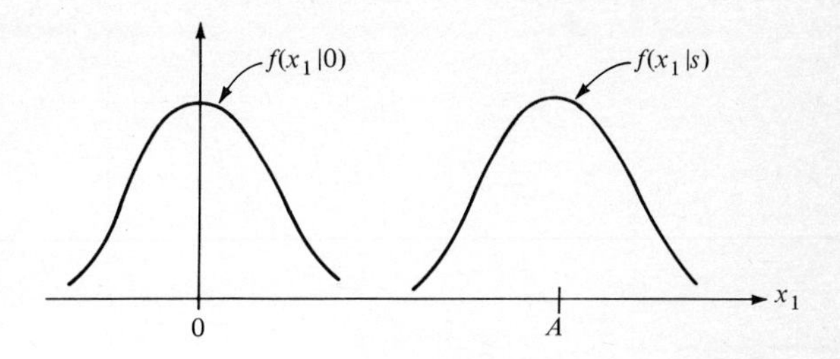

FIGURE 5.4
Density functions, detection problem.

we can find the overall probability of error P_e by adding the two error probabilities after unconditioning them by multiplying by the appropriate probabilities of occurrence of the two events:

$$P_e = P_1 P_{e1} + (1 - P_1)P_{e0} \tag{5.6}$$

Here P_1 is the (assumed known) probability of occurrence of signals.

Obviously both P_{e1} and P_{e0} depend on the respective distributions of x_1 in the two cases of signal present or signal absent, so that the minimization of P_e depends on these statistics as well. Consider a typical case, as shown in Fig. 5.4. With signal absent the statistics of x_1 are identical with those of the noise ($x_1 = n$), so that the probability density function of x_1 in this case is simply

$$f(x_1 \mid 0) = f_n(x_1) \tag{5.7}$$

As an example, if the noise is gaussian (the most common model used),

$$f_n(n) = \frac{e^{-n^2/2\sigma_n^2}}{\sqrt{2\pi\sigma_n^2}} \tag{5.8}$$

The statistics of x_1 are similarly given by

$$f(x_1 \mid 0) = \frac{e^{-x_1^2/2\sigma_n^2}}{\sqrt{2\pi\sigma_n^2}} \tag{5.9}$$

Similarly, with signal present $x_1 = A + n$. The effect of the signal is thus simply to translate the probability density by the fixed value A. This is shown in Fig. 5.4. We thus have

$$f(x_1 \mid s) = f_n(x_1 - A) \tag{5.10}$$

which becomes

$$f(x_1 \mid s) = \frac{e^{-(x_1 - A)^2/2\sigma_n^2}}{\sqrt{2\pi\sigma_n^2}} \tag{5.11}$$

if the noise is gaussian.

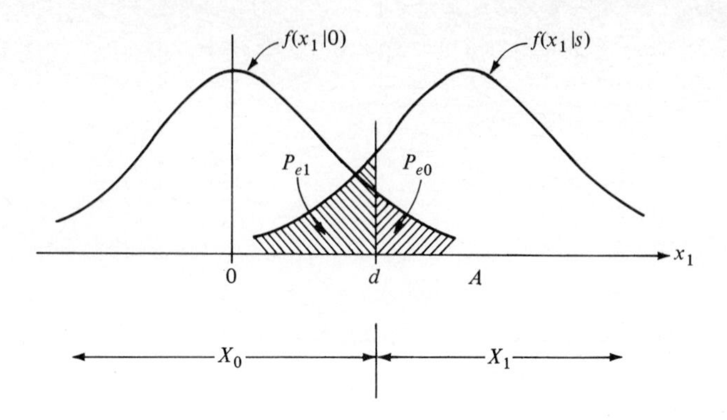

FIGURE 5.5
Decision regions.

How does one now find the two conditional error probabilities and from them the overall error probability? The processor, using one sample x_1, must be a decision rule which specifies two separate ranges of the possible values of x_1, the range of values X_1 corresponding to the signal-present hypothesis and the range of values X_0 corresponding to the signal-absent hypothesis. (For example, X_1 might be the interval $a \leq x_1 \leq b$, and X_0 might be the combined intervals $x_1 < a$ and $x_1 > b$.) The probability of error P_e depends on the particular choice of these two joint regions, and it is the object of our analysis to find those ranges which minimize P_e. The analysis will thus provide processing algorithms guaranteed to minimize P_e.

As an example, assume that the two regions are simply chosen as shown in Fig. 5.5: a decision level d is picked; all values of $x_1 > d$ correspond to signal present (X_1), and all values of $x_1 < d$ correspond to signal absent (X_0). More generally, the two regions X_0 and X_1 incorporate all points on the line corresponding to possible values of x_1 and may cover several decision regions, as shown in Fig. 5.6. By writing expressions for P_{e0} and P_{e1} in terms of the decision regions and adjusting the boundaries until P_e is minimized, we find the solution providing minimum P_e.

It is apparent from the definitions thus far that P_{e0} corresponds to the probability that x_1 will fall into region X_1 given that only noise is present while P_{e1} corresponds to the probability that x_1 will fall into region X_0 given that a signal is present.

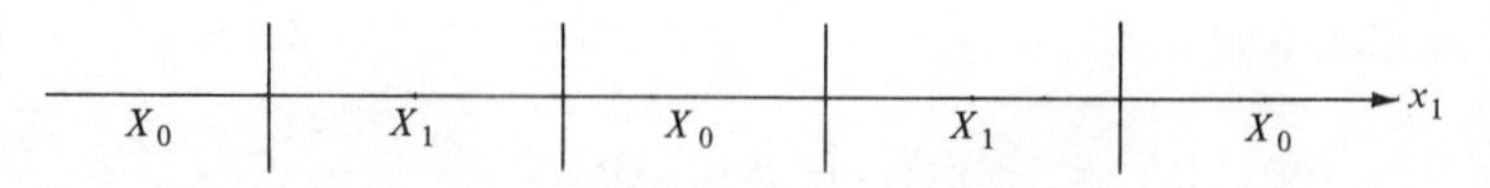

FIGURE 5.6
More general decision regions.

These probabilities in turn are found by integrating the appropriate density functions over the respective regions. An example is shown in Fig. 5.5. Specifically, we have

$$P_{e0} = \int_{X_1} f(x_1 \mid 0)\, dx_1 \tag{5.12}$$

and

$$P_{e1} = \int_{X_0} f(x_1 \mid s)\, dx_1 \tag{5.13}$$

The overall probability of error is then given, from Eq. (5.6), by

$$P_e = P_1 \int_{X_0} f(x_1 \mid s)\, dx_1 + (1 - P_1) \int_{X_1} f(x_1 \mid 0)\, dx_1 \tag{5.14}$$

How does one now find the appropriate X_0 (or X_1) to minimize P_e? In the special case of Fig. 5.5, with the simple boundary $x_1 = d$ to adjust, P_e is a function of d, and one simply differentiates with respect to d to find the minimum, if one exists. More generally, we proceed by the following strategy. Since X_0 and X_1 cover the entire space of x_1, we first eliminate X_0, say, rewriting Eq. (5.14) solely in terms of X_1. Using

$$\int_{X_0 + X_1} f(x_1 \mid s)\, dx_1 = 1$$

we have

$$P_e = P_1 + \int_{X_1} [(1 - P_1) f(x_1 \mid 0) - P_1 f(x_1 \mid s)]\, dx_1 \tag{5.15}$$

What values of x_1 shall we select to include in X_1; that is, which set will make P_e as small as possible? Since P_1, $1 - P_1$, and both density functions are all of necessity positive, a little thought will show that one can do no better than by picking as region X_1 those values of x_1 which satisfy the inequality

$$P_1 f(x_1 \mid s) > (1 - P_1) f(x_1 \mid 0) \tag{5.16}$$

For in this case the integral over X_1 is as negative as one can make it, and hence P_e is as small as possible. The values of x_1 satisfying inequality (5.16) thus correspond to region X_1, while those which reverse the inequality correspond to X_0. More commonly we group expressions involving the data sample x_1 on the left-hand side and write as the solution for region X_1 those values of x_1 for which

$$\ell(x_1) \equiv \frac{f(x_1 \mid s)}{f(x_1 \mid 0)} > \frac{1 - P_1}{P_1} \tag{5.17}$$

The left-hand expression $\ell(x_1)$ is referred to as the *likelihood ratio*. Various other performance criteria aside from minimum probability of error also give rise to an inequality involving the likelihood ratio, as we shall see further by example.

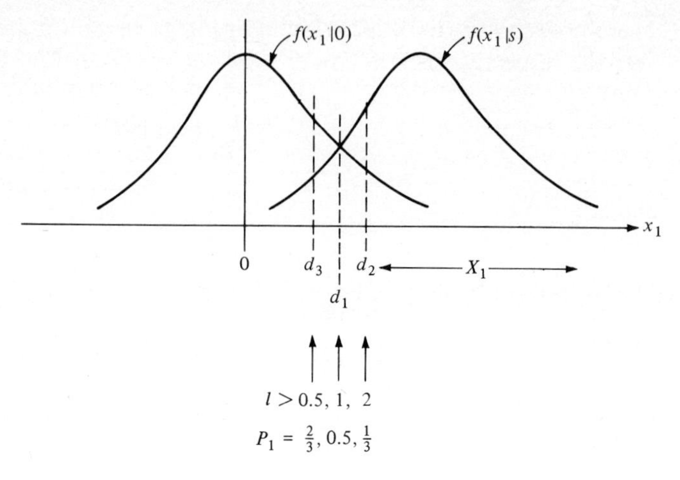

FIGURE 5.7
Example of decision regions for variable P_1.

The interpretation of Eq. (5.17) says simply that the region X_1 corresponds to all values of x_1 for which the ratio of the two conditional density functions is greater than a specified number, $(1 - P_1)/P_1$. As an example say $P_1 = 0.5$. This corresponds to the case where a signal would be expected to be present half the time, on the average. The region X_1 then corresponds to all x_1 for which $\ell(x_1) > 1$ or $f(x_1 \mid s) > f(x_1 \mid 0)$. In the example of Fig. 5.7, X_1 corresponds to all $x_1 > d_1$.

If P_1 increases, the X_1 region increases and the X_0 decreases corresponding to an increased likelihood that signal will be detected. Similarly, as P_1 decreases, the X_1 region is reduced in size. These cases are indicated by example in Fig. 5.7. If the noise in the problem were nongaussian and had a bimodal distribution, as shown in Fig. 5.8, the same kind of analysis would apply but multiple X_1 regions would appear.

In actually implementing the processing algorithm we must solve Eq. (5.17) for the specific values of x_1 corresponding to the inequality. As shown by Figs. 5.7 and

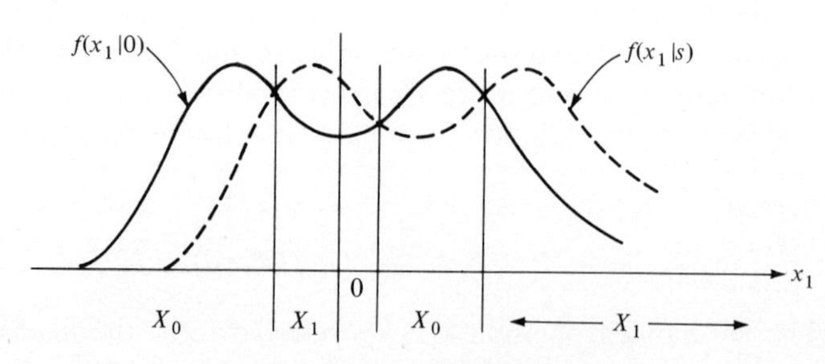

FIGURE 5.8
Another example: $P_1 = 0.5$, $l > 1$.

5.8, we shall then have threshold or decision levels for x_1 between which the system will declare a signal present and outside of which the system decides signal is absent. This processor is then optimum in the sense of minimum probability of error for the signal amplitude, occurrence probability, and noise statistics assumed. Instead of considering specific examples at this point, we move on to the more general case of m samples. (The reader will find other single-sample examples among the problems at the end of this chapter.)

How do we now extend these ideas to multiple samples? Recall the motivating argument above, that a weak radar return should be more effectively detected if we repeat the transmission several times and process the collection of returns. We indicated previously that by picking samples spaced $1/2B$ s apart the noise samples were uncorrelated. In the case of *gaussian* noise this also guarantees *independent* samples. Since many of the applications in real-life situations correspond to additive gaussian noise, we shall henceforth assume independent noise samples. Where the noise is not gaussian (as in some examples to be discussed later, as well as in the problems) we shall assume statistically *independent* samples as well. In the nongaussian case one can generally ensure this by picking the samples far enough apart in time. In these cases the number of samples m is then just whatever number fits into the T-s time interval rather than the $2BT$ factor noted earlier.

With independent noise samples, the received samples x_j are then independent as well. This independence of observations is true in either case: noise or signal-plus-noise. We then look for detection schemes, processing the collective group of m independent samples, that will minimize the overall probability of error. It is instructive to visualize each sample as occupying an orthogonal axis in an m-dimensional space. This space is an extension of the one-dimensional case just considered, and the procedure of selecting an optimum processor generalizes to the choice of the appropriate m-dimensional subspace X_1 corresponding to a decision that the signal is present.

As an example, consider two samples x_1 and x_2. The composite pair (x_1, x_2) must occupy some point in a two-dimensional space, as shown in Fig. 5.9. If this point falls within region X_1, as shown, a signal is declared to be present.

Once we recognize that we are dealing with m random variables $x_1, \ldots, x_j, \ldots, x_m$ and with geometric properties of them, we can invoke the probability theory appropriate to functions of m random variables.† In particular, it is readily shown that one may define an m-dimensional probability density function $f(x_1, x_2, \ldots, x_m)$ integrable over the entire m-dimensional space. In our present case of statistically independent samples, this function is just the *product* $\prod_{j=1}^{m} f(x_j)$ of the individual sample density functions. (Joint probabilities become products of individual probabilities for independent events.) Once an m-dimensional region X_1 is chosen, it is apparent that an error will occur with the signal absent if the composite-sample group falls into region X_1. The probability of this happening is given by the appropriate integration over X_1:

† A. Papoulis, "Probability, Random Variables, and Stochastic Processes," chap. 8, McGraw-Hill, New York, 1965.

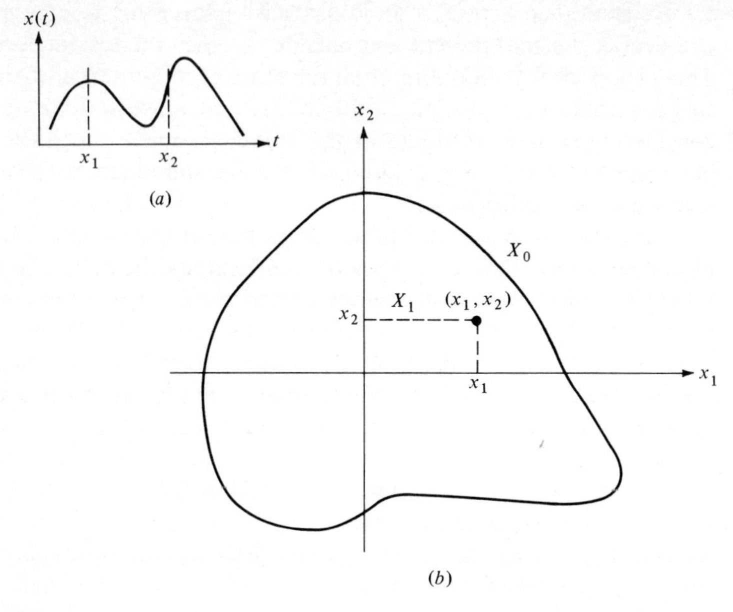

FIGURE 5.9
Two-dimensional detection regions. (*a*) Sampling the data; (*b*) processing the samples.

$$P_{e0} = \int_{X_1} \cdots \int \prod_{j=1}^{m} [f(x_j \mid 0)] \, dx_1 \cdots dx_m \tag{5.18}$$

Note that this is just the m-dimensional extension of Eq. (5.12). Proceeding exactly as in the one-dimensional case, we can show that the detection procedure appropriate to minimizing the overall probability of error consists of deciding the signal is present if

$$\ell(x_1, \ldots, x_m) \equiv \frac{\prod_{j=1}^{m} f(x_j \mid s)}{\prod_{j=1}^{m} f(x_j \mid 0)} > \frac{1 - P_1}{P_1} \tag{5.19}$$

(The actual derivation is left to the reader as an exercise.) A likelihood ratio appears again, this time as the ratio of a joint density of the m samples $x_1, \ldots, x_m$, given s present, to the similar joint density with s absent.

More commonly (because it frequently results in simplification of the processing algorithm) one takes the natural logarithm of both sides of Eq. (5.19), resulting in the detection procedure satisfying

$$\ln \ell \equiv \sum_{j=1}^{m} \ln \frac{f(x_j \mid s)}{f(x_j \mid 0)} > \ln \frac{1 - P_1}{P_1} \tag{5.20}$$

[One can always transform (5.19) into (5.20) since the probabilities make both sides of (5.19) nonnegative and because the logarithm is a monotonically increasing function.] The logarithm-likelihood-ratio test has a rather straightforward interpretation. In general $\ln [f(x_j \mid s)/f(x_j \mid 0)]$ is some known function of the sample value x_j. The detection procedure thus consists of operating on each sample, summing the resultant numbers, and checking to see whether this sum is greater than a specified threshold. Any other detection procedure will be suboptimum in the sense of resulting in a higher probability of error, under the conditions assumed, and may be compared to the optimum procedure of (5.20) on the basis of deterioration in probability of error vs. possible simplification of implementation.

As an example, assume the noise is gaussian. Using the expressions of Eqs. (5.9) and (5.11) to find the ratio of density functions required by Eq. (5.20) and taking the natural logarithm, one has a decision rule which decides that the signal is present when the values of x_j, $j = 1, \ldots, m$, satisfy the inequality

$$\sum_{j=1}^{m} [x_j^2 - (x_j - A)^2] > 2\sigma_n^2 \ln \frac{1 - P_1}{P_1} \tag{5.21}$$

Expanding the term inside the parentheses, canceling, collecting, and rewriting terms, we finally have

$$\sum_{j=1}^{m} x_j > \underbrace{\frac{mA}{2} + \frac{\sigma_n^2}{A} \ln \frac{1 - P_1}{P_1}}_{d} \tag{5.22}$$

The optimum processor in this case thus consists of a summer which simply adds the samples and checks to see whether the sum exceeds a specified level d.

Even more interestingly, if we divide through by the constant m, we have as an alternate optimum processor just the sample mean discussed earlier [Eq. (5.2)]. In this case we decide a signal is present if

$$y = \frac{1}{m} \sum_{j=1}^{m} x_j > \frac{A}{2} + \left(\frac{\sigma_n^2}{mA} \ln \frac{1 - P_1}{P_1} \right) \tag{5.23}$$

Note that when $P_1 = 0.5$, the processor does in fact check to see if $y > A/2$, as suggested earlier. There is some modification to this decision level due to the signal-occurrence statistics, however. If P_1 differs from 0.5, the decision level must be shifted up or down according as the signal-occurrence probability is less or more than 0.5. The relative amount by which one shifts away from $A/2$ depends on the ratio σ_n^2/A^2, the noise variance over signal amplitude squared. It is also reduced by the number of samples m. Even though the signal term A^2 may be small compared with σ_n^2, we shall find it necessary to have $mA^2 > \sigma_n^2$ to ensure reasonably small probabilities of error. (mA^2 plays the role of an effective or equivalent single-sample signal term.) The second term involving P_1 is thus normally small, and we shall neglect it in calculations.

For gaussian noise with fixed signal amplitude A the sample mean is thus the appropriate number to calculate in detecting the presence of the signal. In other

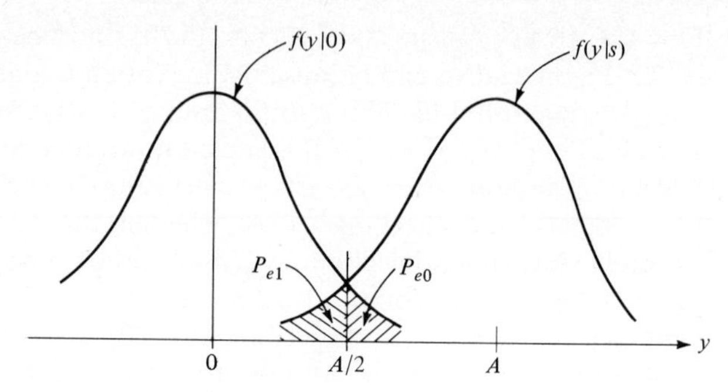

FIGURE 5.10
Performance of optimum detector, gaussian noise.

examples considered later, however, we shall show that other choices for the statistics as well as varying signal amplitude result in different optimum processors. The sample mean is thus suboptimum in those cases.

How well does the sample-mean processor perform in this case? This information is, of course, important in assessing how well the optimum signal-detection scheme performs, in determining how many samples are needed to obtain a reasonable probability of error, and in comparing various other suboptimum schemes to the optimum one. To do this, we actually carry out the probability-of-error calculations. In general, one would have to find P_{e0} and P_{e1} by integrating over the appropriate m-dimensional subregions corresponding to X_1 and X_0, respectively [see Eq. (5.18), for example]. Happily in this case, as in many others, there is no need to do this; for, by Eq. (5.23), we need simply calculate the probability that the random variable y drops below $A/2$ if a signal is present or appears above it if noise alone is present. There is a well-known theorem of probability that the sum of gaussian random variables is still gaussian. Under our assumption of gaussian noise, then, y in Eq. (5.23) is gaussian. In particular, as shown earlier in Eqs. (5.4) and (5.5), its expected value is either A or 0, according as signal is present or absent, and its variance is $\sigma_y^2 = \sigma_n^2/m$. We thus have for signal absent

$$f(y \mid 0) = \frac{e^{-y^2/2\sigma_y^2}}{\sqrt{2\pi\sigma_y^2}} \tag{5.24}$$

and for signal present

$$f(y \mid s) = \frac{e^{-(y-A)^2/2\sigma_y^2}}{\sqrt{2\pi\sigma_y^2}} \tag{5.25}$$

Note that these densities are similar to the ones of Eqs. (5.9) and (5.11) except that there y was equal to x_1 and here y is the average of m samples, with an appropriately modified variance σ_y^2. These two functions are shown sketched in Fig. 5.10. It is apparent from the symmetry of that figure that if we assume $P_1 = 0.5$, so that the

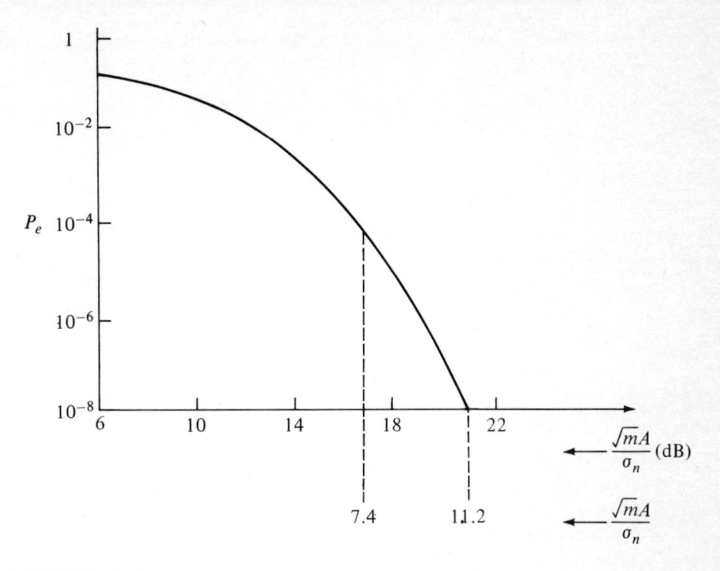

FIGURE 5.11
Performance curve, optimum detector, gaussian noise, $P_1 = 0.5$, m samples.

optimum threshold is $A/2$, $P_{e0} = P_{e1}$. We thus also have $P_e = P_1 P_{e1} + (1 - P_1)P_{e0} = P_{e0}$. In particular

$$P_e = \int_{A/2}^{\infty} f(y|0)\,dy = \int_{A/2}^{\infty} \frac{e^{-y^2/2\sigma_y^2}}{\sqrt{2\pi\sigma_y^2}} = \tfrac{1}{2}\left(1 - \operatorname{erf}\frac{\sqrt{m}\,A}{2\sqrt{2}\,\sigma_n}\right) \tag{5.26}$$

with the error function of x defined as

$$\operatorname{erf} x \equiv \frac{2}{\sqrt{\pi}} \int_0^x e^{-y^2}\,dy \tag{5.27}$$

A table of the error function appears in Chap. 3. Using it, we get the performance curve of Fig. 5.11. Note that if we desire a probability of error P_e less than 10^{-4} we require $\sqrt{m}\,A/\sigma_n > 7.4$. For $\sqrt{m}\,A/\sigma_n = 10$, $P_e = 5 \times 10^{-7}$. Small changes in A/σ_n, the so-called *signal-to-noise ratio*, thus result in significant changes in the probability of error. This is due specifically to the exponentially decreasing tails in the gaussian distribution. It shows why methods for increasing the signal-to-noise ratio are considered so significant in signal-processing systems; one tries to reduce the noise and increase the signal power received as much as possible. If A/σ_n is as large as possible and yet too small to obtain the desired probability of error, one tries to improve the equivalent signal-to-noise ratio by increasing the number of samples m. As an example, if $A/\sigma_n = 1$ (0 dB), $m = 100$ samples are needed to have $\sqrt{m}\,A/\sigma_n = 10$. If $A/\sigma_n = \frac{1}{3}$ (-10 dB), $m \sim 10^3$ samples are needed. Since the samples are spaced a fixed time interval apart to ensure statistical independence ($1/2B$ s for gaussian noise), the time for detection increases with m.

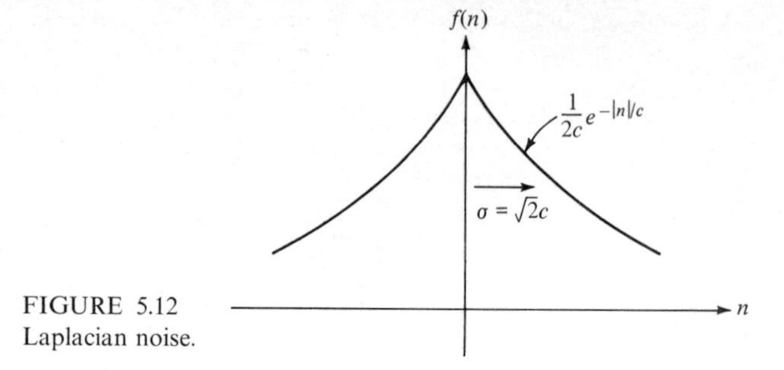

FIGURE 5.12
Laplacian noise.

This example has been discussed in such detail because of the importance and common occurrence of gaussian noise. To show how the detection procedure varies with the noise statistics, consider additive noise whose statistics are laplacian:

$$f_n(n) = \frac{1}{2c} e^{-|n|/c} \tag{5.28}$$

($\sigma_n = \sqrt{2}\,c$ in this case). This is sketched in Fig. 5.12. This density function is sometimes used to model additive impulse or burst-type noise. The simple exponential behavior, rather than the quadratic exponential of the gaussian density function, means that higher amplitudes have correspondingly higher probabilities of appearing, a characteristic of this type of impulse noise. What is the optimum processor, determined from Eq. (5.20), in this case? The two conditional density functions required, $f(x_j \mid s)$ assuming signal present and $f(x_j \mid 0)$ assuming only noise present, are sketched in Fig. 5.13.

In this example, it is convenient to consider each term in the summation of Eq. (5.20) separately. It is apparent from Fig. 5.13 that each term in this sum will have three different mathematical descriptions corresponding to the three ranges of x_j values:

$$x_j < 0 \qquad 0 < x_j < A \qquad x_j > A$$

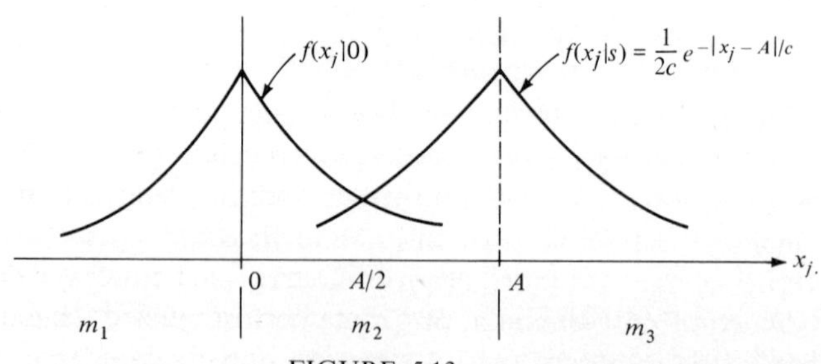

FIGURE 5.13
Conditional density functions, laplacian noise.

Consider these regions one at a time:

$$x_j < 0: \qquad \ln \frac{f(x_j \mid s)}{f(x_j \mid 0)} = \frac{(x_j - A) - x_j}{c} = -\frac{A}{c} \tag{5.29}$$

$$0 < x_j < A: \qquad \ln \frac{f(x_j \mid s)}{f(x_j \mid 0)} = \frac{(x_j - A) + x_j}{c} = \frac{2x_j - A}{c} \tag{5.30}$$

$$x_j > A: \qquad \ln \frac{f(x_j \mid s)}{f(x_j \mid 0)} = \frac{x_j - (x_j - A)}{c} = \frac{A}{c} \tag{5.31}$$

Note that in the first and third cases the dependence on the actual value of x_j cancels out. For samples falling in these two regions, the processor of Eq. (5.20), acting on the samples one at a time as each is received, stores a *count* proportional to A, with sign ($+$ or $-$) depending on the region.

Let m_1 represent the number (of the m total) of samples with negative values, m_2 the number in the range $0 < x_j < A$, and m_3 the number with $x_j > A$. It is apparent that the signal is then declared present if and only if

$$[m_3 - (m_1 + m_2)]\frac{A}{2} + \sum_{0 < x_j < A} x_j > \frac{c}{2} \ln \frac{1 - P_1}{P_1} \tag{5.32}$$

The summation will contain m_2 terms, each satisfying the inequality written below the summation sign. Alternately, since $m = m_1 + m_2 + m_3$, the terms on the left side, which depend on the particular observations, can be simplified:

$$m_3 A + \sum_{0 < x_j < A} x_j > \frac{mA}{2} + \frac{c}{2} \ln \frac{1 - P_1}{P_1} \tag{5.33}$$

A count of A is thus stored every time a sample exceeds A, the counter output is added to the sum of stored sample values in the range 0 to A, and the resultant sum compared to a specified threshold level. The processing scheme incorporates aspects of the gaussian-noise result (pure summing) and of the pure counting mentioned earlier as an ad hoc processing scheme. The pure counter will be discussed later and its performance compared to that of the pure summer for gaussian noise.

5.3 OTHER PERFORMANCE CRITERIA; NEYMAN-PEARSON CRITERION FOR RADAR

The criterion of minimum probability of error, discussed at length in Sec. 5.2, is particularly significant in the digital-communications field, where binary digits represent information to be transmitted and the object is to minimize the number of errors made. We pointed out earlier that in other signal-detection applications different criteria may be more appropriate. In particular we noted that in the radar-type detection problem an appropriate performance criterion (often used in the design of radar-signal processors) is one that maximizes detection probability while keeping the noise, or *false-alarm*, probability fixed at some tolerable value. This criterion will be discussed in detail in this section. In particular we shall use the results obtained to assess the performance of the air-traffic-control radars. But first we consider briefly another criterion which is essentially an extension of the minimum-probability-of-error criterion. The likelihood ratio again arises as the optimum test for the detection of a signal in both cases considered in this section. The choice of criterion simply changes the threshold level to which the likelihood ratio is compared. The optimum processing algorithms, two examples of which were discussed at length in the previous section, remain the same for these other criteria.

The criterion we discuss first before proceeding to the radar example involves average cost; most commonly this is dollars or some other cost function reducible to dollars. We assume that errors, when they occur, involve a specified cost and that correct decisions, when made, involve a negative cost, or gain. (In some applications, this negative-cost term is not included; this simply corresponds to setting the term equal to zero in the discussion below.) Specifically, assume that the failure to detect a signal when it is present costs L_{10} dollars (or some other units), the mistaken detection of noise for signal costs L_{01} dollars, the correct detection of a signal costs L_{11} dollars (generally negative), and the correct decision that signal is absent costs L_{00} dollars (again negative). We then seek a scheme that minimizes the total average cost.

It turns out that this new criterion is a simple extension of the minimum P_e criterion. For the average cost involves exactly the same error expressions used in formulating the error-probability equation [Eq. (5.6) and Eq. (5.14) extended to m samples using Eq. (5.18)]. Specifically, the average cost is given by

$$\bar{L} = L_{01}(1 - P_1)P_{e0} + L_{00}(1 - P_1)(1 - P_{e0}) + L_{10}P_1P_{e1} + L_{11}P_1(1 - P_{e1}) \tag{5.34}$$

[This is derived by defining a discrete random variable $\mathbf{L}$ with four possible values L_{11}, L_{10}, L_{01}, and L_{00}. Then the average cost is

$$\bar{L} = E(\mathbf{L}) = \sum_{i,\,j=0,\,1} L_{ij}P[\mathbf{L} = L_{ij}]$$

giving the expression (5.34).] Here P_{e0} and P_{e1} again represent, respectively, the probability of making an error given that signal is absent and the probability of making an error given that signal is present. Rewriting Eq. (5.34), we have

$$\bar{L} = L_{00}(1 - P_1) + L_{11}P_1 + P_{e0}(1 - P_1)(L_{01} - L_{00}) + P_{e1}P_1(L_{10} - L_{11}) \tag{5.35}$$

Since the cost parameters are all assumed known constants, they simply serve to modify the respective probabilities they multiply. By comparing Eq. (5.35) with Eq. (5.6) it is apparent that except for the first two constant terms, i.e., those independent of the decision rule, they are of the same form. The minimum solutions are identical except that P_1 in Eq. (5.6) is multiplied by $L_{10} - L_{11}$ and $1 - P_1$ multiplied by $L_{01} - L_{00}$. If m independent samples $x_1, x_2, \ldots, x_j, \ldots, x_m$ are processed in this case, it is apparent by comparison with Eq. (5.19) that for minimum average cost a signal is to be assumed present if

$$\ell(x_1, \ldots, x_m) > \frac{(1 - P_1)(L_{01} - L_{00})}{P_1(L_{10} - L_{11})} \tag{5.36}$$

The likelihood ratio ℓ is again the same ratio of products of conditional density functions defined in Eq. (5.19). The minimum-cost detection criterion thus results in the same processing algorithm as that for minimum error probability but with cost parameters modifying the threshold level.

We now turn to the radar-type detection problem and a criterion particularly appropriate for determining processing algorithms in these cases. As noted earlier, this type of problem is encountered when the signal to be detected may appear infrequently, with no predetermined or a priori probability of occurrence. This a priori probability is identically the occurrence probability P_1 defined in the previous section. There are many other detection problems, aside from the radar case, in which it is not possible or makes no sense to define an a priori signal-occurrence probability P_1. An astronomer searching a portion of the sky with a telescope for a particular type of radiating object provides one example. He may have no idea of the frequency of occurrence of the objects. The output of the telescope may perhaps provide a randomly fluctuating wave, as in Fig. 5.1. How does he distinguish the presence of one of the desired objects from the ever-present noise background? A physicist carrying out spectroscopic measurements and searching for the possible presence of particular spectral lines, an engineer processing light scattered by a region of space to determine the presence or absence of certain pollutants—these are also examples where one may have no idea of the frequency of occurrence of the signals to be detected. Sonar and other detection systems using underwater acoustic signals provide similar examples. One may wish to detect seismic signals denoting the occurrence of an earthquake. In all these examples and many similar ones there may be no knowledge of the a priori signal probability P_1. Therefore one cannot define an error probability to be minimized. Alternately the signal may be present relatively rarely (the usual situation in radar and sonar), so that even if P_1 were known, it would be extremely small. The component of probability of error in Eq. (5.6) due to signal being mistaken for noise ($P_1 P_{e1}$) would thus be negligible even if P_{e1} were large. Using the criterion of minimum probability of error, we would thus essentially be minimizing the chance of mistaking noise for signal, with no control over the chance of missing the signal when it appears.

Various criteria have been suggested to handle situations like these. The one already noted earlier, called the *Neyman-Pearson criterion* or *test*, after the two statisticians who first explored its properties, minimizes the probability of signal loss (our parameter P_{e1}) with the probability of mistaking noise for signal, P_{e0}, held fixed at some tolerable level. In radar terminology one would like to maximize the probability of detection $P_d \equiv 1 - P_{e1}$ with a false-alarm probability $P_n \equiv P_{e0}$ specified.

If we again invoke the geometrical picture of an m-dimensional space involving the m independent samples $x_1, x_2, \ldots, x_m$, we have, as in Eq. (5.18),

$$1 - P_{e1} = P_d = \int_{X_1} \cdots \int \prod_{j=1}^{m} [f(x_j|s)] \, dx_1 \cdots dx_m \tag{5.37}$$

and

$$P_{e0} = P_n = \int_{X_1} \cdots \int \prod_{j=1}^{m} [f(x_j|0)] \, dx_1 \cdots dx_m \tag{5.38}$$

It is apparent that we cannot simultaneously adjust the signal region X_1 to maximize

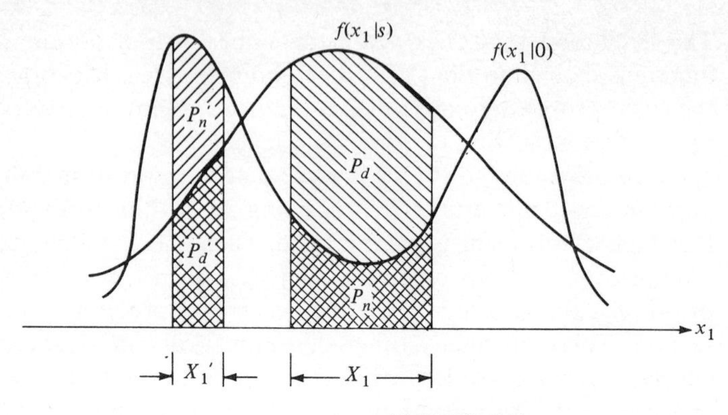

FIGURE 5.14
Neyman-Pearson test, one sample.

the detection probability P_d and minimize the false-alarm probability P_n. The Neyman-Pearson test thus holds P_n fixed at some tolerable level and then searches for the region X_1 that maximizes P_d.

Consider as an example, the one-dimensional case (one sample x_1) of Fig. 5.14. Two regions X_1 and X'_1 are shown, both of which have the same false-alarm probability P_n. [This is the area under the $f(x_1|0)$ curve in each case.] It is apparent that P_d in region X_1 [the area under the $f(x_1 | s)$ curve] is greater than P'_d in region X'_1. It is not too hard to show that the optimum choice of X_1, in the Neyman-Pearson sense in this one-dimensional case, corresponds to all values of x_1 such that $f(x_1 | s) > kf(x_1 | 0)$, with k a constant depending on the choice of P_n.† For if a small strip of X_1 satisfying this relation is replaced by a strip elsewhere violating this relation, with the widths adjusted to keep P_n constant, the area under the $f(x_1|s)$ curve, or P_d, will necessarily decrease. This is apparent from Fig. 5.14. (Notice that this new criterion again gives rise to a likelihood-ratio test.) In the general m-dimensional case, with m independent samples $x_1, \ldots, x_m$ considered, a necessary and sufficient condition for a most powerful test (maximum P_d) can be shown to be just the extension of the one-dimensional case. Thus we decide that the signal is present if

$$\ell = \frac{\prod_{j=1}^{m} f(x_j | s)}{\prod_{j=1}^{m} f(x_j | 0)} > k(P_n) \tag{5.39}$$

with k a constant whose dependence on P_n will be described below. This is exactly the likelihood-ratio test developed earlier under the criterion of minimum error probability [Eq. (5.19)]. The processing algorithm, which depends specifically on the form of the two sets of density functions appearing in the likelihood ratio, is thus also the same in the two cases. To detect a signal in gaussian noise, as an example, one sums

† A. M. Mood and F. A. Graybill, "Introduction to the Theory of Statistics," 2d ed., McGraw-Hill, New York, 1963.

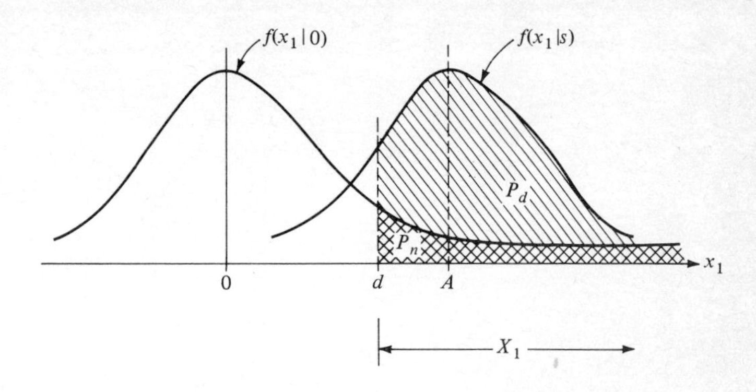

FIGURE 5.15
Neyman-Pearson test, signal in gaussian noise, one sample.

the m samples $x_j, j = 1, \ldots, m$ (or finds the sample mean), and compares the sum with a predetermined threshold. If the signal appears in laplacian noise, one again combines counting and summing, as shown by Eq. (5.33). It is the threshold level, obtained from the right-hand side of the inequality in Eq. (5.39), that varies, depending on the criterion used.

To show specifically how the threshold level is determined in the Neyman-Pearson sense and how it depends on the false-alarm probability P_n, consider again the example of a signal of amplitude A appearing in a gaussian-noise background. Assume for simplicity that only one sample x_1 is taken. Then it is left to the reader to show that the signal is declared present if

$$x_1 > d = \frac{A}{2} + \frac{\sigma_n^2}{A} \ln k \tag{5.40}$$

[Compare with Eq. (5.22) for m samples and minimum-probability-of-error criterion.] This is sketched in Fig. 5.15.

How do we now relate the threshold d (or, equivalently, the constant k) to the false-alarm probability P_n? Assume as a specific example that $P_n = 0.05$. (This implies that noise will be mistaken for signal on the average 5 percent of the times during which no signal is transmitted.) For gaussian noise we have

$$P_n = \int_d^\infty \frac{e^{-x_1^2/2\sigma_n^2}}{\sqrt{2\pi\sigma_n^2}}\, dx_1 = \int_{d/\sqrt{2}\,\sigma_n}^\infty \frac{e^{-x^2}}{\sqrt{\pi}}\, dx = \frac{1}{2}\left(1 - \operatorname{erf}\frac{d}{\sqrt{2}\,\sigma_n}\right) \tag{5.41}$$

upon normalizing, and using the error-function definition introduced earlier [Eq. (5.27)] (see Fig. 5.15). From tables of the error function we get

$$d = 1.65\sigma_n$$

as the threshold level. A different choice of P_n would have changed d. Reducing P_n increases d, and vice versa. Note that in this example the threshold level d is independent of the signal amplitude A. This is apparent from Fig. 5.15. It depends solely on the noise standard deviation. Equation (5.40) is thus somewhat deceptive, indicating

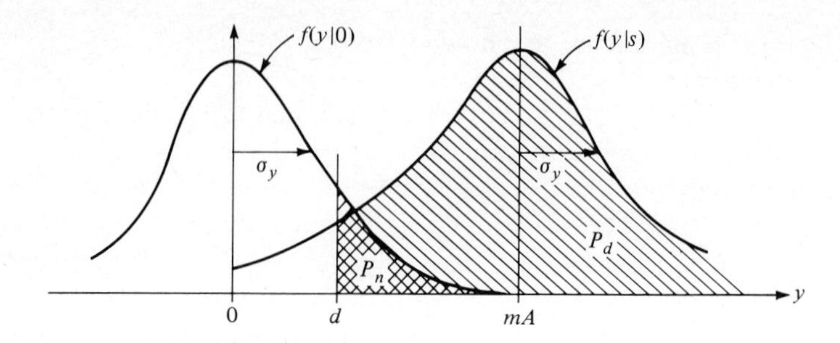

FIGURE 5.16
Neyman-Pearson test, gaussian noise, m independent samples.

as it does a supposed dependence on A. (The constant k obviously must adjust itself to make d independent of A.) This situation differs from that found in the minimum-error-probability case, where the threshold level did depend on A and in fact was close to $A/2$. Thus, although the structure of the processing algorithms is the same for the three performance criteria considered, the threshold levels with which processor outputs are compared do differ.

In the case of m independent samples of a signal of amplitude A appearing in a gaussian-noise background it is apparent that the optimum Neyman-Pearson processor consists of comparing $\sum_{j=1}^{m} x_j$ to some preset threshold or decision level, depending on the false-alarm probability. Thus the signal is declared present if

$$y = \sum_{j=1}^{m} x_j > d \tag{5.42}$$

[Note: this y is not to be confused with the y used in (5.2).] How does the decision level d depend on P_n? As already noted earlier in this chapter, if each sample x_j is gaussian, the sum of the samples is a gaussian random variable as well. Call this random variable y. Then with no signal present, $E(y) = 0$ and $\sigma_y^2 = m\sigma_n^2$ the variance of each of the m noise samples. The density function of y with no signal present is then given by

$$f(y \mid 0) = \frac{e^{-y^2/2\sigma_y{}^2}}{\sqrt{2\pi\sigma_y^2}} \tag{5.43}$$

and, as previously, the false-alarm probability is simply

$$P_n = \int_d^\infty f(y \mid 0)\, dy = \int_c^\infty \frac{e^{-x^2}}{\sqrt{\pi}}\, dx = \tfrac{1}{2}(1 - \operatorname{erf} c) \tag{5.44}$$

using Eq. (5.43) and again normalizing. The parameter c is defined as

$$c \equiv \frac{d}{\sqrt{2}\,\sigma_y} = \frac{d}{\sqrt{2m}\,\sigma_n}$$

The false-alarm probability and its relation to the decision level d is sketched in Fig. 5.16. Note that this is identical to the single-sample case with σ_y replacing σ_n. Although the parameter d is proportional to $\sqrt{m}$, the peak of the $f(y \mid s)$ curve is proportional to m, as shown. Thus as m increases, there is an improvement in detection.

In radar problems $P_n = 0.05$ is much too high. Values of $P_n = 10^{-8}$ to 10^{-10} and even smaller are much more realistic choices because radar signal pulses are commonly of the order of 1 μs in width or less. The m samples to be processed thus occupy 1 μs or less, so that with the signal absent the noise has an opportunity every 1 μs to be mistaken for signal. For $P_n = 10^{-8}$ a false alarm would occur on the average of once every 100 s or, roughly, once every 2 min. For $P_n = 10^{-10}$ the average false-alarm rate is once every 10^4 s or roughly once every 3 h.

Tables of the error function or normal curve generally do not cover values as low as these. A simple integration by parts shows, however, that for small P_n the integral of Eq. (5.44) can be represented by a so-called asymptotic series and for $c \geq 3$ ($P_n \ll 1$) the first term in the series suffices:

$$P_n \approx \frac{e^{-c^2}}{2\sqrt{\pi}\,c} \qquad c > 3 \tag{5.45}$$

Using this equation (taking logarithms and using trial-and-error calculations is the simplest method of evaluation), one finds:

For $P_n = 10^{-10}$: $\qquad c = 4.5 \qquad$ and $\qquad d = 6.35\sigma_y = 6.35\sqrt{m}\,\sigma_n$

and

For $P_n = 10^{-8}$: $\qquad c = 3.98 \qquad$ and $\qquad d = 5.63\sigma_y$

Other decision levels can be found similarly.

The probability of detection P_d can be found by noting that this corresponds to the probability that $y = \sum_{j=1}^{m} x_j > d$ with the signal present. In this case $E(y) = mA$ and $\sigma_y^2 = m\sigma_n^2$, as in the noise-only case. This is indicated in Fig. 5.16. It is left for the reader to demonstrate that P_d is then given by

$$P_d = \int_{c-\sqrt{m}\,A/\sqrt{2}\,\sigma_n}^{\infty} \frac{e^{-x^2}}{\sqrt{\pi}}\,dx = \tfrac{1}{2}\left[1 - \operatorname{erf}\left(c - \frac{\sqrt{m}\,A}{\sqrt{2}\,\sigma_n}\right)\right] \tag{5.46}$$

For $\sqrt{m}\,A/\sigma_n = \sqrt{2}\,c$ the probability of detection is 0.5. As $\sqrt{m}\,A/\sigma_n$ increases beyond this value, P_d increases, approaching 1. For a specified probability of detection, with the false-alarm probability P_n and hence the parameter c fixed, one may then find the required *signal-to-noise ratio* A/σ_n or its equivalent power ratio A^2/σ_n^2. Note that increasing the number of samples m used is equivalent to increasing the effective A^2/σ_n^2 by m as well. The reason for this is apparent from Fig. 5.16. As m increases, the width of the curves shown, or the standard deviation σ_y, increases as $\sqrt{m}$. The decision level d increases as $\sqrt{m}$ as well. The expected value of the variable y with signal increases directly with m, however, so that the ratio of expected value to standard deviation increases as $\sqrt{m}$. The overlap of the two density functions decreases, and the probability of detection thus increases with m.

The physical reason for this is also instructive: as more signal samples are taken, their amplitudes add directly, providing the mA expected value of y. The noise samples being added are random, however, so that their variances add. This is the usual phenomenon in signal-detection problems; adding more and more independent samples makes the signal "rise up out of the noise."

An interesting and useful form for the effective signal-to-noise ratio and for P_d as well can be derived which will recur when we consider the effect of varying signal amplitude. We noted much earlier, at the beginning of this chapter, that the maximum number of uncorrelated noise samples available in a time interval T s long is $m = 2BT$. This result assumes bandlimited white noise, with B the bandwidth prior to sampling and samples taken at a minimum spacing of $1/2B$-s intervals. For gaussian noise these samples are in turn independent. We also noted briefly in passing that the mean noise power in the case of bandlimited white noise is proportional to the bandwidth B. Specifically this boils down to saying that†

$$\sigma^2 = n_0 B$$

with n_0 the so-called white-noise spectral density in watts per hertz. When we let $m = 2BT$ and $\sigma_n^2 = n_0 B$, the effective signal-to-noise ratio mA^2/σ_n^2 turns out to be given simply as $2(A^2T)/n_0$. This has an interesting interpretation. For the rectangular signal pulse of amplitude A and duration T with which we are dealing, A^2 is pulse power (actually as measured across a 1-Ω resistor if A is in units of volts or current) and A^2T the energy E in a single pulse. The detectability of the pulse is thus a function only of the effective signal-to-noise ratio, given by the ratio of its energy E to the noise spectral density n_0:

$$\frac{mA^2}{\sigma_n^2} = \frac{2BTA^2}{n_0 B} = \frac{2E}{n_0} \tag{5.47}$$

This is a well-known and extremely important result in detection theory: one can improve the probability of detection by increasing the received signal *energy* E or by reducing the noise spectral density n_0. The energy is in turn given by the product of the signal power and the pulse duration T. The detectability is thus increased either by increasing signal power (by increasing the height of the rectangular signal pulse, if possible) or by lengthening the duration of the signal pulse. The latter poses problems of signal resolution which will be discussed briefly in the next section.

From this relation connecting signal-to-noise ratio and E/n_0, the probability-of-detection equation (5.46) reduces to the interesting normalized form

$$P_d = \int_{c-\sqrt{E/n_0}}^{\infty} \frac{e^{-x^2}}{\sqrt{\pi}}\, dx = \tfrac{1}{2}\left[1 - \operatorname{erf}\left(c - \sqrt{\frac{E}{n_0}}\right)\right] \tag{5.48}$$

We shall see in a later section that this is a quite general form applicable to the case of varying-amplitude signal pulses as well as the special case of rectangular pulses considered here.

† Schwartz, op. cit., pp. 407 and 597.

5.4 APPLICATION TO AIR-TRAFFIC-CONTROL RADAR

To put all the material discussed in the previous section in context it is instructive to actually carry out some calculations for the air-traffic-control radar mentioned briefly in Chap. 1. The object is to take the known data for these radars and determine how well they might be expected to perform. (In practice one would just do the reverse, however: given some desired performance characteristics, determine some of the significant design parameters.)

As noted in Chap. 1, there are essentially two types of radar in the air-traffic-control system, the airport surveillance (ASR) and air route surveillance (ARSR) systems.† The ARSR radars detect and monitor aircraft en route, and the ASR radar monitors aircraft within 60 mi of the airport at which it is located. Both carry out detection and tracking functions. We shall discuss only the detection capabilities at this point, returning to the tracking function in Chaps. 6 and 7.

In order to discuss the signal-detection characteristics of the radars using the techniques developed in Sec. 5.3 we need some estimate of the power (and from this the energy) reflected from a typical aircraft (or target). For this purpose we first develop the so-called *radar equation.*

Recall from Chap. 1 that a typical surveillance or search radar sends out bursts or pulses of electromagnetic energy at periodic intervals. A portion of this energy, on encountering *any* reflecting material, is reflected back to the radar (Fig. 5.17). This reflected energy indicates the presence of a reflecting object, and the time between transmission and reception of the signal pulse provides the measure of the range or distance of the object from the radar. The radar antenna is designed to provide a narrow beam of energy in azimuth and rotates continuously as it sends out signal pulses. The azimuth position of the antenna at the time of the reception of a pulse thus provides another coordinate of the reflecting object.

To determine the power reflected from an aircraft target let the peak power of the *transmitted* radar pulses be P_T W. (If the high-frequency burst is written $A \cos \omega t$, with $f = \omega/2\pi$ the frequency of transmission over the actual time of transmission, P_T is proportional to $A^2/2$.) If this power were uniformly distributed over all space, the power density at a distance r from the radar would be $P_T/4\pi r^2$. The radar antenna serves to focus the beam in a desired direction, however, this focusing effect being given by its gain G over the uniform (or isotropic) power-density distribution in space. The power density at a distance r is thus $P_T G/4\pi r^2$. Ideally, for an antenna of aperture area A_a m^2, G is given by‡

$$G = \frac{4\pi A_a}{\lambda^2} \tag{5.49}$$

with λ the wavelength of the high-frequency energy. ($\lambda f = c$, where c is the velocity of light in free space.)

† Harry C. Moses, Air Traffic Control Radar Systems Definition Report, *Fed. Aviat. Admin. Rep.* FAA-EM-72-1, March 1972.

‡ D. J. Angelakos and T. E. Everhart, "Microwave Communications," sec. 5-7, McGraw-Hill, New York, 1968.

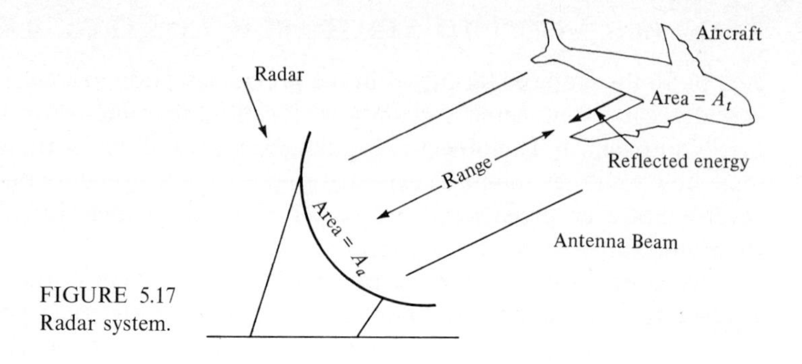

FIGURE 5.17
Radar system.

An ideal reflecting object of area A_t m^2, located r m from the transmitter (see Fig. 5.17), will reflect back to the transmitter $P_T GA_t/4\pi r^2$ W. This power in turn spreads out in space as it moves back to the radar antenna. As seen from the antenna, the reflecting object or target is essentially an isotropic point radiator. The *reflected* power density in watts per square meter back at the antenna is thus the power reflected from the target divided by the surface area $4\pi r^2$ encompassed, or just $(P_T GA_t/4\pi r^2)/4\pi r^2$. Finally, if the receiving antenna has an effective aperture area A_a m^2, the power received from the target is

$$P_R = \frac{P_T GA_t}{4\pi r^2} \frac{A_a}{4\pi r^2} = \frac{P_T G^2 A_t \lambda^2}{(4\pi)^3 r^4} \tag{5.50}$$

putting in expression (5.49) relating antenna gain and aperture size.

Equation (5.50) is the radar equation mentioned earlier. It shows the characteristic $1/r^4$ dependence in distance because of the two-way path taken by the energy. This is actually an *ideal* equation, since it ignores power losses during transmission and reception. For example, the power actually transmitted into space may be less than the power generated because of losses in the transmission link between generator and antenna as well as in the antenna feed and the antenna itself. Equation (5.49) is, as noted, an ideal equation relating antenna aperture and gain. Typically the gain may be 60 to 70 percent of the number calculated from Eq. (5.49), so that the power launched into space is that much less. There may be absorption losses in the transmission medium as well. Equation (5.50) is the so-called *free-space radar equation*, and air and its various constituents provide power absorption not included in that equation. The power absorbed depends on the frequency of transmission, the amount of water vapor present in the air, and other factors as well. We shall ignore these power losses for the time being but put them in as a lumped loss factor later, in interpreting the detection results.

To simplify the power calculations in the use of Eq. (5.50) one often converts to decibels taking 10 $\log_{10}$ of each term in the expression and adding (or subtracting) appropriately. In fact, the gain G is usually given in decibel measure. Using Eq. (5.50), we can now proceed to discuss each of the two air-traffic-control radars. The pertinent specifications for the two radars are given in Table 5.1.

Note that the en route radars are designed to cover almost 4 times the range of the more localized airport radars. They are provided with correspondingly higher-powered transmitters.

Assuming a target to have a 2-m^2 reflecting area at the maximum range (this is a design specification for the two radars), the received powers are

$$P_R = \begin{cases} 2 \times 10^{-13}\ \text{W} & \text{for ASR radar} \\ 0.63 \times 10^{-13}\ \text{W} & \text{for ARSR radar} \end{cases}$$

Note that although the en route radar has 10 times the transmitted power of the airport radar, the much longer distance it has to cover dominates, resulting in a net reduction in the minimum received power.

For detection purposes we must know the received *energy* E and the noise spectral density n_0. Recall from Sec. 5.3 that E is the energy in a rectangular pulse of amplitude A and width T. This is the energy in the so-called *baseband* or *video signal*. This signal, as pointed out earlier, is formed by heterodyning down the so-called rf or high-frequency signal $A \cos \omega t$. The high-frequency energy is proportional to $A^2/2$, while E is proportional to A^2. We must thus multiply the received power by 2 to find the effective E for our detection calculations. We therefore have $E = 2P_R T$, with T the pulse width. The pulse widths for the two radars are, respectively, 0.83 μs for the ASR (airport) radar and 2 μs for the ARSR (en route) radar. This gives for the two video-energy terms

$$E = \begin{cases} 33.2 \times 10^{-20}\ \text{J} & \text{for ASR radar} \\ 25.2 \times 10^{-20}\ \text{J} & \text{for ARSR radar} \end{cases}$$

Note that the wider pulse used for the ARSR radar overcomes the power disadvantage somewhat, increasing its target-detection capability correspondingly. The price paid, however, is that of *range resolution*. A little thought on the part of the reader will show that two targets separated by a radial distance corresponding to less than the pulse width cannot be distinguished by these radars. As the pulse width is increased, more energy is received and the target detectability is increased correspondingly, but the minimum radial spacing of two targets required to distinguish them also goes up correspondingly. Figure 5.18 shows how two airplanes can appear to be a single plane if the difference in propagation time to and from these targets is less than the pulse duration.

Table 5.1 RADAR SPECIFICATIONS

	ASR (airport)	ARSR (en route)
Transmitted power P_T	400 kW	4 MW
Gain G	34 dB	34 dB
Maximum range r	68 mi or 110 km	230 mi or 370 km
Frequency f	2.7 GHz	1.3 GHz
Wavelength λ	0.11 m	0.23 m

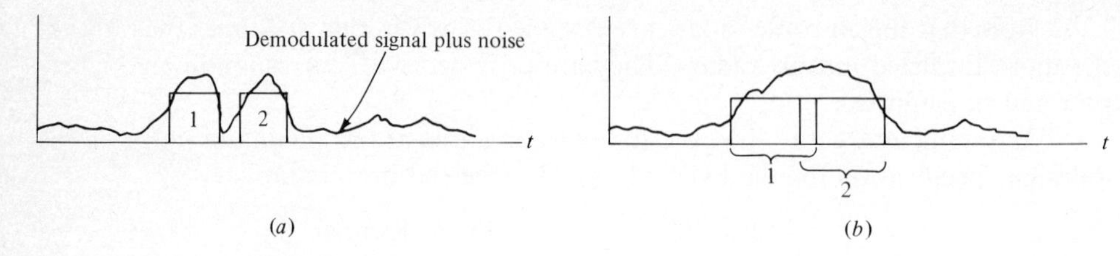

FIGURE 5.18
Radar resolution vs. pulse duration. (*a*) Narrow, separated pulses; (*b*) wide, overlapping pulses.

Some readers may be troubled by the extremely small values of power and energy available at the receiver. Can one really detect something with 10^{-13} W of power? Recall that if it were not for the noise received at the same time (or added in the receiver circuits) *any* signal, no matter how small its power, could be detected, at least theoretically. It is thus not the signal power or energy received that counts but the ratio of signal to noise.

The noise figure for both radar systems is specified to be 4 dB. The noise figure F is essentially a measure of how much additional noise the radar receiving system (from antenna on) introduces over and above the thermal noise picked up by the antenna as it looks into space. In this case $F = 4$ dB or 2.5 means that 150 percent additional noise is introduced. The noise spectral density in terms of F is given by†

$$n_0 = FkT \tag{5.51}$$

with kT the thermal-noise contribution. Here $k = 1.38 \times 10^{-23}$ J/K is the *Boltzmann constant*, and T is the temperature in kelvins (K) of the thermal-noise source, in this case the sky above the radar antennas. As a *conservative* estimate, assume that this temperature is 300K, the usual "room temperature" on the ground. With $F = 4$ dB or 2.5 numerically, we then have $n_0 = 10^{-20}$ W/Hz. The E/n_0 ratios are then, respectively,

$$\frac{E}{n_0} = \begin{cases} 33.2 \text{ or } 15.2 \text{ dB} & \text{for ASR radar} \\ 25.2 \text{ or } 14 \text{ dB} & \text{for ARSR radar} \end{cases}$$

Using the results of Sec. 5.3, we can now determine the performance of the radar systems. Specifically, let the false-alarm probability be 10^{-10} in both cases. Then from Eq. (5.48), with $c = 4.5$ in both cases, we have $P_d = 0.96$ for the ASR radar and 0.76 for the ARSR radar. (Note again how significant even a 1.2-dB difference in E/n_0 is in establishing the probability of detection.)

But this is not the whole picture. We have neglected several important effects in these calculations. First, as noted earlier, we have completely ignored power losses in the systems. This could account for as much as 9 dB in a typical system, reducing the detection probability to extremely small values. As a compensating factor, however, we have also ignored the fact that the radar beam in rotating through space covers a

† Schwartz, op. cit., p. 620.

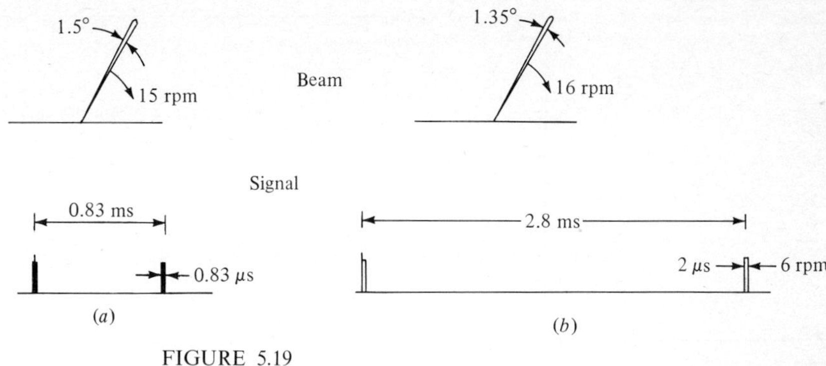

FIGURE 5.19
ATC radar parameters. (*a*) ASR radar—airport: (*b*) ARSR radar—en route.

typical target long enough to have several pulses returned from it. This effect was noted in Chap. 1, in first discussing the air-traffic-control radars. Specifically, the ASR radar rotates at a rate of 15 r/min, completing a 360° scan in 4 s, while the ARSR (en route) radar rotates at 6 r/min, covering a complete scan in 10 s. The two radar beams have azimuth beamwidths of 1.5° and 1.35°, respectively, while the number of signal pulses emitted per second, the so-called *pulse repetition frequency* (PRF), are 1200 and 360 pulses per second, respectively. Figure 5.19 summarizes the parameters used for these two radars. It is left to the reader to show that this corresponds to 20 pulses reflected per target in the ASR case and 13 pulses per target in the ARSR case, providing that many more samples in the signal processing or, equivalently, augmenting the received energy by the same amount. Thus, ideally, the ASR radar energy received should be $20E$ and the ARSR radar energy $13E$. Alternately, this results in a signal-to-noise ratio or E/n_0 improvement of 20 or 13 dB and 13 or 11.1 dB, respectively.

We say *ideally* because in practice target fluctuations introduce random phase variations in the high-frequency signals received back from the target. The high-frequency pulses cannot be simply heterodyned down to the baseband signals and then added, as implied earlier. This is the problem always encountered in adding sine waves of the same frequency but different phase. Sometimes the resultant sum adds up, and sometimes it decreases, because of phase cancellations. The lack of phase coherence pulse to pulse means that the individual pulses must be *envelope-detected* before processing. Figure 5.1 shows the relationship between a high-frequency signal and its envelope. Envelope detection is exactly the procedure employed in home AM receivers for extracting the desired signal from a high-frequency sine wave. It is a nonlinear process, in which the output is proportional to the *envelope* of the input sine wave. Although we shall not analyze this case here (an example is included among the problems at the end of this chapter), one can find the appropriate signal-plus-noise and noise-only statistics at the output of the envelope detector (they are no longer gaussian) and invoke the likelihood ratio to find the optimum processing algorithm, as in the gaussian case. It turns out (as might be expected) that addition of the individual detected pulse amplitudes is called for in this case.

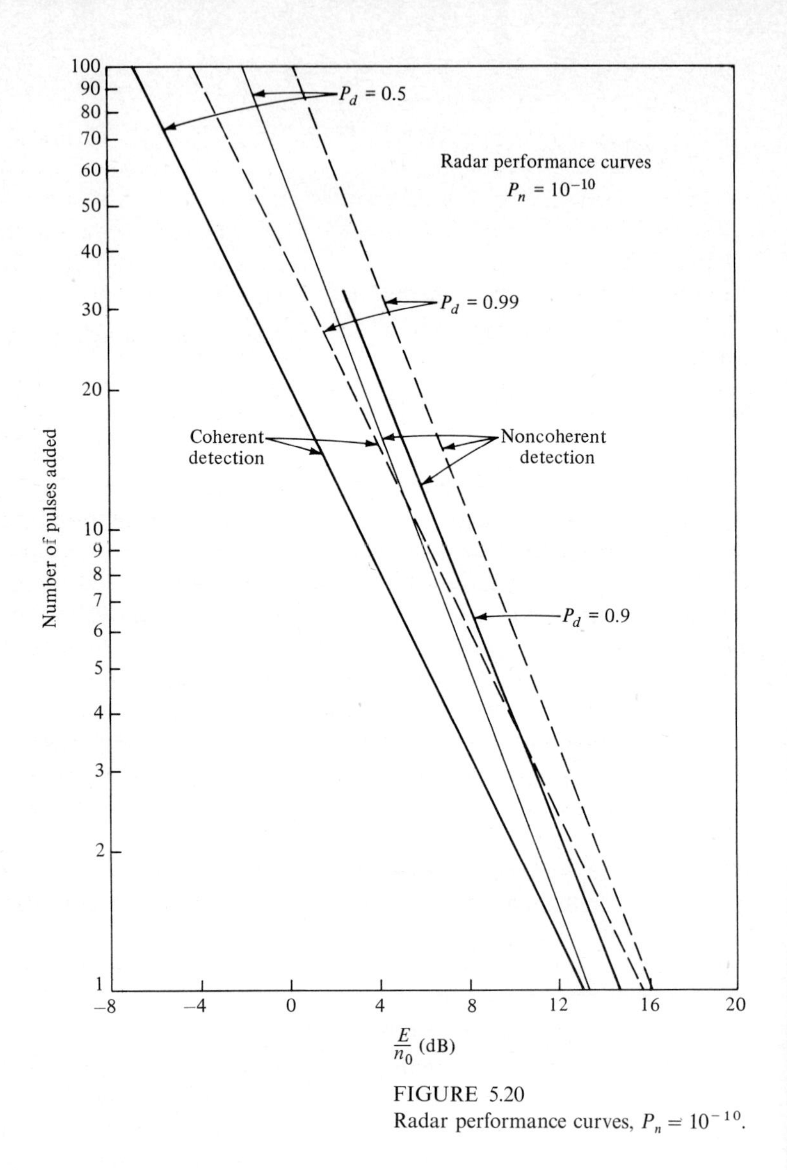

FIGURE 5.20
Radar performance curves, $P_n = 10^{-10}$.

A resultant set of curves describing the performance of this envelope-detector processor is plotted in Fig. 5.20. These curves are labeled *noncoherent detection* and show the signal-energy–noise-spectral-density ratio E/n_0 required per pulse as a function of the number of pulses added for a false-alarm probability $P_n = 10^{-10}$ and $P_d = 0.5$, 0.9, and 0.99, as three examples. Note that a *single* envelope-detected pulse requires an E/n_0 ratio of 13.5 dB for $P_d = 0.5$, while eight pulses added require 6.3 dB each for the same detectability. Also shown on the same figure are two equivalent curves plotted for the *coherent-detection* case: this is the one we have analyzed assuming gaussian statistics. These curves are plots of Eq. (5.48), with $c = 4.5$, as noted earlier, to maintain $P_n = 10^{-10}$.

For one pulse and $P_d = 0.5$, $\sqrt{E/n_0} = 4.5$ from Eq. (5.48). This corresponds to $E/n_0 = 13.1$ dB. Note that the envelope-detection process, made necessary by a lack of phase coherence, has thus resulted in a loss of 0.4 dB for this probability of detection (13.5 dB compared with 13.1 dB). The corresponding loss for $P_d = 0.99$, from Fig. 5.20, is about 0.5 dB (16.2 dB compared with 15.7 dB). When several pulses are added in this case of coherent detection, the energies of the individual pulses are added as well in applying Eq. (5.48), as noted earlier. For the same P_d, then, the energy per pulse may be reduced as the number of pulses added. For example, for 10 pulses added, the energy per pulse may be reduced by 10 dB. This shows up on the curves as an E/n_0 of 3.1 dB instead of 13.1 dB for $P_d = 0.5$ and $P_n = 10^{-10}$.

It is apparent from the curves of Fig. 5.20 that this linear-integration trade-off between number of pulses added and energy required per pulse does not hold for the noncoherent-detection case. Although less energy is required per pulse as more pulses are added, the trade-off is less efficient then in the coherent case, the efficiency decreasing as the number of pulses added increase. Thus, from a 0.4-dB difference between the coherent and noncoherent cases for a single pulse, the difference in E/n_0 required increases to 2.2 dB at 8 pulses and to 5.2 dB at 100 pulses added. This phenomenon, termed an *integration loss*, shows up as an additional loss term in evaluating radar-system performance.

Recapitulating now for the specific case of the two air-traffic-control radars, we have the ASR radar with $E/n_0 = 15.2$ dB on each pulse and 20 pulses added, while the ARSR radar has $E/n_0 = 14$ dB with 13 pulses added. These energy–spectral-noise-density ratios have to be decreased, however, to account for various power losses, as noted earlier. The detection probability will depend on an estimate for these losses. Using three different estimates for the losses (3, 6, and 9 dB) and Fig. 5.20, one can construct Table 5.2. Details are left to the reader.

Table 5.2 DETECTION PROBABILITY ($P_n = 10^{-10}$) FOR ASR (AIRPORT) AND ARSR (EN ROUTE) RADAR

Loss, dB	P_d	
	Coherent detection*	Noncoherent detection
ASR:		
3	≈1	≈1
6	≈1	>0.99
9	≈1	0.99
ARSR:		
3	≈1	≈1
6	≈1	0.99^+
9	0.994	0.5^+

* More precise results can be obtained by using Eq. (5.48) with appropriate account taken of the power losses and gain due to the pulses added.

Note that both radars provide acceptable detection performance ($P_d = 0.99^+$ for the noncoherent detection case) if a power loss of 6 dB is assumed. For a power loss of 9 dB, however, the ARSR radar detection probability drops to a little more than 0.5.

5.5 EXAMPLE OF A SUBOPTIMUM PROCESSOR

Although we have stressed the concept of performance criterion thus far as a means of determining optimum processors for the detection of signals in noise, in practice one often would like to use other processing algorithms. One reason may be that the optimum processor is just too difficult or messy to implement. Economic considerations may dictate implementation with simpler digital circuitry or computer software. In some cases the noise statistics may not be precisely known. (This is one of the unmentioned assumptions in the discussion thus far.) One may thus want to use seat-of-the-pants algorithms that could be applicable over a wide range of noise distributions. (Such techniques are called *robust* or *nonparametric techniques.*) It is of interest, however, to compare these ad hoc techniques with the optimum ones found using known noise statistics. For if the technique suggested is appreciably poorer than the optimum one for the criterion used, it may not pay to use it even though it is simpler or more economic to implement. (The economic or cost benefit is thus illusory in this case. One gains in the cost of implementation at the expense of cost in performance. The overall economics of the problem, incorporating both types of costs, must be considered.) Analysis should be carried out to see the trade-off in performance vs. implementation cost.

We suggested at the beginning of this chapter several ad hoc techniques that appear appropriate for the detection of signals in noise. One, the evaluation of the sample mean and its comparison with a specified threshold, was in fact found optimum for additive gaussian noise. Many other algorithms could be dreamed up, but we shall calculate the performance characteristic of one particular processor as an example of how one goes about comparing it with the optimum. For this purpose we pick the pure counter, mentioned in passing at the beginning of this chapter. We shall then compare results for this case with the Neyman-Pearson processor appropriate to radar-type situations.

Each of the m samples is individually compared with a specified threshold. A counter counts the number exceeding the threshold and declares a signal present if this number exceeds some preset count $k \leq m$. (This technique has been variously called a *double-threshold* or *coincidence procedure.*) Such a scheme is particularly suitable for implementation by simple digital circuitry. How does it compare in performance with the equivalent "optimum" processors? For this purpose we must calculate the two error probabilities, P_{e0} and P_{e1}; note that both appear in the minimum-error-probability and Neyman-Pearson formulation.

Consider the noise-only case. Assume, as previously, that we know the noise distribution. If so, we can readily calculate the probability p_n that an *individual noise sample* will exceed some preset level b. This is given by

$$p_n = \int_b^\infty f(x_j \mid 0)\, dx_j \tag{5.52}$$

P_{e0} is just the probability that at least k of the m samples will each exceed b and is given by the cumulative binomial distribution

$$P_{e0} = \sum_{j=k}^{m} \binom{m}{j} p_n^j (1 - p_n)^{m-j} \tag{5.53}$$

Here $\binom{m}{j}$ represents the number of combinations of m things taken j at a time and is given by

$$\binom{m}{j} = \frac{m!}{j!\,(m-j)!}$$

Similarly, if we call p_s the probability that a signal-plus-noise sample will exceed b when the signal is present, we have

$$1 - P_{e1} = \sum_{j=k}^{m} \binom{m}{j} p_s^j (1 - p_s)^{m-j} \tag{5.54}$$

The probability p_s is of course in turn given by

$$p_s = \int_b^\infty f(x_j \mid s)\, dx_j \tag{5.55}$$

For the error-probability criterion we then calculate

$$P_e = P_1 P_{e1} + (1 - P_1) P_{e0}$$

as noted previously. For the Neyman-Pearson case P_{e0} is equivalent to P_n. We set this at some desired level and calculate P_{e1} or $1 - P_{e1} \equiv P_d$.

How does this scheme of counting the number of samples exceeding a threshold compare with a scheme like that in which the sample mean of all m samples is required to exceed a specified threshold? Consider as an example a detection system for which the Neyman-Pearson procedure is appropriate. Say the false-alarm probability $P_n = P_{e0}$ is chosen at some prescribed level. For m and k specified one can then use Eq. (5.53) to find the appropriate value of p_n, the probability that any one sample will exceed a level b. Knowing the noise-only distribution $f(x_j \mid 0)$, one finds the threshold b using Eq. (5.52). The probability that a single sample containing the signal will exceed b is then given by Eq. (5.55) and the detection probability $P_d = 1 - P_{e1}$ found finally from Eq. (5.54).

As we raise the minimum number of samples k required to exceed the threshold level b, the corresponding value of b must be adjusted to maintain the false-alarm probability P_n at the desired value. The signal amplitude A required to obtain the desired detection probability P_d must then be adjusted as well. One can then obtain a curve relating the signal-to-noise A/σ_n to the parameter k. An example of a set of such curves calculated for a specific radar problem is shown in Figure 5.21, where S/N is

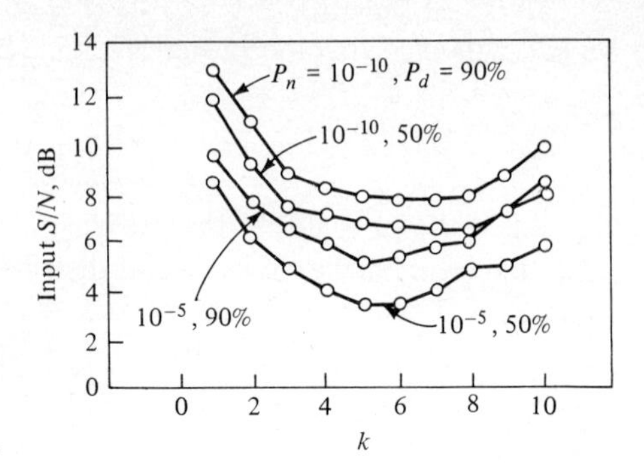

FIGURE 5.21
Signal detection by coincidence ($m = 10$).

used for signal-to-noise ratio and is shown on a decibel scale [dB $\equiv 20 \log_{10} (A/\sigma_n)$]. (Although gaussian noise has been assumed added to a signal of amplitude A in this example, the statistical distributions used in deriving these particular curves were not quite gaussian. They are precisely the same distributions mentioned in the previous section that one obtains at the output of an envelope detector. The curves of Fig. 5.21 are thus appropriate to a noncoherent detection scheme and will in fact be compared shortly to the equivalent noncoherent curves of Fig. 5.20.)

The curves of Fig. 5.21 pertain to the particular case of $m = 10$ samples processed. They show S/N vs. k for various combinations of P_n and P_d. Note that all the curves exhibit a minimum value of S/N at values of k in the vicinity of 5 to 7. This then provides the best value of k to be used since it requires the smallest signal amplitude for given false-alarm probability and probability of detection. Similar curves can be obtained for other values of m. Combining the minimum S/N points of all of these curves for a range of values of m from 1 to 100, one obtains the optimum coincidence curve shown in Fig. 5.22. (Although it is drawn for the specific case

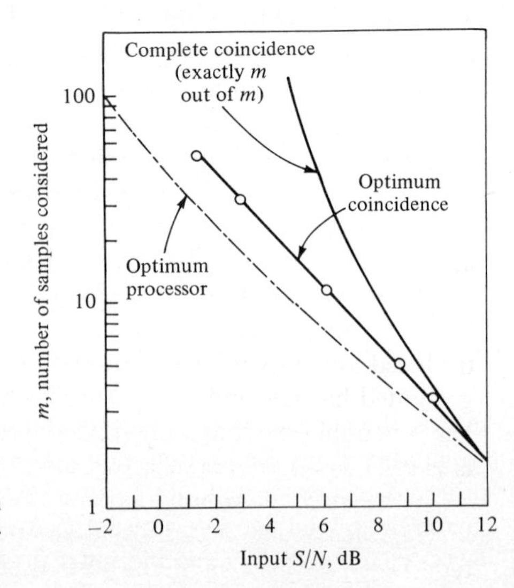

FIGURE 5.22
Comparison of coincidence detection with optimum processor, noncoherent detection ($P_n = 10^{-10}$, $P_d = 0.5$).

$P_n = 10^{-10}$, $P_d = 0.5$, similar curves can be obtained for other combinations of these parameters.) Also shown is a performance curve for the optimum processor in this case. This is exactly the Neyman-Pearson processor for noncoherent detection, and the curve plotted is identically the noncoherent $P_d = 0.5$ curve of Fig. 5.20. (As noted earlier, the derivation of the specific processor appears among the problems for this chapter.) The only difference in the two figures is the use of E/n_0 in Fig. 5.20 and S/N in Fig. 5.22. Also shown for comparison is a performance curve for a processor requiring *all* m samples to exceed a preset threshold. Since the optimum coincidence curve is roughly parallel to the optimum processor curve over most of the range of m, differing from it by about 1.4 dB, one may say that the cost in performance of using the coincidence scheme is to require 1.4 dB more signal power than the smallest power possible. The designer must decide whether the concomitant simplification in the processing algorithm is worth the cost of the additional power required. The reader is encouraged to think through the use of this scheme in the air-traffic-control radar application of the previous section. (The coincidence scheme shows the same 1- to 1.5-dB cost in performance compared with the optimum processor in a variety of other applications in which the noise and signal-plus-noise statistics of the samples are obtained by nonlinearly processing gaussian statistics.)

5.6 DETECTION OF VARIABLE-AMPLITUDE SIGNALS: MATCHED FILTERS

We have assumed throughout the chapter thus far that the signal to be detected is of known, constant amplitude. In many detection problems the signal will vary from sample to sample. Two cases may be distinguished: one in which the signal variation is random, so that the signal whose presence is to be detected may truly be called a *random signal*, and one in which the signal-amplitude variation is assumed known. This case of a signal pulse whose amplitude varies with time is obviously a more realistic representation of actual signal waveshapes than the rectangular, flat-topped pulse we have considered thus far. For all waveshapes, as originally generated, have finite nonzero rise and decay times. They may have some overshoot as well. An example of a typical signal pulse $s(t)$ is shown in Fig. 5.23. Also shown are m samples $s_1, s_2, \ldots, s_j, \ldots, s_m$, of the signal, spaced presumably $1/2B$ s apart, as in the previous section. Because of the time-varying nature of $s(t)$, however, the samples no longer are constant-valued: they now vary in amplitude.

The signal to be detected may have had some amplitude variation built in right from its initial generation, or additional amplitude variations may have been introduced along its transmission path to the receiver. For example, a radar pulse as generated has rise and decay times as noted above. The transmitting antenna beam has a nonuniform shape that introduces some additional amplitude variation. (This is usually negligible because the microsecond pulse widths used are short compared to the relatively slow antenna scan rate measured in seconds. But successive pulses leaving the antenna and directed at a particular target will vary somewhat in amplitude because of the antenna rotation and nonuniform antenna beam shape.) Slight

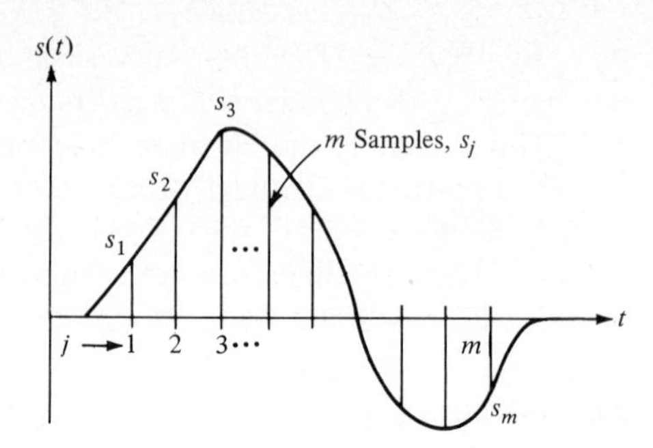

FIGURE 5.23
Typical signal amplitude variations.

rotations of the target may introduce amplitude variations into the reflected signal. The receiving antenna may also introduce some amplitude variation (although again this is usually negligible).

In a pulse communications system the medium over which the signals are sent may spread the pulses out or distort them considerably. This is particularly true of digital communications over cable or wire, as in the typical telephone system. In an astronomical example a stellar object to be detected may have its light or radio-wave output fluctuating in time. Spectral lines under investigation in spectroscopic analysis typically show line broadening, which when probed with high enough resolution shows up as a rise and decay.

In these various examples and others that might be cited, the signal variations may sometimes be modeled as *known amplitude variations* and in other cases as *random variations*. In this section we consider the known-amplitude-variation case. We thus assume that the form of $s(t)$ and hence the samples s_j, $j = 1, \ldots, m$, as in Fig. 5.23, is known. In the next section we briefly consider the random-signal case.

With either type of signal variation, a likelihood-ratio algorithm will again be obtained as the optimum processor. The optimum processing algorithms in both cases arise from simple generalizations of the constant-amplitude signal-detection problem studied thus far.

The optimum processor for a signal with known amplitude variation to be detected in the presence of additive gaussian noise is particularly interesting. We shall find it consisting of a device that *weights* the received-signal-plus-noise samples by the corresponding known (and stored) signal amplitudes and adds them together. This weighting operation is called a *matched filter*. The matched-filter output is then compared with a decision level, as in previous sections. (If the signal sample values are all the same, as with constant signal amplitude, the weighting may of course be ignored and the result reduces to the optimum processor already discussed.) This is intuitively satisfying since it simply says that a received sample whose signal component (if present) is known to be large would be given more weight than a sample at a time when the signal component is small.

Not only is the processor for this case of a known signal in gaussian noise a simple extension of the constant-amplitude case, but the expressions for the probability of error, or detection probability in a radar-type situation, turn out to be identical to those of the constant-amplitude case obtained earlier.

The ratio of signal energy to noise spectral density E/n_0 is again found to play the fundamental role in determining the detectability of a signal in gaussian noise. The energy E is now in this more general case given by $E = \int_0^T s^2(t)\,dt$, the energy in the time-varying signal. As an example, Eq. (5.48) for the probability of detection of a radar signal in gaussian noise (assuming phase-coherent detection) still applies in the case of varying-amplitude signals. These extensions of the constant-amplitude-signal case will be developed in the paragraphs following.

Ad Hoc Formulation: Filtering of Signals in Noise

To obtain quantitative results for the variable-signal case we first use an empirical approach, postulating a particular form of processor. We then consider optimum approaches utilizing the likelihood ratio. The reader may recall that this was how we handled the constant-amplitude case as well.

While this ad hoc formulation does not minimize the probability of a detection error, it is very useful. This approach yields a processor which maximizes a signal-to-noise ratio for a wide variety of noise distributions. Later we shall see that if the noise is gaussian, the same processor is optimal in the sense of minimizing P_e.

As a generalization of our previous discussion, then, assume that when the signal is present, the received signal-plus-noise samples are given by

$$x_j = s_j + n_j \qquad j = 1, 2, \ldots, m \tag{5.56}$$

The values of the m signal samples $s_j, j = 1, \ldots, m$, are assumed known and obtained by sampling some known waveshape $s(t)$, as in Fig. 5.23.

How does one now process the m samples x_j to determine the presence or absence of a signal? Simply adding the samples together, as suggested at the beginning of this chapter, does not appear appropriate since it weights the noise samples the same whether the signal sample is small or large. A more appropriate procedure might be to provide less weight to x_j when s_j is small, greater weight when s_j is large. (This is exactly the optimum result we shall obtain later in the case of gaussian noise, as pointed out earlier. At this point no particular noise statistics are specified.) We thus propose as a possible processor a weighted-averaging device

$$y(m) = \sum_{j=1}^{m} h_j x_j \tag{5.57}$$

$y(m)$ is then compared with some preset threshold, as proposed earlier. Here the known constants $h_j, j = 1, \ldots, m$, are presumably chosen small when s_j is small, large when s_j is large. (When $h_j = 1/m$, all j, we have the previous sample-mean case.)

This use of weighted sums to pick out a signal of desired shape corresponds to the signal-approximation schemes mentioned in Chap. 2 where linearly increasing weights were used to obtain the slope of a straight line fit to data, and the coefficient of a harmonic term in a Fourier series was found by integrating the data with a weighting function of the same shape.

Note, moreover, that this processor providing a weighted sum of the x_j's is just a special case of the moving-average or nonrecursive *filter* discussed in Chap. 2. We are in fact suggesting by Eq. (5.57) that one possible way of detecting the presence of the signal is to pass the successive samples x_j, $j = 1, \ldots, m$, through a *linear filter* given by Eq. (5.57) and to compare the output with a threshold. Recall that in Chap. 2 we pointed out that the averaging of two successive samples provides signal smoothing in the presence of noise and that this is the special case of a linear filter smoothing many samples. (In Chap. 2 we emphasized the problem of estimating a signal in noise. Smoothing was found to improve the estimation. In this chapter we have focused thus far on the equivalent *signal-detection* problem. We shall consider more general linear filtering for estimation purposes later in Chap. 6.)

How does one quantitatively specify the weighting or filter coefficients h_j? Various seat-of-the-pants approaches are of course possible. One intuitively satisfying approach is to maximize the difference between $y(m)$ with signal present and $y(m)$ with signal absent. Since the noise samples n_j are random, $y(m)$ is a random variable. We must therefore talk of the difference in an expected-value or mean sense. One would expect filter coefficients chosen on this basis to provide enhanced signal detectability [we are "separating" the two possible values of $y(m)$ as far apart as possible]. As noted above, we shall in fact show that the resultant linear filter is identical to the optimum processor obtained when the noise is assumed gaussian and minimum error probability (or maximum detection probability in a Neyman-Pearson sense) is desired. The filter also has the property of weighting the received samples x_j proportionally to the respective signal sample s_j, exactly the procedure suggested earlier. Such a filter, to be discussed in detail later, is called a matched filter, as we have noted.

Specifically, we propose finding the filter coefficients h_j that maximize a parameter λ given by

$$\lambda = \frac{\left(E[y]\big|_{\text{signal}} - E[y]\big|_{\text{no signal}}\right)^2}{\sigma_y^2\big|_{\text{no signal}}} \tag{5.58}$$

The denominator term is introduced for normalization purposes, and the numerator is squared to keep the parameter positive. λ may from its definition be interpreted as a signal-to-noise parameter. With the signal absent, $x_j = n_j$ and $y(m) = \sum_{j=1}^{m} h_j n_j$. With the signal present, $x_j = s_j + n_j$ and $y(m) = \sum_{j=1}^{m} h_j(s_j + n_j)$. If we assume that the noise samples are independent and have zero mean, $E[n_j] = 0$, we get the following results for the three parameters comprised by λ:

$$E[y]\Big|_{\text{no signal}} = 0 \qquad E[y]\Big|_{\text{signal}} = \sum_{j=1}^{m} h_j s_j$$

and

$$\sigma_y^2\Big|_{\text{no signal}} = \sigma_n^2 \sum_{j=1}^{m} h_j^2$$

Here σ_n^2 is the variance of each noise sample. (Details of the calculations are left to the reader.) Then

$$\lambda = \frac{\left(\sum_{j=1}^{m} h_j s_j\right)^2}{\sigma_n^2 \sum_{j=1}^{m} h_j^2} \tag{5.59}$$

The choice of filter coefficients h_j to maximize λ is readily found through the use of the Schwarz inequality

$$\left(\sum_j a_j b_j\right)^2 \leq \left(\sum_j a_j^2\right)\left(\sum_j b_j^2\right) \tag{5.60}$$

Here a_j and b_j are in general any real numbers. In particular, equality occurs when

$$a_j = kb_j \qquad j = 1, 2, \ldots m \tag{5.61}$$

k an arbitrary constant.

In our problem the samples s_j $(j = 1, \ldots, m)$ are assumed known. Thus dividing λ in Eq. (5.59) through by the constant $\sum_{j=1}^{m} s_j^2$ cannot change the maximization, but in that case the numerator and denominator of Eq. (5.59) are exactly in the form of Eq. (5.60). Hence λ is maximized when $h_j = ks_j$, just the result we are seeking. As noted earlier, a filter with this property is a special case of a matched filter (the filter coefficients are *matched* to the signal). The filter weights each input sample x_j by the known value of the signal s_j. Those samples possibly containing smaller values of signal are thus weighted less heavily than those possibly containing larger values. Note that the maximum value of λ is in particular given by

$$\lambda\Big|_{\max} = \frac{\sum_{j=1}^{m} s_j^2}{\sigma_n^2} \tag{5.62}$$

This again has the form of a signal-to-noise ratio. We shall show later that this is just $2E/n_0$ in the notation used earlier. This comes from Eq. (2.66), relating energy E and the sum of squares of samples. As a matter of fact, when $s_j = A$, all j, just the case discussed exhaustively earlier, $\lambda = mA^2/\sigma_n^2 = 2E/n_0$ with $m = 2BT$, $E = A^2T$, $\sigma_n^2 = Bn_0$, as previously. This is of course exactly the parameter appearing earlier in the probability-of-error calculations. One would like to make $\lambda|_{\max}$ as large as possible. This can obviously be done by reducing the noise variance σ_n^2 as much as possible, by increasing the size of the signal samples s_j as much as possible, or by increasing the number of independent samples m used. This was noted earlier as well. Obviously $\lambda|_{\max}$ can be made as large as possible when $s_j = A$ and A is made as large as possible. The fact that the signal samples may vary in amplitude thus provides deterioration of the processor performance over the equal-sample case, but if this situation cannot be helped (the usual real-life situation), a matched filter provides the best detectability for *the criterion chosen* (maximum λ). Looked at another way, if the signal varies in amplitude, its energy given by $E = \int_0^T s^2(t)\, dt$ is less than that which would be obtained for the constant-amplitude case $s = A$, with A the *maximum* value of $s(t)$. The decrease in energy accounts for the deterioration in performance as measured by λ, the signal-to-noise ratio at the detector. We shall see in the case of gaussian noise that the probability of error (or probability of detection in the radar-type detection problem) will also deteriorate if the signal amplitude varies. Signals, whether with varying amplitudes or constant amplitude, with the same value of $\sum_{j=1}^{m} s_j^2$ or energy E, however, will be detected with the *same* probability of error.

The unequal m-sample case can be phrased much more succinctly by using m-dimensional vector notation. Since this notation will also simplify much of the

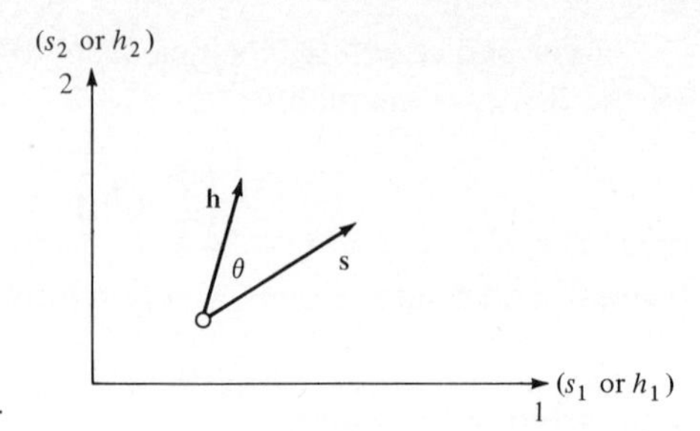

FIGURE 5.24
Example of two-dimensional space.

discussion following, we digress to introduce the notation and repeat some of the formulation above in vector form. Specifically, consider column vectors **h** and **s** representing the respective arrays of numbers $h_1, h_2, \ldots, h_m$ and $s_1, s_2, \ldots, s_m$. In an m-dimensional space one can define the euclidean length of a vector **h** as an extension of the equivalent three-dimensional length by $\sqrt{\sum_{j=1}^{m} h_j^2}$. This is called the norm $\|\mathbf{h}\|$ of the vector. The sum of squares of vector components $\sum_{j=1}^{m} h_j^2$ is just the square of the norm and is also given by

$$\sum_{j=1}^{m} h_j^2 = \mathbf{h}^T\mathbf{h} \equiv (\mathbf{h}, \mathbf{h}) \tag{5.63}$$

The parenthetical notation is often used to represent the "product" of **h** with itself, just the vector interpretation of the norm squared.

The dot, or inner, product of the vector in an m-dimensional euclidean space can be defined, just as in the three-dimensional case, as the projection of one vector on the other. Thus

$$(\mathbf{h}, \mathbf{s}) = \mathbf{h}^T\mathbf{s} = \mathbf{s}^T\mathbf{h} = \sum_{j=1}^{m} h_j s_j \tag{5.64}$$

Consider the angle θ between the two vectors. (See the two-dimensional case of Fig. 5.24.) Then it is readily shown that θ is defined by

$$\cos\theta = \frac{(\mathbf{h}, \mathbf{s})}{\sqrt{(\mathbf{h}, \mathbf{h})} \cdot \sqrt{(\mathbf{s}, \mathbf{s})}} \leq 1 \tag{5.65}$$

But notice that this is just the Schwarz inequality of Eq. (5.60), rewritten in vector form. In particular the equality holds when the two vectors are collinear and θ is zero.

Optimum Formulation

We now discuss the optimum processing for the detection of an arbitrary but known signal in the presence of noise. This is just an extension of the previous approach for fixed-amplitude samples. In the absence of signal we have, as earlier,

$$x_j = n_j \qquad j = 1, \ldots, m$$

and the density function $f(x_j \mid 0) = f(n)$ can readily be found. With signal present we have

$$x_j = s_j + n_j \qquad j = 1, \ldots, m$$

and again $f(x_j \mid s_j) = f_n(x_j - s_j)$. In this case the density function differs for each sample because of the variation in signal level. Consider the received samples $x_j, j = 1, \ldots, m$. When the vector $\mathbf{x}$ is used to denote the m samples, the density functions, in the case of independent noise samples, are given respectively by

$$f(\mathbf{x} \mid 0) = \prod_{j=1}^{m} f(x_j \mid 0) \tag{5.66}$$

and

$$f(\mathbf{x} \mid \mathbf{s}) = \prod_{j=1}^{m} f(x_j \mid s_j) \tag{5.67}$$

Here we have used the symbol $\mathbf{s}$ to denote the presence of the signal vector. Proceeding exactly as in the equal-amplitude signal case, one finds that the optimum procedure for declaring a signal present in the sense of minimum probability of error is to have

$$\ell(\mathbf{x}) = \frac{\prod_{j=1}^{m} f(x_j \mid s_j)}{\prod_{j=1}^{m} f(x_j \mid 0)} > \frac{1 - P_1}{P_1} \tag{5.68}$$

Alternately, taking logarithms, we get the logarithm-likelihood-ratio test

$$\log \ell(\mathbf{x}) = \sum_{j=1}^{m} \log \frac{f(x_j|s_j)}{f(x_j|0)} > \log \frac{1 - P_1}{P_1} \tag{5.69}$$

There is no essential difference between this result and that obtained previously. We simply have to calculate the appropriate conditional density function for each signal sample s_j, take logarithms, and sum accordingly. The actual processing algorithm again depends on the form of the noise statistics.

Consider now in particular gaussian noise. Here we have

$$f(x_j \mid s_j) = \frac{e^{-(x_j - s_j)^2/2\sigma_n{}^2}}{\sqrt{2\pi\sigma_n^2}} \tag{5.70}$$

with s_j as the mean value of the jth observation x_j. The logarithm of the ratio of the respective density functions in Eq. (5.70) is given by

$$\frac{1}{2\sigma_n^2}[x_j^2 - (x_j - s_j)^2] = \frac{1}{2\sigma_n^2}(2s_j x_j - s_j^2)$$

Substituting into Eq. (5.70) and collecting the terms involving the measured samples x_j on the left and all other terms on the right, we get as the rule for declaring a signal present

$$\sum_{j=1}^{m} s_j x_j > \frac{1}{2} \sum_{j=1}^{m} s_j^2 + \sigma_n^2 \log \frac{1 - P_1}{P_1} \tag{5.71}$$

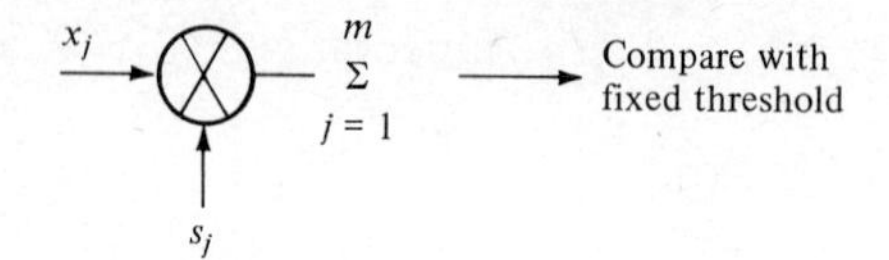

FIGURE 5.25
Optimum processor, gaussian noise.

Note that if $s_j = A$, all j, we again have the optimum processing algorithm derived earlier as Eq. (5.22). Note specifically the appearance of the *matched filter* in this case. Because of the form of the filter, this processor is often called a *correlation detector*: it cross-correlates samples of the incoming signals x_j with stored samples s_j of the known signal to be detected. The same form of detector is obviously obtained for the Neyman-Pearson and minimum-average-cost criteria discussed earlier. The right-hand side of Eq. (5.71), the decision level, is the only quantity changed in these two cases. The optimum processor is shown schematically in Fig. 5.25.

In vector notation the processing algorithm can be visualized as forming the inner (dot) product of **x** and **s** and comparing with a threshold depending on the squared length of **s** (**s**, **s**) as well as the signal-occurrence probability P_1

$$(\mathbf{s}, \mathbf{x}) > \tfrac{1}{2}(\mathbf{s}, \mathbf{s}) + \sigma_n^2 \log \frac{1 - P_1}{P_1} \tag{5.72}$$

This analysis and that preceding for the case of fixed-amplitude signals can readily be extended to the case of deciding which of two signals is present when one or the other is always present and corrupted by noise.† This is particularly significant in digital communications, where two different signal waveshapes are used to carry the appropriate binary information. The likelihood ratio again appears, and in the special case of gaussian noise one finds the processor consisting of two matched filters, one matched to each signal. The outputs of these two filters are then compared to determine which is larger.

We noted at the start of this section on the detection of varying-amplitude signals that the expression for the probability of error (or probability of detection in the Neyman-Pearson case) turns out to be the same as that obtained in the constant-amplitude case. This is readily demonstrated with the use of the identity, Eq. (2.66), relating the energy E in a bandlimited function to the sum of the squares of its sample values. Specifically, recall that it was shown there that any time function $s(t)$ whose frequency components lie within a finite bandwidth B Hz is related in the following way to its Nyquist samples s_j, $j = 0, \pm 1, \pm 2, \ldots$, spaced $T_s = 1/2B$ s apart:

$$E = \int_{-\infty}^{\infty} s^2(t)\, dt = \frac{1}{2B} \sum_{j=-\infty}^{\infty} s_j^2 \tag{5.73}$$

It was also indicated that this relation connecting the energy E to the sum of the uares of the sample values is valid to a good approximation for finite-time signals, ., those constrained to occupy a finite-time interval T rather than the infinite :erval of Eq. (5.73), providing we chose $2BT \gg 1$. This comes from the fact that

† Schwartz, op. cit., chap. 8.

such signals constrained in time cannot be constrained to occupy a finite bandwidth in the frequency domain as well. To a good approximation, however, essentially all the signal energy will lie in a frequency range $B \gg 1/T$. The relation $2BT \gg 1$ ensures that the bandwidth we have chosen is large enough. Since $2BT$ is also the number of samples in T s, it also ensures that enough samples are used in approximating the infinite sum of Eq. (5.73). For smooth pulses, 10 samples will ensure the validity of this approximation, so that $2BT > 10$ suffices to satisfy the relation $2BT \gg 1$. In the case of a time-limited signal we also have, to a good approximation,

$$E = \int_0^T s^2(t)\,dt \approx \frac{1}{2B} \sum_{j=1}^{m=2BT} s_j^2 \tag{5.74}$$

with $m = 2BT \gg 1$. This expression enables us to connect the performance of the matched-filter processor with that of the constant-amplitude signal processor discussed in previous sections.

Specifically, note that the output of the matched-filter processor is, as shown in Eq. (5.71), just $\sum_{j=1}^m s_j x_j$. This is just the random variable $y(m)$ introduced in Eq. (5.57) with the filter coefficient $h_j = s_j$. (This is the reason for the use of the name matched filter: the filter coefficients are *matched* to the signal.) Dropping the reference to m and letting $y(m) = y$ for ease of notation, we note that $y = \sum_{j=1}^m s_j x_j$ has two different statistics according as the signal is present or absent. In these two cases we have $x_j = s_j + n_j$ or n_j, respectively. In the first case we have $E(y) = \sum_{j=1}^m s_j^2$ and $\sigma_y^2 = \sigma_n^2 \sum_{j=1}^m s_j^2$; while in the second case, $E(y) = 0$ and σ_y^2 is the same. These results are left to the reader to justify. Since n_j and hence x_j are gaussian, y is gaussian as well. In particular the two density functions for y in the two cases of signal present and signal absent, respectively, are precisely those used earlier in handling the equal-amplitude case; see Eqs. (5.24), (5.25), and (5.43) and the discussion following each equation.

The key fact emerging is that the quantity $\sum_{j=1}^m s_j^2$ appears in the expressions for $E(y)$ and σ_y^2. From Eq. (5.74), however, this is very nearly $2BE$, with $E = \int_0^T s^2(t)\,dt$ the energy in the signal. Using this equivalence and also the relation $\sigma_n^2 = n_0 B$, with n_0 the noise spectral density introduced in previous sections, it is left to the reader to show that the probability of error in the equally likely binary-transmission case is given by

$$P_e = \tfrac{1}{2}\left[1 - \operatorname{erf}\left(\frac{1}{2}\sqrt{\frac{E}{n_0}}\right)\right] \tag{5.75}$$

This is a generalization of Eq. (5.26), previously obtained for the constant-amplitude signal case. [As a check, if we let $E = A^2 T$, and write $E/n_0 = BE/n_0 B = mA^2/2\sigma_n^2$, Eq. (5.75) becomes exactly Eq. (5.26).] If binary signals of the form $\pm s(t)$ (bipolar signals) are used in place of $s(t)$ and 0 (*on-off signals*), as discussed here, it is readily shown that the probability of error becomes $P_e = \frac{1}{2}(1 - \operatorname{erf}\sqrt{E/n_0})$.† For a given probability of error such a set of binary signals requires only one-quarter the energy of the equivalent on-off signals (6 dB less). They are thus preferred where energy

† Ibid, p. 606.

(power) is at a premium. (Since the on-off case requires no power when a 0 is transmitted, the *average* power improvement is 2/1 or 3 dB.) It is left to the reader to show similarly that in the radar-detection problem, the probability of detection is precisely that obtained previously [Eq. (5.48)] for constant-amplitude signals:

$$P_d = \int_{c-\sqrt{E/n_0}}^{\infty} \frac{e^{-x^2}}{\sqrt{\pi}}\,dx = \tfrac{1}{2}\left[1 - \operatorname{erf}\left(c - \sqrt{\frac{E}{n_0}}\right)\right] \tag{5.76}$$

The constant c depends (as previously) on the false-alarm probability deemed tolerable [Eqs. (5.44) and (5.45)].

Matched filters or approximations to them are used in all modern digital-communication systems and radar systems to provide the appropriate smoothing against noise. In past years they were often implemented by using resistors and capacitors to form *analog-filter* equivalents to the digital filter discussed here, but the reduced cost of digital filtering brought about by modern integrated-circuit technology has led to the increased use of digital filters.

Although we had not previously touched on the effect of signal-amplitude variations in discussing the air-traffic-control radar, the results obtained there are still valid since we used power and then energy directly in evaluating the radar performances. The transmitted power used in the calculations already had amplitude variations taken into account. The only other factors that might be considered in modifying the results obtained previously would be the effect of random received signal variations due to random aircraft motion as the signal is reflected from it and the effect of a nonuniform antenna beam. The random signal effect is considered briefly in the next section. The effect of a nonuniform antenna beam appears primarily in processing the multiple pulses reflected from the aircraft as the beam rotates past it. Pulses transmitted and received at the edges of the beam particularly may have substantially less power and energy in them than those from the beam center. This means that the multiple pulses assumed added in the air-traffic-control radars (20 pulses for the ASR or airport radar, 13 for the ARSR or en route radar) have different energies associated with them and so do not provide the improvement expected. The reduction in expected improvement, often referred to as *beam-shape loss*, may be added into the overall losses listed earlier.

5.7 DETECTION OF RANDOM SIGNALS

So far in this chapter we have considered the signals being detected to have known amplitude, possibly time-varying. An obvious extension is the area where the signals themselves can be defined only probabilistically. This is the class of random signals introduced in Chap. 3. One example might be a gaussian signal. A sample of such a signal then has a gaussian distribution with specified mean and variance. Two successive samples are jointly gaussian with correlation between them dependent on the sampling interval. A group of m such samples is represented by an m-dimensional gaussian random variable. The autocorrelation function, as defined in Chap. 3, pro-

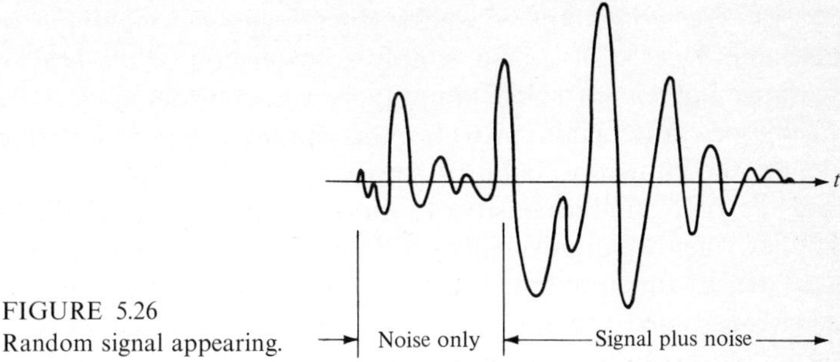

FIGURE 5.26
Random signal appearing.

vides the desired correlation terms, pair by pair, needed in describing the probability distribution of this random variable.

The gaussian noise we have used over and over again in the examples given in this chapter is, of course, an example of such a signal. We call it noise only when its presence is not desired. Random signal variations introduced by the turbulent motion of the aircraft in the radar-detection problem, as noted in the previous section, provide another example. If the measured statistics of the reflected signal turn out to be gaussian, the signal part of the received signal-plus-noise may be modeled as a gaussian signal. (This is not always the case, however; other statistical distributions may arise as well). Another specific example of a gaussian signal arises when one searches the sky with a radio telescope looking for a particular thermal-noise source. Ever-present noise provides a continuous output at the receiver. When an additional noise source (the "signal") is detected, the receiver output may fluctuate more widely, as shown in Fig. 5.26. The problem is then that of distinguishing the presence of the random signal: is it there in the time record being analyzed?

As already noted, a random signal does not have to have gaussian statistics. A common nongaussian example is the sine wave $s(t) = a \cos(\omega_0 t + \theta)$. If any (or all) of the parameters a, ω_0, and θ are random, $s(t)$ is a random signal. If such a signal appears obscured by noise of some kind, we say that we are attempting to detect a periodic random signal in the presence of the noise.

Our basic assumption here will be that the probability distribution of the signal samples is known. We shall further assume, as in the case thus far of the noise samples, that successive signal samples are statistically independent and possess the same statistical distribution. (The random signal is said to be *white* and *stationary*.) This rules out random signals which are slowly varying in relation to the sampling intervals. (An extreme case of a slowly varying signal is one with fixed but unknown amplitude. The detection procedure for this situation is not difficult to handle; it consists of conditioning on the constant signal level, leading to exactly the likelihood approach discussed earlier, and then averaging over the allowable signal levels. We shall not pursue this case here, however. This example of a very slowly varying random signal will be discussed in Chap. 6.)

Consider then the m measured samples x_j, $j = 1, \ldots, m$. They come from either of two distributions, one corresponding to signal present and the other to signal absent. We would like to decide which, again using some specified criterion for decision. This is, of course, simply an extension of the previous sections, where a variable but *known* signal amplitude was considered. It is thus apparent that a likelihood ratio again provides the optimum processor. Specifically, let $f(x_j \mid s)$ denote the known probability density of the jth sample with signal assumed present and $f(x_j \mid 0)$ again the density of x_j in the absence of signal. (Exactly the same analysis applies when we are deciding on the presence of either one of two random signals. One density function represents the conditional density function with one of the two signals assumed present, the other with the other signal assumed present. In the terminology of statistics we are dealing here with binary hypothesis testing: setting up a test to determine which of two possible hypotheses to choose.)

We decide that the signal is present if

$$\sum_{j=1}^{m} \ln \frac{f(x_j \mid s)}{f(x_j \mid 0)} > K \tag{5.77}$$

where K is some specified threshold. [Compare with Eq. (5.20), the logarithm of Eq. (5.19), and Eq. (5.69).] Previous derivations of this kind of test apply here also. The only difference between this random-signal case and previous signal models is in the evaluation of the densities in (5.77).

Consider as an example the gaussian signal case mentioned earlier. Let

$$f(s_j) = \frac{e^{-s_j^2/2\sigma_s^2}}{\sqrt{2\pi\sigma_s^2}} \tag{5.78}$$

We are thus assuming that each signal sample has zero mean and variance σ_s^2. Successive signal samples are independent. (Actually the condition of uncorrelated signal samples suffices in the gaussian case.) Let zero-mean gaussian noise of variance σ_n^2 be continuously present. Then if the signal is present, $x_j = s_j + n_j$ is also zero-mean gaussian with variance $\sigma_s^2 + \sigma_n^2$. The two conditional density functions are

$$f(x_j \mid s) = \frac{e^{-x_j^2/2\sigma^2}}{\sqrt{2\pi\sigma^2}} \tag{5.79}$$

with

$$\sigma^2 = \sigma_n^2 + \sigma_s^2$$

and

$$f(x_j \mid 0) = \frac{e^{-x_j^2/2\sigma_n^2}}{\sqrt{2\pi\sigma_n^2}} \tag{5.80}$$

These are sketched in Fig. 5.27. Our decision rule in this case is thus one for testing between two gaussian random variables of different variance.

From Eq. (5.77) we have, after some simplification,

$$\sum_{j=1}^{m} \frac{x_j^2}{2}\left(\frac{1}{\sigma_n^2} - \frac{1}{\sigma^2}\right) > \frac{m}{2} \ln \frac{\sigma^2}{\sigma_n^2} + K \tag{5.81}$$

or

$$\sum_{j=1}^{m} x_j^2 > d^2 \equiv \frac{2}{1/\sigma_n^2 - 1/\sigma^2}\left(\frac{m}{2} \ln \frac{\sigma^2}{\sigma_n^2} + K\right) \tag{5.82}$$

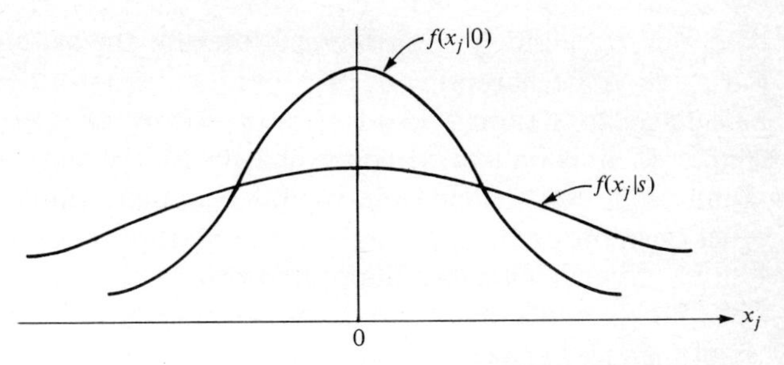

FIGURE 5.27
Gaussian signal plus noise.

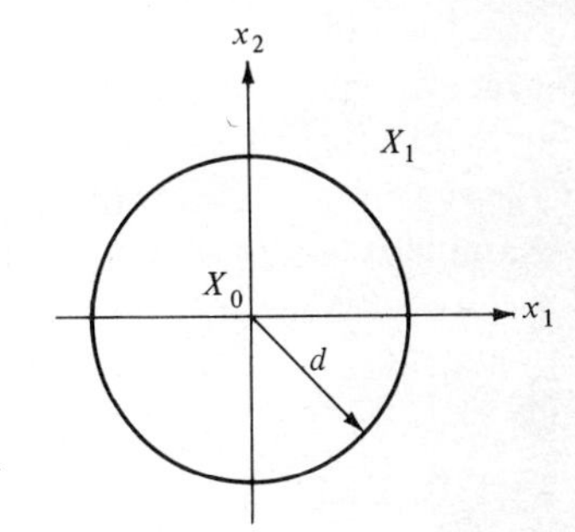

FIGURE 5.28
Two-sample decision regions, gaussian signal.

Note that the rule here simply says to square each sample, sum, and compare with some specified threshold. This has some interesting interpretations. First, the fact that the signal and noise are both zero-mean gaussian, equally likely to be positive or negative, should rule out dependence on the polarity of the x_j samples for determining the presence of the signal. The processor squaring operation of Eq. (5.82) in fact eliminates the polarity dependence. Second, x_j with the signal present has a larger variance and is more likely to have larger negative and positive expressions (Figs. 5.26 and 5.27). One would thus expect a test that looks for signals that are in absolute value above a specified threshold. For two samples the decision-present region X_1, as shown in Fig. 5.28, consists of all points *outside* the circle of radius d [Eq. (5.82)] surrounding the origin. In an m-dimensional space, X_1 corresponds to all points outside the hypersphere of radius d surrounding the origin.

5.8 SUMMARY

In this chapter we have considered a particularly simple signal-processing problem, that of detecting the presence of a signal in noise. This is obviously a prelude to any further signal processing to be carried out. We stressed the case where the statistics of the signal-plus-noise are known, showing that a likelihood ratio arises in determining the optimum digital processor. Where the noise statistics are gaussian and the signal deterministic, although possibly time-varying, the matched filter followed by threshold arises quite naturally as the optimum signal-processing operation. Optimum here refers to a rather broad class of performance criteria. The examples in this chapter focused on two criteria primarily: that of minimizing the probability of error (appropriate, for example, in digital communications) and the Neyman-Pearson criterion appropriate to radar, sonar, and other systems in which the a priori probability of receiving a signal is not known.

The air-traffic-control radar was invoked as a prime example of the applicability of the ideas presented here.

In the next chapter we focus on the follow-up signal-processing problem, the estimation of desired signal parameters. For example, in the radar case, once we have detected a signal denoting a target, we may be interested in estimating the target location and its velocity, among other quantities. This may be useful in tracking the target (the aircraft) to ensure that it is following an appropriate sky path, to prevent air collisions, or to help in the approach to an airport. We thus explore several signal-estimation techniques, again using the air-traffic-control problem as a specific example, culminating in Chap. 7 in the discussion of digital processors, called Kalman filters, which have been found particularly useful in signal-estimation problems.

PROBLEMS

5.1 A single observation $x = s + n$ consists of zero-mean gaussian noise plus either $s^{(1)} = 1$ or $s^{(2)} = 2$, with the two possible signals equally likely to be present.
(*a*) Show that the best minimum P_e decision rule says that $s^{(1)}$ is present if $x \leq 1.5$ and $s^{(2)}$ is present if $x > 1.5$.
(*b*) Find the corresponding P_e if $\sigma_n^2 = 1$.

5.2 Prove that the detection procedure for minimizing the probability of error when m independent data samples are available is given by Eq. (5.19).

5.3 Verify Eqs. (5.4) and (5.5) for the sample-mean detector of Eq. (5.2).

5.4 Consider the following detection problem. $x = s + n$, where s is either $+1$ or -1 with equal probability and n has the probability density function shown in Fig. P5.4. A single

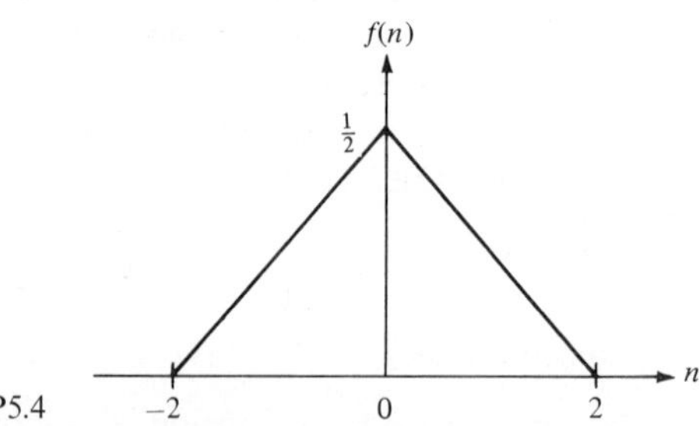

FIGURE P5.4

sample is taken and compared with a threshold set at 0:

$$\text{If } x \geq 0 \qquad +1 \text{ is declared present}$$
$$\text{If } x < 0 \qquad -1 \text{ is declared present}$$

Calculate P_e, the probability of error for this detection scheme.

5.5 Equally likely bipolar binary digits (amplitude $\pm A$) are transmitted as shown in Fig. P5.5. They are received with noise added. To determine whether a $+A$ or a $-A$ was

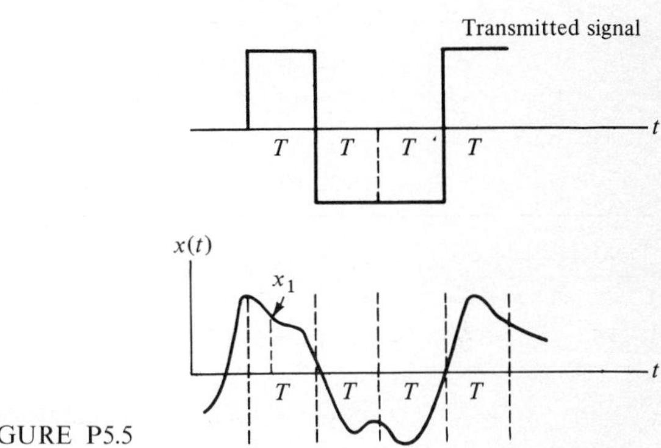

FIGURE P5.5

transmitted in each time interval T, one sample x_1 is taken, as shown. Then $x_1 = A + n$, *or* $x_1 = -A + n_1$, with n_1 the noise sample. The processing procedure is as follows:

$$\text{If } x_1 > 0 \qquad +A \text{ is deemed present}$$

$$\text{If } x_1 < 0 \qquad -A \text{ is deemed present}$$

Assume the noise has gaussian statistics. Then

$$f(n_1) = \frac{e^{-n_1{}^2/2\sigma_n{}^2}}{\sqrt{2\pi\sigma_n^2}}$$

Find the probability of error in the two cases $A/\sigma_n = 1$ and $A/\sigma_n = \sqrt{3}$. Use the table of the error functions in Chap. 3 or tables from any statistics or probability book.

Ans. 0.159, 0.041

5.6 Three independent samples, x_1, x_2, x_3, of the received signal of Prob. 5.5 are taken in each T-s interval.

(*a*) The three samples are added. Their sum is checked for polarity: if positive, $+A$ is declared present; if negative, $-A$ is declared present. Find the probability of error if $A/\sigma_n = 1$ ($\equiv 0$ dB signal-to-noise ratio). Compare with Prob. 5.5. What is the effect of either increasing A (the signal amplitude) *or* increasing the number of samples? *Hint:* The sum of gaussian random variables is still gaussian. *Ans.* $P_e = 0.041$

(*b*) An alternative processing procedure is to be investigated. The polarity of each of the three samples is checked, using logic circuitry. If positive, a 1 is stored in an up-down counter. If negative, a -1 is counted. If the net count for the three samples is positive, a $+A$ is declared present. If the net count < 0, $-A$ is declared present. (A majority-rule test gives the same result: declare $+A$ present if the majority of the polarities are positive.) Calculate the probability of error for $A/\sigma_n = 1$ and compare with part (*a*), where the samples were added. Which processing technique is better? Why? *Hint:* Assume $-A$ present. An error occurs if two or three of the samples are positive. Let the probability of a sample being positive $= p$. Then the probability that two are positive is $3p^2(1-p)$.

Why? The probability that three are positive is p^3. Why? What is the probability that two *or* three samples are positive? *Ans.* 0.068

5.7 Consider the same situation as in Prob. 5.5. The noise now has a *triangular* density function, as shown in Fig. P5.7.

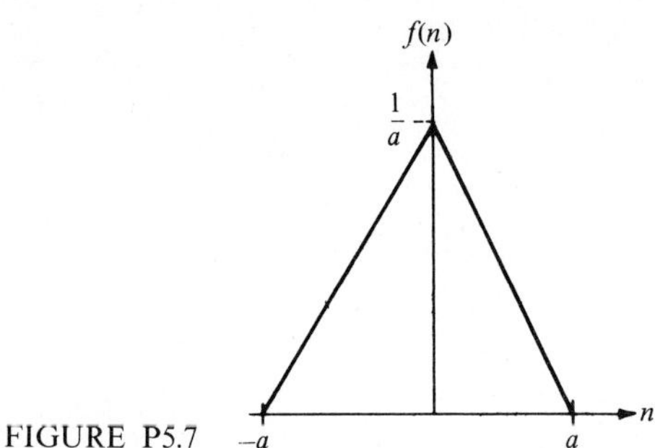

FIGURE P5.7

(*a*) Sketch the two density functions $f(x_1 \mid A)$ and $f(x_1 \mid -A)$ corresponding to $x_1 = +A + n$ and $x_1 = -A + n_1$, respectively, for $a > A$. Indicate the two regions X_1 and X_0 corresponding to $+A$ and $-A$, respectively. Remember that the signals are equally likely.
(*b*) $a = 1$. Calculate the *minimum* probability of error if $A = \frac{1}{8}, \frac{1}{4}, \frac{1}{2}$.

5.8 Consider $x_j = \pm A + n_j$, with the density function of n_j given by the laplacian density function (Eq. 5.28)

$$f(n_j) = \frac{e^{-|n_j|/c}}{2c}$$

Find the optimum processing procedure for n samples of x_j if $+A$ and $-A$ are equally liked to be transmitted (optimum here means minimum probability of error).

5.9 The binary signals 0 or A are transmitted with equal probability. They are received in additive noise with probability density

$$f(n) = \frac{1}{\pi c} \frac{1}{1 + (n/c)^2} \qquad -\infty < n < \infty$$

(*a*) One sample of signal plus noise, $x = s + n$, $s = 0$ or A, is taken. Show that the decision rule for minimum probability of error corresponds to signal A declared present if $x \geq A/2$. Show that the probability of error is $\frac{1}{2} - (1/\pi) \tan^{-1} (A/2c)$.
(*b*) If m independent samples, $x_1, x_2, \ldots, x_m$, are now taken, find the decision rule for processing the m samples that guarantees minimum probability of error. *Do not attempt to simplify.*

5.10 Refer to Fig. P5.10. One sample of x is taken. If $x > d$, signal is declared present.
(*a*) Show that if $d = 2$, the false-alarm probability is 0.135 (probability that noise alone is mistaken for signal).
(*b*) Find the probability of detection (probability that signal, when present, is correctly detected) for $d = 2$.

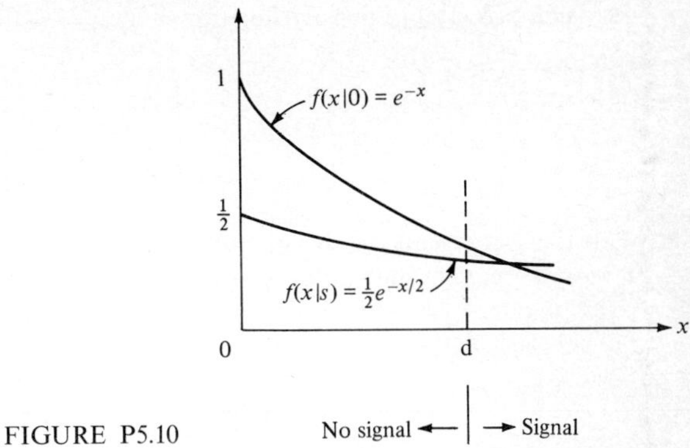

FIGURE P5.10

5.11 Using the Neyman-Pearson criterion, show that the decision procedure for declaring a signal of amplitude A present in a gaussian-noise background is given by Eq. (5.40) for one sample and Eq. (5.42) for m independent samples.

5.12 The output of a radar receiver is sampled m times in T s (the expected width of a received signal pulse that would appear if a target were present). The individual samples correspond to zero-mean independent gaussian-noise samples of variance σ_n^2 in the no-signal case or to an amplitude A plus noise in the signal-present case. The individual samples are summed as in Eq. (5.42), and a signal is declared present if the sum $> d$, a decision level.

(*a*) Show that the false-alarm probability is given by Eq. (5.44), with $c = d/\sqrt{2m}\,\sigma_n$, and the error function defined by Eq. (5.27).

(*b*) Show that the probability of detection is given by Eq. (5.46).

(*c*) $P_n = 10^{-10}$ ($c = 4.5$), and 10 samples are taken. Find the signal-to-noise ratio A/σ_n required if $P_d = 0.5$. Repeat for $P_d = 0.9$.

(*d*) $m = 2BT$ samples are now taken; Eq. (5.48) is now applicable. Consider the ASR radar and show that $P_R = 2 \times 10^{-13}$ W, using Eq. (5.50).

(*e*) Then $E/n_0 = 33.2$. Show that if the decision level is chosen to have $P_n = 10^{-10}$ ($c = 4.5$), $P_d = 0.96$. [Use Eq. (5.48).] What happens if the energy E is halved (or noise spectral density n_0 is doubled)? What if E is doubled? [*Note*: $\operatorname{erf}(-z) = -\operatorname{erf} z$].

5.13 Consider the ASR (airport) and ARSR (en route) radars discussed in Sec. 5.4.

(*a*) Show that for the transmitted power, target reflecting area, and maximum range indicated for each, the received powers are respectively 2×10^{-13} and 0.63×10^{-13} W. What happens to these powers if the transmitted powers are halved? If the range is halved?

(*b*) The noise figure for the ARSR radar is 4 dB. Show that for a false-alarm probability of 10^{-10} the probability of detection, using Eq. (5.48), is 0.76. This assumes that only one pulse is reflected back from a reflecting aircraft.

(*c*) Verify the numbers shown in Table 5.2 when the appropriate number of pulses reflected back is included in the calculations.

5.14 Consider high-frequency gaussian noise with zero mean and variance σ^2 (Fig. 5.1*b*). It can be shown that the *envelope* r of this noise when sampled is a random variable with the

Rayleigh probability density function

$$f(r) = \begin{cases} \dfrac{re^{-r^2/2\sigma^2}}{\sigma^2} & r \geq 0 \\ 0 & r < 0 \end{cases} \tag{i}$$

This then represents the density function of samples at the output of an *envelope detector* if high-frequency gaussian noise appears at the input. Similarly, if a high-frequency signal of amplitude A (say $A \cos \omega_0 t$) is added to the gaussian noise (Fig. 5.1*b*), the density of the *envelope* of the sum is given by the Rician function

$$f(r|s) = \begin{cases} \dfrac{re^{-A^2/\sigma^2}}{\sigma^2} e^{-A^2/2\sigma^2} I_0\left(\dfrac{rA}{\sigma^2}\right) & r \geq 0 \\ 0 & r < 0 \end{cases} \tag{ii}$$

where $I_0(x)$ is the modified Bessel function. Equations (i) and (ii) are the equations used in constructing the noncoherent radar curves of Fig. 5.20.

(*a*) m independent samples $r_1, r_2, \ldots, r_m$ of the envelope-detector output are taken. Find the form of the optimum processor.

(*b*) For the case of high signal-to-noise ratio ($A^2 \gg \sigma^2$) Eq. (ii) may be approximated by a gaussian function of variance σ^2 and average value A. Sketch Eqs. (i) and (ii) in this case and show graphically what the detection procedure is for *one* sample.

5.15 *Coherent Radar Detection*

(*a*) Show how one obtains the coherent-detection curves of Fig. 5.20. *Hint:* Use Eq. (5.48) and see the discussion of linear integration at the end of Sec. 5.4.

(*b*) Each pulse at a radar receiver is separately detected and compared to a threshold. Say 10 pulses are received. If k of these exceed the threshold, a signal is declared to be present. The threshold is chosen to have the false-alarm probability $P_n = 10^{-10}$. The desired probability of detection $P_d = 0.5$. Find E/n_0 required per pulse for $k = 3, 5, 7$ and compare with the appropriate point on the $P_d = 0.5$ coherent-detection curve of Fig. 5.20. *Note:* Assume gaussian statistics in this case; Eq. (5.48) then applies for *each* pulse. Use the technique outlined in Sec. 5.5: Eqs. (5.52) to (5.55) are appropriate. Tables of the cumulative binomial distribution are available, or you may solve the problem by computer.

5.16 Starting with Eq. (5.58), show that one gets Eqs. (5.59) and (5.62).

5.17 Verify Eqs. (5.75) and (5.76) for the optimum (matched-filter) detection of a signal of energy E in the presence of gaussian noise.

5.18 Refer to Fig. P5.18.

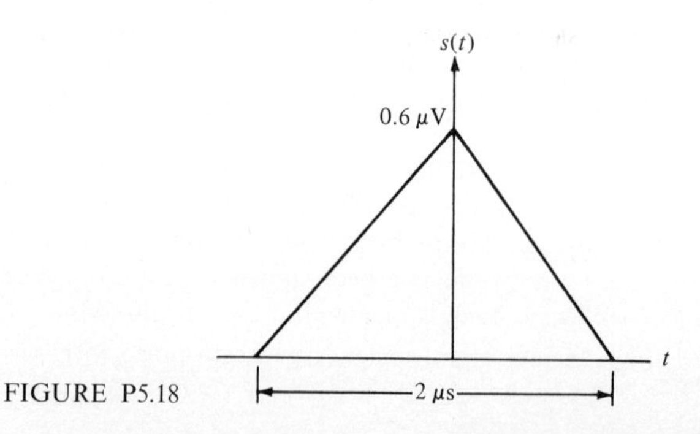

FIGURE P5.18

(*a*) Show that the energy in $s(t)$ is $E = 24 \times 10^{-20}\ \mathrm{V^2 \cdot s}$.
(*b*) A pulse of this type is received from a target in a radar system. Gaussian noise of spectral density $n_0 = \frac{2}{3} \times 10^{-20}\ \mathrm{V^2/Hz}$ is added. Find the probability of detection if the false-alarm probability is fixed at 10^{-10}. *Hint:* Use (5.76), $c = 4.5$ here.

Ans. 0.983

5.19 Nine independent observations $x_i = s + n_i$, $i = 1, 2, 3, \ldots, 9$, are available with $s = 0$ or $\frac{1}{3}$, and $n_1, n_2, \ldots$ mutually independent, gaussian, zero mean, and each with variance $\sigma^2 = 0.09$. We shall say $s = \frac{1}{3}$ if $y = \sum_1^9 x_i > d$. Find d so that the false-alarm probability $P_{FA} = 10^{-8}$. What is the corresponding detection probability P_d? Sketch $f(y \mid 0)$ and $f(y \mid \frac{1}{3})$. If x_i are the samples indicated on the $x(t)$ shown in Fig. P5.19, what decision should we make ($s = 0$ or $\frac{1}{3}$)?

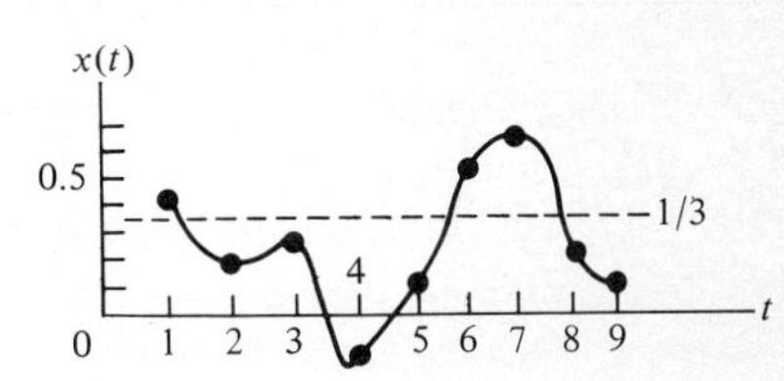

FIGURE P5.19

5.20 Eight samples are either

$$x_i = s_1(t_i) + n_i \qquad \text{or} \qquad x_i = s_0(t_i) + n_i$$

where $s_1(t)$ and $s_0(t)$ are as shown in Fig. P5.20. The two signals are equally likely, and the noise samples are independent, gaussian, zero mean, with variance σ_n^2. The minimum probability of error rule is used. What is the decision when $\{x_i\} = \{-2, -6, 1, 6, 4, 2, 0, -7\}$? Find the distribution of $\sum_1^8 s_1(t_i) n_i$ and determine P [declare $s_1 \mid s_0$ is present].

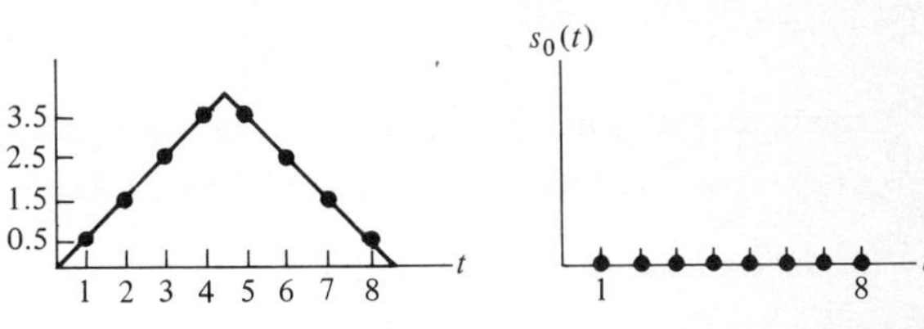

FIGURE P5.20

5.21 A radar transmits the triangular pulse $s(t)$ of Fig. P5.21. If an aircraft is present to reflect the signal, a triangle is returned with zero-mean gaussian noise added. If no target is

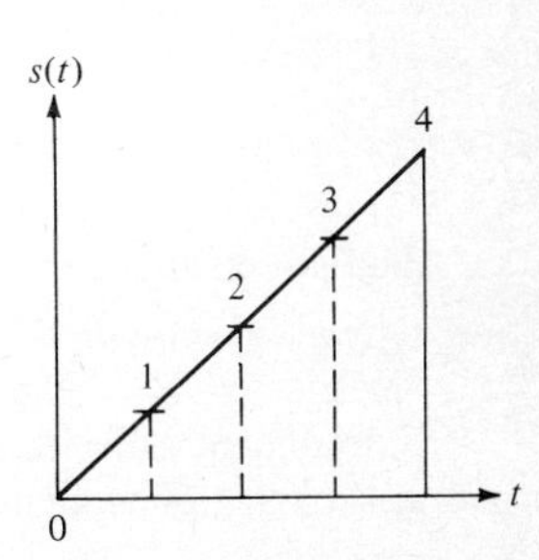

FIGURE P5.21

present, noise alone appears. Thus

$$x_j = s_j + n_j \qquad \text{or} \qquad x_j = n_j$$

$$s_1 = 1 \qquad s_2 = 2 \qquad s_3 = 3 \qquad s_4 = 4$$

$$\sigma_n^2 = 2$$

(*a*) Show that the use of the likelihood ratio results in the following optimum processor. Signal declared present if

$$y = \sum_{j=1}^{4} s_j x_j > d$$

(This is then a matched filter followed by a threshold level.)
(*b*) The false-alarm probability is to be kept at 10^{-8}. Derive the following equation from which the threshold level d can be found:

$$10^{-8} = \int_{d/\sqrt{120}}^{\infty} \frac{e^{-w^2}\,dw}{\sqrt{\pi}}$$

(*c*) Show that the probability of detection is given by

$$P_d = \int_{(d-30)/\sqrt{120}}^{\infty} \frac{e^{-w^2}\,dw}{\sqrt{\pi}}$$

5.22 Two equally likely, bipolar, signals $s(t)$ and $-s(t)$ are detected in the presence of zero-mean gaussian noise, $\sigma_n^2 = 1$. Thus,

$$x(t) = s(t) + n(t) \qquad \text{or} \qquad x(t) = -s(t) + n(t)$$

$s(t)$ is declared present if $\sum_{j=1}^{m} s_j x_j > 0$, s_j the jth sample of $s(t)$.
(*a*) $m = 3$ independent samples of $x(t)$ are taken and found to be

$$x_1 = -1 \qquad x_2 = +1 \qquad x_3 = +3$$

The corresponding samples of $s(t)$ are

$$s_1 = +1 \qquad s_2 = -1 \qquad s_3 = +1$$

Which signal, $s(t)$ or $-s(t)$, is declared present?
(*b*) Let $y = \sum_{j=1}^{m} s_j x_j$, $m = 3$, as in part (*a*).
(1) $s(t)$ is *known* to have been transmitted. Show that $E[y] = 3$ and $V(y) = 3$ (variance of y).
(2) $-s(t)$ is *known* to have been transmitted. Show that $E[y] = -3$ and $V(y) = 3$.
(3) Sketch $f(y|s)$ and $f(y|-s)$. Indicate on the sketches how one calculates the probability of error.

5.23 One of two equally likely signals is transmitted and received in additive gaussian noise of variance $\sigma^2 = 1$. The signals are $s_1(t) = 1 = -s_2(t)$, in a 1-ms binary interval. Eight independent samples, $x_i = s_{1i}$ or $s_{2i} + n_i$, $i = 1, 2, \ldots, 8$, are taken.
(*a*) For the following sets of observed data, which signal would you decide was sent?

$$\{x_i\} = \{1.0, 0.5, -0.6, 1.5, -2.0, -1.0, 0.0, -0.3\} \qquad \text{(i)}$$

Ans. s_2

$$\{x_i\} = \{-1.0, -2.0, -1.5, -0.5, -0.4, -0.1, -0.2, 6.0\} \quad \text{(ii)}$$

Ans. s_1

(*b*) What is the probability of error? *Ans.* 0.00234

5.24 $s_1(t)$ is the triangle shown in Fig. P5.24, and $s_2(t) = 0$. The two signals are equally likely to be transmitted. The received signal $x(t)$ consists of $s_1(t)$ or $s_2(t)$ plus gaussian noise.

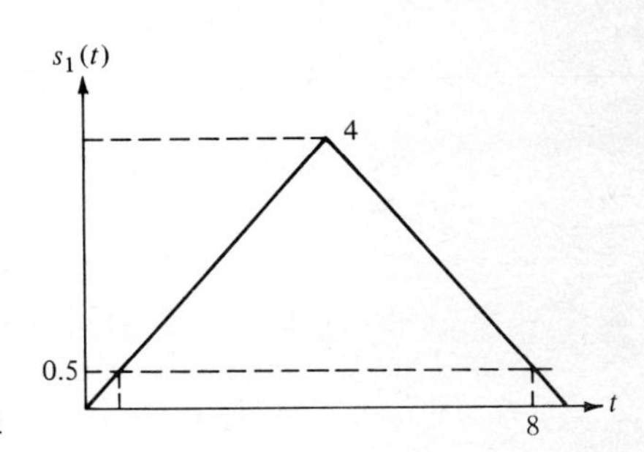

FIGURE P5.24

$x(t)$ is sampled eight times, starting at the point $s_1 = 0.5$ and ending at $s_1 = 0.5$, as shown. The samples are assumed statistically independent. For the following sets of samples, which signal would you decide was sent?

$$\{x_i\} = \{-2, -6, +1, +5, +6, +2, +1, -8\}$$ *Ans.* s_1

$$\{x_i\} = \{0, -2, -4, +5, +6, +2, -4, -2\}$$ *Ans.* s_1

5.25 Five samples $x_i = s_i + n_i$ are observed, with all $s_i = s_i^{(0)} = 0$ (signal 0 present) or $s_1^{(1)} = 1$, $s_2^{(1)} = 2$, $s_3^{(1)} = 3$, $s_4^{(1)} = 2$, $s_5^{(1)} = 1$ (signal 1 present). A matched filter computes $y = \sum_1^5 s_i x_i$. The n_i are independent, zero mean, gaussian, with $E[n_i^2] = \sigma_n^2$.
(*a*) Find the mean and variance of y given $s^{(0)}$ present.
(*b*) Compute the false-alarm probability if we declare $s^{(1)}$ present when $y > \frac{1}{2} \sum_1^5 (s_i^{(1)})^2$. Take $\sigma_n^2 = \frac{8}{19}$.

5.26 A random variable x is observed, coming from one of two possible densities, as sketched in Fig. P5.26. $f(x \mid s)$ says your grade is more likely to be 100 if you study; $f(x \mid 0)$ corresponds to the case of lower grades likely for lazy students.

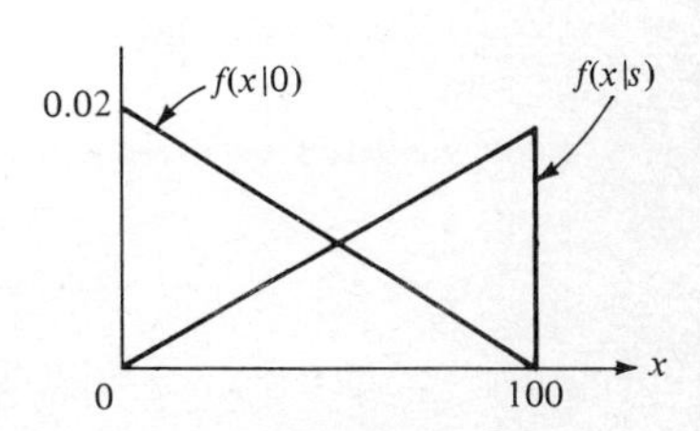

FIGURE P5.26

(*a*) Sketch the likelihood ratio $\ell(x)$ and show that $\ell(x) < K$ corresponds to $x < d$ for some d.
(*b*) Find a d to maximize my probability of correctly detecting that you studied, with a probability of 0.09 that I mistakenly think you studied when you really did not.
(*c*) What is the detection probability using the threshold from part (*b*)?

5.27 A message is transmitted by selecting between two noise generators whose outputs have the densities shown in Fig. P5.27.

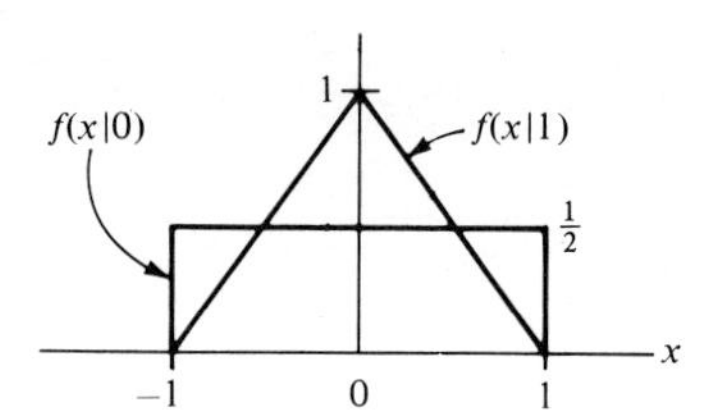

FIGURE P5.27

(*a*) Find the decision rule for maximizing the probability of detecting a 1 (triangular density) but keeping the probability of mistaking a 0 for a 1 equal to 0.1. *Hint:* Guess a K and find the x-region corresponding to $\ell(x) > K$.

Ans. $s = 1$ when $|x| < 0.1$

(*b*) What is the probability of correctly detecting a 1?

5.28 Two signals, s_1 and s_2, are equally likely to be received. If s_1 is present, the probability density of a data sample is given by

$$f(x \mid 1) = \frac{xe^{-x^2/2\sigma_1^2}}{\sigma_1^2} \qquad x \geq 0$$

If s_2 is present, the probability density of a data sample is given by

$$f(x \mid 2) = \frac{xe^{-x^2/2\sigma_2^2}}{\sigma_2^2} \qquad x \geq 0 \quad \sigma_2 > \sigma_1$$

(*a*) Sketch the two density functions and indicate *graphically* the region corresponding respectively to s_1 present and s_2 present if *one* data sample is used. (Probability of error is to be minimized.)
(*b*) Calculate the probability of error in this case.
(*c*) m independent samples $x_1, x_2, \ldots, x_m$ are now taken. Find the optimum processing algorithm for the m samples in this case. (Probability of error is to be minimized.) Compare the special case of $m = 1$ with the result of part (*a*).

5.29 m independent samples of $x(t)$ are to be used to determine the presence of a signal. Verify the processors for each of the following examples. (x_i is the ith sample.)

$$f(x_i \mid s) = \frac{e^{-x_i^2/2\sigma_s^2}}{\sqrt{2\pi\sigma_s^2}} \qquad f(x_i \mid 0) = \frac{e^{-x_i^2/2\sigma_n^2}}{\sqrt{2\pi\sigma_n^2}} \qquad \sigma_s^2 > \sigma_n^2 \qquad \text{(i)}$$

Then s is declared present if $\sum_{i=1}^{m} x_i^2 > d$, a prescribed threshold level.

$$f(x_i|s) = \frac{x_i e^{-x_i^2/2S}}{S} \qquad f(x_i|0) = \frac{x_i e^{-x_i^2/2N}}{N} \qquad x_i \geq 0 \quad S > N \qquad \text{(ii)}$$

Then s is declared present if $\sum_{i=1}^{m} x_i^2 > b$, a prescribed threshold level.

5.30 In a radio-astronomy experiment a specified region of the sky is continually scanned by a radio telescope tuned to 1 GHz. Energy at that frequency is always present in the form of sky noise and receiver noise. In this experiment a change in the ambient-noise level at 1 GHz due to a radio star transmitting is to be detected.

(*a*) A sample x_1 of the system output is taken. If ambient noise alone is present, it is known that x_1 is a positive random variable with density function

$$f(x_1 \mid 0) = \begin{cases} \dfrac{e^{-x_1/N}}{N} & x_1 \geqslant 0 \\ 0 & x_1 < 0 \end{cases}$$

(N is a measure of the ambient-noise intensity.) If an additional noise source is present (the "signal" in this case), the density function of x_1 is

$$f(x_1 \mid s) = \begin{cases} \dfrac{e^{-x_1/N_T}}{N_T} & x_1 \geqslant 0 \\ 0 & x_1 < 0 \end{cases}$$

Here $N_T = N + S$, with S the intensity of the new source. Use the likelihood ratio to show that the optimum way of processing x_1, in the sense of maximizing the probability of detection P_s of a source if present with the false-alarm probability fixed at some value P_n, is to check whether $x_1 > d$. Sketch $f(x_1|0)$ and $f(x_1|s)$ and justify this graphically.

(*b*) $P_n = 10^{-8}$. Show that $d = 18.4N$. Show that S/N = 25.6 (14.1 dB) is the required ratio of "signal" source to ambient-noise source if $P_s = 0.5$. Show S/N = 175 (22.4 dB) if $P_s = 0.9$.

(*c*) Show that $\sum_{j=1}^{m} x_j > d$ is the optimum processor if m independent samples are processed.

6

ESTIMATION OF SIGNALS IN NOISE

6.1 INTRODUCTION

In Chap. 5 we concentrated on the problem of signal detection: given a set of data samples, how does one process them to determine whether a signal of known characteristics is present or not? The signal could be completely defined, i.e., deterministic, or random, with known statistics. We found that the optimum processor consists of comparing the likelihood ratio (or its logarithm) with a specified threshold.

We now move on to a somewhat different problem, that of estimating the values of a signal from the given data samples. Here we *know* the signal is present but because of noise, inaccuracies in the data samples, limited precision of instruments (these are often all modeled as additive noise), or other obscuring artifacts, the data samples scatter about the actual signal values. The question now is: How does one process the data samples to obtain the "best estimate" of the signal?

The problem of signal estimation was first introduced at the beginning of Chap. 2, where the data samples were assumed to represent a signal of known type, e.g., a straight line of the form $c_0 + nc_1$, or some other polynomial, with n the sample number, and least-squares estimates of the desired parameters were obtained. This extremely important problem is essentially one of curve fitting: given a prescribed analytical form of a curve (this is then the "signal"), how does one best fit the data to

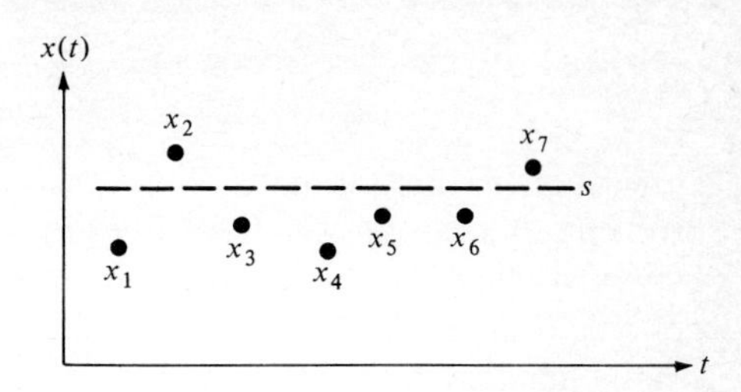

FIGURE 6.1
Estimation of signal parameter s from data samples.

it? This problem often occurs in trajectory calculations, in estimating vehicle velocity and acceleration for control purposes, in astronomical-orbit calculations, etc.

The air-traffic-control radar discussed in Chap. 5 is an example of a processor that carries out signal estimation. The estimation problem in this case is that of *tracking* the aircraft, once detected. The tracking function requires *predicted* estimates of the vehicle range as well as estimates of the vehicle angular velocity. These are updated on each scan of the radar, using past estimates and new data for the necessary processing. Details of this application appear in Chap. 7, following the introductory treatment of estimation in this chapter.

In this chapter and the next, we approach the estimation problem in a step-by-step fashion. As noted above, the estimation of random signals is an extension of the deterministic case considered in Chap. 2. We first take up estimating a signal of fixed but unknown amplitude in this chapter and then extend it in Chap. 7 to more generalized situations, e.g., those encountered in the radar tracking function. (There the signal, representing range and velocity, varies with time.) In saying that the signal is fixed, unvarying in time, we are implicitly assuming it to be a random variable. (Once the signal is allowed to take on a sequence of values in time, each defined probabilistically, as in Sec. 5.7 on the detection of random signals, we begin to deal with sequences of random variables.) This differs from the case treated in Chap. 2 simply because the signal may have *any* value within a range specified by its probability density function. Thus we know that a signal is present, and we can estimate its value, but at some other time we would expect to find another value present.

Again, for simplicity, assume that m data samples containing a signal parameter s and additive noise are to be processed. We thus have

$$x_j = s + n_j \qquad j = 1, 2, \ldots, m \tag{6.1}$$

The signal s is a random variable with some expected value $E[s] = s_0$ and variance σ_s^2. Assume that the noise samples are zero mean with identical variance σ_n^2 and are uncorrelated. (This latter condition is not really necessary but is introduced to simplify the analysis later.) An example of such a set of data samples is shown in Fig. 6.1. How do we now process these samples to obtain an estimate of s?

It is apparent from our previous discussions on detection and from the signal-estimation examples discussed in Chap. 2 that the sample mean should serve as a good estimator in this case. Specifically, then, using the symbol $\hat{s}$ to denote the estimate, we have

$$\hat{s} = \frac{1}{m} \sum_{j=1}^{m} x_j \tag{6.2}$$

How good is this estimate? How "close" does it come to s? As in the detection problem, one can answer these questions only after defining a measure of goodness. Obviously, one would like the probability that $\hat{s}$ deviates from s to be as small as possible. But unlike the detection case, probabilistic measures of goodness are not readily handled in the estimation case. Instead, it is more common to use the square of the error $(\hat{s} - s)^2$ as a measure of the deviation between the signal and its estimate. This is similar to the approach adopted in Chap. 2. Because s and $\hat{s}$ are random variables, however, the error $\varepsilon = \hat{s} - s$ is a random variable as well, and a small error in one measurement could very well be followed by a very large error at another time. A better approach is to average the square error over all possible values of $\hat{s}$ and s. This gives the mean-squared error $E[\varepsilon^2] = E[(\hat{s} - s)^2]$ as the measure of goodness of the estimate. It is the most commonly used performance measure in estimation problems.

In this particular case with $x_j = s + n_j$, we have

$$\varepsilon = \hat{s} - s = \frac{1}{m} \sum_{j=1}^{m} n_j$$

Note that the sample-mean estimator from (6.2) results in an error depending only on the noise samples; s cancels out. The expectation over $\hat{s}$ and s is thus the same as averaging over the m uncorrelated noise samples. It is thus apparent that

$$E(\varepsilon^2) = E[(\hat{s} - s)^2] = \frac{\sigma_n^2}{m} \tag{6.3}$$

As the number of samples m increases the mean-squared error decreases, an obviously good property. The sample mean is thus a good estimate of s in this sense.

The sample mean has one other interesting property in estimating s. If we take the expectation of $\hat{s}$ in Eq. (6.2), and recall that $E[n_j] = 0$, $E[s] = s_0$, we have

$$E[\hat{s}] = E[s] = s_0 \tag{6.4}$$

The estimate of s, on the average, is the same as the average value of the estimate. An estimator with this property is called an *unbiased estimator*. "On the average" it produces the desired result.

But notice from Eq. (6.4) that the random value $\varepsilon = \hat{s} - s$ has zero mean. The mean-squared error $E[\varepsilon^2] = E[(\hat{s} - s)^2]$ is thus the variance of $\hat{s} - s$; and, by the Tschebychev inequality of Chap. 3, the probability that the error $\varepsilon = \hat{s} - s$ will deviate beyond some specified value is bounded by the variance of ε. The mean-squared error in this case thus provides a bound on the probability that the estimate $\hat{s}$ deviates from s. In particular, as the number of samples m is increased, reducing the mean-squared error [Eq. (6.3)], the probability that $\hat{s}$ deviates from s is reduced as well. This provides further justification for the utility of the mean-squared error as a measure of goodness in estimation problems.

6.2 LINEAR MEAN-SQUARED ESTIMATION

Since we have evaluated the sample mean as a possible signal estimator in terms of the mean-squared error, why not go all the way and use that error as a performance criterion for determining optimum processors in the signal-estimation case? This is exactly what we shall do in much of this chapter, although other measures of performance will be considered as well. Specifically, we investigate signal estimates $\hat{s} = r(x_1, x_2, \ldots, x_m)$ with the function $r(\cdot)$ of the m samples chosen to minimize $E[(\hat{s} - s)^2]$. We shall show in a later section that nonlinear functions of the samples generally provide the "best" mean-squared estimates of a signal s. But at this point we shall specialize to linear estimates for simplicity. We thus assume

$$\hat{s} = \sum_{j=1}^{m} h_j x_j \tag{6.5}$$

and ask for the choice of the m coefficients $h_j, j = 1, \ldots, m$, that minimizes $E[\varepsilon^2] = E[(\hat{s} - s)^2]$. Note that this linear processor is our old friend the nonrecursive filter. We found filtering useful in the signal-detection problem; the sample mean, whose good estimation properties were just discussed and were found useful as well for smoothing purposes in Chap. 2, is a special case ($h_j = 1/m$).

To set up the problem we write

$$E[\varepsilon^2] = E[(\hat{s} - s)^2] = E\left[\sum_{j=1}^{m} h_j x_j - s\right]^2 \tag{6.6}$$

We differentiate $E[\varepsilon^2]$ with respect to each of the m parameters h_j, $j = 1, \ldots, m$, setting each partial derivative equal to zero to obtain the required m equations from which to find the values of h_j that minimize the mean-squared error. As an example, consider the equation involving the ith parameter h_i. It is proved in books on probability that the operations of expectation and differentiation can be interchanged. We thus have

$$\frac{\partial E[\varepsilon^2]}{\partial h_i} = E\left[\frac{\partial \varepsilon^2}{\partial h_i}\right] = 2E\left[\varepsilon \frac{\partial \varepsilon}{\partial h_i}\right] = 2E[\varepsilon x_i] = 0 \tag{6.7}$$

from Eq. (6.6). The set of equations that must be solved to find the optimum choice of $h_j, j = 1, \ldots, m$, is thus given by

$$E[\varepsilon x_i] = 0 \qquad i = 1, 2, \ldots, m \tag{6.8}$$

This is called the *orthogonality principle* in works on estimation.† By this we mean that the product of the error $\varepsilon = \hat{s} - s$ with each of the measured samples x_i is set equal to zero in an expected-value sense. (This comes from the concept of orthogonality in an m-dimensional space: two vectors are orthogonal, or "perpendicular," if the projection of one onto the other is zero.)

† See, for example, A. Papoulis, "Probability, Random Variables, and Stochastic Processes," chap. 8, McGraw-Hill, New York, 1965.

Alternately, substituting the actual expression for $\varepsilon = \sum_{j=1}^{m} h_j x_j - s$ into Eq. (6.8), we have as the set of equations to be solved

$$E\left[\left(\sum_{j=1}^{m} h_j x_j - s\right) x_i\right] = 0 \qquad i = 1, 2, \ldots \tag{6.9}$$

This can be put into more succinct form by recognizing that $R_{ij} = E[x_i x_j]$ is the autocorrelation between x_i and x_j (see Chap. 3)† and defining the parameter $g_i \equiv E[sx_i]$, the correlation between the random variables s and x_i. Then the set of equations is also given by

$$\sum_{j=1}^{m} h_j R_{ij} = g_i \qquad i = 1, 2, \ldots, m \tag{6.10}$$

In expanded form this set of equations corresponds to

$$\begin{aligned} R_{11}h_1 + R_{12}h_2 + \cdots + R_{1m}h_m &= g_1 \\ R_{21}h_1 + R_{22}h_2 + \cdots + R_{2m}h_m &= g_2 \\ \cdot \quad \cdot \quad \cdot \quad \cdot \quad \cdot \quad \cdot \quad \cdot \quad \cdot \quad \cdot \quad & \cdot \quad \cdot \\ R_{m1}h_1 + R_{m2}h_2 + \cdots + R_{mm}h_m &= g_m \end{aligned} \tag{6.11}$$

More compactly yet, if we define the $m \times m$ correlation matrix R with elements R_{ij}, i, $j = 1, 2, \ldots, m$, a column vector $\mathbf{h}$ with elements $h_1, h_2, \ldots, h_m$, and a column vector $\mathbf{g}$ with elements $g_1, g_2, \ldots, g_m$, we have as the set of equations in matrix-vector form,

$$R\mathbf{h} = \mathbf{g} \tag{6.12}$$

[The reader may check for himself that this vector equation does in fact correspond to the set of equations given by Eq. (6.11).]

It can readily be proved that R is a positive definite matrix and is invertible. The formal solution to Eq. (6.12) is thus

$$\mathbf{h} = R^{-1}\mathbf{g} \tag{6.13}$$

with R^{-1} the inverse matrix of R. This is the formal solution to the problem we have outlined, the set of filter coefficients in Eq. (6.5) by which the m data samples $x_1, x_2, \ldots, x_m$ are to be multiplied and then added to obtain the "best" linear estimate of the signal s in the least mean-squared error sense. The solution vector $\mathbf{h}$ represents the optimum linear filter through which the data samples are to be passed. This corresponds, in the estimation case, to the matched filter of the detection case.

A filter of this type is often called a *Wiener filter*, after the American mathematician who was among the first to discuss its properties. If the reader will recapitulate the steps that led to the derivation of the Wiener filter, he will find that nowhere was

† In the notation of Chaps. 3 and 4 this would be $R_x(i - j)$. We use the notation R_{ij} to simplify the writing and to also allow us to handle nonstationary cases, where the autocorrelation does depend on i, j and not just on their difference $i - j$.

the fact that the data samples $x_j, j = 1, \ldots, m$, are given by $s + n_j$ invoked. The result is much more general. It states that if the data samples x_j, $j = 1, 2, \ldots, m$, "somehow" contain the unknown random variable s (the signal), the best *linear*-filter operation on the samples to estimate s is given by the Wiener filter. This is true whether s is the random amplitude (or phase, say) of a sine wave that is to be estimated, the random slope of a linear trajectory, the random velocity (or acceleration) of a vehicle that is to be estimated from the data, etc. In all these cases the best linear filter, if linear filtering is to provide the processing, will have its coefficients given by the solution of the set of equations (6.11). The dependence on the particular problem involved appears in the various correlation terms of Eq. (6.11) to be calculated. The utility of the linear filter, whether or not it provides an adequate estimate, also depends on the specific problem involved. (It is not hard to find examples in which the smallest mean-squared error still provides large errors and hence poor estimates of s. This simply means that the linear filter is a poor processor for these cases.) One difficulty with the Wiener filter result, aside from the obvious one that it requires previous knowledge (or stored estimates) of the autocorrelation matrix R, is that one must specify the number of data samples m to be used in the processing beforehand. If m is changed for any reason (more data may become available), the calculations must be done all over again. In addition, it requires the inversion of the $m \times m$ matrix R. If m is large, this can take substantial computer time. To allow updating of the estimate as more information becomes available and to save on digital-processing cost, it would be of interest to develop a processing scheme that continually generates a new estimate from the previous stored one plus the next data sample as it comes in. Such a processing scheme is called a *recursive processor*. The Kalman filter, to be discussed in Chap. 7, is exactly of this recursive type and continually updates the estimate to keep the mean-squared error as small as possible. We discuss implementation of the Kalman filter in the air-traffic-control radar tracking problem as well as in other examples.

At this point, however, we would like to consider some specific examples of the Wiener filter derived in Eq. (6.13). First consider the particular case of signal-plus-noise with which we introduced this discussion of least mean-squared filtering. What does the Wiener filter look like, and how does it compare with the sample-mean estimator of s discussed previously?

To find the filter in this case we must first calculate the various correlation terms and then solve Eqs. (6.11). Specifically, let $x_j = s + n_j$, as in Eq. (6.1). Assume, as previously, that the noise samples are zero mean, with variance σ_n^2, uncorrelated with each other and with the signal s as well. We thus assume that

$$E[n_i n_j] = \begin{cases} 0 & i \neq j \\ \sigma_n^2 & i = j \end{cases} \qquad \text{and} \qquad E[sn_j] = 0$$

Also, let $E[s^2]$, the second moment of s, be a constant S. Assume further that $E[s] = 0$ to simplify the results. We then have, for this example,

$$R_{ij} = E[x_i x_j] = E[(n_i + s)(n_j + s)] = S + \sigma_n^2 \delta_{ij} \tag{6.14}$$

with δ_{ij} the Kronecker delta

$$\delta_{ij} = \begin{cases} 1 & i = j \\ 0 & i \neq j \end{cases}$$

Also
$$g_i = E[sx_i] = E[s^2] = S \tag{6.15}$$

The system of equations to be solved is thus

$$S \sum_{j=1}^{m} h_j + \sigma_n^2 h_i = S \qquad i = 1, 2, \ldots, m \tag{6.16}$$

A little thought will indicate that the solution has all filter coefficients equal: $h_1 = h_2 = \cdots = h_m$. Then from Eq. (6.16),

$$h_1 = h_2 = \cdots = h_m = \frac{S}{mS + \sigma_n^2} = \frac{1}{m + b} \tag{6.17}$$

with $b \equiv \sigma_n^2/S$ a noise-to-signal ratio. The linear least mean-squared estimate is thus given by

$$\hat{s} = \frac{1}{m + b} \sum_{j=1}^{m} x_j \tag{6.18}$$

Note that this is very much like the sample-mean estimate of s. In particular, for large signal-to-noise ratio, $b \ll m$, it does approximate the sample mean.

It is left to the reader to show that the mean-squared error between the signal s and its estimate $\hat{s}$ is given by

$$E[\varepsilon^2] = E[(\hat{s} - s)^2] = \frac{\sigma_n^2}{m + b} \tag{6.19}$$

This error is thus less than that provided by the sample mean in (6.2). For large signal-to-noise ratio ($b \ll m$) the two errors are about the same and decrease as $1/m$, as noted earlier.

The observant reader may have noted that this discussion of the choice of linear-filter coefficients to minimize the mean-squared error between some unknown parameter s and the best linear estimate $\hat{s}$ for it is similar to the least-squares-fit discussion of Chap. 2. For example, the best least-squares fit of the intercept c_0 of a straight line was found in Chap. 2 to be just the average of the data points. This is the same as the sample-mean result here, for the case of large signal-to-noise ratio. Although the emphasis there was on least-squares fitting of data to curves of known or specified shape, in contrast to the mean-squared estimation in a *statistical* sense discussed here, the approaches are similar. The data points were in fact implicitly assumed statistically independent so that the squares of the fitting errors at each data point could be summed. This is, of course, exactly the assumption made in the example just considered. We shall again find analogous results when estimating the (unknown) random slope of a line.

Consider another example now. Assume again that we have the problem of estimating a random (although fixed) signal s in noise but the noise is now *correlated* from sample to sample. An example might be a radar tracking an airplane. Wind gusts hitting the radar antenna cause its azimuthal position to be randomly disturbed so that the measurement of azimuth (one of the necessary signals in this case) has a noise component added. If the random motion of the antenna is slow compared with the signal-sampling rate, successive noise samples are correlated. One would expect the signal estimate to be less accurate with a given number of samples because the dependence of the successive noise samples reduces the effect of smoothing attainable by linearly filtering the signal plus noise. (In the extreme case, where each noise sample is the same as the previous one—and the noise is thus assumed to be unvarying from sample to sample, processing more than one sample will provide no additional information. To provide an improved estimate of a signal one needs a sequence of samples in which the noise scatters essentially independently about the signal level so that the effects of the noise can be averaged or smoothed out.) As a simple model to illustrate the essence of this material, assume the noise to have been generated by passing white noise through a first-order *recursive filter:*

$$n_j = an_{j-1} + z_j \tag{6.20}$$

Here a is some specified constant ($|a| < 1$), and z_j is a sample of a zero-mean white-noise (or orthogonal-increment) process of variance σ_z^2. As noted in Chap. 3, the process n_j is often called an autoregressive process of the first order. In Chap. 3 we showed the autocorrelation function for the process of (6.20) to be given by

$$R_n(j + k, j) \equiv E[n_{j+k}n_j] = R_n(j - k, j) = a^{|k|}\sigma_n^2 \tag{6.21}$$

with $\sigma_n^2 = \sigma_z^2/(1 - a^2)$. A sketch of this autocorrelation function is shown in Fig. 6.2 for two cases, a negative and a positive. Note that the autocorrelation function decreases exponentially ($|a| < 1$) with the separation k between samples. For $a = 0$ we obviously have the uncorrelated-noise example previously considered. For $|a| \to 1$ the noise samples become more and more correlated.

Consider now samples x_j to be processed that are made up, as previously, of the signal s and the correlated noise n_j just discussed. What is the best estimate of s in the least mean-squared error sense based on these samples? To simplify the algebra we shall consider two samples only. We also consider linear processing only. We thus look for the best choice of the two coefficients h_1 and h_2 so that the estimate

$$\hat{s} = h_1 x_1 + h_2 x_2 \tag{6.22}$$

is as close to s as possible. The answer is obviously given by solving the set of Eqs. (6.11). Recalling the definition of the various parameters in Eqs. (6.11), we have, for this example,

$$g_i = E[sx_i] = E[s^2] = S \qquad i = 1, 2 \tag{6.23}$$

as previously, and

$$R_{11} = E[x_1^2] = E[(s + n_1)^2] = E[s^2] + E[n_1^2] = S + \sigma_n^2 \tag{6.24}$$

again assuming that n_1 and s are uncorrelated.

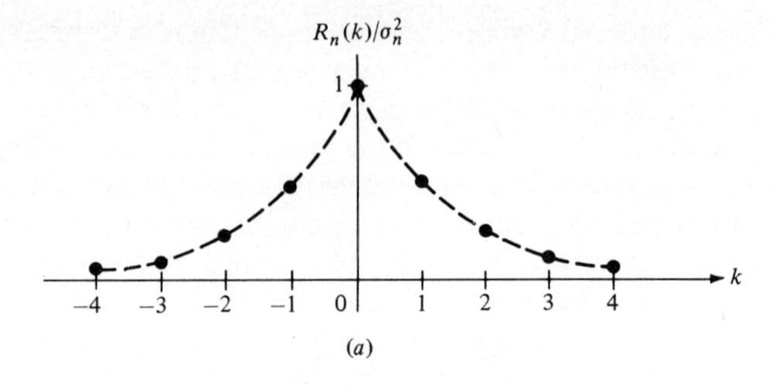

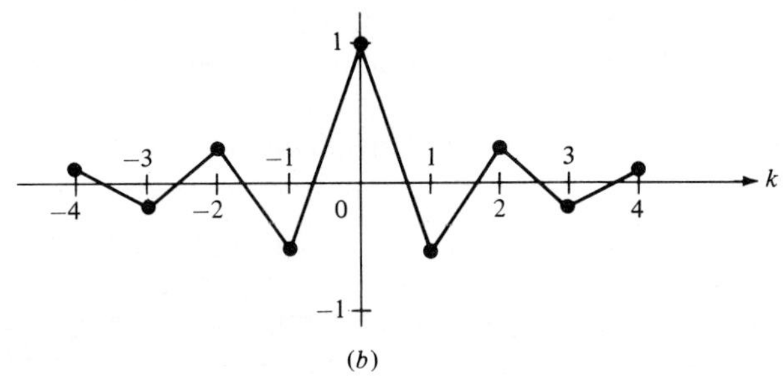

FIGURE 6.2
Autocorrelation function, autoregressive process.
(a) $a = 0.5$. (b) $a = -0.5$.

A little thought will indicate that $R_{22} = E[x_2^2] = R_{11}$ and that

$$R_{12} = R_{21} = E[x_1 x_2] = S + a\sigma_n^2 \tag{6.25}$$

This is where the noise correlation enters, via the parameter a. It is left for the reader to show that the solution to the pair of equations

$$R_{11}h_1 + R_{12}h_2 = g_1 \qquad \text{and} \qquad R_{21}h_1 + R_{22}h_2 = g_2$$

is given by

$$h = h_1 = h_2 = \frac{1}{2 + b(1 + a)} \tag{6.26}$$

with $b = \sigma_n^2/S$ again a noise-to-signal ratio. Note that with $a = 0$ (uncorrelated noise samples) we have just the result found previously, with $m = 2$ samples used [see (6.17)]. The linear least mean-squared estimate is thus given by

$$\hat{s} = h(x_1 + x_2) \tag{6.27}$$

with h defined by Eq. (6.26).

What is the mean-squared error actually resulting, and how does it compare with the sample mean or other processor errors? The mean-squared error is in general

$$E[\varepsilon^2] = E[(\hat{s} - s)^2] = E\left[\left(\sum_{i=1}^{m} h_i x_i - s\right)^2\right] \tag{6.28}$$

Letting $m = 2$, with $x_i = s + n_i$, we have, after some manipulation,

$$E[\varepsilon^2] = S(h_1 + h_2 - 1)^2 + h_1^2 E[n_1^2] + h_2^2 E[N_2^2] + 2h_1 h_2 E[n_1 n_2] \tag{6.29}$$

Setting $h_1 = h_2 = h$, and using the results found previously for the statistics of n_j, we have, finally,

$$\frac{E[\varepsilon^2]}{\sigma_n^2} = \frac{(2h - 1)^2}{b} + 2h^2(1 + a) \tag{6.30}$$

In this form we can calculate the mean-squared error not only for the optimum value of h given by Eq. (6.26) but for other values as well. For example, let $h = \frac{1}{2}$. This is just the sample-mean processor. Then

$$\left.\frac{E[\varepsilon^2]}{\sigma_n^2}\right|_{\text{sample mean}} = \tfrac{1}{2}(1 + a) \tag{6.31}$$

For $0 < a < 1$, the presence of correlated noise samples *increases* the error or decreases the effectiveness of the processing. In particular, if the noise samples are completely correlated, with $a = 1$, using two samples provides no improvement over one sample. This is exactly the situation pointed out earlier. If the noise samples are correlated, the effectiveness of smoothing or filtering is reduced. Notice also an additional interesting result: if $a = -1$, the error goes to zero. (The same result will appear as well with the optimum processor, as it should.) This is a rather special and artificial situation; here the noise samples are assumed *negatively* correlated. A positive noise sample is thus *highly likely* to be followed by a negative sample and vice versa. (It is important to caution at this point that since the measure of correlation is a statistical parameter, one can talk only of the relation between successive sample values in a *statistical sense.* A high measure of correlation does *not* mean two successive samples are deterministically related. There is presumably a high probability that one is related to the other.) If two successive signal-plus-noise samples are averaged, the noise samples will most often tend to cancel, improving the signal estimation. One must caution, however, that such situations are rather artificial and that many "noise-cancellation schemes" invented by overeager engineers and scientists end up doing just the opposite.

If we now insert the optimum weighting factor of Eq. (6.26) into Eq. (6.30), we obtain, after some manipulation,

$$\left.\frac{E[\varepsilon^2]}{\sigma_n^2}\right|_{\min} = \frac{(1 + a)/2}{1 + b(1 + a)/2} \tag{6.32}$$

Since $(1 + a)/2 > 0$, the denominator of Eq. (6.32) is greater than 1 and the mean-squared error is *smaller* than that given by the sample-mean processor [Eq. (6.31)], as expected. Note again, as in the previous example, that as the signal-to-noise ratio increases (b decreases), the optimum processor and the sample-mean processor approach one another.

The examples thus far have assumed a constant signal amplitude s plus additive noise to be processed. The optimum filter has been one which weights each sample the same. This is the same result as obtained in the detection case. What if the signal component *varies* from sample to sample? This might, for example, be the situation where one is trying to estimate the amplitude s of a sine wave $s[\cos(2\pi j/m)]$ with j the jth sample out of m in a complete period. An even simpler example is that of determining the slope of a straight line. Such a problem occurs, for example, when estimating the velocity of a vehicle from measured displacement data corrupted by noise. This is analogous to the filtering problem discussed in Chap. 2, where *least-squares* fits of straight lines to data points were considered. We shall find, in working this problem out for independent noise samples, that the result obtained is also analogous. Specifically, our result will be the same as that of Eq. (2.11), for the case of high signal-to-noise ratio. This will correspond in our case to a filter whose coefficients increase linearly with sample number.

Note that these results are analogous only for the two cases of estimating a constant and the slope of a straight line. Linear estimates are appropriate in these cases. The more general case of nonlinear mean-squared estimation, to be discussed immediately following, must be invoked when comparing estimation with fitting of higher-order polynomials.

For the final example, then, with linear processing assumed to provide the estimate of a signal, let the m samples

$$x_j = js + n_j \qquad j = 1, 2, \ldots, m \tag{6.33}$$

be measurements of a linearly increasing signal plus noise. As noted above, we are thus estimating the slope s of a known straight line. We again desire the h_j coefficients of the optimum linear processor

$$\hat{s} = \sum_{j=1}^{m} h_j x_j \tag{6.34}$$

The answer is again obtained by solving the system of equations (6.11).

Assuming that the noise samples are uncorrelated, for simplicity, with variance σ_n^2, one readily finds that a typical parameter R_{ij} in this set of equations is given by

$$R_{ij} = E[x_i x_j] = ijS + \sigma_n^2 \delta_{ij} \tag{6.35}$$

with δ_{ij} the Kronecker delta and $S = E[s^2]$. Similarly,

$$g_i = E[sx_i] = jS \qquad i = 1, 2, \ldots, m \tag{6.36}$$

Consider the case of $m = 2$ samples. The appropriate weighting factors h_1 and h_2 are given by the solution of the two equations

$$(S + \sigma_n^2)h_1 + 2Sh_2 = S \qquad 2Sh_1 + (4S + \sigma_n^2)h_2 = 2S \tag{6.37}$$

Dividing through by S and again letting $b = \sigma_n^2/S$, we have

$$(1 + b)h_1 + 2h_2 = 1 \qquad 2h_1 + (4 + b)h_2 = 2 \tag{6.38}$$

It is left to the reader to show that the solutions are

$$h_1 = \frac{1}{5 + b} \qquad h_2 = 2h_1 = \frac{2}{5 + b} \tag{6.39}$$

and the estimate of s is given by

$$\hat{s} = \frac{x_1 + 2x_2}{5 + b} \tag{6.40}$$

Note that if $b = 0$ (zero noise-to-signal ratio), this agrees with the least-squares fitting solution of Eq. (2.11). (The lower limit of the sums in that equation must be taken as 1. Then $\sum_{n=1}^{2} n^2 = 5$, as in the case here.)

Similarly, it is left to the reader to show that the filter weights for $m = 3$ samples are given by

$$h_1 = \frac{1}{14 + b} \qquad h_2 = 2h_1 \qquad h_3 = 3h_1 \tag{6.41}$$

The estimate of s is then given by

$$\hat{s} = \frac{x_1 + 2x_2 + 3x_3}{14 + b} \tag{6.42}$$

[Compare with Eq. (2.11).]

More generally, say

$$x_j = f_j s + n_j \qquad j = 1, 2, \ldots, m \tag{6.43}$$

$f_1, f_2, \ldots, f_m$ are *known* numbers, while s is a random variable to be estimated. (Other examples in addition to the straight line js, just discussed, whose slope s is to be estimated, include $x_j = s \cos j\pi/L + n_j$, the estimate of the amplitude of a sine wave; $x_j = sj^2 + n_j$, the estimate of an acceleration parameter s; etc.) Again assume $E[s] = E[n_j] = 0$, $E[s^2] = S$, $E[n_j^2] = \sigma_n^2$, $E[n_i n_j] = 0$. White noise is thus assumed added to the random term whose parameter s is to be estimated. It is then left to the reader as an exercise to show that the optimum *linear* estimator is our old friend the matched filter. For with

$$\hat{s} = \sum_{j=1}^{m} h_j x_j \tag{6.44}$$

the values of h_j that provide the smallest mean-squared error $E[(\hat{s} - s)^2]$ are found to be given by

$$h_1 = \alpha f_1, \qquad h_2 = \alpha f_2, \qquad \ldots, \qquad h_j = \alpha f_j \tag{6.45}$$

with

$$\alpha = \frac{S}{\sigma_n^2 + S \sum_{j=1}^{m} f_j^2} \tag{6.46}$$

The filter coefficients h_j are thus matched to the respective signal terms $f_j s$, $j = 1, 2, \ldots, m$. The appropriate filter is one that again weights each received sample in accordance with the relative amplitude of the signal in that sample.

As a check, let $f_j = j$, just the case of Eq. (6.33), in which the slope of a straight line is to be estimated. It is apparent that (6.45) and (6.46) agree with (6.39) for $m = 2$ samples and (6.41) for $m = 3$ samples.

6.3 BAYES ESTIMATORS: NONLINEAR ESTIMATES

We have stressed *linear* mean-squared estimation in the last section because of its relative simplicity and because of its connection with linear filtering. It is apparent that linear processing, although simple to carry out, may not always provide the best possible performance. We shall in fact show in this section that nonlinear processors are often called for, both for least mean-squared estimates and for other performance criteria as well. This has already been alluded to in the discussion of polynomial least-squares fitting in Chap. 2. The rub, however, is that we shall have to assume knowledge of the probability density function of the data samples and the signal s to be estimated. This was not the case with linear estimation. There we simply had to have some knowledge of the respective *second moments* (correlation functions). The more general estimation procedures to be discussed in this section thus require more statistical information.

To be more specific and to develop some feeling for the estimates to be discussed, assume for simplicity that we are dealing with a single-sample situation. We measure x_1, which depends in some way on the signal parameter s, and from this measurement attempt to estimate s. Since we assume knowledge of the statistics of x_1 and s, we can generate the a posteriori density function $f(s \mid x_1)$ from the measurement of x_1. (We shall show how to do this later.) This is the probability density function of s *conditioned* on knowing x_1. A sketch of such a density function is shown in Fig. 6.3 for two different values of x_1. Say $x_1 = 5$ so that we operate along the left-hand curve. What value shall we select for the estimate $\hat{s}$ of s? One possible value is the peak or mode of the curve, shown at $s = 3$. Another valid estimate is the mean. Another possible estimate might be the median. There are various estimates that might be used, depending on the particular problem, the performance criterion to be adopted, etc.

The peak of the curve serves as a particularly satisfying estimate: it represents the *most probable* value of s. If the measured value of x_1 is $x_1 = 10$, say, then we operate along the right-hand curve and the most probable value of s is now $\hat{s} = 10$. The estimate of s based on the peak value of the a posteriori density function is called the *maximum a posteriori* (MAP) estimate, written $\hat{s}_{\text{MAP}}$. We shall discuss it in further detail later in this section. The mean of the distribution, $\int sf(s \mid x_1)\, ds$, will also be discussed in detail since we shall show that it corresponds to the least mean-squared estimate $\hat{s}_{\text{MS}}$. Since it is the mean of the a posteriori distribution *conditioned* on knowing the sample x_1, it is called the *conditional mean.*

How does one determine the a posteriori density function on which both the MAP and conditional-mean estimates are based? As indicated earlier, this requires knowledge of the statistics of both x_1 and s. (We are still assuming one-sample estimation here for simplicity, but this will be extended shortly to the more realistic

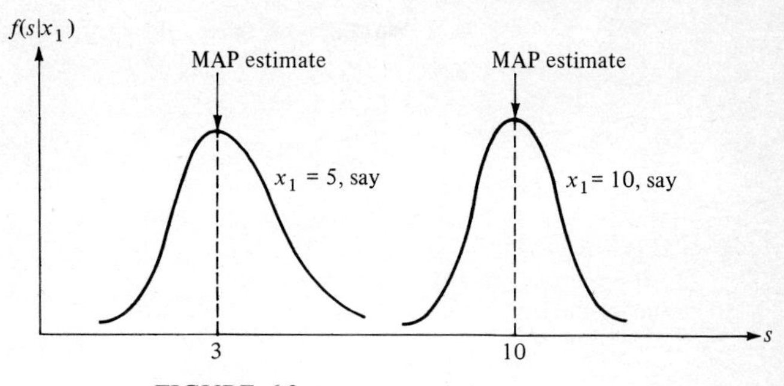

FIGURE 6.3
A posteriori density function, $f(s \mid x_1)$.

case of m-sample estimation.) Specifically, assume $f(s)$ known as well as $f(x_1 \mid s)$, just the conditional density function encountered previously in our discussion of detection. From the familiar law relating joint and conditional density functions we have

$$f(s, x_1) = f(s \mid x_1) f(x_1) = f(x_1 \mid s) f(s) \tag{6.47}$$

It is thus apparent that the desired a posteriori density function is simply given by

$$f(s \mid x_1) = \frac{f(x_1 \mid s) f(s)}{f(x_1)} \tag{6.48}$$

Alternately, since $\int f(s \mid x_1)\, ds = 1$, as must be true for any valid probability density function,

$$f(s \mid x_1) = \frac{f(x_1 \mid s) f(s)}{\int f(x_1 \mid s) f(s)\, ds} = K f(x_1 \mid s) f(s) \tag{6.49}$$

with

$$K = \frac{1}{\int f(x_1 \mid s) f(s)\, ds} \tag{6.50}$$

a constant independent of the signal s. This constant essentially serves the role of normalizing $f(s \mid x_1)$ to have its area equal unity.

This can be checked by noting that

$$f(x_1) = \int f(x_1, s)\, ds \tag{6.51}$$

i.e., the *marginal* density is found by integrating over the two-dimensional density. From (6.47), however,

$$f(x_1, s) = f(x_1 \mid s) f(s)$$

and (6.49) and (6.50) follow. Often the effort to evaluate K can be avoided, since its value will not change the location of the peak of $f(s \mid x_1)$ as a function of s.

As an example, say that we have a signal received in additive noise, just the model used in much of the discussion thus far in this chapter, as well as in Chap. 5.

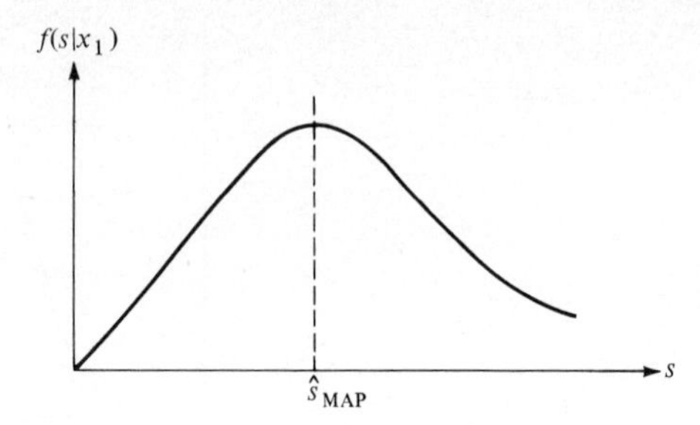

FIGURE 6.4
A posteriori density function, additive noise, Rayleigh signal.

Thus,

$$x_1 = s + n_1 \tag{6.52}$$

Say n_1 is zero-mean gaussian with variance σ_n^2. The conditional density of the observation x_1, given s, is then the shifted noise-density function

$$f(x_1 \mid s) = \frac{1}{\sqrt{2\pi\sigma_n^2}} e^{-(x_1 - s)^2/2\sigma_n^2} \tag{6.53}$$

We assume that the signal s has a Rayleigh distribution [see (3.100)]

$$f(s) = \begin{cases} \dfrac{se^{-s^2/2\sigma_s^2}}{\sigma_s^2} & s \geq 0 \\ 0 & s < 0 \end{cases} \tag{6.54}$$

since this is a very common model for random amplitude fluctuations and occurs frequently in the study of signals transmitted through random media. A little bit of algebra provides the desired a posteriori distribution

$$f(s \mid x_1) = Kf(s)f(x_1 \mid s) = Cse^{-(a/2\sigma_n^2)(s - x_1/a)^2} \tag{6.55}$$

where all constants not involving s have been lumped into the constant C and $a \equiv 1 + \sigma_n^2/\sigma_s^2$. (Details are left to the reader.) A typical plot is sketched in Fig. 6.4. The peak value, found by differentiating, occurs at $s = (x_1/2a)(1 + \sqrt{1 + 4a\sigma_n^2/x_1^2})$. This is the MAP estimate $\hat{s}_{MAP}$ and is so indicated in the figure. Note that this is in general a nonlinear estimate. It is only for $x_1 \gg 2\sigma_n\sqrt{a}$ that the estimate is linearly related to x_1. This particular example is discussed again, in more general form, in Example 4 of the next section.

As another example *not* involving a random signal in additive noise, assume that

$$f(x_1 \mid s) = se^{-sx_1} \qquad \begin{matrix} x_1 \geq 0 \\ s \geq 0 \end{matrix} \tag{6.56}$$

(Here $E[x_1 \mid s] = 1/s$. As the parameter s *decreases*, the average value of x *increases*.) Assume that s itself follows a similar exponential distribution

$$f(s) = \lambda e^{-\lambda s} \qquad \begin{matrix} s \geq 0 \\ \lambda \geq 0 \end{matrix} \tag{6.57}$$

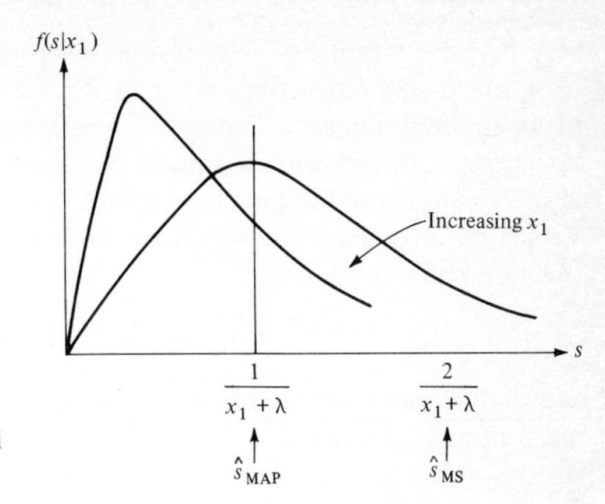

FIGURE 6.5
Another example of an a posteriori density function.

Thus s itself is an exponentially distributed random variable with average value $1/\lambda$. Such an example might occur in a traffic situation in a city. Say s is the mean rate of arrival of vehicles at an intersection. This parameter is to be estimated from a one-sample measurement. It is known that for this city as a whole the mean arrival rate of vehicles is exponentially distributed with average value $E[s] = 1/\lambda$. Let x_1 be the observed time between the arrival of two successive vehicles. From this one measurement, knowing λ, we would like to find an estimate for s.

It is left for the reader to show that the a posteriori distribution in this case is given by

$$f(s \mid x_1) = (x_1 + \lambda)^2 s e^{-s(\lambda + x_1)} \qquad s \geq 0 \tag{6.58}$$

Typical sketches are shown in Fig. 6.5. It is left to the reader as well to show that the peak of the distribution occurs at $1/(x_1 + \lambda)$ and that its mean value is given by $2/(x_1 + \lambda)$. These points are indicated in Fig. 6.5 and are labeled, respectively, $\hat{s}_{\text{MAP}}$ and $\hat{s}_{\text{MS}}$, for MAP and least mean-squared (MS) estimates. Note that in this example the MAP and MS estimates differ by a factor of 2. Which one to pick depends ultimately on the designer of the experiment and the processing equipment to go with it. We shall see shortly that the MAP estimate can be related to minimizing a particular cost function. The choice will thus depend on which cost function it is desired to implement. Typically, in much more complex estimation problems, the designer may be looking for any processing algorithm that can readily be implemented and has some connection with a valid cost criterion. The one chosen may thus be the one that is most readily implemented at least dollar cost.

We have noted several times in passing that the two estimates up to now, the MAP estimate and the conditional mean, can be shown to represent minima of two different cost functions. We would now like to prove this statement, for it enables us to demonstrate the validity of these estimates in a more concrete and less intuitive way and to extend the estimation procedures to the multisample case.

We shall do this first by considering the least mean-squared estimate and then extending it to other, more general cost functions. Again consider the single-sample case first, for simplicity. The results are then readily generalized to m samples. By least mean-squared estimate we mean the estimate $\hat{s} = g(x_1)$ that minimizes the mean-squared difference between $\hat{s}$ and the desired "signal" parameter s. This is of course the same criterion discussed previously in the linear-estimation case. There we chose $\hat{s} = h_1 x_1$ and found the value of h_1 to minimize the mean-squared error. Here we search for a generally nonlinear function $g(x_1)$. Specifically, then, we have as the function to be minimized

$$E[\varepsilon^2] = E[(\hat{s} - s)^2] = E\{[g(x_1) - s]^2\} \tag{6.59}$$

We shall in fact show that the optimum processor in this case is just the conditional mean already discussed:

$$\hat{s}_{\text{MS}} \equiv g(x_1) = E[s \mid x_1] = \int_{-\infty}^{\infty} sf(s \mid x_1)\, ds \tag{6.60}$$

To demonstrate this we write the mean-squared error to be minimized as

$$E[(\hat{s} - s)^2] = \iint_{-\infty}^{\infty} (\hat{s} - s)^2 f(s, x_1)\, ds\, dx_1 \tag{6.61}$$

with $f(s, x_1)$ the joint density function already introduced earlier. We are assuming that x_1 and s are *continuous* random variables to simplify the analysis. (In the examples considered previously the random variables were also assumed continuous.) But the result is actually more general, and we shall in fact consider examples in which x_1 or s, or both, are discrete-valued variables.

Rewriting the joint density function in terms of the conditional density function, we have

$$E[(\hat{s} - s)^2] = \iint_{-\infty}^{\infty} (\hat{s} - s)^2 f(s \mid x_1) f(x_1)\, dx_1\, ds$$

$$= \int_{-\infty}^{\infty} f(x_1) \left[\int_{-\infty}^{\infty} (\hat{s} - s)^2 f(s \mid x_1)\, ds \right] dx_1 \tag{6.61a}$$

carrying out the necessary integration with respect to s first. Note that $\hat{s}$ in the expression above is a function of the data sample x_1 and so can properly be written $\hat{s}(x_1)$.

The smallest mean-squared error will be obtained when $\hat{s}(x_1)$ is adjusted to make the quantity inside the brackets as small as possible for each x_1. This is due to the *positive* character of the probability density function $f(x_1)$. [If $f(x_1)$ were negative for some x_1 values, we would want the bracketed quantity to be big for those values. Since $f(x_1)$ is positive for all x_1, however, we want the inner integral to be as small as possible for *all* x_1, independent of $f(x_1)$.] The bracketed quantity represents the second moment of s about some point $\hat{s}$. It is well known from probability theory that the point with respect to which the second moment is minimum is just the mean or expected value. (This can easily be checked by differentiating with respect to $\hat{s}$. The resultant second moment is just the variance and is often called the second central moment as well.) The best choice of $\hat{s}$, in the sense of minimizing the mean-squared error, is thus

$$\hat{s}_{MS} = \int_{-\infty}^{\infty} sf(s \mid x_1)\, ds \tag{6.62}$$

just the result we cited previously.

As an example, consider the a posteriori distribution of Eq. (6.58) and Fig. 6.5. For this case

$$\hat{s}_{MS} = E[s \mid x_1] = \int_{-\infty}^{\infty} (x_1 + \lambda)^2 s^2 e^{-s(\lambda + x_1)}\, ds \tag{6.63}$$

A direct calculation shows that this is just $2/(x_1 + \lambda)$, as noted earlier. As the data sample x_1 varies, the estimate $\hat{s}_{MS}$ varies as well.

We have shown that the mean of the a posteriori distribution, the so-called *conditional* mean $E[s \mid x_1]$, is the best estimate of the parameter s, given the data sample x_1, in the least mean-squared sense. We previously noted that the *peak* of the a posteriori distribution is another perfectly valid estimate and is in fact called the maximum a posteriori (MAP) estimate. Is there any way of justifying the use of this latter estimate other than stating that it appears to provide a valid estimate? We can in fact show that it does minimize a particular function of the error between s and its estimate $\hat{s}$. To do this, we generalize the concept of mean-squared error to include more generalized cost functions. The MS and MAP estimates then correspond to two particular (and commonly used) cases of these cost functions. They are examples of a class of estimates called *Bayes estimates*, designed to minimize the expectation of a specified cost function.

Specifically, assume some cost function $C(s, \hat{s})$ associated with the desired signal parameter s and its estimate $\hat{s}$. Examples of possible cost functions are shown in Fig. 6.6. The first uses an absolute-error criterion. The second tolerates any error within $\Delta/2$ of the desired signal parameter and weights all errors outside this range equally. The third is the squared-error criterion already discussed, and the fourth is a normalized power of the error. The Bayes estimate is then designed to minimize the average or expected value of this cost. When these or other cost functions are used, the average cost to be minimized is given by

$$R = \iint C(\hat{s}, s) f(x_1, s)\, ds\, dx_1 \tag{6.64}$$

in the case of a one-sample estimate. More generally, if m samples $x_j, j = 1, 2, \ldots, m$, are available, we have as an obvious extension

$$R = \iint C(\hat{s}, s) f(\mathbf{x}, s)\, ds\, d\mathbf{x} \tag{6.65}$$

The vector $\mathbf{x}$ denotes the ensemble of m samples, the joint density function $f(\mathbf{x}, s)$ is a function of the m samples and the signal parameter s, and $d\mathbf{x}$ is the shorthand notation representing $dx_1, dx_2, \ldots, dx_m$. The integration is then over the $(m + 1)$-dimensional space representing the range of values of the samples and s.

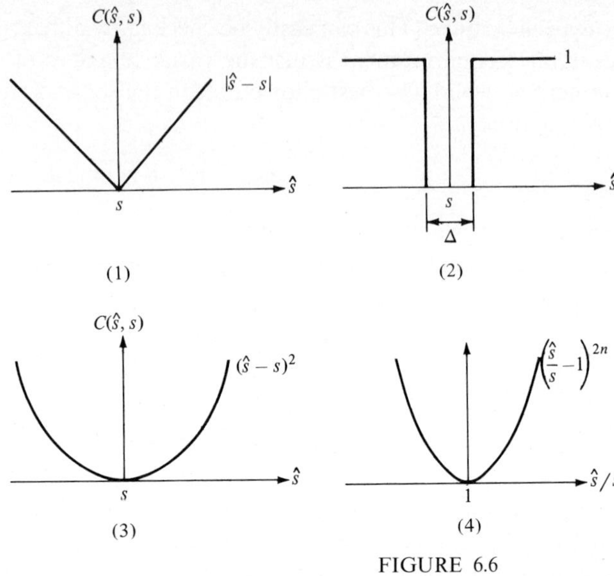

FIGURE 6.6
Examples of cost functions.

Consider now in particular cost function 2 in Fig. 6.6. We shall show that this is the one that leads to the MAP estimate. As already noted, this is a criterion which allows small errors to accumulate but weights all others equally. This criterion is sometimes called "hit-or-miss" or a "miss is as good as a mile," by analogy to hunting. This may apply to a radar signal which indicates the position of a runway. If the radar gets the pilot close enough to the runway, he can complete the landing visually. What is the best choice of the signal estimate $\hat{s}$ to minimize the average cost under this criterion? To answer this question we rewrite the average cost R, as was done earlier in Eqs. (6.61) and (6.61a):

$$R = \iint C(\hat{s}, s) f(\mathbf{x}) f(s \mid \mathbf{x})\, ds\, d\mathbf{x} = \int f(\mathbf{x}) \left[\int C(s, \hat{s}) f(s \mid \mathbf{x})\, ds \right] d\mathbf{x} \tag{6.66}$$

Consider the inner integral. For the cost function 2 of Fig. 6.6, this is zero in the range $\pm\Delta/2$ about $\hat{s}$. Since the integral of $f(s \mid \mathbf{x})$ must be 1 over all values of s, R then takes on the value

$$R = \int f(\mathbf{x}) \left[1 - \int_{\hat{s}-\Delta/2}^{\hat{s}+\Delta/2} f(s \mid \mathbf{x})\, ds \right] d\mathbf{x} \tag{6.67}$$

Now let $\Delta \to 0$; that is, we tolerate errors only in the limit, when they are very small. The integral is then given by $\Delta f(\hat{s} \mid \mathbf{x})$. As in the case of the squared-error criterion, minimum R corresponds to minimizing the expression inside the brackets. [Recall again that $f(\mathbf{x})$ is everywhere positive and independent of $\hat{s}$.] A little thought will indicate that this is accomplished by choosing $\hat{s}$ at the largest value of $f(s \mid \mathbf{x})$. It is

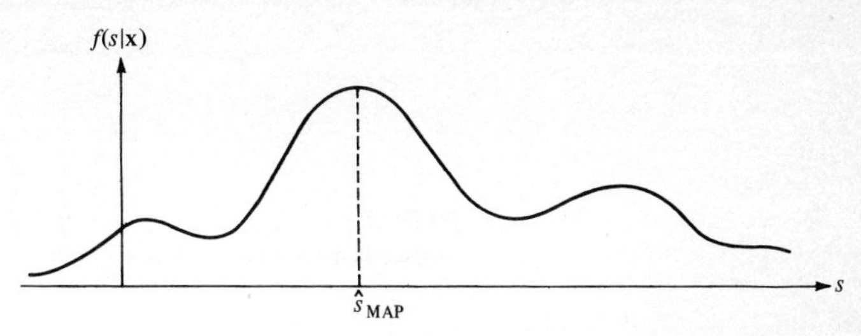

FIGURE 6.7
Maximum a posteriori (MAP) estimate.

thus exactly the MAP estimate considered previously. An example of such an estimate for an arbitrary function is sketched in Fig. 6.7. Other examples previously noted appear in Figs. 6.3 and 6.5. Assuming that the maximum value of $f(s \mid \mathbf{x})$ occurs at a point where the function is continuous in s (this may not always be the case), the MAP estimate is given by the solution of the equation

$$\frac{\partial f(s \mid \mathbf{x})}{\partial s} = 0 \tag{6.68}$$

A further comparison with the least mean-squared estimate is in order. As noted earlier, that estimate is just the mean or expected value of the a posteriori probability [Eq. (6.62)]. The MAP estimate is the peak value of the density function. Where the peak value and the mean coincide (gaussian density functions as well as other symmetrical density functions provide the obvious examples), the two estimates will be the same. In many situations it is simpler to use Eq. (6.68) to compute the MAP estimate than to compute the conditional mean of the a posteriori density; i.e., the Bayes estimate for hit or miss is easier to compute, in those cases, than the one for least mean-squared error.

Two alternate equations for determining the MAP estimate can be found from Eq. (6.68) and are often simpler to use. Recall from the relation between joint and conditional probabilities [Eqs. (6.47) and (6.48)] that

$$f(s \mid \mathbf{x}) = \frac{f(s)f(\mathbf{x} \mid s)}{f(\mathbf{x})} \tag{6.69}$$

Differentiating this with respect to s and noting that $f(\mathbf{x})$ is independent of s, we have the MAP estimate also given by

$$\frac{\partial}{\partial s} f(s)f(\mathbf{x} \mid s) = 0 \tag{6.70}$$

Alternately, since the logarithm of a function varies monotonically with that function, another, often simpler, equation for the MAP estimate is given by

$$\frac{\partial}{\partial s} \ln f(s \mid \mathbf{x}) = 0 = \frac{\partial}{\partial s} \ln f(s) + \frac{\partial}{\partial s} \ln f(\mathbf{x} \mid s) \tag{6.71}$$

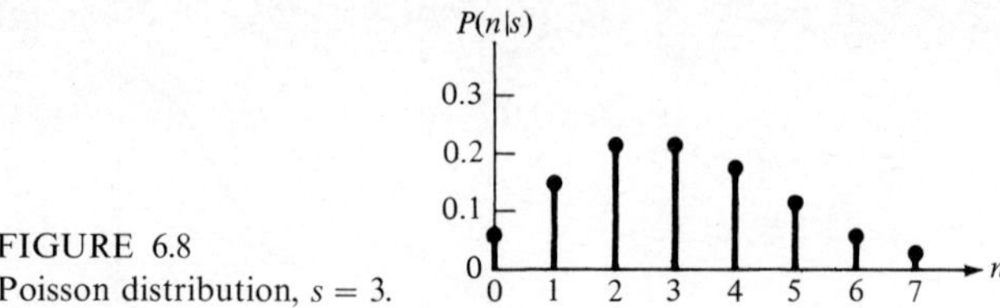

FIGURE 6.8
Poisson distribution, $s = 3$.

As a specific example of the applicability of the MS and MAP estimates and comparison between them, consider the following hypothetical situation (other examples are included in the next section). It is desired to estimate the average number of telephone calls per hour coming into a particular telephone exchange. Call this number s. Assume in this hypothetical situation that it has been well established from many tests performed throughout the telephone system that the actual number of calls per hour follows the Poisson distribution. Thus, the probability of n calls per hour being received at the exchange is given by

$$P[n \mid s] = \frac{s^n e^{-s}}{n!} \qquad n = 0, 1, 2, \ldots \tag{6.72}$$

(see Fig. 6.8 for a typical plot of this distribution). This distribution has the average value s, the parameter to be estimated. In addition, say that it is also known from experience that the average number of calls at various exchanges throughout the system itself obeys an exponential distribution,

$$f(s) = \lambda e^{-\lambda s} \qquad s \geq 0 \tag{6.73}$$

with average value $1/\lambda$. This thus represents the average number throughout the entire system. (Note that although the actual number of calls must be an integer, the average does not have to be and may in fact be approximated by a continuous random variable, as assumed here.)

Measurements of the actual number of calls are made during one particular hourly period. Say the number measured is N. (This corresponds to the variable x_1 used previously.) From this one measurement, what can one say about the average number s? We assume that we want to estimate s in both the least mean-squared sense and in the MAP sense. To find $\hat{s}$ we must first find the a posteriori density function $f(s \mid N)$. As previously, we write

$$f(s, N) = f(s \mid N)P[N] = P[N \mid s]f(s) \tag{6.74}$$

But since N is an integer value, we must use probabilities here rather than density functions in describing N. We thus have

$$f(s \mid N) = \frac{P[N \mid s]f(s)}{P[N]} \tag{6.75}$$

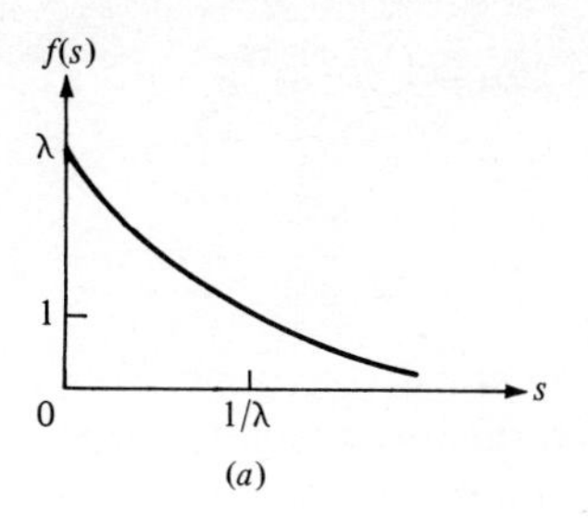

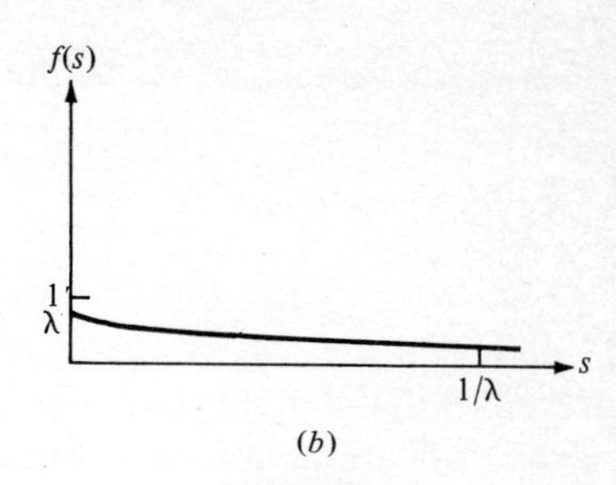

FIGURE 6.9
Exponential distribution.

Alternately, since $\int_0^\infty f(s \mid N)\, ds = 1$, as must be true for any valid probability density function,

$$f(s \mid N) = kP[N \mid s]f(s) \tag{6.75a}$$

with k an appropriately chosen normalizing constant. The probability $P[N \mid s]$ of getting N calls is given by Eq. (6.72) and the density function $f(s)$ for s by Eq. (6.73). Introducing these into Eq. (6.75a), we have

$$f(s \mid N) = \frac{k\lambda s^N e^{-s(\lambda+1)}}{N!} \tag{6.75b}$$

From integral tables, we find that $\int_0^\infty s^N e^{-s(\lambda+1)}\, ds = N!/(\lambda + 1)^{N+1}$. Using the condition that $\int_0^\infty f(s \mid N)\, ds = 1$ to evaluate k, we finally have

$$f(s \mid N) = \frac{\lambda + 1}{N!}[s(\lambda + 1)]^N e^{-s(\lambda+1)} \tag{6.75c}$$

It is then readily found from integral tables that

$$\hat{s}_{\text{MS}} = \int_0^\infty sf(s \mid N)\, ds = \frac{N + 1}{\lambda + 1} \tag{6.76}$$

This is the desired least mean-squared estimate of s in terms of the measured sample N, as well as the system average $1/\lambda$. It has an interesting interpretation. Consider first the extreme case $\lambda \ll 1$. This corresponds to an average number of calls $1/\lambda \gg 1$. Most exchanges thus have large numbers of calls per hour, on the average (see Fig. 6.9b). In this case the least mean-squared estimate of s is just the number $N + 1$ measured. Now consider the other extreme case, $\lambda \gg 1$. This corresponds to few calls, on the average. (This is presumably highly uneconomical and hence obviously a poor model for a typical telephone system.) Figure 6.9a portrays this case. If $N \gg 1$ is measured, this corresponds to an unlikely situation and the estimate accounts for this by reducing $N + 1$ by $\lambda + 1$. If $N = 0$ is measured, the estimate is actually given by $1/(\lambda + 1)$. This large-λ case could also correspond to an observation interval which is too short, e.g., measuring the number of phone calls *per second*.

What about the MAP estimate now? Differentiation of $f(s \mid N)$ to find the maximum value (the logarithm is simpler to work with in this case) provides the MAP estimate

$$\hat{s}_{\text{MAP}} = \frac{N}{\lambda + 1} \tag{6.77}$$

For $N \gg 1$ the two estimates are obviously the same. For $N = 0$ or 1, however, the results will differ substantially. (See the discussion following the mean-squared estimate evaluation, however.) Both estimates are biased. (This should be checked by the reader.)

How else do the two results compare? One approach is to actually evaluate the mean-squared error $E[(\hat{s} - s)^2]$ in the two estimates and compare. Carrying out the required algebra in this example, we find for the least mean-squared (MS) error,

MS:
$$E[(\hat{s} - s)^2] = \frac{1}{\lambda(\lambda + 1)} \tag{6.78}$$

and for the MAP estimate,

MAP:
$$E[(\hat{s} - s)^2] = \frac{1}{\lambda^2} > \frac{1}{\lambda(\lambda + 1)} \tag{6.79}$$

Here $1/\lambda$ represents the average number of calls per hour, averaged over all exchanges in the system. For $1/\lambda \gg 1$, the two mean-squared errors differ radically. But this is just the case where even the least mean-squared (MS) error is quite large and hence possibly meaningless as a measure of processing performance. It is only for $1/\lambda \ll 1$ that both mean-squared errors are the same. A similar comparison on the basis of a mean hit or miss would obviously show the MAP estimate to be superior for large $1/\lambda$, where the two estimates differ substantially.

Before considering other examples it is appropriate at this point to introduce still a third common parameter estimator. This is the *maximum-likelihood* (ML) estimator, discussed briefly in Chap. 3, in connection with sample estimates of mean and variance, and again in Sec. 4.10, in connection with parametric estimates of autocorrelation function and power spectral density. Here we consider it in connection with estimates of a *signal* parameter s. The maximum-likelihood estimator in this case is developed simply in terms of the MAP equations of (6.70) and (6.71). Note that the MAP estimate involves prior knowledge of $f(s)$, the density function of the signal parameter being estimated. In some cases s may be concentrated about some specified value, s_0 say. A situation like this is shown in Fig. 6.10*a*. It is apparent that this biases the MAP estimate of s to within the vicinity of s_0. [Recall that the MAP estimate is at the peak of the product of $f(s)$ and $f(\mathbf{x} \mid s)$. In the limit as $f(s)$ approaches a delta function about s_0, s becomes precisely known independent of any knowledge garnered from the sample vector $\mathbf{x}$.] In many situations, however, $f(s)$ may be rather broadly distributed (Fig. 6.10*b*). The MAP estimate will then lie close to the peak of $f(\mathbf{x} \mid s)$. In particular, if s is *uniformly* distributed over some broad range, $f(s)$ has no influence (except possibly in delimiting the range of s) and the MAP estimate *does* occur at the peak of $f(\mathbf{x} \mid s)$. This peak value of $f(\mathbf{x} \mid s)$ provides

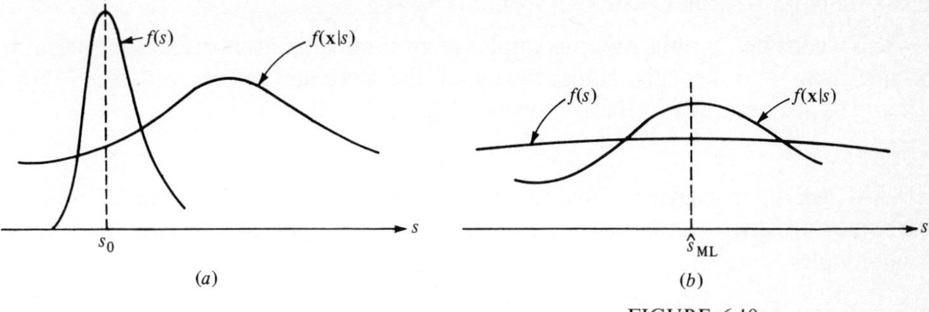

FIGURE 6.10
Effect of a priori distribution of s.

the ML estimate $\hat{s}_{ML}$ of s. This is so indicated in Fig. 6.10*b*. In mathematical terms the ML estimate is given by the solution of

$$\frac{\partial}{\partial s} f(\mathbf{x} \mid s) = 0 \tag{6.80}$$

or

$$\frac{\partial}{\partial s} \ln f(\mathbf{x} \mid s) = 0 \tag{6.81}$$

Because the ML estimate ignores a priori information about s, it is often simpler to find. It is also obviously an estimate to use in the event of complete lack of knowledge about s. Since $f(s)$ is not known, one might as well ignore it (this is equivalent to assuming s uniformly distributed over any feasible range) and use the information garnered from the sample vector $\mathbf{x}$ only, with no prior knowledge of s or its form thrown in. This is exactly what the ML estimate does. The ML estimate is also an appropriate one to use when the signal parameter s is not a random variable but some deterministic although unknown number. There then is no density function $f(s)$, and one can only work with the sample density function $f(\mathbf{x} \mid s)$.

The telephone-call problem again serves to illustrate these points. Recall from Fig. 6.9 that the $\lambda \ll 1$ case is the one for which s is broadly distributed over a wide range of values. (This corresponds to a large number of calls per hour on the average.) One would thus expect the MAP and ML estimates to coincide in this case. From Eq. (6.77) this should be

$$\hat{s}_{ML} = N \tag{6.82}$$

We can check this out by recalling that the distribution of calls was assumed Poisson with average value s [Eq. (6.72)]. Differentiation of this distribution with respect to s does yield $\hat{s} = N$ as the peak value. Note, incidentally, that since s is the average value of the Poisson distribution, $E(N) = s$. The ML estimate is thus *unbiased* in this case.

6.4 EXAMPLES OF BAYES ESTIMATES

We now consider various other examples of all three estimators discussed thus far to relate them and provide comparisons of the different results obtained. Other examples are provided in the problems at the end of this chapter.

EXAMPLE 1 Consider a signal s received in gaussian noise. The signal is known to be uniformly distributed between $-s_M$ and s_M. We would like to estimate s, using one sample

$$x_1 = s + n_1 \tag{6.83}$$

ML ESTIMATE Since n_1 is zero-mean gaussian, we have

$$f(x_1 \mid s) = \frac{e^{-(x_1 - s)^2/2\sigma_n{}^2}}{\sqrt{2\pi\sigma_n^2}} \tag{6.84}$$

The maximum-likelihood estimate, the peak of this density function, is thus the sample value x_1. This estimate *ignores* the fact that s is constrained to lie within the range $\pm s_M$.

MAP ESTIMATE Since $f(s) = 1/2s_M$, $-s_M < s < s_M$, the peak value of $f(s)f(x_1 \mid s)$ is the same as that of $f(x_1 \mid s)$ *providing* $-s_M < x_1 < s_M$. For $x_1 > s_M$ and $x_1 < -s_M$, however, the peak values are s_M and $-s_M$, respectively. (The reader should check this for himself.) The MAP estimate is thus

$$\hat{s}_{\text{MAP}} = \begin{cases} x_1 & -s_M < x_1 < s_M \\ s_M & x_1 \geq s_M \\ -s_M & x_1 \leq -s_M \end{cases} \tag{6.85}$$

This estimate, as well as the ML estimate, is shown in Fig. 6.11. It is obviously a nonlinear estimate.

MS ESTIMATE The least mean-squared estimate is given by the conditional mean, the mean value of $f(s \mid x_1)$. In this case this is just

$$\begin{aligned} \hat{s}_{\text{MS}} &= \int_{-\infty}^{\infty} sf(s \mid x_1)\, ds = K \int_{-\infty}^{\infty} sf(s)f(x_1 \mid s)\, ds \\ &= \frac{K}{2s_M} \int_{-s_M}^{s_M} \frac{se^{-(s-x_1)^2/2\sigma_n{}^2}}{\sqrt{2\pi\sigma_n^2}}\, ds \end{aligned} \tag{6.86}$$

The constant K is to be chosen so that $\int_{-\infty}^{\infty} f(s \mid x_1)\, ds = 1$. Alternately, one may write

$$\hat{s}_{\text{MS}} = \frac{\int_{-\infty}^{\infty} sf(s)f(x_1 \mid s)\, ds}{\int_{-\infty}^{\infty} f(s)f(x_1 \mid s)\, ds} \tag{6.86a}$$

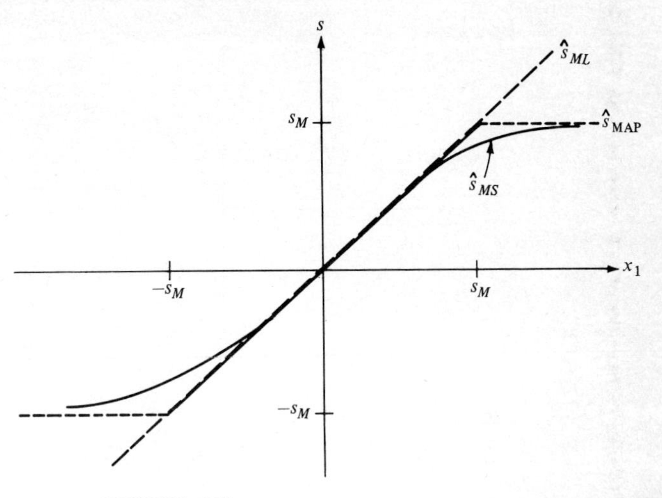

FIGURE 6.11
Estimation of uniformly distributed signal in gaussian noise.

The integration of the numerator is readily carried out. The denominator can only be written in terms of the function $\varphi(y)$ (related to the error function of Chap. 3)

$$\varphi(y) = \frac{1}{\sqrt{2\pi}} \int_0^y e^{-x^2/2}\, dx \tag{6.87}$$

This function approaches $\pm\frac{1}{2}$ respectively for y very large positively and negatively (see Fig. 6.12). It is left to the reader as an exercise to show that the estimate $\hat{s}_{MS}$ is given by

$$\hat{s}_{MS} = x_1 - \frac{\sigma_n}{\sqrt{2\pi}\, e(z)} \underbrace{\left(e^{-(1/2)(a-z)^2} - e^{-(1/2)(a+z)^2} \right)}_{d(z)} \tag{6.86b}$$

with the function $e(z)$ given by

$$e(z) \equiv \varphi(a - z) + \varphi(a + z)$$

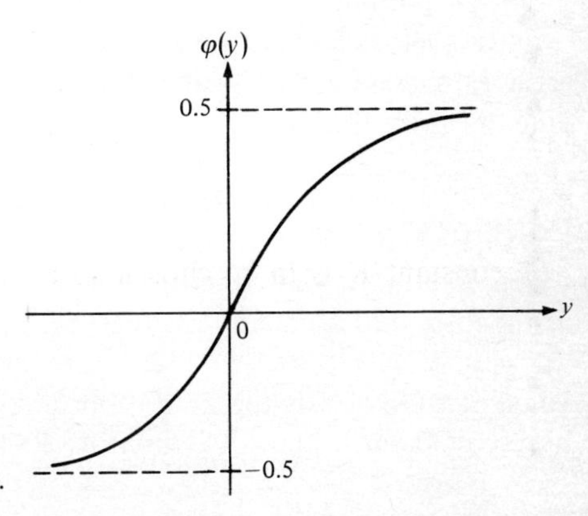

FIGURE 6.12
Probability function $\varphi(y)$.

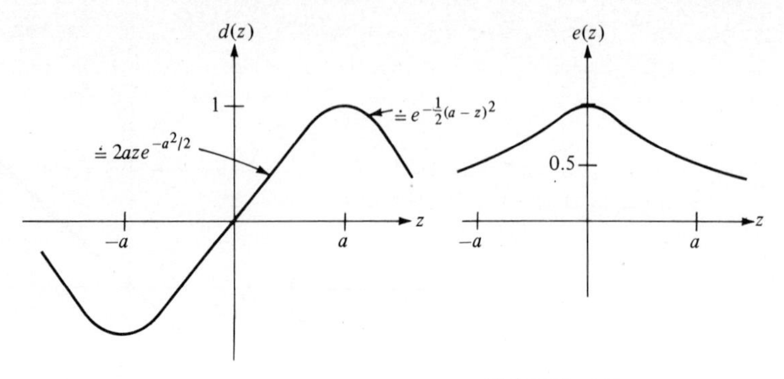

FIGURE 6.13
Functions $d(z)$ and $e(z)$, $a > 3$.

The parameter z is just x_1/σ_n and represents the sample value normalized to the noise standard deviation σ_n. a is a signal-to-noise ratio, $a = s_M/\sigma_n$. Note that, except for the first term in the estimate, the expression is highly nonlinear. Some idea of the variation of the estimate $\hat{s}_{MS}$ with sample value x_1 (or its normalized value z) can be obtained by sketching each of the functions defined, $d(z)$ and $e(z)$, separately and combining on one curve. Since the estimate depends on the signal-to-noise ratio a, a set of curves for different values of a would normally have to be plotted. For $a > 3$ or 4, however, for which $\varphi(a) \approx 0.5$, the expressions simplify considerably. In particular,

$$d(z) \approx \begin{cases} 2aze^{-a^2/2} & \text{for } |z| \ll a \text{ or } |x_1| \ll s_M \\ 1 & \text{for } z = a \\ e^{-(a-z)^2/2} & \text{for } z \gg a \end{cases}$$

and

(see Fig. 6.13). $e(z)$ can be similarly approximated by simple expressions and is also sketched in Fig. 6.13. The resultant composite curve for the least mean-squared estimate, shown in Fig. 6.11, is very much like the one for the MAP estimate. For large values of the sample x_1 ($x_1 > s_M$), $\hat{s}_{MS} \approx s_M$. It is thus a smoothed version of the MAP estimate.

It is left as an exercise for the reader to show that the *linear* least mean-squared error estimate for this problem is a straight-line function of x_1 but with a slope different from that of $\hat{s}_{ML}$. ////

EXAMPLE 2 We now consider the binary-detection problem, first discussed in Chap. 5, but now viewed in the form of a parameter-estimation problem. Specifically, assume a signal s, equally likely to be $\pm s_0$, received with additive gaussian noise. m samples, $x_j = s + n_j$, $j = 1, 2, \ldots, m$, are taken, and it is desired to estimate the value of s. Obviously the limitation of s to two possible values makes the optimum binary detector a kind of estimator of s in this case. We can in fact show that MAP estimation is identifiable with minimum probability of error binary detection in this case. For consider the probability density functions $f(s)$ and $f(\mathbf{x} \mid s)$. Here

$$f(s) = \tfrac{1}{2}\delta(s + s_0) + \tfrac{1}{2}\delta(s - s_0) \tag{6.88}$$

with $\delta(s)$ the delta function, because of the equally likely binary-signal assumption.

Assume that the noise samples n_j, $j = 1, 2, \ldots, m$, are independent. Then

$$f(\mathbf{n}) = \frac{\exp\left(-\sum_{i=1}^{m} \frac{n_i^2}{2\sigma_n^2}\right)}{(2\pi\sigma_n^2)^{m/2}} \tag{6.89}$$

and

$$f(\mathbf{x}|s) = \frac{\exp\left(-\sum_{i=1}^{m} \frac{(x_i - s)^2}{2\sigma_n^2}\right)}{(2\pi\sigma_n^2)^{m/2}} \tag{6.90}$$

The product $f(s)f(\mathbf{x} \mid s)$ must then also be a pair of delta functions, one at $s = s_0$, with magnitude $f(\mathbf{x} \mid s_0)$, the other at $s = -s_0$, with magnitude $f(\mathbf{x} \mid -s_0)$. The MAP estimate is in this case simply the larger of the two. The estimate is thus

$$\hat{s}_{\text{MAP}} = \begin{cases} +s_0, & \text{if } \ell(\mathbf{x}) = \dfrac{f(\mathbf{x} \mid s_0)}{f(\mathbf{x} \mid -s_0)} > 1 \\ -s_0, & \text{if } \ell(\mathbf{x}) < 1 \end{cases} \tag{6.91}$$

This is just the binary detection rule written, as in Chap. 5, in terms of the likelihood ratio $\ell(\mathbf{x})$. It is left to the reader to show that in this case

$$\ell(\mathbf{x}) = \exp \frac{2m\bar{x}s_0}{\sigma_n^2}$$

with $\bar{x} \equiv 1/m \sum_{i=1}^{m} x_i$ just the sample mean introduced early in this chapter. It is apparent that the MAP estimate is equally well given by $\hat{s}_{\text{MAP}} = +s_0$ if $\bar{x}$ or $\sum_{i=1}^{m} x_i > 0$. Similarly, if $\sum_{i=1}^{m} x_i < 0$, $\hat{s}_{\text{MAP}} = -s_0$. This is thus a highly nonlinear estimate, as shown in Fig. 6.14. (Is this estimate biased or unbiased?)

Why the MAP estimate is identical with the minimum-probability-of-error detection rule is easily explained. Visualize receiving and storing the m-element sample vector $\mathbf{x}$. We use this to calculate two probabilities, the *a posteriori probability* $P[+s_0 \mid \mathbf{x}]$ that $+s_0$ was present as the signal and the *a posteriori probability* $P[-s_0 \mid \mathbf{x}]$ that $-s_0$ was present. A little thought will indicate that the choice of the larger probability will guarantee the minimum probability of error over the long run. We thus pick $+s_0$ if

$$P[+s_0 \mid \mathbf{x}] > P[-s_0 \mid \mathbf{x}]$$

But we have

$$P[+s_0|\mathbf{x}]f(\mathbf{x}) = P[+s_0]f(\mathbf{x}|s_0)$$

and

$$P[-s_0 \mid \mathbf{x}]f(\mathbf{x}) = P[-s_0]f(\mathbf{x} \mid -s_0)$$

from the Bayes theorem relating joint and conditional probabilities. Here $P[+s_0]$ and

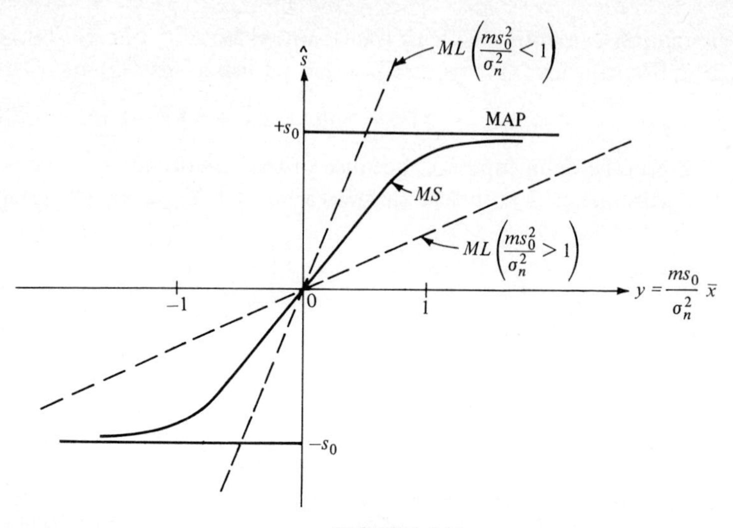

FIGURE 6.14
Estimates, binary signals in gaussian noise.

$P[-s_0]$ are the respective a priori probabilities of transmitting $+s_0$ and $-s_0$. Substituting into the inequality, we again have the likelihood-ratio rule. The interesting thing about this approach is that it is easily extended to the case of the detection of one of M possible signals transmitted.†

What is the maximum-likelihood estimate in this case? Here we simply find the peak value of $f(\mathbf{x} \mid s)$, independent of the statistics of the signal s. The maximum-likelihood estimate would thus be the same for binary signals (as in this case), a uniformly distributed signal (as in the previous example), a gaussian signal, or deterministic signal of unknown amplitude. Specifically, differentiating the logarithm of $f(\mathbf{x} \mid s)$ with respect to s and setting this equal to zero, we find

$$\hat{s}_{\text{ML}} = \bar{x} = \frac{1}{m} \sum_{i=1}^{m} x_i \tag{6.92}$$

This is obviously independent of the signal amplitude s_0, since the maximum-likelihood estimate makes no use of this information. The sample mean is thus the maximum-likelihood estimate in *any* situation involving a constant signal parameter plus additive independent gaussian-noise samples. (Is this estimate biased or unbiased?)

The MS estimate in this example is also of interest. Recall that this is just

$$\hat{s}_{\text{MS}} = \frac{\int s f(s) f(\mathbf{x} \mid s)\, ds}{\int f(s) f(\mathbf{x} \mid s)\, ds} \tag{6.93}$$

† M. Schwartz, W. R. Bennett, and S. Stein, "Communication Systems and Techniques," p. 87, McGraw-Hill, New York, 1966.

Plugging in the appropriate expressions for the two density functions in this case and using the relation $\int \delta(x - x_0) f(x)\, dx = f(x_0)$, we find

$$\hat{s}_{\text{MS}} = s_0 \frac{f(\mathbf{x} \mid s_0) - f(\mathbf{x} \mid -s_0)}{f(\mathbf{x} \mid s_0) + f(\mathbf{x} \mid -s_0)} \tag{6.94}$$

(Again the details are left to the reader.)

Alternatively, using the definition of the likelihood ratio, $\ell(\mathbf{x}) = f(\mathbf{x} \mid s_0)/f(\mathbf{x} \mid -s_0)$, we have

$$\hat{s}_{\text{MS}} = s_0 \frac{\ell(\mathbf{x}) - 1}{\ell(\mathbf{x}) + 1} \tag{6.94a}$$

Instead of the two-valued estimate involving the likelihood ratio that we obtained for the MAP estimator, we have here an estimate varying continuously with the sample values. [Note only that $\ell(\mathbf{x}) > 1$ ensures a positive estimate. In the MAP or detection case, this gave the precise value $+s_0$; see Fig. 6.14.] If we actually use the expression for the likelihood ratio written above, we find, after some algebraic manipulation, that the MS estimate is given by

$$\hat{s}_{\text{MS}} = s_0 \tanh y \tag{6.94b}$$

with

$$y \equiv \frac{ms_0}{\sigma_n^2} \bar{x}$$

a dimensionless parameter proportional to the sample mean $\bar{x}$. The hyperbolic tangent is for small values of the argument just the argument itself:

$$\tanh y \approx y \qquad y \ll 1$$

For large values it approaches ± 1 asymptotically

$$\tanh y \approx \begin{cases} 1 & y \gg 1 \\ -1 & y \ll -1 \end{cases}$$

This estimate is sketched in Fig. 6.14, together with the MAP estimate. We can also superimpose the ML estimate $\hat{s}_{\text{ML}} = \bar{x}$ by rewriting in terms of y. We then have

$$\hat{s}_{\text{ML}} = \frac{s_0 y}{ms_0^2/\sigma_n^2} \tag{6.92a}$$

The signal-to-noise parameter ms_0^2/σ_n^2 was encountered earlier in this chapter. Two lines representing the maximum-likelihood estimate are shown in Fig. 6.14, one for large and the other for small signal-to-noise ratio. ////

EXAMPLE 3 As an extension of Example 2, consider a binary signal again received in the presence of gaussian noise but now a known waveshape is used to carry the signal information. We thus have

$$x_j = as_j + n_j \qquad j = 1, 2, \ldots, m \tag{6.95}$$

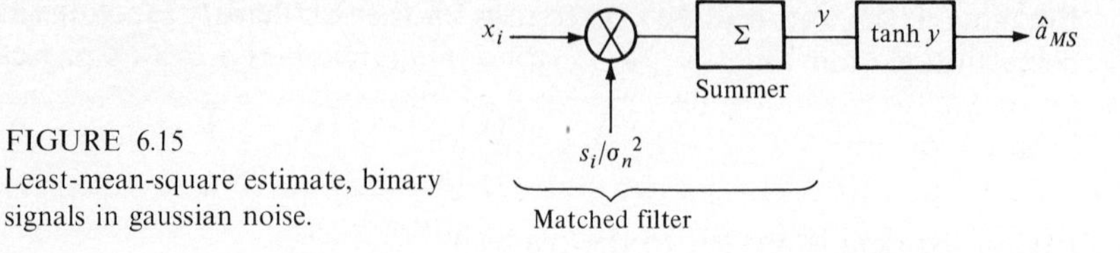

FIGURE 6.15
Least-mean-square estimate, binary signals in gaussian noise.

with s_j *known* signal samples and a a random parameter to be estimated that is equally likely to be ± 1. This is of course the variable-signal-amplitude problem considered previously in the binary-detection case of Chap. 5. This is the problem that led to the matched filter. It is left to the reader to show, as simple extensions of Example 2, that the various estimates of a are given by

$$\hat{a}_{\text{MAP}} = \begin{cases} +1 & \text{if } \sum_{i=1}^{m} s_i x_i > 0 \\ -1 & \text{if } \sum_{i=1}^{m} s_i x_i < 0 \end{cases} \tag{6.96}$$

$$\hat{a}_{\text{ML}} = \frac{\sum_{i=1}^{m} s_i x_i}{\sum_{i=1}^{m} s_i^2} \tag{6.97}$$

and

$$\hat{a}_{\text{MS}} = \tanh y \qquad y = \frac{1}{\sigma_n^2} \sum_{i=1}^{m} s_i x_i \tag{6.98}$$

Note that these are the same as the previous results, with the appropriate redefinition of the dimensionless parameter y. Note also that *matched filtering* is called for in all three cases. As an example, the MS processor is shown schematically in Fig. 6.15. If the tanh y box is approximated by a perfect limiter, the MAP estimate or optimum binary-detection processor is obtained. ////

EXAMPLE 4 The next example is essentially a composite of Examples 1 and 3. Assume that a known signal waveshape has been transmitted. The signal is received in additive noise with the same shape but with a random amplitude factor appended. Thus, we have

$$x_j = as_j + n_j \qquad j = 1, 2, \ldots, m \tag{6.99}$$

Here s_j is known, all j, n_j is gaussian, and the unknown amplitude a is to be estimated.

This example might represent a simplified model of a radar or sonar received signal, whose target reflections and variation in distance cause a random amplitude factor to appear. It could represent a simple AM receiver situation: the s_j samples represent samples of a known sine-wave carrier, the amplitude parameter a representing the information transmitted. It could represent a model of a received pulse-amplitude modulation (PAM) signal sometimes used in telemetry systems, in which the heights (amplitudes) of narrow signal pulses (the *carrier*) are varied in accordance with the information to be transmitted. It might also represent a processing situation in an on-line pollution test. A signal of known waveshape (perhaps at microwave, infrared, or optical frequencies) is transmitted through a medium under test (air or water). The energy transmitted through the medium, or in some cases scattered back from it, is a measure of the concentration of pollutants. This energy is in turn proportional to the square of a.

What are some means of processing the m samples x_j to extract an estimate of a? First consider the maximum-likelihood processor. This requires no knowledge of the statistics of a. A little thought will indicate that this is identical with the maximum-likelihood estimate of Example 3 and given by

$$\hat{a}_{\mathrm{ML}} = \frac{\sum_{i=1}^{m} s_i x_i}{\sum_{i=1}^{m} s_i^2} \tag{6.97}$$

Matched-filter processing is again called for.

Now assume that the density function $f(a)$ is known. Recall that the MAP estimate is the solution of the equation

$$\frac{\partial}{\partial a} \ln f(a) + \frac{\partial}{\partial a} \ln f(\mathbf{x} \mid a) = 0 \tag{6.100}$$

In this example we have

$$f(\mathbf{x} \mid a) = \frac{\exp - \sum_{i=1}^{m} \frac{(x_i - as_i)^2}{2\sigma_n^2}}{(2\pi\sigma_n^2)^{m/2}} \tag{6.101}$$

(Can you justify this?)

Taking the logarithm and differentiating, as required, we have

$$\frac{\partial}{\partial a} \ln f(\mathbf{x}|a) = \frac{1}{\sigma_n^2} \sum_{i=1}^{m} s_i(x_i - as_i) \tag{6.102}$$

To complete the problem we must assign some known form to $f(a)$. Two interesting cases follow.

GAUSSIAN STATISTICS Let

$$f(a) = \frac{e^{-(a-a_0)^2/2\sigma_a^2}}{\sqrt{2\pi\sigma_a^2}} \tag{6.103}$$

Here a_0 is the average value of a, and σ_a^2 its variance. Then

$$\frac{\partial}{\partial a} \ln f(a) = -\frac{a - a_0}{\sigma_a^2} \tag{6.104}$$

It is left to the reader to show that the MAP estimate is then just

$$\hat{a}_{\text{MAP}} = \frac{a_0 + \dfrac{\sigma_a^2}{\sigma_n^2} \sum_{i=1}^{m} s_i x_i}{1 + \dfrac{\sigma_a^2}{\sigma_n^2} \sum_{i=1}^{m} s_i^2} \tag{6.105}$$

Note that if a_0, the average value of the unknown amplitude factor, were zero, a *linear* estimate of the form $\sum_{i=1}^{m} h_i x_i$ would be obtained. It is left to the reader to show that this MAP estimate is the *same* in this case as the least mean-squared (MS) estimate. This is specifically due to the *gaussian* statistics assumed for both the signal and noise: the a posteriori probability $f(s \mid \mathbf{x})$ [actually $f(a \mid \mathbf{x})$ in this case] is then also gaussian. Its mean and mode (maximum point) are then the same, leading to identical MS and MAP estimates. [Any cost function $C(s, \hat{s})$ which is symmetric about s, that is, $C(s, \hat{s})$ is of the form $C(s - \hat{s}) = C(\hat{s} - s)$, will lead to the same result as with gaussian statistics. The examples in Fig. 6.6 are examples of exactly this type. This means that the absolute-error criterion of Fig. 6.6(1), for example, will also provide the same estimate on minimization.] In the case of zero-mean signal this optimum estimate is also the optimum least-squared linear estimate obtained by assuming $\hat{a} = \sum_{i=1}^{m} h_i x_i$ and finding the h_i's to minimize $E(\hat{a} - a)^2$ directly. This is of course the problem solved in Sec. 6.2. With zero-mean gaussian statistics the optimum MS estimate is linear rather than nonlinear and agrees with that found by assuming that a linear processor is to be used.

One can show generally that zero-mean gaussian statistics *always* result in linear estimates. Note also that a *matched filter* again appears as the appropriate processor (Fig. 6.17*a*). The term $(\sigma_a^2/\sigma_n^2) \sum s_i^2$ is essentially a signal-to-noise ratio and provides appropriate weighting to the matched-filter output. If we now have $\sigma_a \gg \sigma_n$, so that the signal parameter is *broadly* distributed compared to the noise statistics and $\sum_{i=1}^{m} s_i^2$ is not too small,

$$\hat{a}_{\text{MAP}} \approx \frac{\sum_{i=1}^{m} s_i x_i}{\sum_{i=1}^{m} s_i^2} \tag{6.105a}$$

This is just the maximum-likelihood result quoted earlier and agrees with our earlier statement that if the parameter to be estimated has a broad distribution, maximum-likelihood and MAP estimates are the same.

RAYLEIGH STATISTICS One very common model for random amplitude variations is to assume the Rayleigh distribution

$$f(a) = \frac{a e^{-a^2/2\sigma_a^2}}{\sigma_a^2} \qquad a \geq 0 \tag{6.106}$$

as already noted briefly in passing [see Eq. (6.54)]. This distribution is sketched in Fig. 6.16. It has been found to be a fairly accurate representation of amplitude statistics in many radar and sonar situations. It is also often a good model for the

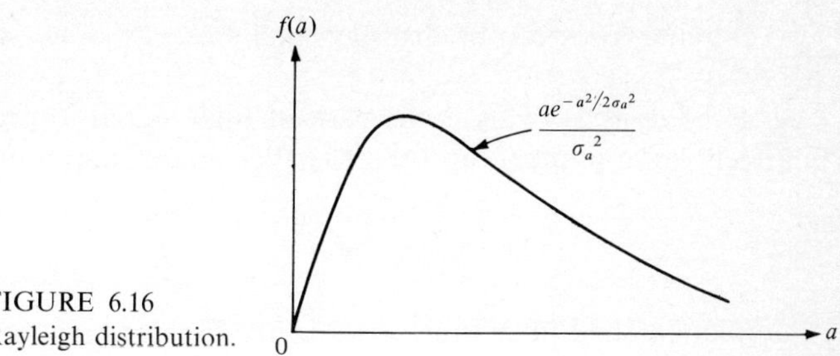

FIGURE 6.16
Rayleigh distribution.

statistics of signals transmitted through so-called *fading media*. (Shortwave reflection from the ionosphere is one classic example.)

It is again left for the reader to show that the MAP estimate in this case is given by

$$\hat{a}_{\text{MAP}} = \frac{y}{2P}\left(1 + \sqrt{1 + \frac{4P}{y^2}}\right) \tag{6.107}$$

with

$$y \equiv \frac{1}{\sigma_n^2} \sum_{i=1}^{m} s_i x_i$$

the same parameter introduced previously in Example 3, and

$$P \equiv \frac{1}{\sigma_n^2} \sum_{i=1}^{m} s_i^2 + \frac{1}{\sigma_a^2}$$

a known signal- and noise-weighting parameter. This is in general a highly nonlinear function of the samples, although matched filtering is again called for (Fig. 6.17*b*).

Since the received signal terms are as_i, we may let s_i be a dimensionless quantity and incorporate the actual signal dimensions in a. We can then normalize by letting $\sum_{i=1}^{m} s_i^2 = 1$. Actual amplitude variations are then incorporated in σ_a^2. The parameter $y^2/4P$ that appears in the expression for $\hat{a}_{\text{MAP}}$ is then given by

$$\frac{y^2}{4P} = \frac{\left(\sum_{i=1}^{m} \frac{s_i x_i}{2\sigma_n}\right)^2}{1 + \sigma_n^2/\sigma_a^2}$$

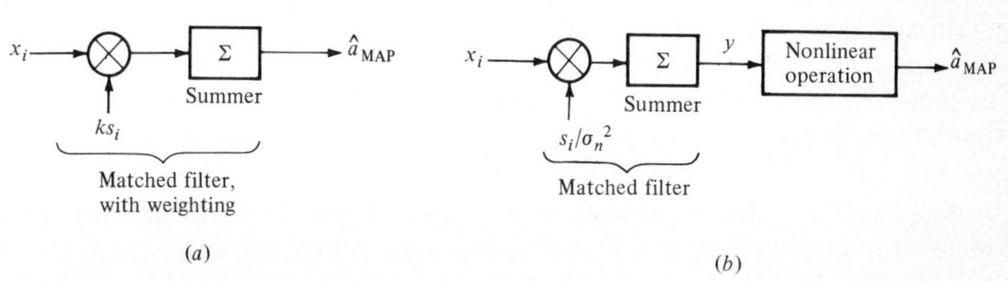

FIGURE 6.17
MAP signal processors, amplitude signal variations. (*a*) Gaussian signal statistics (k a constant); (*b*) Rayleigh signal statistics.

Two extreme cases can be distinguished.

CASE 1: $\sigma_n^2/\sigma_a^2 \ll 1$ This is the case of high signal-to-noise ratio; $y^2/4P$ is then a relatively large number, and the MAP estimate is approximately

$$\hat{a}_{\text{MAP}} \approx \frac{y}{P} \approx \sum_{i=1}^{m} s_i x_i \tag{6.107a}$$

A simple matched filter thus arises in the case of high signal-to-noise ratio.

CASE 2: $\sigma_n^2/\sigma_a^2 \gg 1$ This is the case of low signal-to-noise ratio; $y^2/4P$ is a small number, and the estimate for this case becomes

$$\hat{a}_{\text{MAP}} \approx \frac{1}{\sqrt{P}} = \sigma_a \tag{6.107b}$$

The best MAP estimate is thus just the signal standard deviation σ_a *independent of sample values measured.* Said another way, if the signal received is small compared with the noise, one should disregard the samples taken. (They carry no useful information.) Signal processing is not possible in this situation. ////

6.5 MAXIMUM-LIKELIHOOD ESTIMATION OF PARAMETERS OF LINEAR SYSTEMS

In Chap. 2 we discussed at length the time- and frequency-domain properties of linear filters. In particular, the two important classes of digital filters, nonrecursive and recursive, were handled in some detail. They are of course used extensively in practice for their frequency-selective properties, for their smoothing capability, as estimators of signals in the presence of noise, etc. These are just the signal-processing objectives that are the theme of this book.

In most physical situations additional filtering is introduced by the medium or the system through which signals are transmitted. A digital signal transmitted over the wires and cables of the telephone plant is smeared out in time by the distributed capacitance of the lines; a driver attempting to get the feel of the road through his manipulation of the steering wheel has the transfer function or filtering properties of the system from the car wheels to the steering wheel interposed between himself and the road; instruments and transducers used to measure temperature, pressure, and other physical parameters varying with time (the "signals") introduce their own dynamics into the problem. Conceptually, we denote this by saying that some input signal $x(t)$ is converted to another signal $y(t)$ after passing through the physical system. This is indicated in Fig. 6.18. We can characterize the medium or physical system by saying that a pulselike input signal (approximating an impulse) is converted into a wider output signal (the impulse response of the system).

Obviously, if we are attempting to measure the input $x(t)$, we must know the characteristics of the medium. For simple systems or media the transfer function or impulse response can be determined quite readily, in a straightforward way. In such

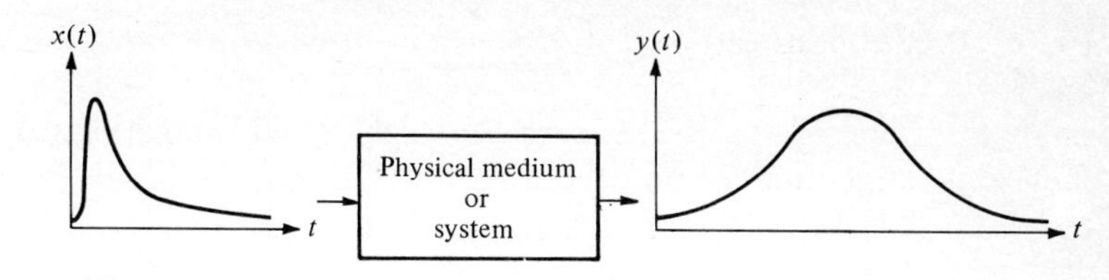

FIGURE 6.18
Effect of physical medium on input signal—signal is dispersed or spread out in time.

cases the effect of the medium is easily taken into account. In much more complicated situations, e.g., an entire telephone system over which signals must be conveyed or water through which underwater signals must travel, physical considerations are not as readily applied in determining the characteristics of the medium. In addition, the medium or signal may be changing its characteristics with time. (Each time you dial a long-distance call, for example, the message may take a different route to reach the same number.) How does one *measure* the system characteristics? Various possibilities exist. One method is simply to shock-excite the system, measuring the resultant impulse response; another is to measure its frequency transfer function by actually finding the response to a sequence of input sinusoids. The method we discuss here, as an application of maximum-likelihood estimation, is to assume a model of the system under study, with certain parameters unknown, and use measurements to determine these parameters. Thus this is another example of parametric estimation.

In Sec. 4.10 we discussed parametric spectral estimation by means of maximum-likelihood techniques. Specifically we assumed there that the data samples x_k had been generated by a first-order autoregressive process, or higher-order processes if deemed necessary [see Eqs. (4.98) and (4.119), for example]. The objective was then to estimate the (unknown) parameters of these processes. The discussion of that section is directly applicable here if the medium or linear system we have been discussing can be modeled as a recursive filter of first or higher order. The parameter a of (4.98) or the parameters a_1 and a_2 of (4.119) are thus the parameters that characterize the linear system under test, whose values are to be estimated by measurements of the output-data samples x_k. Very often in tests of this type the input, or driving-force, samples are samples taken from a white-noise process or an approximation of such a process. These would then be the samples u_k of (4.98). In this case the parameters b or b_1 and b_2 of (4.98) and (4.119) would presumably be known and would not have to be estimated.

An alternate representation of a linear system is of course that of a nonrecursive digital filter. Since the recursive-filter model has already been discussed in Sec. 4.10, we focus here on the nonrecursive representation. (Linear systems may of course be represented by continuous-time models as well; the *RC* filter, for example, is the simplest such model and is the continuous equivalent of the first-order recursive filter. But we are focusing in this book on digital signal processing and are therefore interested in discrete-time models, in which signal samples are measured only at

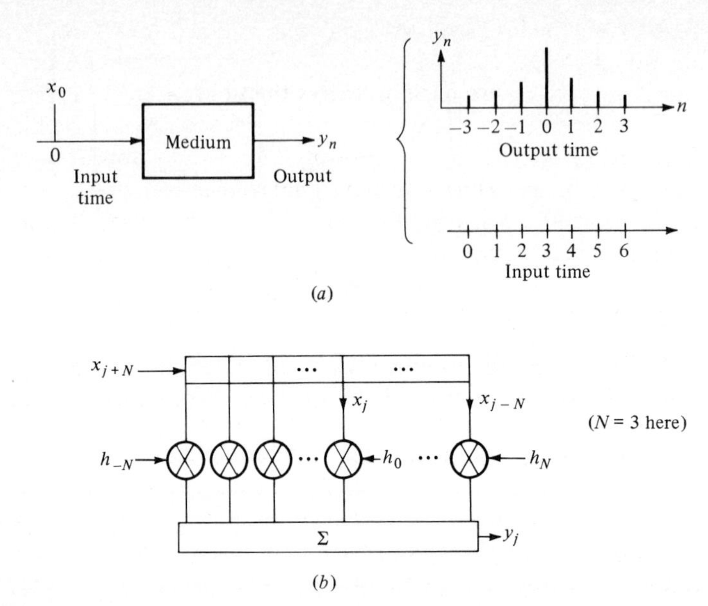

FIGURE 6.19
Linear representation of physical system or medium. (*a*) Medium impulse response; (*b*) nonrecursive-filter model of medium (output time).

prescribed instants of time rather than at all times.) We thus assume in this section that the medium or linear system under study or test can be modeled as a nonrecursive digital filter. Our objective is then to estimate the filter (or system) coefficients. Specifically, if the input-signal samples are x_j, the samples at the output of the system under consideration are given by

$$y_j = \sum_{n=0}^{m} h_n x_{j-n} \tag{6.108}$$

We thus are interested in determining the $m + 1$ coefficients h_n, $n = 0, \ldots, m$, that characterize the system. Roughly speaking, the number of coefficients needed, $m + 1$ in this case, is determined by the expected spread or dispersion introduced by the system. As an example, consider the solitary pulse x_0 introduced at time $t = 0$ in Fig. 6.19. The output is shown spread out over seven time slots. One would then expect the model to need at least seven coefficients for appropriate characterization.

In many physical situations the system impulse response builds up to a peak value some time units after the impulse is applied and then decays back to zero. It simplifies the analysis to measure the time of the output samples from this peak output value. We shall also assume, for simplicity, that the spread or dispersive effect of the medium is equally distributed about the peak value. (One can always add 0 values to ensure this.) The output samples can then equally well be written

$$y_j = \sum_{n=-N}^{N} h_n x_{j-n} \tag{6.109}$$

with time now measured with reference to the output samples and $2N + 1 = m + 1$ the number of coefficients needed. Figure 6.19*a* compares output with input time for that example. Figure 6.19*b* portrays the nonrecursive filter of Eq. (6.109).

One problem with measuring the desired $2N + 1$ coefficients h_n, $n = -N, \ldots, 0, \ldots, N$, directly is that the y_j samples normally appear corrupted with noise. This noise may be introduced during signal transmission, or it may represent inaccuracies in the measurements or the model representation itself. The actual received information is thus given by

$$y_j = \sum_{n=-N}^{N} h_n x_{j-n} + n_j \tag{6.110}$$

with the desired coefficients now to be determined from noisy measurements. For simplicity we assume that the noise samples are uncorrelated, zero-mean, gaussian variables, of variance σ^2.

How does one now estimate the h_n's from the measured samples y_j? One common method is that of sending a known signal sequence $\{x_j\}$, measuring the error $y_j - x_j$ introduced by the medium and the additive noise, and using the sequence of errors to find the h_n's. As already noted, one signal sequence of particular interest for this purpose is a white-noise process. One particularly simple embodiment of such a sequence, particularly simple to implement for digital processing, is the so-called *pseudorandom* pulse sequence. This is a relatively long sequence of binary pulses, ± 1, one per time slot or sampling interval, that can be made to approximate truly random (equally likely and independent) binary symbols. The pulses are commonly generated with shift registers and digital logic. Call this sequence $\{a_j\}$. The output signal sequence is then

$$y_j = \sum_{n=-N}^{N} h_n a_{j-n} + n_j \tag{6.111}$$

The $\{a_j\}$ sequence thus corresponds to the $\{u_k\}$ sequence of Sec. 4.10.

If there were no dispersion but simply innate time delay N units long, the medium output would be $h_0 a_j + n_j$. The rest of the terms in y_j represent distortion due to the medium. Since we assume a_j known (the pseudorandom sequence is made available at the output), we can measure the error $\varepsilon_j = y_j - a_j$ and use it to estimate the medium coefficients. (One could use y_j directly for estimation purposes, but a calculation indicates that the variance of the estimates is proportional to h_0, the coefficient corresponding to the undispersed but delayed output. Generally h_0 is greater than the other coefficients to be found; as noted earlier, the system impulse response gradually rises to a peak and then decays. The variance of the estimates can then be rather large. By using $\varepsilon_j = y_j - a_j$, we find below that we estimate $h_0 - 1$, a number comparable to the other coefficients to be estimated, resulting in a substantially smaller variance.)

It is apparent that the error is given by

$$\begin{aligned} \varepsilon_j = y_j - a_j &= h_{-N} a_{j+N} + \cdots + (h_0 - 1)a_j + \cdots + h_N a_{j-N} + n_j \\ &= \sum_{n=-N}^{N} h'_n a_{j-n} + n_j \end{aligned} \tag{6.112}$$

with h'_n indicating that $h'_0 = h_0 - 1$. Assume that we now measure K such terms in succession, say $\varepsilon_1, \varepsilon_2, \ldots, \varepsilon_K$, and use these measured errors to estimate the h_n coefficients.

What processing shall we use to estimate the h_n's? Here, because of its simplicity, we assume maximum-likelihood estimation is to be used. Specifically, with the noise samples assumed gaussian and independent, the ensemble of K samples $\varepsilon_1, \varepsilon_2, \ldots, \varepsilon_K$, denoted by the vector $\boldsymbol{\varepsilon}$ and conditioned on the coefficients $h_{-N}, \ldots, h_0 - 1, \ldots, h_N$, to be estimated, is itself jointly gaussian, the density function $f(\boldsymbol{\varepsilon} \mid h_{-N} \cdots)$ being given by the *product* of the individual density functions

$$f(\boldsymbol{\varepsilon} \mid h_{-N} \cdots) = \frac{\exp\left[-\sum_{j=1}^{K} \frac{(\varepsilon_j - \sum_n h'_n a_{j-n})^2}{2\sigma^2}\right]}{(2\pi\sigma^2)^{K/2}} \tag{6.113}$$

The maximum-likelihood estimate is now found by setting the derivative of the logarithm of $f(\cdot)$ equal to zero

$$\frac{\partial}{\partial h'_i} \ln f(\boldsymbol{\varepsilon} \mid h_{-N} \cdots) = 0 = \sum_{j=1}^{K} \left(\varepsilon_j - \sum_n \hat{h}'_n a_{j-n}\right) a_{j-i} \qquad i = -N, \ldots, N \tag{6.114}$$

The hat notation is again used to denote the maximum-likelihood estimate.

Rewriting this set of equations, we have

$$\sum_{j=1}^{K} a_{j-i}\varepsilon_j = \sum_{j=1}^{K} \sum_{n=-N}^{N} \hat{h}'_n a_{j-n} a_{j-i} \qquad i = -N, \ldots, N \tag{6.115}$$

We thus have to solve this set of $2N + 1$ equations simultaneously to find the estimates $\hat{h}'_n$. They look rather formidable but can be put in less forbidding form by defining the coefficients

$$g_i \equiv \sum_{j=1}^{K} a_{j-i}\varepsilon_j \tag{6.116}$$

$$R_{in} \equiv \sum_{j=1}^{K} a_{j-n} a_{j-i} \tag{6.117}$$

The set of equations (6.115) is then written much more simply as

$$g_i = \sum_{n=-n}^{N} \hat{h}'_n R_{in} \qquad i = -N, \ldots, N \tag{6.118}$$

Note, however, that this is exactly of the form of the equations obtained in Sec. 6.2 in using linear mean-squared estimation to estimate a random signal s [see Eqs. (6.5) and (6.10)]. As a matter of fact, the R_{ij} coefficients defined there were the correlation coefficients; here they are *sample* correlation coefficients. As there, the set of equations (6.118) here can be written in still simpler form by using vector notation: let g_i be the ith component of a $(2N + 1)$-dimensional vector $\mathbf{g}$, while R_{in} is a typical term in a $(2N + 1) \times (2N + 1)$ matrix R. We then have

$$\mathbf{g} = R\hat{\mathbf{h}} \tag{6.119}$$

with $\hat{\mathbf{h}}$ the vector consisting of the $2N + 1$ unknown coefficients to be found. The vector solution is thus, of course,

$$\hat{\mathbf{h}} = R^{-1}\mathbf{g} \tag{6.120}$$

As noted earlier, such vector equations can be easily manipulated but they can be costly (in terms of computer time) to solve if the order of the matrices [here $(2N + 1) \times (2N + 1)$] is large. Iterative techniques have been developed for solving such equations on a computer. Here we simply point out another estimate for the h_n's, developed from the maximum-likelihood approach just considered, that obviates the need for simultaneous solution of $2N + 1$ equations. Specifically, in this case we stipulated that the pseudorandom input signals $\{a_i\}$ were to be binary and essentially independent. This implies that over a long enough interval of time, $a_i a_j$, $i \neq j$, is equally likely to be positive or negative. Then the correlation coefficient R_{in}, $i \neq n$, is very nearly zero

$$R_{in} = \sum_{j=1}^{K} a_{j-n} a_{j-i} \approx 0 \tag{6.121}$$

for $i \neq n$, and K large. Stated another way, the matrix R defined above may be expected to be nearly diagonal. With this assumption and recalling that $a_i^2 = 1$ (the symbols are ± 1), Eq. (6.115) reduces simply to

$$\hat{h}_i'\bigg|_{\text{subopt}} \approx \frac{1}{K} \sum_{j=1}^{K} a_{j-i} \varepsilon_j \qquad i = -N, \ldots, N \tag{6.122}$$

We use suboptimum to denote that this estimate is suboptimum in a maximum-likelihood sense, but it is of course a perfectly valid estimate in its own right.

This estimate has an interesting interpretation: the measured errors ε_j are to be stored and then *cross-correlated* with, or multiplied by, the time-shifted stored binary signals a_{j-i}. A cross-correlation operation again appears as the appropriate signal processor. The stored products are then accumulated to provide the derived estimates. These operations can readily be instrumented using digital circuitry in a hard-wired digital processor.

How "good" is the estimate of Eq. (6.122)? If we take the expectation of $\hat{h}_i'$ with respect to the noise n_j that appears in the error term ε_j and again assume $\sum_{j=1}^{K} a_{j-k} a_{j-n} \approx 0$ unless $i = n$, it is readily shown that

$$E\left[\hat{h}_i'\bigg|_{\text{subopt}}\right] = \begin{cases} h_i' = h_0 - 1 & i = 0 \\ h_i & i \neq 0 \end{cases} \tag{6.123}$$

(Can you show this?) These estimates are then *unbiased*.

The variance $V(\hat{h}_i')$ of this estimate is not so readily obtained, principally because sums of terms such as $a_k a_n a_j a_l$ appear in calculating it. It is not clear how to handle these terms. A similar problem arose in Chap. 4 in determining the variance of the autocorrelation-function estimate [see (4.12)]. Under some simplifying assumptions, however, one can show

$$\frac{V(\hat{h}_i')}{E^2(\hat{h}_i')} \sim \frac{1}{K} \tag{6.124}$$

The variance thus reduces with the number of accumulated error terms ε_j measured. For large enough K, then, the estimates of Eq. (6.122) represent fairly good approximations to the medium-model coefficients h_i that were to be estimated.

This technique for estimating the coefficients of a linear-model representation of some physical system has other applications than that of simply representing a system for processing purposes. It is also used in studying media whose physical properties may be of great interest in their own right. For example, consider the problem of searching for minerals underground in a given region. If one shock-excites the ground, as is frequently done in seismic and geophysical exploration, the ground response at some distance from the excitation must depend on the physical characteristics of the earth (the medium) between. A cross correlation of input signal and output signal should help in the identification of the region underground: What types of mineral does it contain? Is it stratified? Are there oil-bearing rocks underneath? One approach is obviously that of providing a catalog of different types of nonrecursive models for each of the different characteristic media. The identification of thc most likely model thus serves to characterize the medium.

PROBLEMS

6.1 A voltage s which is known to have a density function $f(s) = (1/2\sqrt{2\pi})e^{-s^2/8}$ is to be estimated by the linear combination of several measurements. Two expensive meters give readings x_1 and x_2, which are the true voltage plus zero-mean gaussian errors. The errors are independent with variances $\sigma_n^2 = 2\ \text{V}^2$.
(*a*) What is the best (mean-squared error) estimate of the form $\hat{s} = h_1 x_1 + h_2 x_2$? That is, find the best h_1 and h_2.
(*b*) Repeat when four cheap meters produce readings y_1, y_2, y_3, y_4 with bigger errors having variances $\sigma_n^2 = 4\ \text{V}^2$. That is, find the best $h_1, \ldots, h_4$ in

$$\hat{s} = h_1 y_1 + h_2 y_2 + h_3 y_3 + h_4 y_4$$

(*c*) Which is the better procedure: two expensive meters or four cheap ones?

6.2 Consider the signal-estimation example given by Eqs. (6.14) and (6.15).
(*a*) Show that Eq. (6.17) satisfies the set of equations (6.16), thus providing the smallest linear mean-squared estimate.
(*b*) Show that with the signal estimate of Eq. (6.18), the mean-squared error is given by Eq. (6.19).

6.3 A velocity sensor mounted on a radar tracking a space vehicle is used to measure the radial velocity of the vehicle. Five independent sensor readings, measured with respect to a nominal average velocity of 10,000 mi/h, are taken. These turn out to be $x_1 = 980$ mi/h, $x_2 = 550$ mi/h, $x_3 = 1100$ mi/h, $x_4 = 1050$ mi/h, $x_5 = 760$ mi/h. The wide dispersion in the sample values is assumed due to noise added in the system: noise picked up by the radar antennas, noise in the tracking and instrument sensors, etc. The five samples are to be processed to form an estimate of vehicle velocity v_r, measured with respect to the average 10,000 mi/h velocity. Thus $v_r = v - 10{,}000$, with v the vehicle velocity, and $E[v] = 10{,}000$. (When an estimate $\hat{v}_r$ of v_r is found, the estimate of v becomes $\hat{v} = \hat{v}_r +$

10,000.) As a reasonable model one may assume the noise additive, with noise samples independent and zero mean. Thus

$$x_j = v_r + n_j \qquad j = 1, 2, \ldots, 5$$

It is known that $\sigma_v = 1000$ mi/h, and the noise standard deviation σ_n is the same.
(*a*) The samples are added together and averaged as an estimate of v_r. What is $\hat{v}_r$ in this case and the corresponding mean-squared error $E([v_r - \hat{v}_r]^2)$?

Ans. 888 mi/h, $(1000)^2/5$

(*b*) The samples are linearly processed to *minimize* $E([v_r - \hat{v}_r]^2)$. Find the estimate $\hat{v}_r = \sum_{j=1}^{5} h_j x_j$ and the corresponding mean-squared error.

Ans. 740 mi/h, $(1000)^2/6$

6.4 Refer to the linear-estimation example with correlated noise samples represented by Eqs. (6.23) and (6.24).
(*a*) Verify Eq. (6.25) and then show that the optimum filter coefficients are given by Eq. (6.26).
(*b*) Show that the least mean-squared error is given by Eqs. (6.30) and (6.32).

6.5 Two data samples are taken:

$$x_1 = s + n_1 \qquad x_2 = 2s + n_2$$

$$E[s] = E[n] = 0 \qquad E[s^2] = 3 \qquad E[n_1^2] = E[n_2^2] = 2$$

The signal and noise are uncorrelated, but $E[n_1 n_2] = 1$. Find $R_{11} = E[x_1^2]$, $R_{22} = E[x_2^2]$, $R_{12} = E[x_1 x_2]$, $g_1 = E[x_1 s]$, and $g_2 = E[x_2 s]$, used in finding the best linear estimate of s.

6.6 Consider the signal-estimation example of Eq. (6.33).
(*a*) Show that the R_{ij} and g_i parameters are given by Eqs. (6.35) and (6.36), respectively.
(*b*) Show that the optimum linear estimates are given by Eq. (6.40) for two samples and Eq. (6.42) for three samples.

6.7 A mass in free fall on a strange planet drops a distance $s(t) = \frac{1}{2}gt^2$ in t s. The gravitational acceleration g is to be estimated from observations by a noisy instrument:

$$x_j = g_\varepsilon \frac{j^2}{2} + n_j \qquad j = 1, 2, \ldots$$

in which the mean of g has been subtracted out. We thus actually estimate g_ε, which has zero mean and variance $1(\mathrm{m/s^2})^2$. The noise samples are also zero mean. They have an autocorrelation $R_n(k) = E[n_j n_{j+k}] = (\frac{1}{2})^k$ and are uncorrelated with the signal; $E[g_\varepsilon n_j] = 0$.
(*a*) One sample $x_1 = g_\varepsilon/2 + n$ is taken. Show that the linear estimate of g_ε that minimizes the mean-squared error is $\hat{g}_\varepsilon = \frac{2}{5}x_1$.
(*b*) Two samples x_1 and x_2 are taken. Show that the corresponding linear estimate is now $\hat{g}_\varepsilon = -(x_1/8) + \frac{7}{16}x_2$. Can you relate the negative sign in $\hat{g}_\varepsilon$ to the noise-correlation properties?

6.8 Consider $x(t) = s \cos \omega_0 t + n(t)$. The random parameter s is to be estimated by sampling $x(t)$ and linearly processing the samples. The frequency ω_0 is known. [This is a very common processing problem where periodicities of known frequency but unknown amplitude appear. One obvious example is an AM signal, and the processor is then one form of a digital AM receiver. Any application where the signal to be estimated is a sine wave in noise is equally valid. The additive noise $n(t)$ accounts not only for receiver noise but for inaccuracies of instruments, as well as any other scattering about the periodic signal term.] Specifically, let $x(t)$ be sampled at $\omega_0 t = 0$, $\omega_0 t = \pi/4$, giving x_1 and x_2. Assume $E[s] = 0$. $E[n_1 n_2] = 0$ (white noise). $E[s^2] = S$, $E[n_1^2] = E[n_2^2] = \sigma_n^2$. Find the estimate $\hat{s} = h_1 x_1 + h_2 x_2$ such that $E[(s - \hat{s})^2]$ is minimum. Can you satisfy yourself that this is a two-sample matched filter?

Ans. $h_1 = S/(3S/2 + \sigma_n^2)$, $h_2 = h_1/\sqrt{2}$

6.9 Say $x_j = f_j s + n_j$, $j = 1, 2, \ldots, m$, $f_1, f_2, \ldots, f_m$, are *known* numbers. s is a *random* variable to be estimated:

$$E[s] = E[n_j] = 0 \qquad E[s^2] = S$$

$$E[n_j^2] = \sigma_n^2 \qquad E[n_i n_j] = 0 \qquad i \neq j$$

(*a*) Show that if we use linear estimation, $\hat{s} = \sum_{j=1}^{m} h_j x_j$, the following values of h_j provide the smallest mean-squared error $E[(\hat{s} - s)^2]$:

$$h_1 = \alpha f_1, h_2 = \alpha f_2, \ldots, h_j = \alpha f_j, \ldots$$

$$\alpha = \frac{S}{\sigma_n^2 + S \sum_{j=1}^{m} f_j^2}$$

(*b*) Let $f_1 = 1$, $f_2 = 1/\sqrt{2}$. Check the answer of Prob. 6.8.

6.10 *Drill with Conditional Probability Density Functions*

(*a*) $f(x \mid s) = se^{-sx}$ $x \geq 0$, $s \geq 0$, x a random variable, s a parameter. Show that $E[x \mid s] = 1/s$. Thus, as the parameter s decreases, the average value of x increases. Verify this by plotting se^{-sx} for $s = \frac{1}{2}, 1, 2$.

(*b*) $f(s) = \lambda e^{-\lambda s}$, $s \geq 0$, $\lambda \geq 0$. Thus s itself is a random variable with average value $1/\lambda$. Then the joint density function is $f(x, s) = f(x \mid s) f(s) = \lambda s e^{-\lambda s} e^{-sx}$. Show that $f(x) = \lambda/(x + \lambda)^2$.

(*c*) $f(x, s) = f(s \mid x) f(x)$ also. Show that $f(s \mid x) = (x + \lambda)^2 s e^{-s(\lambda + x)}$. This is called the a posteriori distribution. It represents the probability distribution of s, *given* x. Sketch $f(s \mid x)$. Show that its peak or maximum value occurs at $1/(x + \lambda)$. This is called the maximum a posteriori (MAP) value. It is given by solving $\partial f(s \mid x)/\partial s = 0$. It represents the *most probable value* of s, given x. Show as well that the conditional mean $E[s \mid x] = \int s f(s \mid x)\, ds$ is given by $2/(x + \lambda)$.

6.11 *Application of Prob. 6.10 to Estimation* s is a signal (random parameter) that is to be estimated by making one measurement x. As in Prob. 6.10, we know the statistics of x, *given* s: $f(x \mid s) = se^{-sx}$. (x is thus inversely related to s.) We also know the statistics of s: $f(s) = 2e^{-2s}$, so that $E[s] = \frac{1}{2}$.

(*a*) The measurement produces $x = 2$. What is the most probable estimate $\hat{s}$ of s? (This is the MAP estimate.) What is the estimate $\hat{s}$ that minimizes the mean-squared error $E[(s - \hat{s})^2]$? (This is shown in the text to be the conditional mean: $\hat{s} = E(s \mid x)$.]

Ans. $\hat{s} = \frac{1}{4}, \frac{1}{2}$

(*b*) Repeat part (*a*) for $x = 4$. *Ans.* $\hat{s} = \frac{1}{6}, \frac{1}{3}$

6.12 A high-frequency sine wave is transmitted to a distant receiver by reflection from the ionosphere (an ionized layer surrounding the earth). The ionosphere introduces random variations in the amplitude of the sine wave. The amplitude as received is thus a random variable with the Rayleigh probability density function of Eq. (6.54). Zero-mean gaussian noise is added during transmission as well. Show that the a posteriori distribution is given by Eq. (6.55). Find the MAP estimate of a single sample x_1 of the amplitude and show that it has the value given in the text.

6.13 You were driving in a 55 mi/h zone but had not looked at your speedometer for several minutes; your speed was uniformly distributed between 50 and 60 mi/h. A policeman in a car behind you says that you were going 60 mi/h. You say his speedometer error is gaussian, with mean = 0 and standard deviation = 5 mi/h.
(*a*) Explain to the judge why the linear least mean-squared estimate of your speed is 56.25 mi/h.
(*b*) Explain why the policeman's estimate is the MAP estimate of your speed.
(*c*) Your father (who is also mayor) was following the police car, with his equally accurate speedometer. He says that you were going exactly 55 mi/h. Explain why the least mean-squared estimate using *both* his measurement and the policeman's still exceeds the speed limit.

6.14 A single observation

$$x = s + n$$

is the sum of a gaussian signal and independent uniformly distributed noise, with the densities shown in Fig. P6.14.
(*a*) Sketch $f(s|x)$ for an x in each of the ranges $x > 1$, $x < -1$, and $|x| < 1$.
(*b*) Find $\hat{s}_{\text{MAP}}$ for each of the three cases.
(*c*) For $x > 1$, express the least mean-squared estimate $\hat{s}_{\text{MS}}$ as a ratio of two integrals.

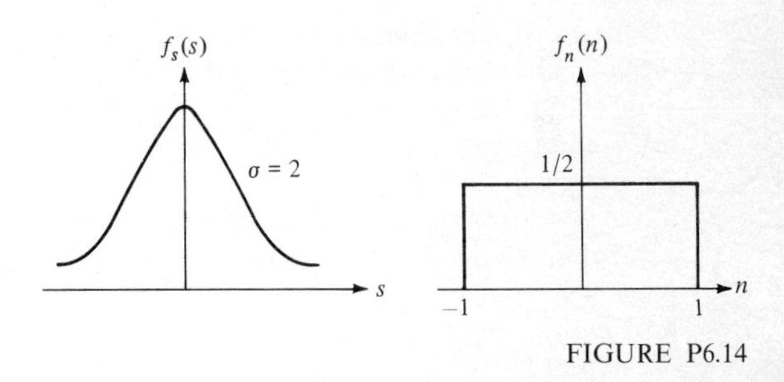

FIGURE P6.14

6.15 The azimuth angle of an antenna is to be estimated from noisy observations. Before making observations, it is known that the angle s is equally likely to take any value between -1 and 1 mrad. The noise samples n_i are independent of each other and of s, with the triangular density shown in Fig. P6.15. The observation samples are given by $x_i = s + n_i$.

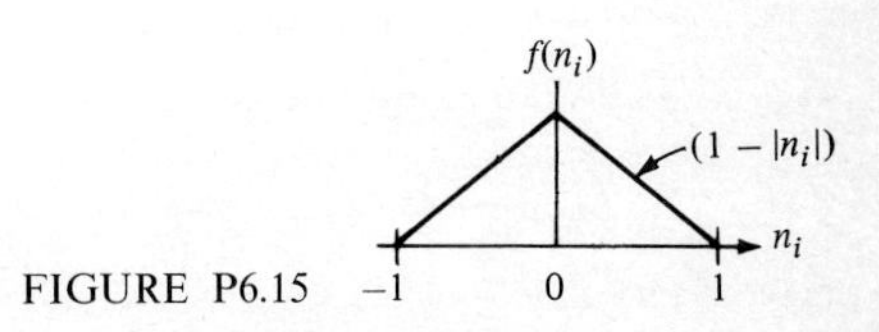

FIGURE P6.15

(*a*) Explain why the assumptions imply that $E[sn_i] = 0$ and $E[n_k n_j] = 0, j \neq k$. Find the least mean-squared linear estimate *and* the corresponding mean-squared error when

$$x_1 = 0.9 \qquad x_2 = -0.1 \qquad x_3 = 0.4 \qquad x_4 = 0.7 \qquad x_5 = 0.6$$

Note that

$$\int_{-1}^{1} n^2(1 - |n|)\, dn = \tfrac{1}{6}$$

(*b*) Find the nonlinear least mean-squared estimate for a single observation $x_1 = 1.5$.
(*c*) Find $\hat{s}_{\mathrm{MAP}}(x_1)$ when $x_1 = 1.5$ and interpret the result in terms of the hit-or-miss cost function associated with the MAP estimator.
(*d*) A new observation technique allows noise reduction in the second observation; i.e., new observations are

$$y_1 = s + n_1 \qquad y_2 = s + \frac{n_2}{2}$$

where the s and n_i properties are unchanged. Find the best h_1 and h_2 for a least mean-squared estimate of the form

$$\hat{s}_L = h_1 y_1 + h_2 y_2$$

6.16 Three traffic counters are used to measure the average speed of vehicles on an expressway. Each counter averages over the effects of several vehicles and produces a number x_j for the speed. The expressway speed is known (or assumed) to be uniformly distributed ± 5 mi/h about 35 mi/h and the counters all have accuracies of ± 5 mi/h. We observe the numbers 27, 39, and 33 mi/h, respectively, on the three meters.
(*a*) Find the least-squared estimate of the expressway speed.
(*b*) Assuming that the meter errors are uncorrelated and additive, estimate the expressway *speed deviation* from 35 mi/h in the linear least mean-squared error sense. When doing this, use the deviations of the measurements from 35 mi/h as the observations to be linearly combined.
(*c*) Find the mean-squared errors in both cases.
(*d*) Repeat parts (*b*) and (*c*) assuming different accuracies for the three counters, of ± 1, ± 3, and ± 5 mi/h, respectively.
Parts (*a*) to (*c*) can be answered using formulas in the chapter, but (*d*) requires a new derivation starting from the orthogonality conditions.

6.17 We are interested in the least mean-squared estimate of s based on a single observation

$$x = s + n$$

with additive noise. The signal and noise are independent with the uniform densities shown in Fig. P6.17*a* and *b*. The desired estimate is the conditional mean value

$$\hat{s}(x) = \int_{-\infty}^{\infty} s f(s \mid x)\, ds$$

in which

$$f(s \mid x) = \frac{f(x \mid s) f(s)}{f(x)} \tag{i}$$

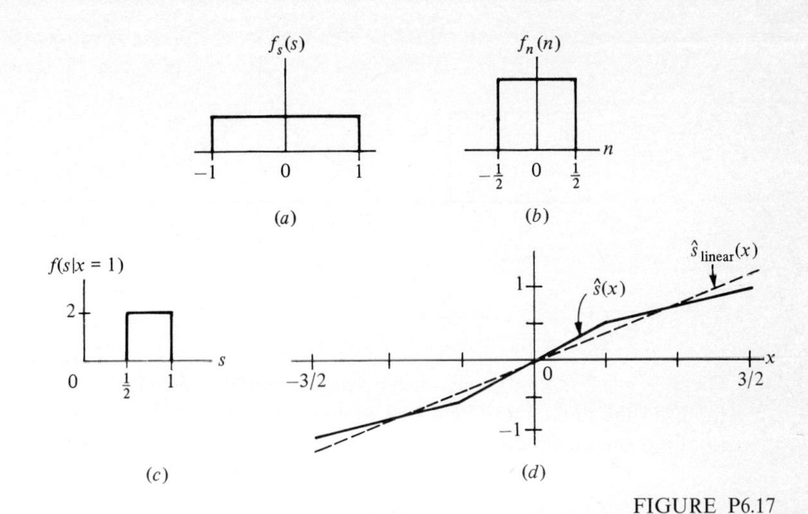

FIGURE P6.17

(a) Show that

$$f(x \mid s) = f_n(x - s) = 1 \qquad |x - s| < \tfrac{1}{2}$$

and sketch this as a function of s for $x = 1$.
(b) Show that combination of part (a) with $f(s)$ yields the $f(s \mid x = 1)$ shown in Fig. P6.17c. [We need not compute $f(x)$ in (i) since it serves only to normalize, and the proper height of the uniform density is easily determined to make the area equal to unity.] Explain why this implies $\hat{s}_{\text{MS}}(1) = 3/4$.
(c) Repeat the preceding argument for other values of x to get the nonlinear estimator curve shown in Fig. P6.17d.
(d) Show that the best linear least mean-squared estimate is $\hat{s}_{\text{linear}} = 0.8x$ and that such an estimate has the undesirable property of saying that $\hat{s}_{\text{linear}}(\frac{3}{2}) = 1.2$ when the largest possible signal value is $s = 1$.

6.18 Refer to Example 1 of Sec. 6.4. Show that the least mean-squared estimate is given by Eq. (6.86b). This is sketched in Fig. 6.11. Find the *linear* least mean-squared estimate using one sample as well and compare with the various estimates shown in Fig. 6.11.

6.19 One data sample x is used to estimate a random signal s

$$x = 2s + n$$

n is a sample of noise. s and n have probability density functions as shown in Fig. P6.19a and b.
(a) Ignore $f(s)$. Find $f(x \mid s)$, sketch it as a function of s, and show that its peak value occurs at $s = x/2$. This is the maximum-likelihood estimate $\hat{s}_{\text{ML}}$.
(b) Include $f(s)$. Sketch $f(s)f(x|s)$ as a function of s for various values of x. From this deduce the MAP estimate and show that it is given by the curve of Fig. P6.19c.
(c) Superimpose $\hat{s}_{\text{ML}}$ and $\hat{s}_{\text{MAP}}$ on the same figure and compare.

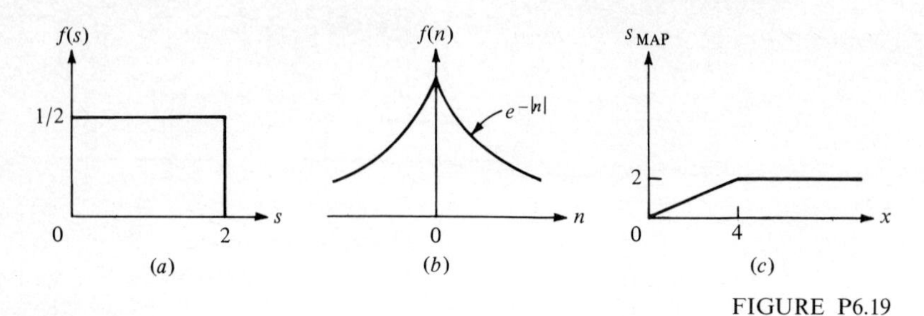

FIGURE P6.19

6.20 Given $x = s/2 + n$. n is zero-mean gaussian with a variance of 1.

(a) Find the maximum-likelihood estimate $\hat{s}_{ML}$ of x [this is the maximum of $f(x \mid s)$].

(b) Find the maximum a posteriori estimate $\hat{s}_{MAP}$ [maximum of $f(s)f(x \mid s)$] for

$$f(s) = \begin{cases} \frac{1}{4}e^{-s/4} & s \geq 0 \\ 0 & s < 0 \end{cases}$$

(c) Sketch $\hat{s}_{MAP}$ and $\hat{s}_{ML}$ vs. the data sample x, and compare.

6.21 Refer to Example 2 of Sec. 6.4. Carry out the details of the derivations of all three estimates discussed and show that they are given by the curves of Fig. 6.14.

6.22 The acceleration a of a vehicle is to be estimated from measurements of displacement. The measurements are noisy, so that the actual data samples are of the form

$$x_j = aj^2 + n_j \qquad j = 1, 2, \ldots$$

We know that $E[a] = 0$, $E[a^2] = K$, $E[n_j] = 0$, $E[n_j^2] = \sigma_n^2$, and $E[an_j] = 0$. The noise samples are uncorrelated: $E[n_i n_j] = 0$, $i \neq j$.

(a) *Two* data samples, $x_1 = a + n_1$ and $x_2 = 4a + n_2$, are taken. Show that the least mean-squared linear estimate is

$$\hat{a} = \tfrac{1}{18}x_1 + \tfrac{4}{18}x_2 \qquad \text{if } \frac{\sigma_n^2}{K} = 1$$

(b) The noise samples n_j are now assumed to be gaussian:

$$f(n_j) = \frac{e^{-n_j^2/2\sigma_n^2}}{\sqrt{2\pi\sigma_n^2}}$$

Hence

$$f(x_j \mid a) = \frac{e^{-(x_j - j^2 a)^2/2\sigma_n^2}}{\sqrt{2\pi\sigma_n^2}}$$

Show that the maximum-likelihood estimate based on *two* samples,

$$x_1 = a + n_1 \qquad x_2 = 4a + n_2$$

is given by

$$\hat{a}_{ML} = \frac{x_1}{17} + \frac{4x_2}{17}$$

(*c*) Using the same two samples, show that the maximum a posteriori estimate is given by

$$\hat{a}_{\text{MAP}} = \tfrac{1}{18}x_1 + \tfrac{4}{18}x_2$$

if a is assumed gaussian, n_j gaussian, $E[a] = 0$, $E[a^2] = K = \sigma_n^2$.

6.23 Refer to Example 3 in Sec. 6.4. Here a signal a is to be estimated, with a equally likely to be $+1$ or -1. It thus has a probability density function $f(a) = \frac{1}{2}\delta(a + 1) + \frac{1}{2}\delta(a - 1)$ (see Example 2 in the same section). The received signal samples are $x_j = as_j + n_j, j = 1, 2, \ldots, m$. The s_j are *known* numbers (samples of a *known* waveshape). Show that the various estimates of a are given by (6.96) to (6.98). Verify Fig. 6.15 for $\hat{a}_{\text{MS}}$.

6.24 A signal *known* to have gaussian statistics (zero mean, variance σ_s^2) is detected in the presence of gaussian noise (zero mean, variance σ_n^2).

(*a*) One sample, $x_1 = s + n_1$, is taken. Show that the MAP estimate is

$$\hat{s} = \frac{x_1}{1 + b} \qquad b = \frac{\sigma_n^2}{\sigma_s^2}$$

(*b*) Show that the ML estimate is $\hat{s} = x_1$. [Here there is no knowledge of the statistics of s. As a check, let $\sigma_s^2 \gg \sigma_n^2$ in part (*a*). The result there agrees with this result. Why should this be so?]

(*c*) m samples, $x_j = s + n_j, j = 1, 2, \ldots, m$, are taken. The noise samples are independent. Show that

$$f(x_1, x_2, \ldots, x_m \mid s) \equiv f(\mathbf{x} \mid s) = \frac{\exp\left[-\sum_{j=1}^{m} \frac{(x_j - s)^2}{2\sigma_n^2}\right]}{(2\pi\sigma_n^2)^{m/2}}$$

Show from this that the MAP estimate is $\hat{s}_{\text{MAP}} = 1/(m + b) \sum_{j=1}^{m} x_j$, $b = \sigma_n^2/\sigma_s^2$ [s is still gaussian as in part (*a*). How does this result compare with the result on *linear* least mean-squared estimation?] Show that the ML estimate is $\hat{s}_{\text{ML}} = (1/m) \sum_{j=1}^{m} x_j$.

6.25 Consider Example 4 in Sec. 6.4. Two cases of the statistics of the random parameter a to be estimated are examined.

(*a*) With gaussian statistics assumed for a, show that both the MAP and mean-squared estimates of a, using m received data samples, are given by the matched-filter forms of Eqs. (6.105) and (6.105*a*).

(*b*) The random parameter a now is assumed to obey the Rayleigh statistics of Eq. (6.106). Verify that the MAP estimate is now given by Eq. (6.107). Show that the special cases of Eqs. (6.107*a*) and (6.107*b*) arise under the signal-to-noise conditions shown.

7

RECURSIVE LINEAR MEAN-SQUARED ESTIMATION: TIME-VARYING SIGNALS AND KALMAN FILTERING

7.1 INTRODUCTION: ESTIMATION OF A SIGNAL PARAMETER

We have discussed (Chap. 6) various techniques for estimating random signals in the presence of noise, and in doing so we have come across the need for solving sets of algebraic equations simultaneously. This corresponds to inverting a matrix whose order is that of the number of simultaneous equations involved. Examples include Eqs. (6.10), (6.11), and (6.13) and the maximum-likelihood estimation of linear systems considered in Sec. 6.5. Many computationally efficient techniques exist for carrying out the necessary matrix inversion. One problem with these techniques, and with the estimation techniques discussed thus far, is that they rely on a specified number of data samples to be processed. Should new data become available, the matrix inversion has to be done over, and each successive matrix to be inverted has more rows and columns. One would like a technique in which previously determined estimates are simply updated as new data come in, rather than solving the problem all over again. The recursive-estimation technique introduced in this section and discussed in detail in the sections following is exactly such a scheme. We shall also show in Sec. 7.2 that recursive estimation is readily extended to include the estima-

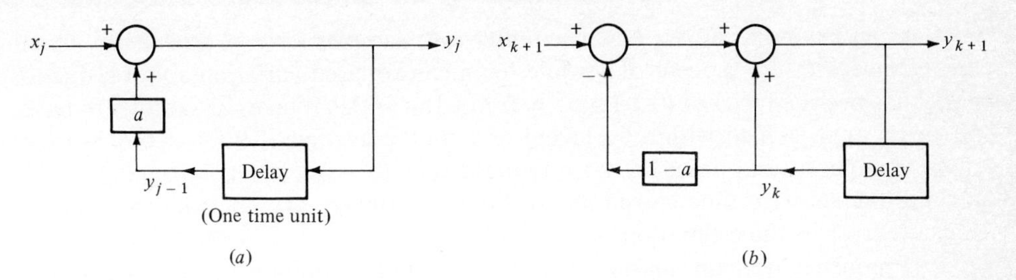

FIGURE 7.1
Recursive filter: two equivalent forms.

tion of time-varying random signals or random processes. It is thus a powerful computational tool for estimating dynamically varying desired signals. It has been used for making space-orbit measurements, for tracking vehicle velocity and acceleration in the presence of disturbances (noise), for real-time navigation, and for a host of other applications. In Sec. 7.3 we shall apply the recursive techniques specifically to the tracking function of radars such as the air-traffic-control radars discussed previously. Other examples and applications are considered briefly in Sec. 7.4.

The recursive filter discussed at length in Chap. 2 is an example of a recursive estimator in a rather special sense. For assume that successive data samples $x_j = s + n_j$ contain a signal term s with $E[s^2] = S$ to be estimated, as well as additive uncorrelated noise samples n_j with variance σ_n^2, as considered in detail in previous chapters. Both s and n_j are assumed to be zero-mean random variables. Then the output of a first-order recursive filter taking in successive x_j samples at the input is given by

$$y_j = ay_{j-1} + x_j \qquad |a| < 1 \tag{7.1}$$

This filter continually updates, adding a new data sample x_j, to a fraction of the previous output. A block diagram of this recursive operation is shown in Fig. 7.1*a*. What are the properties of this filter? Say we start with the first input sample x_1 and allow the filter to operate until the kth output sample y_k. We then have

$$y_k = a^{k-1}x_1 + a^{k-2}x_2 + \cdots + ax_{k-1} + x_k \tag{7.2}$$

Separating the signal and noise terms, we have

$$\begin{aligned} y_k &= (a^{k-1} + a^{k-2} + \cdots + a + 1)s + a^{k-1}n_1 + \cdots + an_{k-1} + n_k \\ &= s\frac{1-a^k}{1-a} + \sum_{j=1}^{k} a^{k-j}n_j \end{aligned} \tag{7.3}$$

For large k ($|a|^k \ll 1$), the signal part of y_k approaches $s/(1-a)$, while the *variance* due to the *noise* approaches $\sigma_n^2/(1-a^2)$. If one multiplies y_k by $1-a$, one then finds $(1-a)y_k \to s$, for k very large, while the variance of $(1-a)y_k \to \sigma_n^2(1-a)/(1+a)$. Thus this device does provide some smoothing, as expected, producing an estimate of s [$\hat{s} = (1-a)y_k$] that has a smaller variance or mean-squared variation about s than would be the case with one sample.

As an example, let $a = 0.5$. Then after four samples ($a^k = \frac{1}{16}$), $(1 - a)y_k$ is on the average within 6 percent of s, while the mean-squared variation about s, due to noise, has been reduced to $(0.5/1.5)\sigma_n^2 = 0.3\sigma_n^2$. If $a = 0.9$, it takes 27 samples to have $(1 - a)y_k$ approach to within 6 percent of s, on the average ($0.9^{27} = 0.06$), but the noise variance at that point has been reduced to $\sigma_n^2(1 - a)/(1 + a) = 0.053\sigma_n^2$. So by making the effective time constant a of the filter longer, we can reduce the mean-squared error in the estimation of s.

[Frequency-domain analysis also helps in understanding this filtering action. The noise is white, with a constant spectral density over all ω. The filter has a low-pass characteristic whose passband becomes narrower, letting less noise power through, as the coefficient a takes values closer to unity (see Fig. 3.32).]

But this recursive device, although simple, is not the best estimator of s. The averaging of m samples provides a mean-squared error reduction to σ_n^2/m. For 27 samples this is $0.037\sigma_n^2$. We showed in Sec. 6.2 that the best mean-squared linear estimator of s, $\hat{s} = (m + b)^{-1} \sum_{j=1}^{m} x_j$, reduced the error to $\sigma_n^2/(m + b)$, with b the noise-to-signal ratio. Is it possible to obtain a recursive estimate that automatically provides the smallest mean-squared error? We shall in fact do so, comparing it with the simple recursive filter just discussed above. The beauty of the approach leading to the recursive estimator with the smallest mean-squared error, however, is that it is readily extended to much more complex situations, as already noted. These include cases in which the signal and noise vary dynamically, several possibly related signals are to be estimated simultaneously, signals are to be predicted ahead, etc. A discussion of this more general case of recursive estimation appears in Secs. 7.2 to 7.4.

To demonstrate the recursive estimation of a signal parameter at this point we use exactly the problem just discussed in connection with the first-order recursive filter: given successive samples $x_j = s + n_j$, provide a *linear* estimator $\hat{s}_k = \sum_{j=1}^{k} h_j x_j$ such that the mean-squared error $E[(\hat{s}_k - s)^2]$ is as small as possible. We already know the nonrecursive solution to this problem. As shown in Sec. 6.2 [Eq. (6.17)], we must set $h_j = 1/(k + b)$ with $b \equiv \sigma_n^2/S$. But now we would like to obtain the same solution recursively. As noted above, such a recursive approach reduces the computational requirements. It also leads directly to the more general recursive-estimation case appropriate to situations in which the statistics of the signal and noise may be varying with time.

Specifically, we have as the best linear mean-squared estimator, with k samples available,

$$\hat{s}_k = \sum_{j=1}^{k} h_j x_j \qquad h_j = \frac{1}{k + b} \tag{7.4}$$

while the corresponding mean-squared error is

$$e_k = E[[\hat{s}_k - s]^2] = \frac{1}{k + b} \sigma_n^2 \tag{7.5}$$

[see Eqs. (6.17) and (6.19)]. For $k + 1$ samples the estimate and corresponding mean-squared error would be

$$\hat{s}_{k+1} = \sum_{j=1}^{k+1} h_j x_j \qquad h_j = \frac{1}{(k+1)+b} \tag{7.6}$$

$$e_{k+1} = E[[\hat{s}_{k+1} - s]^2] = \frac{1}{(k+1)+b}\sigma_n^2 \tag{7.7}$$

Is it possible to obtain $\hat{s}_{k+1}$ *recursively* in terms of the already calculated $\hat{s}_k$ and the new data sample x_{k+1}?

Note that we should actually write $h_j(k) = 1/(k + b)$ since the coefficients change with the number of samples used. A comparison of Eqs. (7.4) and (7.5) makes it apparent that for this special signal-estimation case we have

$$h_j(k) = \frac{e_k}{\sigma_n^2} = p_k \tag{7.8}$$

We have used the symbol p_k here to represent the normalized mean-squared errors. Similarly, we have

$$h_j(k+1) = \frac{e_{k+1}}{\sigma_n^2} = p_{k+1} \tag{7.9}$$

Using these last two equations plus Eqs. (7.4) and (7.5) for the filter coefficients, we obtain the following expression relating the normalized errors:

$$\frac{p_{k+1}}{p_k} = \frac{k+b}{k+1+b} = \frac{1}{p_k + 1} \tag{7.10}$$

Thus given p_k, we can find p_{k+1}, then p_{k+2}, etc. So we already have a simple algorithm for finding the variation of mean-squared error with sample size.

Now consider the $(k + 1)$st estimate of the signal parameter $\hat{s}_{k+1}$. For this special problem we may write

$$\begin{aligned}\hat{s}_{k+1} &= \frac{1}{k+1+b}\sum_{j=1}^{k} x_j + \frac{1}{k+1+b}x_{k+1} \\ &= \frac{k+b}{k+1+b}\hat{s}_k + \frac{1}{k+1+b}x_{k+1}\end{aligned} \tag{7.11}$$

using (7.4). From Eq. (7.10) we also have

$$\hat{s}_{k+1} = \frac{p_{k+1}}{p_k}\hat{s}_k + p_{k+1}x_{k+1} \tag{7.11a}$$

This is the desired recursive form. We use Eq. (7.10) to find p_{k+1} in terms of p_k. Then from the stored, previous value $\hat{s}_k$ plus the new data sample x_{k+1}, we can calculate $\hat{s}_{k+1}$. The first estimate $\hat{s}_1$ based on a single observation must be computed separately by nonrecursive methods to get this recursion started. This procedure has the property that it *continually* generates the best linear mean-squared estimator of s, while the normalized mean-squared error p_{k+1} is also available if desired. From the error relation of (7.10) we have $p_k \to 0$ as k increases, in agreement with Eq. (7.5).

As an example, say that the signal-to-noise ratio $E[s^2]/\sigma_n^2 = S/\sigma_n^2 = 1/b = 1$. The first estimate of s, using the first data sample, is just half that sample, as determined from the nonrecursive estimate $\hat{s}_k = 1/(k+b)\sum_{j=1}^{k} x_j$, with $k = 1$, $b = 1$. The corresponding mean-squared error is $p_1 = \frac{1}{2}$. The recursion begins by using (7.10) to get $p_2 = p_1/(1 + p_1) = \frac{1}{3}$. Then $\hat{s}_2 = \frac{2}{3}\hat{s}_1 + \frac{1}{3}x_2$. (As a check, since $\hat{s}_1 = x_1/2$, $\hat{s}_2 = \frac{1}{3}x_1 + \frac{1}{3}x_2$, agreeing with the nonrecursive result.) Similarly, $p_3 = p_2/(1 + p_2) = \frac{1}{4}$. Then $\hat{s}_3 = \frac{3}{4}\hat{s}_2 + \frac{1}{4}x_2$, and the process continues.

If we now compare the recursive relationship of Eq. (7.11*a*) with the recursive filter of Eq. (7.1), we note that it is of the same form, but with time-varying coefficients. Thus we can write

$$\hat{s}_{k+1} = a_{k+1}\hat{s}_k + b_{k+1}x_{k+1} \tag{7.12}$$

with two parameters apparently replacing the one parameter a used previously. Actually, the two parameters may be replaced by one, and the recursive estimates rewritten in a rather instructive form. For from Eq. (7.11) or from Eq. (7.11*a*) using Eq. (7.10), we have

$$a_{k+1} = \frac{p_{k+1}}{p_k} = \frac{1}{1 + p_k} = 1 - b_{k+1} \tag{7.13}$$

and

$$b_{k+1} = p_{k+1} = \frac{p_k}{1 + p_k} \tag{7.14}$$

Similar relationships will be encountered in the additional recursive estimators we shall discuss later. Other examples of such time-varying linear processors are the recursive mean and variance estimators in Prob. 3.9.

With this relationship, the form of the recursive estimator can be expressed as

$$\hat{s}_{k+1} = \hat{s}_k + b_{k+1}(x_{k+1} - \hat{s}_k) \tag{7.12a}$$

The interpretation of this result is quite interesting: the $(k + 1)$st estimate of s is the same as the previous kth estimate plus a *correction* term involving the difference between the new sample value x_{k+1} and the previous estimate. This correction term is multiplied by a variable-gain factor $b_{k+1} = p_{k+1} = 1/(k + 1 + b)$ that continually *decreases* with k. So ultimately, the estimate stabilizes at some value depending on the data samples read in. This estimate will be modified only if a new sample x_{k+1} differs *considerably* from the previous estimate. The two forms of the recursive estimator, corresponding respectively to Eqs. (7.12) and (7.12*a*), are shown in Fig. 7.2*a* and *b*.

The time-invariant recursive filter can be expressed in the same fashion as Eq. (7.12*a*) for comparison

$$y_{k+1} = ay_k + x_{k+1} = y_k + [x_{k+1} - (1 - a)y_k] \tag{7.15}$$

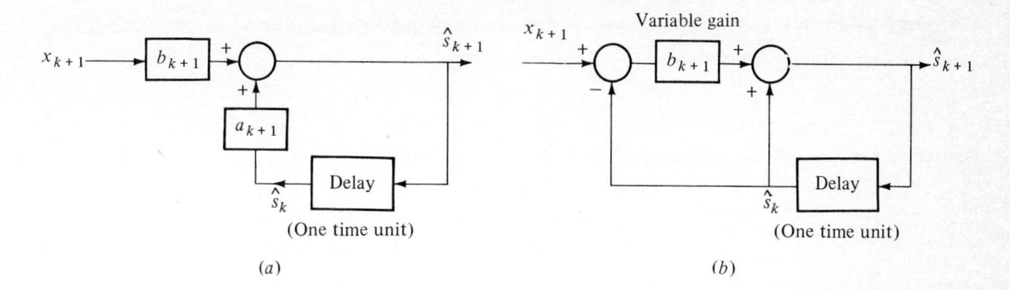

FIGURE 7.2
Recursive estimator, signal in additive white noise; two equivalent forms.

Here the correction term to be added to the previous estimate is the difference between the new sample value and $1 - a$ times the previous estimate. The block diagram depicting this form of the filter is shown in Fig. 7.1*b*.

The recursive estimator for a signal parameter in additive uncorrelated (white) noise given by Eqs. (7.12) and (7.12*a*) was obtained by a trick technique: we took the known solution for the least mean-squared estimate and *rewrote* it essentially in a recursive form. This served to demonstrate the *possibility* of obtaining a recursive estimate. But this approach has two obvious faults: (1) It requires solving the m-sample case first, just the process we wished to avoid since it normally involves inverting an $m \times m$ matrix. (This problem does not arise in the example we picked.) (2) It is not clear how one generally converts the m-sample estimate result to the desired recursive solution. Thus the approach just used does not allow simple extension to other, possibly more complex, problems. We therefore repeat the problem, using a *systematic* procedure for obtaining the recursive estimate. This approach is then readily extended to a variety of other signal-estimation cases.

This systematic approach *assumes* a recursive form of the type of Eq. (7.12); i.e., it assumes the structure of the recursive estimator to be that of Fig. 7.2*a*. It then uses the orthogonality principle [Eq. (6.8)] discussed in Sec. 6.2 to find the appropriate recursive relations for the gain factors a_{k+1} and b_{k+1} as well. We shall demonstrate the technique by starting with first one sample, then two, and then, by extension, k samples. Specifically, let $x_k = s + n_k$, $k = 1, 2, \ldots$, with n_k uncorrelated zero-mean noise samples, and s a zero-mean random-signal parameter to be estimated. (Nonzero-mean signals s^* are included in the analysis that follows if we let s here be $s = s^* - E[s^*]$ and $\hat{s}_k^* = \hat{s}_k + E[s^*]$.) If the first sample x_1 is taken, it is apparent that our only *linear* choice for the estimate $\hat{s}_1$ is

$$\hat{s}_1 = b_1 x_1 \tag{7.16}$$

with b_1 the parameter to be found. The appropriate value of b_1 is the one that minimizes the mean-squared error $e_1 = E[\varepsilon_1^2] = E[(s - \hat{s}_1)^2]$. As demonstrated in

Sec. 6.2 [Eq. (6.8)], the minimum $E[\varepsilon_1^2]$ corresponds to having $\varepsilon_1 = s - \hat{s}_1$ orthogonal to the data sample x_1:

$$E[\varepsilon_1 x_1] = E[(s - \hat{s}_1)x_1] = 0 \tag{7.17}$$

Since $x_1 = s + n_1$, $\hat{s}_1 = b_1 x_1$, this immediately gives

$$b_1 = \frac{S}{S + \sigma_n^2} = \frac{1}{1 + b} \tag{7.18}$$

with $E[s^2] = S$, $E[n_1^2] = \sigma_n^2$, and $b = \sigma_n^2/S$, as previously defined.

For this particular choice of b_1 the mean-squared error is given by

$$e_1 = E[\varepsilon_1^2] = E[(s - \hat{s}_1)(s - \hat{s}_1)] = E[(s - \hat{s}_1)s] \tag{7.19}$$

since $\hat{s}_1 = b_1 x_1$ and $s - \hat{s}_1$ is orthogonal to x_1 [Eq. (7.17)]. Carrying out the averaging indicated in Eq. (7.19) and substituting in the appropriate value for b_1, we have

$$e_1 = S(1 - b_1) = \sigma_n^2 b_1 = \frac{\sigma_n^2}{1 + b} \tag{7.20}$$

We thus also have

$$b_1 = \frac{e_1}{\sigma_n^2} = p_1 \tag{7.18a}$$

Now consider a second data sample x_2 available to be processed. The recursive procedure involves using the estimate $\hat{s}_1$ already calculated and adding to it, in a weighted fashion, x_2. We thus take as an appropriate second estimate

$$\hat{s}_2 = a_2 \hat{s}_1 + b_2 x_2 \tag{7.21}$$

with a_2 and b_2 the two parameters that must be found. We shall first show that for this particular example, a_2 and b_2 are simply related. Specifically, we shall prove, as already shown in Eq. (7.13), that

$$a_2 = 1 - b_2 \tag{7.22}$$

We shall show that a_2 and b_2, besides being related, can be found recursively in terms of the normalized mean-squared error p_2. More generally, a_{k+1} and b_{k+1} will be found to be simply related [Eq. (7.13)] and in turn dependent on the mean-squared error p_{k+1}. This is itself dependent on p_k, the error obtained using k data samples. The sequence of mean-squared errors thus plays an important role in the estimation recursion. These steps also apply to the more general recursive procedure we shall discuss in the next section, in dealing with the estimation of random signals varying with time: one first updates an expression for the mean-squared error and then uses it to find the new weighting parameters for an updated estimate of the signal, making use of a new data sample.

To demonstrate all this quantitatively for two samples, we again rely on the *orthogonality* conditions of Eq. (6.8). For the second-step minimum mean-squared error $E[\varepsilon_2^2] = E[(s - \hat{s}_2)^2]$, we must have

$$E[\varepsilon_2 x_i] = 0 \qquad i = 1, 2 \tag{7.23}$$

There are thus *two* equations involved here, rather than the one used in the single-sample case [Eq. (7.17)]. Specifically, from the equation involving x_1 we have

$$E[\varepsilon_2 x_1] = 0 = E[(s - \hat{s}_2)x_1] \tag{7.24}$$

Now $\hat{s}_2 = a_2 \hat{s}_1 + b_2 x_2$. Then $s - \hat{s}_2$ may be written, after adding and subtracting $a_2 s$, in the following form:

$$\varepsilon_2 = s - \hat{s}_2 = a_2(s - \hat{s}_1) + s(1 - a_2) - b_2 x_2 \tag{7.25}$$

The reason for writing ε_2 in this form is apparent: the term $s - \hat{s}_1 = \varepsilon_1$ is orthogonal to x_1 [Eq. (7.17)], so that by using Eq. (7.25) the orthogonality condition [Eq. (7.24)] becomes

$$(1 - a_2)E[sx_1] - b_2 E[x_1 x_2] = 0 \tag{7.26}$$

Now with $x_1 = s + n_1$, $x_2 = s + n_2$, and $E[n_1 n_2] = 0$, we have $E[sx_1] = S$ and $E[x_1 x_2] = S$. (The reader should check this for himself.) Rearranging Eq. (7.26), we finally get

$$a_2 = 1 - b_2 \tag{7.22}$$

as desired.

To find parameter b_2 we invoke the second of the two orthogonal equations $E[\varepsilon_2 x_2] = 0$. Using $\hat{s}_2 = a_2 \hat{s}_1 + b_2 x_2$, we have

$$E[\varepsilon_2 x_2] = E[(s - \hat{s}_2)x_2] = 0 = E[sx_2] - a_2 E[\hat{s}_1 x_2] - b_2 E[x_2^2] \tag{7.27}$$

It is left to the reader to show that $E[sx_2] = S$ and $E[x_2^2] = S + \sigma_n^2$. Equation (7.27) can thus be written

$$b_2 \sigma_n^2 = S - Sb_2 - a_2 E[\hat{s}_1 x_2] \tag{7.28}$$

All that remains now is to evaluate $E[\hat{s}_1 x_2]$. This is simple to do using Eqs. (7.16) and (7.18), but we would like to involve the mean-squared errors. For this purpose we introduce the mean-squared error $e_2 = E[(s - \hat{s}_2)^2]$. As noted earlier, our final result will thus involve an expression for b_2 in terms of e_2 (or its normalized form $p_2 = e_2/\sigma_n^2$). The mean-squared error e_2 will in turn be found recursively from e_1. We have

$$e_2 = E[(s - \hat{s}_2)^2] = E[(s - \hat{s}_2)s] \tag{7.29}$$

since $\hat{s}_2$ is a linear function of x_1 and x_2 and $\varepsilon_2 = s - \hat{s}_2$ is orthogonal to each of these [Eq. (7.23)]. Putting in $\hat{s}_2 = a_2 \hat{s}_1 + b_2 x_2$ and noting that $E[s^2] = S$, $E[x_2 s] = S$, we have

$$e_2 = S - Sb_2 - a_2 E[\hat{s}_1 s] \tag{7.29a}$$

Note how similar this is to Eq. (7.28). As a matter of fact, writing $s = x_2 - n_2$ and noting that $E[\hat{s}_1 n_2] = 0$ ($\hat{s}_1 = b_1 x_1$, and both s_1 and n_1 are uncorrelated with n_2), gives

$$e_2 = b_2 \sigma_n^2 \tag{7.29b}$$

In the normalized form, then,

$$b_2 = p_2 = \frac{e_2}{\sigma_n^2} \tag{7.30}$$

as expected.

To actually find the desired recursive relation involving p_2 and p_1 we must evaluate $E[\hat{s}_1 s]$ in Eq. (7.29*a*). It is left to the reader to show that this is given by $b_1 S$. Combining all the equations thus far $[p_1 = b_1 = 1/(1 + b), p_2 = b_2, a_2 = 1 - b_2]$ and applying a little algebra, we get, finally,

$$p_2 = \frac{p_1}{1 + p_1} \tag{7.31}$$

as expected [see Eq. (7.10)].

It is apparent that this recursive approach for two samples is much more tedious algebraically than the straightforward calculations for linear mean-squared filtering of Sec. 6.2. We obviously would not attempt it if we were just interested in processing the two samples. The beauty is that the same algebra allows the results to be extended to any number of samples processed, while the same type of technique allows extension to more complex signal-processing problems. Specifically, assume now that we are at the $(k + 1)$st sample x_{k+1}. We want to use this plus the previously calculated kth estimate $\hat{s}_k$ to obtain the $(k + 1)$st estimate $\hat{s}_{k+1}$ of s. As already noted, we do this by writing formally

$$\hat{s}_{k+1} = a_{k+1}\hat{s}_k + b_{k+1}x_{k+1} \tag{7.12}$$

This is of course possible since $\hat{s}_k$ is in turn the weighted sum of $\hat{s}_{k-1}$ and x_k. Upon working back to the first estimate $\hat{s}_1 = b_1 x_1$, it is apparent that $\hat{s}_{k+1}$ can be rewritten as the weighted linear sum of all $k + 1$ data samples, $x_1, x_2, \ldots, x_{k+1}$. The linear estimation leading to the Wiener filter of Sec. 6.2 and the recursive-estimation approach of Eq. (7.12) are thus the same and must lead to the same results.

To find the two coefficients a_{k+1} and b_{k+1} we simply repeat the process carried out for a_2 and b_2. Since our criterion is that of minimizing the mean-squared error $e_{k+1} = E[\varepsilon_{k+1}^2] = E[(s - \hat{s}_{k+1})^2]$, it is apparent that the orthogonality relations of Eq. (7.23), extended to $k + 1$ samples, must hold here as well:

$$E[\varepsilon_{k+1}x_i] = 0 \qquad i = 1, 2, \ldots, k + 1 \tag{7.32}$$

Consider in particular the orthogonality relations $E[\varepsilon_{k+1}x_i] = 0$, $i = 1, 2, \ldots, k$. A little thought will indicate that *all* k of these conditions must lead to the same result obtained using $E[\varepsilon_2 x_1] = 0$ [Eq. (7.24)] previously. Thus, replacing the index 2 by $k + 1$ and the index 1 by k, we have, paralleling Eq. (7.22),

$$a_{k+1} = 1 - b_{k+1} \tag{7.33}$$

Parameter b_{k+1} is in turn found by invoking the orthogonality relation $E[\varepsilon_{k+1} x_{k+1}] = 0$. It is left for the reader to show that all the algebraic manipulations used to find b_2 previously are appropriate here as well, with index 2 replaced by $k + 1$ and 1 replaced by k. The recursive relations relating b_{k+1}, $p_{k+1} = e_{k+1}/\sigma_n^2$

(normalized mean-squared error for $k + 1$ samples), and $p_k = e_k/\sigma_n^2$ can thus be written down immediately. From Eqs. (7.30) and (7.31) we have

$$b_{k+1} = p_{k+1} \tag{7.34}$$

and
$$p_{k+1} = \frac{p_k}{1 + p_k} \tag{7.35}$$

Knowing p_k, we calculate p_{k+1}. This then gives us b_{k+1}, from which in turn we get the $(k + 1)$st estimate of s:

$$\hat{s}_{k+1} = \hat{s}_k + b_{k+1}(x_{k+1} - \hat{s}_k) \tag{7.36}$$

These results obviously check with those obtained previously as portrayed in block-diagram form in Fig. 7.2*b*.

A further note about initiation of recursive-estimation algorithms is in order. Instead of starting with a nonrecursive estimate of $\hat{s}_1$ based on the first observation x_1, we could try to get an estimate $\hat{s}_0$ based on *no* observations. From this viewpoint, $\hat{s}_0$ is the number which minimizes

$$e_0 = E[(s - \hat{s}_0)^2]$$

Setting the derivative of e_0 with respect to $\hat{s}_0$ equal to zero, we find that the best $\hat{s}_0$ is the mean value of s

$$\hat{s}_0 = E[s]$$

This choice of an estimate seems quite reasonable when all we know about s is its density function. The corresponding mean-squared error is clearly the variance of s

$$e_0 = \sigma_s^2$$

In our present problem s was assumed to have a zero mean and variance of S, and so the recursion could be started with

$$\hat{s}_0 = 0 \qquad p_0 = \frac{S}{\sigma_n^2} = 1/b$$

Future recursive estimators of zero-mean signals will also be initiated with estimates of $\hat{s}_0 = 0$ and corresponding mean-squared errors.

In the next section we extend this technique to time-varying signals. A further extension, using vector notation, will enable us to consider simultaneous recursive estimation of several signals, just the case one encounters, for example, in tracking radars.

7.2 RECURSIVE ESTIMATION OF TIME-VARYING SIGNALS: KALMAN FILTERING

In the material on estimation in Sec. 7.1 and throughout Chap. 6, we have considered the estimation of a random *parameter*. This parameter, whether a signal corrupted by additive noise, the time of arrival of a radar pulse, the power of a laser beam scattered by concentrated pollutants in the air, etc., was assumed to be nonvarying with time and hence a *random variable*. Commonly, however, signals vary randomly with time. Voice or picture information transmitted by radio varies randomly. (Otherwise why transmit the information?) An airplane being tracked by radar is of course in constant motion. It may accelerate, and it may roll and pitch. Thus estimates of range and velocity must vary with time. The velocity and acceleration of a space vehicle being tracked on the ground for control purposes will vary irregularly with time. If these quantities are to be estimated in the presence of noise (a common occurrence), one cannot assume they are non-time-varying parameters.

Signals will thus generally vary randomly with time, and if we are interested in following this time variation, we must consider processors that provide time-varying estimates. The signal to be estimated is thus an example of a *random process*. Various possibilities may be envisioned. If the signal appears corrupted by noise, we want a processor that smooths the signal-plus-noise, providing a better estimate than at the input to the processor. This is of course the *filtering operation* discussed extensively in connection with estimating constant-signal parameters. Here we want the filter to track the current value of a changing signal. Another possibility, of particular importance in data transmission, statistical time-series forecasting, or vehicle control, among other applications, is that of *predicting* the signal, i.e., estimation of a future value of the signal, using past samples. The utility of such predictive techniques for economic forecasting, vehicle control, weather forecasting, etc., requires no explanation. (The basic problem of course is that of providing techniques that have some measure of accuracy yet do not require inordinate computer time and instrumentation.) An application to data transmission is in the area of data compression; here future data samples that are deemed to be predicted accurately enough need not be transmitted. It is thus possible to reduce system data requirements.

These recursive-estimation techniques were developed around 1960, most notably by Rudolph E. Kalman. For this reason, the processors devised at that time (as well as the wide variety of generalizations and extensions to time-varying statistics, continuous-time signals, nonlinear dynamics, etc.) are referred to as *Kalman filters*.

In this section we discuss a simple recursive approach to the filtering and prediction of random processes. For simplicity we assume the processes to be stationary with time. (The underlying structure generating the signals does not vary with time.) The approaches used can be extended to nonstationary cases as well, however.

Signal Model

The recursive-estimation technique assumes that the random signal to be estimated can be modeled as a *first-order* recursive process driven by zero-mean white noise. Recall that such a model was also assumed in Sec. 4.10. The output y_j of the recursive filter of Eq. (7.1) or Fig. 7.1 is exactly such a signal if the input x_j is a sample of zero-mean white noise ($E[x_i x_j] = 0,\ i \neq j$). To simplify the development of more general relations later we change the notation somewhat, using $x(k)$ in place of x_k,

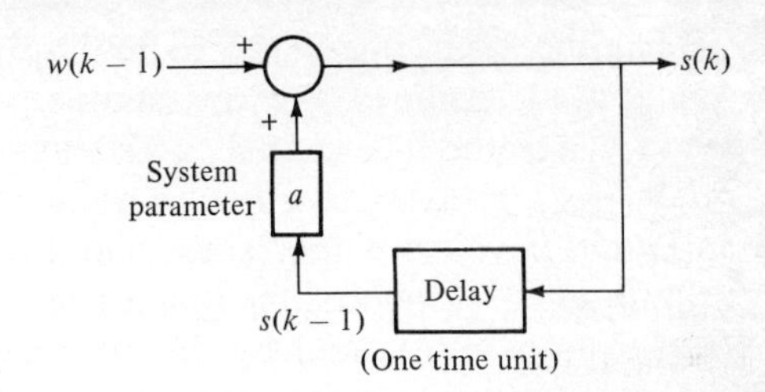

FIGURE 7.3
Model of random-signal process.

$y(k)$ in place of y_k, etc. Then the random signal $s(k)$ to be estimated will be assumed to evolve in time according to the dynamical equation

$$s(k) = as(k-1) + w(k-1) \tag{7.37}$$

As noted above, $w(k)$ must then be a sample of a zero-mean white-noise process

$$E[w(k)] = 0$$

$$E[w(k)w(j)] = \begin{cases} 0 & k \neq j \\ \sigma_w^2 & k = j \end{cases} \tag{7.38}$$

Figure 7.3 is a block-diagram representation of Eq. (7.37). If $\sigma_w^2 = 0$, so that the white-noise process disappears, and $a = 1$, then $s(k) = s(k-1)$ and we are left with the random-signal *parameter* s considered thus far in this chapter and in Chap. 6. We shall extend this dynamical representation of $s(k)$ to more complex processes with the use of vector representation later in this section. Recall from Chap. 3 that a random process like that defined by Eq. (7.37) is said to be an *autoregressive process* of the first order. As shown there and as repeated in Chap. 6 [Eq. (6.21)], the pertinent statistical parameters of $s(k)$ are

$$E[s(k)] = 0 \qquad \sigma_s^2 = \frac{\sigma_w^2}{1-a^2}$$

$$R_s(j+l, j) = R_s(l) = a^{|l|}\sigma_s^2 \tag{7.39}$$

Here $R_s(l) = E[s(k)s(k+l)]$ is the autocorrelation of two samples of $s(k)$ spaced l units apart. σ_s^2 is of course the signal variance.

The parameter a (more exactly, its logarithm) plays the role of a time constant of the process: the larger a is (approaching 1), the more sluggish the process is, requiring a longer time interval (in terms of units of the characteristic sample spacing T_s) to change significantly from its current value.

In the next section we shall consider as specific examples of such a signal the deviations from the average range, azimuth, range velocity, and azimuthal velocity of an aircraft as tracked by an air-traffic radar. The motivation for the use of the dynamic model of Eq. (7.37), aside from its relative simplicity, is clear. The model states simply that the signal has a characteristic time constant, indicating the rate at which it changes. Samples taken closer together than this time constant will not differ very much. Samples spaced farther apart will show significant changes. As their spacing increases, the changes show up more and more. The changes are in turn randomly determined. As an example, visualize an automobile moving at some velocity in traffic. The car may accelerate or decelerate randomly, as the driver steps on the gas or brakes according to traffic conditions on the road ahead. The resultant variations in velocity (the "signal" in this case) thus depend on two parameters, the overall system response time (the driver and car combined) and the random velocity perturbations introduced by the random variations in acceleration. Similarly $s(k)$ could represent an airplane's speed at time t_k, and $w(1), w(2), \ldots$ could be wind gusts which accelerate the plane. Successive velocities are correlated via the parameter a representing air resistance and the plane's inertia. The model of Eq. (7.37) is the *simplest* that could be written to represent such a random dynamical process. More complex processes will be considered later.

How does one determine the two parameters a and σ_s^2 that, according to Eq. (7.39), characterize the process? Various procedures are available, e.g., using one's knowledge of the physical process. If previous signal measurements are available, one can use the maximum-likelihood technique of Sec. 4.10 and the other techniques of Chap. 6. Alternately, one may use the autocorrelation and spectral-density measurement techniques discussed in Chap. 4 to find the spectral density of $s(k)$ and then approximate this in some sense by the "best" first-order autoregressive process. The use of physical reasoning as one approach in the modeling of the signal process by an autoregressive process of the form of Eq. (7.37) will be discussed briefly in connection with the air-traffic-control tracking function described in Sec. 7.3.

Filtering of Signal in Noise

With the signal dynamics modeled according to Eq. (7.37), we now assume, as previously, that the current signal sample $s(k)$ appears obscured by additive white noise $n(k)$ with variance σ_n^2. [This is sometimes called *measurement* or *observation noise* to avoid confusion with the white-noise term $w(k-1)$ appearing in the signal-dynamics model.] We thus have as the data sample $x(k)$ from which an estimate of $s(k)$ is to be made

$$x(k) = s(k) + n(k) \tag{7.40}$$

The problem is now one of determining the "best" recursive linear estimate of $s(k)$. By "best" we again mean in the least mean-squared sense.

We shall not present the detailed algebra involved in deriving the optimum linear recursive estimator. Suffice it to say that the approach is similar to that used in deriving the recursive estimator for a random-signal parameter, as carried out in the previous section. Thus we again assume for the form of the estimator

$$\hat{s}(k) = a_k \hat{s}(k-1) + b_k x(k) \tag{7.41}$$

It uses the weighted sum of the previous estimate $\hat{s}(k-1)$ and the new data sample $x(k)$. [See Eq. (7.12) and recall that $\hat{s}_k$ there was the kth estimate of the parameter s. Here $\hat{s}(k)$ is the estimate of the signal process $s(k)$ at time k. The change in notation was introduced specifically to avoid confusion between these two different quantities.] The two time-varying gain terms a_k and b_k are then to be found such that the mean-squared error

$$e_k = E[[\hat{s}(k) - s(k)]^2]$$

is minimized. We again use the orthogonality relations $E[[\hat{s}(k) - s(k)]x(j)] = 0$, $j = 1, 2, \ldots, k$, which correspond to minimizing e_k, to find the appropriate values for a_k and b_k. Specifically, the orthogonality relation $E[[\hat{s}(k) - s(k)]x(k - 1)] = 0$ is first used to eliminate a_k in terms of b_k. This is exactly the procedure adopted in the previous section. Introducing the two expressions (7.37) and (7.40) modeling the signal process and the noisy-data sample, respectively, into the orthogonality relation and carrying out the algebra indicated, we find

$$a_k = a(1 - b_k) \tag{7.42}$$

Note that this is a generalization of Eq. (7.33) for the signal parameter a_k. Equation (7.41) for the recursive estimate of the kth sample of the signal process can thus be written

$$\hat{s}(k) = a\hat{s}(k - 1) + b_k[x(k) - a\hat{s}(k - 1)] \tag{7.43}$$

The first term, $a\hat{s}(k - 1)$, represents the best estimate of $s(k)$ *without* any additional information and is therefore a *prediction* based on past observations. The second term is a *correction term* involving the difference between the new data sample and the updated estimate $a\hat{s}(k - 1)$, with a variable gain factor b_k appended. This predictor-corrector form for the recursive estimate is intuitively very satisfying. It looks very much like the model, Eq. (7.37), originally assumed for the signal itself. We shall find that generalizations of recursive estimation to vector signals retain this predictor-corrector form.

The results thus far are summarized in Fig. 7.4, where Fig. 7.4*a* shows the autoregressive model assumed for the random-signal process with noise then added to form the observation or data sample $x(k)$ and Fig. 7.4*b* shows the optimum processing equation (7.43) in block-diagram form. A comparison of the two parts, particularly those portions involving the generation of the signal $s(k)$ and its estimate $\hat{s}(k)$, shows how much the form of the estimator resembles the model assumed.

How is the time-varying gain parameter b_k now determined? As in the previous section, we make use of the orthogonality relation $E[[s(k) - \hat{s}(k)]x(k)] = 0$. This expression plus that for the mean-squared error

$$e_k = E[[s(k) - \hat{s}(k)]^2] = E[[s(k) - \hat{s}(k)]s_k]$$

(the justification for this is left to the reader) enable us to establish the following equation relating the gain b_k and the mean-squared error e_k:

$$b_k = \frac{e_k}{\sigma_n^2} \equiv p_k \tag{7.44}$$

Note that this is identical with the relation derived in the previous section [Eq. (7.34)]. We have again used the symbol p_k to represent the mean-squared error e_k

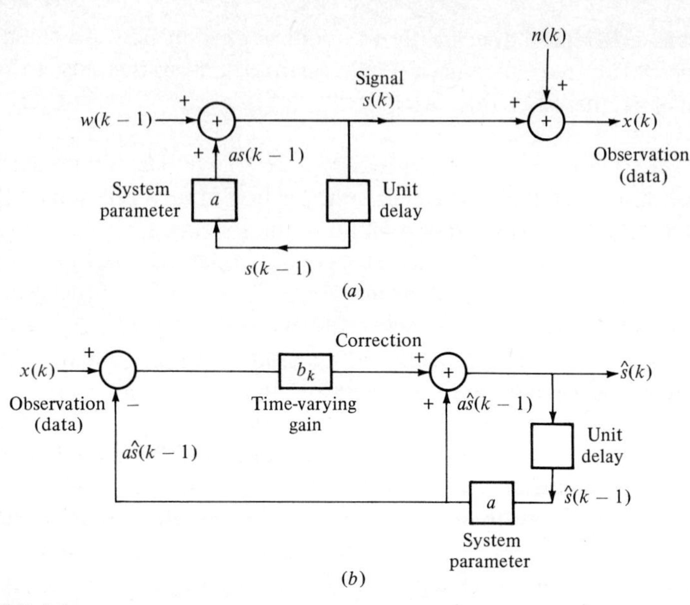

FIGURE 7.4
Model for, and estimator of, random-signal process. (*a*) Model of signal and observation processes; (*b*) optimum recursive estimator.

normalized to the observation-noise variance σ_n^2. In addition to Eq. (7.44) the orthogonality relation and the mean-squared error expression can be combined to provide the following recursive form for the normalized mean-squared error:

$$p_k = \frac{A + a^2 p_{k-1}}{A + 1 + a^2 p_{k-1}} \tag{7.45}$$

The parameter A is just σ_w^2/σ_n^2, the ratio of the variance of the white-noise part of the signal to the variance of the observation noise. It is essentially a signal-to-noise parameter. As A increases (due either to a reduction in the additive observation noise or to an increase in the signal fluctuations), $b_k \to 1$. In this limiting case $\hat{s}(k) = x(k)$, and all past estimates of the signal are disregarded. For with the additive noise very small $x(k) \approx \hat{s}(k)$, and the best estimate of $s(k)$ is just the data sample $x(k)$. At the other extreme, if $A = 0$ ($\sigma_w^2 = 0$), $b_k = p_k = a^2 p_{k-1}/(1 + a^2 p_{k-1})$. In particular, with $a = 1$, $s(k) = s$, just the signal *parameter* of the previous section. The varying gain is then $b_k = p_{k-1}/(1 + p_{k-1})$, in agreement with Eqs. (7.34) and (7.35).

Equation (7.45) for the normalized mean-squared error indicates that a steady-state value for both this error and for the varying gain b_k is approached for large k. This value of the steady-state error and gain can be found in terms of the parameter A by letting $p_k = p_{k-1} = p$ and solving the resultant quadratic equation for p. As an example, let $A = 1$, $a^2 = \frac{1}{2}$. It is left to the reader to show that the limiting mean-squared error and gain is $b_\infty = p = 0.56$.

How does the recursive estimator operate, and how is the limiting gain value approached? We must first initialize the system, as in the previous section [Eqs. (7.16) to (7.18)]. Specifically, with the first data sample $x(1)$ we can only write $\hat{s}(1) = b_1 x(1)$, since $\hat{s}(0) = 0$. (Recall that we are dealing with *zero-mean* signals. The best estimate with no data is then just 0.) To find b_1 we use the orthogonality relation $E[[s(1) - \hat{s}(1)]x(1)] = 0$. Writing $x(1) = s(1) + n(1)$, taking the expectation as called for, and solving for b_1, we find

$$b_1 = p_1 = \frac{\sigma_s^2}{\sigma_s^2 + \sigma_n^2} = \frac{1}{1 + b} \tag{7.46}$$

exactly as in Eq. (7.18). (There we used S to represent the variance of the random-signal *parameter*. Here we used σ_s^2 to represent the variance of the signal *process*. $b \equiv \sigma_n^2/\sigma_s^2$ is again a noise-to-signal ratio.) Actually we could have used the previous initialization results directly, for with one sample signal dynamics do not play a role and signal-parameter and signal-process estimates should be the same.

As an example, let $A = 1$, $a^2 = \frac{1}{2}$, as noted above. Since $A = \sigma_w^2/\sigma_n^2 = (1 - a^2)\sigma_s^2/\sigma_n^2 = (1 - a^2)/b$, we have $b = \frac{1}{2}$ and $b_1 = p_1 = \frac{2}{3}$. Then $b_2 = p_2 = (A + a^2 p_1)/(A + 1 + a^2 p_1) = 0.63$. Continuing, $b_3 = p_3 = 0.57$, and we are already close to the limiting value of 0.56.

One-Step Signal Prediction

The discussion thus far has concentrated on the estimation of the *current* value of a random-signal process in additive white noise. This is often referred to as a filtering problem, and the recursive estimator of Fig. 7.4 thus provides the best linear *filter* for $s(k)$ in the least mean-squared sense. In many real-life situations, particularly those involving control or data compression, one would like to *predict* ahead, if possible. We have already noted this in passing. In the next few paragraphs we therefore outline the approach to recursive signal prediction. This is then taken up further in more detail in connection with the prediction of *vector*-signal processes, with application to the radar tracking problem considered in Sec. 7.3.

Prediction is often referred to as one-step, two-step, or l-step prediction, depending upon how many time units into the future we would like to predict. Obviously the further in the future we look the larger the prediction error we shall experience. In this book we discuss only the one-step case. Reference is then made to the literature for l-step prediction results. Specifically, with the signal again modeled as a first-order autoregressive process [Eq. (7.37)] and white noise assumed added during the measurement [Eq. (7.40)], we would like to know the "best" linear estimate of $s(k + 1)$ (the signal at time $k + 1$) *given* the data and previous estimate at time k. We call this one-step prediction estimate $\hat{s}(k + 1 \mid k)$. By "best" we mean the predictor that minimizes the mean-squared prediction error $v(k + 1 \mid k) \equiv E[[s(k + 1) - \hat{s}(k + 1 \mid k)]^2]$. This is comparable to the mean-squared error $e_k = E[[s(k) - \hat{s}(k)]^2]$ in the filtering case. [In a common notation a filtered estimate would be $\hat{s}(k \mid k)$.]

By extension of the previous discussion on filtering it is apparent that the one-step linear predictor will be of the form

$$\hat{s}(k+1 \mid k) = \alpha_k \hat{s}(k \mid k-1) + \beta_k x(k) \tag{7.47}$$

We can proceed to find the varying gains α_k and β_k (comparable to a_k and b_k in the filtering case) just as previously by using the appropriate orthogonality relations again. Since the algebra involved is straightforward, albeit tedious, we dispense with the derivations and write the final results directly. Specifically, we find that the final expression for the recursive predictor is given by

$$\hat{s}(k+1 \mid k) = a\hat{s}(k \mid k-1) + \beta_k[x(k) - \hat{s}(k \mid k-1)] \tag{7.48}$$

The variable-gain term β_k is in turn found recursively from the normalized mean-squared prediction error p'_k, as in the filtering case just considered:

$$\beta_k = \frac{ap'_{k-1}}{1 + p'_{k-1}} \tag{7.49}$$

with

$$p'_k \equiv \frac{v(k \mid k-1)}{\sigma_n^2} = A + \frac{a^2 p'_{k-1}}{1 + p'_{k-1}} = A + a\beta_k \tag{7.50}$$

The parameter A is again defined to be σ_w^2/σ_n^2. Note that with $A = 0$ and $a = 1$, the two equations (7.48) and (7.49) again reduce to those found previously in estimating a random-signal parameter recursively [Eqs. (7.34) to (7.36)].

Again the optimum processor consists of simply multiplying the previous estimate by a, in agreement with the signal-process model assumed, and then adding a weighted correction term. Note that the correction term consists of the difference between the new data sample $x(k)$ and the previous prediction estimate $\hat{s}(k \mid k-1)$ directly. In the filtering problem considered previously the correction term involved $x(k)$ less a times the previous estimate [Eq. (7.43)]. The two gain factors b_k and β_k are apparently different as well. Does this mean that filtering and one-step prediction must be done independently if both are desired? Actually it turns out the two are related, for the one-step predicted value in the least mean-squared sense is just a times the filtered value. This again is in agreement with the model assumed for the signal: in predicting ahead simply take a times the present estimated value. A similar result will be found to hold, more generally, in the vector-signal case. Specifically, then,

$$\hat{s}(k+1 \mid k) = a\hat{s}(k) \tag{7.51}$$

It is also found that

$$\beta_k = ab_k \tag{7.52}$$

i.e., the prediction gain and filtering gain are actually related by the parameter a as well. The filtering and prediction operations are thus intimately connected.

Another interesting special case of one-step prediction corresponds to letting the observation noise vanish by setting $\sigma_n^2 = 0$. This amounts to estimating $s(k+1)$ given perfect observations of signal samples $s(k), s(k-1), \ldots, s(1)$. The resulting best predictor is simply a times the most recent observation

$$\hat{s}(k+1 \mid k) = as(k)$$

with an error which is simply the next zero-mean unpredictable white-noise sample $w(k)$ and

$$v(k+1 \mid k) \doteq \sigma_w^2$$

It is precisely this simple prediction property for the time-varying signal which allows the best linear mean-squared estimators from noisy observations to be put into the recursive form. A second-order signal model

$$s(k) = a_1 s(k-1) + a_2 s(k-2) + w(k)$$

could not be estimated or predicted by the same kind of recursive scheme. We shall see later how such a signal can be estimated recursively by viewing it as the output of a first-order *vector* equation with a white-noise input.

As a check, if we divide Eq. (7.48) through by a and use Eqs. (7.51) and (7.52), we get exactly Eq. (7.43) for the recursive filtering operation. In addition, the mean-squared estimation error e_k and the prediction error $v(k+1 \mid k)$ can be shown quite readily to be directly related. For

$$\begin{aligned} v(k+1|k) &= E[[s(k+1) - \hat{s}(k+1|k)]^2] = E[[as(k) + w(k) - a\hat{s}(k)]^2] \\ &= E[[a[s(k) - \hat{s}(k)] + w(k)]^2] \end{aligned}$$

putting in the known dynamics of $s(k+1)$ and using Eq. (7.51). Since $w(k)$ is a sample of a white-noise process, it is uncorrelated with the error term $s(k) - \hat{s}(k)$. Hence on expansion we get

$$v(k+1 \mid k) = a^2 e_k + \sigma_w^2 \tag{7.53}$$

It is left for the reader to show that the two recursive equations for the mean-squared errors [Eqs. (7.45) and (7.50)] are in agreement with Eq. (7.53). As a matter of fact, Eq. (7.52), relating the prediction and filtering gains, can readily be derived using Eq. (7.53) and the recursive expressions for the mean-squared errors.

The optimum one-step predictor of Eq. (7.48) is sketched in block-diagram form in Fig. 7.5*a*. Because of the intimate connection between filtering and prediction, the predicted value can be obtained as well by tapping off at the appropriate point in the optimum filter. The resultant predictor-estimator combination is outlined in Fig. 7.5*b*. Note that this device is the same as that of Fig. 7.4*b* except for an inconsequential interchange of the unit delay and a-gain box. It is left for the reader to show, in the light of the discussion above, that the two block diagrams of Fig. 7.5 do in fact both provide the one-step prediction $\hat{s}(k+1 \mid k)$.

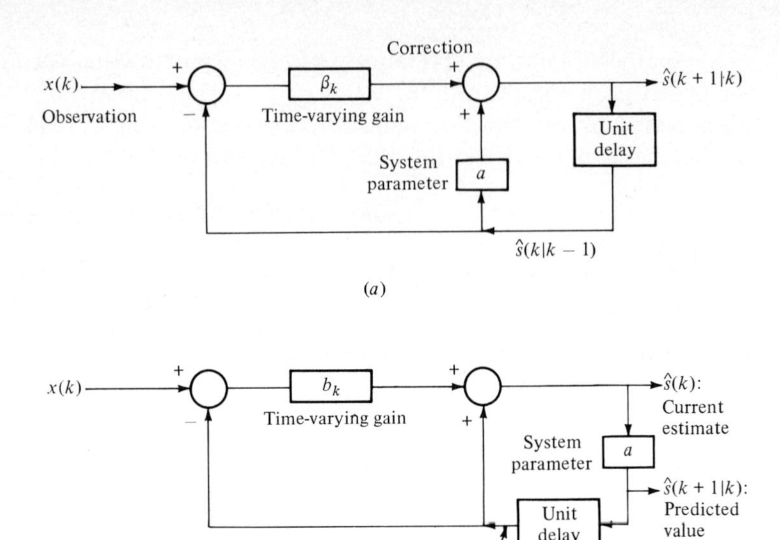

FIGURE 7.5
One-step prediction of random-signal process. (*a*) One-step predictor; (*b*) filtering and prediction simultaneously.

Vector Signals†

We have concentrated thus far on the filtering and prediction of random signals assumed generated by a first-order autoregressive process. It would obviously be worthwhile to extend the process to broader classes of signals. In addition one often handles the estimation of several signals simultaneously (components of a three-dimensional displacement vector, a velocity vector, etc.). Both cases (more complex signal classes and simultaneous estimation of possibly correlated signals) are readily treated by using vector notation. In place of simple gain parameters we then get matrix operations on vectors.

To demonstrate the procedure assume first that we have N-independent signals to be estimated or predicted simultaneously. Call the samples of these at time k $s_1(k)$, $s_2(k)$, ..., $s_N(k)$. Assume that each one is generated according to its own first-order autoregressive process. The jth signal, for example, is then formed according to the equation

$$s_j(k) = a_j s_j(k-1) + w_j(k-1) \qquad j = 1, \ldots, N \tag{7.54}$$

Each of the w_j processes is assumed white, zero mean, and independent of all the others. We can define N-dimensional vectors made up of the N signals and N white-noise driving processes:

† See Sec. 3.7.

$$\mathbf{s}(k) \equiv \begin{bmatrix} s_1(k) \\ s_2(k) \\ \vdots \\ s_N(k) \end{bmatrix} \quad \text{and} \quad \mathbf{w}(k) \equiv \begin{bmatrix} w_1(k) \\ w_2(k) \\ \vdots \\ w_N(k) \end{bmatrix} \tag{7.55}$$

In terms of these defined vectors the N equations (7.54) can then be written as the one vector equation

$$\mathbf{s}(k) = \Phi\mathbf{s}(k-1) + \mathbf{w}(k-1) \tag{7.56}$$

with the $N \times N$ matrix Φ a diagonal matrix given by

$$\Phi = \begin{bmatrix} a_1 & 0 & 0 & \cdots & 0 \\ 0 & a_2 & 0 & \cdots & 0 \\ 0 & 0 & a_3 & \cdots & 0 \\ \cdots & \cdots & \cdots & \cdots & \cdots \\ 0 & 0 & 0 & \cdots & a_N \end{bmatrix} \tag{7.57}$$

It is left to the reader to demonstrate that this is in fact the case.

By letting Φ become a nondiagonal matrix all kinds of other situations can be handled. As an example, say that the signal $s(k)$ obeys not a first-order but a *second-order* equation. Thus, let

$$s(k) = as(k-1) + bs(k-2) + w(k-1) \tag{7.58}$$

This enables us immediately to handle much more complex signals—those in which second-order dynamics are known from the physics of the situation to play a role or those in which the spectral density of a first-order autoregressive process, for example, just does not fit the measured spectral density of $s(k)$. (There are now two parameters, a and b, to adjust rather than the simple one parameter previously.) This case is easily handled by defining two components of a signal vector. (This is a special case of the state-space approach to linear system analysis.) Specifically, in Eq. (7.58) let $s_1(k)$ be $s(k)$ and $s_2(k)$ be $s(k-1) = s_1(k-1)$. We then have, corresponding to Eq. (7.58), the *two* equations

$$s_1(k) = as_1(k-1) + bs_2(k-1) + w(k-1)$$

and
$$s_2(k) = s_1(k-1) \tag{7.58a}$$

When the two-dimensional vector

$$\mathbf{s}(k) \equiv \begin{bmatrix} s_1(k) \\ s_2(k) \end{bmatrix}$$

is formed, the two equations can be combined to form the single vector equation

$$\begin{bmatrix} s_1(k) \\ s_2(k) \end{bmatrix} = \begin{bmatrix} a & b \\ 1 & 0 \end{bmatrix} \begin{bmatrix} s_1(k-1) \\ s_2(k-1) \end{bmatrix} + \begin{bmatrix} w(k-1) \\ 0 \end{bmatrix} \tag{7.58b}$$

Note that this is exactly of the form of Eq. (7.56) with

$$\Phi = \begin{bmatrix} a & b \\ 1 & 0 \end{bmatrix} \quad \text{and} \quad \mathbf{w}(k) = \begin{bmatrix} w(k) \\ 0 \end{bmatrix}$$

It is apparent that this process can be continued indefinitely. Various types of signals and coupling between them can be handled by the appropriate definition of a new *vector* signal. The system of equations relating the original signals is then converted into the *first-order vector equation*

$$\mathbf{s}(k) = \Phi\mathbf{s}(k-1) + \mathbf{w}(k-1) \tag{7.56}$$

Equations with time-varying coefficients can be handled as well by defining a time-varying matrix $\Phi(k)$. (The reader is referred to the literature for these more general approaches to recursive estimation.†)

Consider another example which comes specifically from the radar tracking problem to be treated in more detail in the next section. Say that a vehicle being tracked is at range $R + r(k)$ at time k and at range $R + r(k+1)$ at time $k+1$, T s later. (T thus represents the spacing between data samples. We use T rather than the symbol T_s used up to now to avoid confusion between more densely packed samples in a pulse used to detect the presence of a target, as in the previous chapter, and those representing pulses spaced a scan apart, as is the case here.) $r(k)$ and $r(k+1)$ are thus deviations away from some average range R. We are interested in estimating these deviations, assumed statistically random with zero-mean value. To a first approximation, if the vehicle is traveling at radial velocity $\dot{r}(k)$ and T is not too large,

$$r(k+1) = r(k) + T\dot{r}(k) \tag{7.59}$$

Similarly, the radial acceleration $u(k)$ is given by

$$Tu(k) = \dot{r}(k+1) - \dot{r}(k) \tag{7.60}$$

Assume that $u(k)$ is a zero-mean, stationary *white-noise process*. The acceleration is thus, on the average, zero and uncorrelated from interval to interval: $E[u(k+1)u(k)] = 0$. It has some known variance $\sigma_u^2 = E[u^2(k)]$, however. Such accelerations might be caused by sudden wind gusts or short-term irregularities in engine thrust.

Then $u_1(k) \equiv Tu(k)$ is a white-noise process as well, and we have, in place of Eq. (7.60),

$$\dot{r}(k+1) = \dot{r}(k) + u_1(k) \tag{7.61}$$

Equations (7.59) and (7.61) together define a *second-order* dynamic process. For, from Eq. (7.59), we have

$$T\dot{r}(k+1) = r(k+2) - r(k+1) \tag{7.62}$$

† See, for example, A. P. Sage and J. L. Melsa, "Estimation Theory with Applications to Communications and Control," chap. 7, McGraw-Hill, New York, 1971.

From Eq. (7.61), multiplying through by T and substituting in Eq. (7.59) for $T\dot{r}(k)$, we also have

$$T\dot{r}(k+1) = r(k+1) - r(k) + Tu_1(k) \tag{7.63}$$

Equating these two expressions, we have finally

$$r(k+2) = 2r(k+1) - r(k) + Tu_1(k) \tag{7.64}$$

the equation of a second-order autoregressive process. The random range variations in this model are thus related directly to the (assumed) known statistics of the radial acceleration. The white-noise acceleration process turns out to be the forcing function or driving force for the random range process.

If we now define a two-component signal vector $\mathbf{s}(k)$ with one component the range $r(k)$ and the other the radial velocity $\dot{r}(k)$, that is, $s_1(k) = r(k)$ and $s_2(k) = \dot{r}(k)$, the two equations (7.59) and (7.61) *or* the one equivalent equation (7.64) can be written in the first-order vector-equation form of Eq. (7.56). It is left for the reader to demonstrate that for this example

$$\Phi = \begin{bmatrix} 1 & T \\ 0 & 1 \end{bmatrix} \quad \text{and} \quad \mathbf{w}(k) = \begin{bmatrix} 0 \\ u_1(k) \end{bmatrix} \tag{7.65}$$

We shall have more to say about this example in connection with radar tracking in the next section.

Kalman Filters: One-Step Prediction

With the extension of the signal model to vector signals to include multiple signal measurements and signals with more complex dynamic behavior, we are now in a position to handle much more realistic problems involving the recursive estimation of signals in additive noise. We shall again simply present the resultant optimum mean-squared filter equations, in vector form, to avoid carrying out the algebraic derivations here. Optimum mean-squared recursive estimators for signal models of this form (including filters and predictors, with extensions to time-varying situations and nonlinear cases as well) are called Kalman filters, as noted earlier.

The Kalman filter results we are about to discuss require a model for the measurement process of the vector signal in addition to the model for the generation of the signal already discussed. Assume that in estimating the signal vector $\mathbf{s}(k)$ we make M simultaneous noisy measurements at time k. The M measurements may be of a selected number of the N signal-vector components directly or of combinations (weighted sums) of the signals. We thus have $M \le N$. Let these measurement samples be labeled $x_1(k), x_2(k), \ldots, x_M(k)$. As an example, if these measurements are of the first M of the N signal components, we have

$$\begin{aligned} x_1(k) &= s_1(k) + n_1(k) \\ x_2(k) &= s_2(k) + n_2(k) \\ &\cdots\cdots\cdots \\ x_M(k) &= s_M(k) + n_M(k) \end{aligned} \tag{7.66}$$

The $n_j(k)$ terms represent additive noise.

Equation (7.66) can be put into vector form by defining M-component vectors $\mathbf{x}(k)$ and $\mathbf{n}(k)$. In terms of the previously defined N-component signal vector $\mathbf{s}(k)$, we then have

$$\mathbf{x}(k) = H\mathbf{s}(k) + \mathbf{n}(k) \tag{7.67}$$

with H the $M \times N$ matrix

$$H = \overset{\rightarrow N}{\begin{bmatrix} 1 & 0 & 0 & \cdots & 0 \\ 0 & 1 & 0 & \cdots & 0 \\ 0 & 0 & 1 & \cdots & 0 \\ & & \vdots & & \\ 0 & 0 & \cdots & 1 \cdots & 0 \end{bmatrix}} \downarrow M$$

It is left for the reader to show that Eq. (7.67) is appropriate as well for measurements involving two or more components of the signal $s(k)$ coupled together. One simply defines H appropriately. [What is H, for example, if $x_1(k) = 2s_1(k) + 3s_2(k) + n_1(k)$, with the other x_j's given by Eq. (7.66)?]

For the specific example of the previous radar tracking case, say that the signals to be estimated are range $r(k)$, radial velocity $\dot{r}(k)$, bearing (azimuth) $\theta(k)$, and angular velocity $\dot{\theta}(k)$. Then $\mathbf{s}(k)$ is the four-component vector

$$\mathbf{s}(k) = \begin{bmatrix} r(k) \\ \dot{r}(k) \\ \theta(k) \\ \dot{\theta}(k) \end{bmatrix}$$

Measurements of only range and bearing are made, however, with additive noise $n_1(k)$ and $n_2(k)$, respectively. We thus have $N = 4$, $M = 2$ in this example [the velocities are then found in terms of these using equations such as Eq. (7.59)]. The matrix H is here given by

$$H = \begin{bmatrix} 1 & 0 & 0 & 0 \\ 0 & 0 & 1 & 0 \end{bmatrix}$$

Returning to the basic problem, we have a signal vector $\mathbf{s}(k)$ obeying a known dynamical equation

$$\mathbf{s}(k+1) = \Phi\mathbf{s}(k) + \mathbf{w}(k) \tag{7.56a}$$

to be extracted from a noisy-measurement vector $\mathbf{x}(k)$:

$$\mathbf{x}(k) = H\mathbf{s}(k) + \mathbf{n}(k) \tag{7.67}$$

How do we form the "best" linear estimate (filtered value) $\hat{\mathbf{s}}(k)$ of $\mathbf{s}(k)$ and its "best" predicted value $\hat{\mathbf{s}}(k \mid k-1)$? Note that the problem is formally the same as that stated previously in dealing with single time-varying signals obeying a first-order dynamical equation. Here we simply look for the simultaneous estimates (filtered or predicted) of N signal components. By "best" we now mean estimators that minimize the mean-squared error of *each* signal component simultaneously. In the filtering operation each mean-squared error $E[[s_j(k) - \hat{s}_j(k)]^2]$, $j = 1, 2, \ldots, N$, is to be minimized. In the one-step prediction each prediction error $E[[s_j(k) - \hat{s}_j(k|k-1)]^2]$, $j = 1, 2, \ldots, N$, is to be minimized.

As noted earlier, we shall not carry out the derivations of the optimum estimators here. Suffice it to say they are extensions of the scalar case outlined previously. Readers interested in the details are referred to the literature.† We shall outline only the results here and discuss them as in the scalar case. Specifically, in the one-step prediction case one finds the optimum prediction vector to be given by the recursive vector equation

$$\hat{\mathbf{s}}(k+1 \mid k) = \Phi\hat{\mathbf{s}}(k \mid k-1) + G_{k+1}[\mathbf{x}(k) - H\hat{\mathbf{s}}(k \mid k-1)] \tag{7.68}$$

Note how similar this is to the one-step predictor for the scalar signal case [Eq. (7.48)]. Note also how the form of the solution parallels the form of the vector equation (7.56) representing the signal dynamics. As in the scalar case, the best prediction of $\mathbf{s}(k+1)$, using current data $[\mathbf{x}(k)]$, is formed by extrapolating the previous predicted vector (multiplication by the signal-dynamics matrix Φ is equivalent to multiplying by the constant a in the first-order scalar-signal case), and then adding a correction term formed by subtracting the previous estimate from the new observation vector $\mathbf{x}(k)$. In place of the previous time-varying scalar gain β_k in Eq. (7.48) we have a time-varying gain matrix G_k. The Kalman one-step predictor and the corresponding signal-plus-noise model are portrayed in Fig. 7.6. Equation (7.68) for the one-step predictor can readily be generalized to nonstationary signals, with the signal-dynamics matrix Φ changing with time in some known manner. The corresponding time-varying matrix Φ_k then appears in Eq. (7.68).‡

For the scalar signal the time-varying gain was found recursively from the mean-squared prediction error [Eq. (7.49)]. In this vector-signal case the time-varying gain matrix G_k is found to be related recursively to the *covariance matrix* of the predicted estimate. Calling this quantity $V(k+1 \mid k)$, we have as its defining relationship

$$V(k+1|k) \equiv E[[\mathbf{s}(k+1) - \hat{\mathbf{s}}(k+1|k)][\hat{\mathbf{s}}(k+1) - \hat{\mathbf{s}}(k+1|k)]^T] \tag{7.69}$$

Note that this is a square $N \times N$ matrix. The diagonal terms in fact represent the individual mean-squared prediction errors that are to be simultaneously minimized, while the off-diagonal terms represent the covariances between the various signal components. [A typical element of the matrix is $v_{ij} = E[[s_i(k+1) - \hat{s}_i(k+1 \mid k)] \times [s_j(k+1) - \hat{s}_j(k+1|k)]]$.] The recursive expression for the time-varying gain matrix in terms of the covariance matrix turns out to be the matrix equation

$$G_{k+1} = \Phi V(k|k-1)H^T[HV(k|k-1)H^T + R_k]^{-1} \tag{7.70}$$

The matrix R_k appearing is just the $M \times M$ additive-noise covariance matrix $R_k \equiv E[\mathbf{n}(k)\mathbf{n}(k)^T]$. [In terms of the previous example with two simultaneous measurements $x_1(k)$ and $x_2(k)$ made, there are two additive-noise terms $n_1(k)$ and $n_2(k)$. R_k

† Sage and Melsa, op. cit., pp. 253–269.

‡ Ibid.

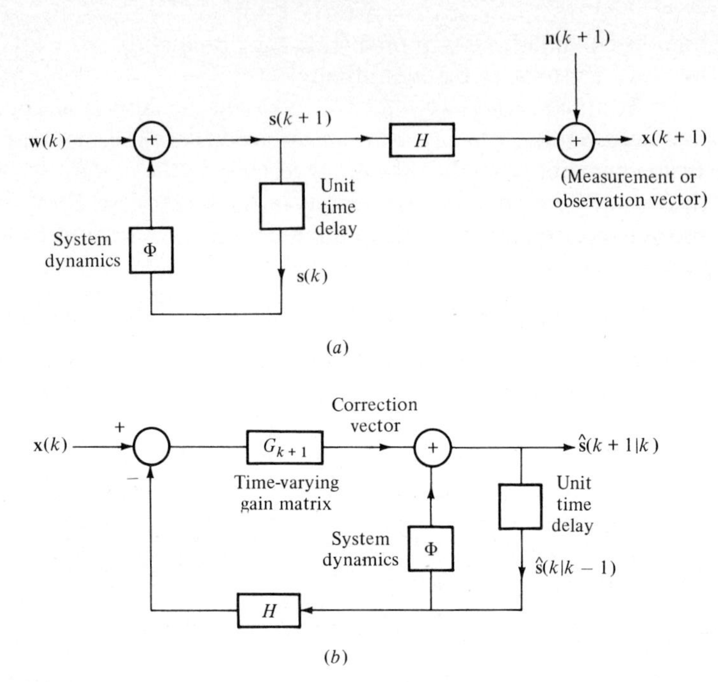

FIGURE 7.6
Kalman one-step predictor. (*a*) Model of signal and observation processes; (*b*) optimum one-step predictor.

then has *four* components, namely $E[n_1^2(k)]$ and $E[n_2^2(k)]$ along the diagonal and $E[n_1(k)n_2(k)]$ in the off-diagonal slots, as a measure of the correlation between the noise processes. If the noises are uncorrelated, as is usually assumed to be the case, the off-diagonal terms are zero. Most commonly the noise processes are assumed to be stationary; i.e., their statistics do not vary with time. R_k is then a matrix R independent of time k.]

The covariance matrix of the error in the predicted estimate $V(k \mid k - 1)$ is itself found recursively by means of the matrix equation

$$V(k + 1 \mid k) = (\Phi - G_{k+1}H)V(k \mid k - 1)\Phi^T + U_k \tag{7.71}$$

Here $U_k \equiv E[\mathbf{w}(k)\mathbf{w}(k)^T]$ is the $N \times N$ signal driving-noise covariance matrix. (For stationary noise processes, as usually assumed, this is just a constant matrix U.)

Summarizing, the complete prediction operation consists of calculating the appropriate gain matrix G_{k+1} at time $k + 1$ from Eq. (7.70) and using it to update the predicted estimate [Eq. (7.68) or Fig. 7.6*b*]. The covariance matrix of the predicted estimate is then itself updated by means of Eq. (7.71), which in turn is used to find a new gain matrix G_{k+2}, and the process repeats itself. The initialization necessary to start the procedure is discussed in the next section in terms of a specific example. The recursive equation for the covariance of the predicted estimate is particularly useful since it enables one to keep track of the prediction error.

As checks on the two recursive equations (7.70) and (7.71) written down and discussed thus far, it is left for the reader to show that the two equations produce, respectively, an $N \times M$ matrix G_{k+1} and an $N \times N$ matrix $V(k+1|k)$. As a further check assume that the signal vector $\mathbf{s}(k)$ and observation vector $\mathbf{x}(k)$ have one component only. The problem should then reduce to the one-step prediction of a time-varying scalar signal discussed previously. In particular, let the $N \times N$ matrix Φ become the parameter a, let the gain matrix G_k become the gain parameter β_k, let $V(k\,|\,k-1)$ become the mean-squared prediction error $v(k\,|\,k-1)$, and let R_k become σ_n^2, the additive-noise variance. It is left to the reader to show that the three vector equations, (7.68), (7.70), and (7.71), describing the Kalman one-step predictor do in fact reduce to the corresponding three equations for the scalar-signal predictor, (7.48), (7.49), and (7.50).

Kalman Filters: Signal Filtering

To complete this discussion of the recursive estimation of vector signals we summarize the results for the direct estimation or filtering of a signal from a set of noisy measurements. These results are almost obvious now and follow directly from the scalar-estimation results discussed previously, as well as the vector-prediction case just considered.

Specifically, we again assume that the signal or set of signals to be estimated obeys the first-order vector difference equation

$$\mathbf{s}(k+1) = \Phi\mathbf{s}(k) + \mathbf{w}(k) \qquad (7.56a)$$

Here $\mathbf{w}(k)$ is a white-noise forcing function. $\mathbf{s}(k)$ is an N-component vector. We have available a set of M noisy measurements, which we denote by the vector $\mathbf{x}(k)$. This vector in turn is given by the vector equation

$$\mathbf{x}(k) = H\mathbf{s}(k) + \mathbf{n}(k) \qquad (7.67)$$

We now ask for the recursive algorithm (or filter or estimator) that provides an estimate $\hat{\mathbf{s}}(k)$ of $\mathbf{s}(k)$ that is as close to it as possible in a mean-squared sense. By this we mean an estimate that simultaneously minimizes *each* mean-squared error term $E[[s_j(k) - \hat{s}_j(k)]^2]$, $j = 1, 2, \ldots, N$. (Compare this with the corresponding prediction error $E[[s_j(k+1) - \hat{s}_j(k+1\,|\,k)]^2]$.)

The answer turns out to be available in a variety of equivalent forms.† The simplest, and the one directly analogous to Eq. (7.43) for the scalar-signal estimator, is given by the vector estimator equation

$$\hat{\mathbf{s}}(k) = \Phi\hat{\mathbf{s}}(k-1) + K_k[\mathbf{x}(k) - H\Phi\hat{\mathbf{s}}(k-1)] \qquad (7.72)$$

The variable-gain matrix K_k of course reduces to the gain parameter b_k in the scalar case [Eq. (7.43)], while the signal-dynamics matrix Φ reduces to the parameter a in

† Ibid., pp. 265–270.

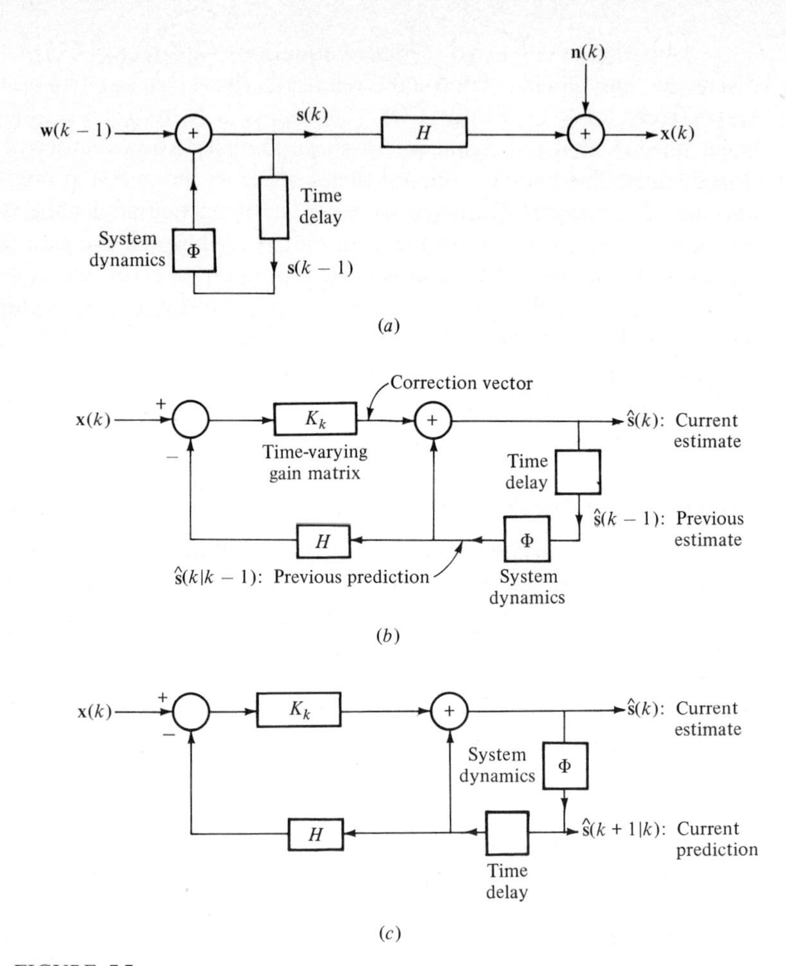

FIGURE 7.7
Kalman filter. (*a*) Signal and observation (measurement) model; (*b*) optimum Kalman filter; (*c*) simultaneous estimation and prediction.

the scalar case. This form of the optimum Kalman filter is again simply explained, as the scalar-filter and vector-predictor results were. Comparing with Eq. (7.56), it is apparent that the first portion of the estimate is provided by a linear operation on (or extrapolation of) the previous estimate. Added to this is a correction term involving the weighted difference between the new data sample and the extrapolated estimate. The signal model, observation model, and this version of the Kalman filter appear in Fig. 7.7*a* and *b*.

As in the scalar case, one can again prove the following interesting and intuitively satisfying connection between estimated and predicted signal vectors:

$$\hat{\mathbf{s}}(k+1 \mid k) = \mathbf{\Phi}\hat{\mathbf{s}}(k) \tag{7.73}$$

Given the filtered signal, the best estimate of the signal one step in the future ignores the noise and assumes that the signal-dynamics matrix Φ operates only on the estimate. The best prediction in the least mean-squared sense simply extrapolates the best filtered estimate into the future using the known signal time variation. Equation (7.72) for the filtered estimate can thus be rewritten in terms of the predicted estimate if so desired. The relation between filtered and predicted signal estimates is shown pictorially in Fig. 7.7. An alternate version of the Kalman filter, obtained by interchanging a Φ box and a time-delay element, is shown in Fig. 7.7*c*. This allows simultaneous measurement of the filtered and predicted estimates of the signal $\mathbf{s}(k)$ and is of course an extended (vector) version of Fig. 7.5*b* for the scalar-signal case.

How does one now compute the variable-gain matrix K_k? Various recursive relations are available.† A simple one involves the use of the three matrix equations

$$K_k = V(k \mid k-1)H^T[HV(k \mid k-1)H^T + R_k]^{-1} \tag{7.74}$$

$$V(k \mid k) = V(k \mid k-1) - K_k HV(k \mid k-1) \tag{7.75}$$

$$V(k+1 \mid k) = \Phi V(k \mid k)\Phi^T + U_k \tag{7.76}$$

All the matrices here have been previously defined except for $V(k \mid k) \equiv E[[\mathbf{s}(k) - \hat{\mathbf{s}}(k)][\mathbf{s}(k) - \hat{\mathbf{s}}(k)]^T]$. This is just the covariance matrix of the filtered estimate. The reader can check for himself that the diagonal terms of this matrix are just the mean-squared estimation errors that are to be minimized by this filtering operation. As in the one-step prediction case previously discussed, one can keep track of the mean-squared errors as the recursive estimation proceeds.

Note again, as in the scalar case, the close connection between the one-step predictor and filter results. It is apparent from Eqs. (7.70) and (7.74) that the two gain matrices are simply related: $G_{k+1} = \Phi K_k$. This should also be apparent from a comparison of Figs. 7.6 and 7.7. The three matrix equations used to find K_k recursively can thus be reduced to two, as in the one-step predictor case. Equation (7.75) has been kept intact, however, to provide the running measure of the mean-squared error noted above.

The operation of the Kalman filter is apparent from Eqs. (7.72) and (7.74) to (7.76). A new value of K_k is found from Eq. (7.74) and then used to update the estimate of the signal vector, using Eq. (7.72). The covariance matrix $V(k \mid k)$ is then calculated, using Eq. (7.75), and $V(k+1 \mid k)$ is then updated, using Eq. (7.76). A new value of the gain can then be calculated, and the process repeated again.

Note that the gain matrices G_{k+1} and K_k could be calculated *before* estimation and prediction are carried out. They do not depend at all on the measurements. This approach requires storing the calculated vectors for each recursion and feeding them out as needed. In the approach above, in which the gains are updated recursively as the estimation proceeds, there is no need to store all gain values. The previous value is the only one necessary.

† Ibid., p. 268.

7.3 APPLICATION TO AIR-TRAFFIC-CONTROL RADAR TRACKING†

To clarify the discussion of recursive estimation and the Kalman filters developed in the previous section, we show in this section how one might apply the recursive techniques to the tracking function of an air-traffic-control radar. In Chap. 5 we discussed the detection function of these radars. We pointed out there that as the radars rotate, they continuously send out pulses of electromagnetic energy. Pulses intercepting an airplane in space are reflected back to the radar. The return pulses with noise mixed in must then be processed to show the presence of the aircraft. The time delay between transmission and reception of the pulses provides an estimate of the aircraft range (radial distance), while the location of the antenna beam at the time of detection provides the aircraft bearing (azimuth).

For purposes of control one would like to track the aircraft as it moves through space, either en route or as it approaches a particular destination. In tracking the aircraft one may be interested simply in providing improved estimates of its range and bearing at regular intervals of time or in predicting ahead to provide better control capability. Recall that the relatively short-range airport surveillance radar (ASR) rotates at a scan rate of 15 r/min while the longer-range air-route surveillance radar (ARSR) rotates at 6 r/min. The ASR system is thus capable of providing new range and bearing estimates every 4 s, while the ARSR system can provide them every 10 s. The tracking filters in the two systems are thus updated at these respective time intervals, corresponding to the T-s time interval assumed in the previous section.

Assume now that the only measurements available are those of range r and bearing θ and that from them, estimates and predictions are to be made, not only of range and bearing, but of aircraft range rate or radial velocity $\dot{r}$ and bearing rate $\dot{\theta}$ as well. This is exactly the example noted in passing in the last section. (In some radars the range rate $\dot{r}$ is directly determined from doppler measurements while elevation measurements are made as well.) By assuming reasonable models for the aircraft flight characteristics one can come up with the appropriate Kalman filters for providing the desired recursive estimates and predicted values.

Specifically, the simplest relatively reasonable signal model that one can assume relating range, range rate, bearing, and bearing rate at T-s intervals is an extension of that already cited as an example in the previous section [Eqs. (7.59) and (7.60)]:‡

$$r(k+1) = r(k) + T\dot{r}(k) \tag{7.77}$$

$$\dot{r}(k+1) = \dot{r}(k) + u_1(k) \tag{7.78}$$

$$\theta(k+1) = \theta(k) + T\dot{\theta}(k) \tag{7.79}$$

$$\dot{\theta}(k+1) = \dot{\theta}(k) + u_2(k) \tag{7.80}$$

† This section draws heavily on R. A. Singer and K. W. Behnke, Real-Time Tracking Filter Evaluation and Selection for Tactical Applications, *IEEE Trans. Aerosp. Electron. Syst.*, vol. AES-7, no. 1, pp. 100–110, January 1971.

‡ Ibid., p. 101. Recall that we are again modeling variations about average values.

The terms $u_1(k)$ and $u_2(k)$ represent, respectively, the change in radial velocity and bearing rate over the T-s recursion interval. They are each T times the radial and angular acceleration, respectively [see Eq. (7.60)]. We would thus expect them to increase with T in some sense. The model for random maneuver acceleration we shall use will in fact have the appropriate standard deviations increasing with T. We assume $u_1(k)$ and $u_2(k)$ to be random with zero average values. We shall also assume they are uncorrelated with each other. In addition we assume, in order to use the recursive-estimation approach, that they are white-noise variables: $E[u_1(k+1)u_1(k)] = 0$ and $E[u_2(k+1)u_2(k)] = 0$. They are thus individually uncorrelated from one time interval to the next. (We shall indicate later how to introduce the effect of correlation by augmenting the signal-vector model.) The two terms $u_1(k)$ and $u_2(k)$ thus represent *maneuver noise*, in the r and θ directions, respectively, that accounts for vehicle deviation away from a constant-velocity trajectory. Under these assumptions the vehicle accelerations in both the radial and azimuthal directions are random and uncorrelated from scan to scan. The two accelerations are assumed uncorrelated with each other as well.

With the vector notation of the previous section, Eqs. (7.77) to (7.80) may be written exactly as there in the succinct form

$$\mathbf{s}(k+1) = \Phi\mathbf{s}(k) + \mathbf{w}(k) \tag{7.81}$$

with the signal vector $\mathbf{s}(k)$ and maneuver-noise vector $\mathbf{w}(k)$ given by

$$\mathbf{s}(k) = \begin{bmatrix} r(k) \\ \dot{r}(k) \\ \theta(k) \\ \dot{\theta}(k) \end{bmatrix} \quad \text{and} \quad \mathbf{w}(k) = \begin{bmatrix} 0 \\ u_1(k) \\ 0 \\ u_2(k) \end{bmatrix} \tag{7.82}$$

The state transition matrix Φ, augmented from the 2×2 example of the previous section, is in turn given by

$$\Phi = \begin{bmatrix} 1 & T & 0 & 0 \\ 0 & 1 & 0 & 0 \\ 0 & 0 & 1 & T \\ 0 & 0 & 0 & 1 \end{bmatrix} \tag{7.83}$$

As indicated earlier, the radar sensors are assumed to provide noisy estimates of the range $r(k)$ and bearing $\theta(k)$ at T-s intervals. At time k, the two sensor outputs are then

$$x_1(k) = r(k) + n_1(k) \tag{7.84}$$

and

$$x_2(k) = \theta(k) + n_2(k) \tag{7.85}$$

In matrix form, as in the previous section,

$$\mathbf{x}(k) = H\mathbf{s}(k) + \mathbf{n}(k) \tag{7.86}$$

with the matrix H given by

$$H = \begin{bmatrix} 1 & 0 & 0 & 0 \\ 0 & 0 & 1 & 0 \end{bmatrix} \tag{7.87}$$

(Note that $N = 4$ and $M = 2$ in the notation of the previous section.) The two additive noises are assumed zero-mean white, uncorrelated with each other, and with variances $\sigma_r^2(k)$ and $\sigma_\theta^2(k)$, respectively. (Although we shall assume the noise to be stationary in the material following, so that the variances do not vary with k, the nonstationary case is easily handled as well. For example, angular measurement errors might be due to wind gusts twisting the antenna, and these gusts might be bigger on the average during certain times of the day.) The noise covariance matrix, needed for the Kalman filter implementation, is thus given by

$$R_k \equiv E[\mathbf{n}(k)\mathbf{n}(k)^T] = \begin{bmatrix} \sigma_r^2(k) & 0 \\ 0 & \sigma_\theta^2(k) \end{bmatrix} \tag{7.88}$$

The basic recursive-estimation problem, already discussed in detail in the previous section, is thus one of estimating the signal vector $\mathbf{s}(k)$ and its one-step predicted value $\hat{\mathbf{s}}(k + 1 \mid k)$ from the noisy sensor data $\mathbf{x}(k)$ as well as (stored) previous values.

The formal Kalman filter solution to this problem is exactly that of the previous section (Fig. 7.7) and may be written down directly. Two additional things remain to be done, however. First, we must specifically evaluate the signal driving noise or maneuver-noise covariance matrix U_k [Eq. (7.71)] and then initialize the filter.

The maneuver-noise covariance matrix is formally defined as $U_k \equiv E[\mathbf{w}(k)\mathbf{w}(k)^T]$. Using the definition of the noise vector $\mathbf{w}(k)$ and the assumption that the two maneuver-noise terms $u_1(k)$ and $u_2(k)$ are uncorrelated, we find

$$U_k = \begin{bmatrix} 0 & 0 & 0 & 0 \\ 0 & \sigma_1^2 & 0 & 0 \\ 0 & 0 & 0 & 0 \\ 0 & 0 & 0 & \sigma_2^2 \end{bmatrix} \tag{7.89}$$

Here $\sigma_1^2 = E[u_1^2]$ and $\sigma_2^2 = E[u_2^2]$ represent the variance, respectively, of T times the radial and angular acceleration.

Specific numbers must of course be put in for those variances in order to define the Kalman filter numerically. To do this we use a model for the vehicle acceleration that is simple and appears reasonable on physical grounds.† We assume that the vehicle acceleration u in either of the two orthogonal directions (r and θ) is random and equally likely to be positive or negative with some maximum value A. (As an example, A might be $\frac{1}{3}g$ or approximately 3 m/s^2.) Because these two acceleration terms are actually the projections of the vector acceleration onto the two coordinates r and θ, it seems reasonable to assume the acceleration to be uniformly distributed between $\pm A$. The probability density function of the acceleration in either direction (r or θ) is thus assumed to have the form of Fig. 7.8. Three impulse functions representing discrete probabilities at $\pm A$ and 0 acceleration have been superimposed to make the model a little more flexible. These then simply say that there is a probability P_2 that the aircraft will proceed at constant radial and angular velocities,

† Ibid., pp. 101, 102.

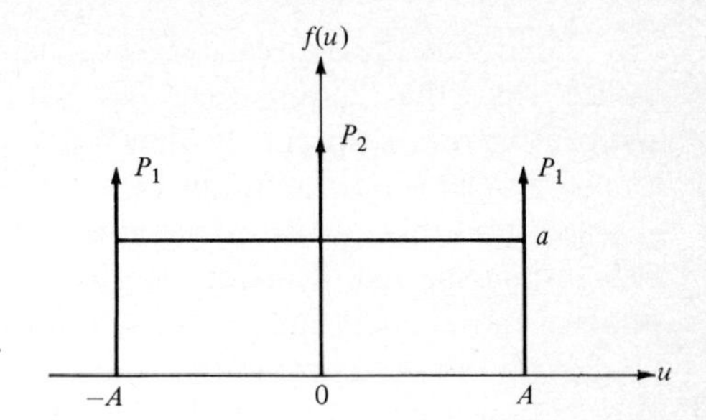

FIGURE 7.8
Assumed probability distribution of maneuver noise.

while there is a probability P_1 that its acceleration (deceleration) in either direction is at the maximum value A. It is left to the reader to show that the height of the uniform distribution is just $a = (1 - 2P_1 - P_2)/2A$ and that the variance of the random variable u is given by

$$\sigma_u^2 = \frac{A^2}{3}(1 + 4P_1 - P_2) \tag{7.90}$$

As an example, if $P_1 = 0.1$ and $P_2 = 0.3$ (there is then a 50 percent chance that the aircraft will be at 0 or maximum acceleration), $\sigma_u^2 = 1.1A^2/3$. If $A = \frac{1}{3}g$, $\sigma_u^2 = 3.9$ and $\sigma_u = 2 \text{ m/s}^2$.

To find σ_1^2 and σ_2^2, the variances of the maneuver-noise terms u_1 and u_2, respectively, and the components of matrix U_k [Eq. (7.89)], we note that $\sigma_1 = T\sigma_u$. We thus get

$$\sigma_1^2 = T^2\sigma_u^2 = \frac{A^2T^2}{3}(1 + 4P_1 - P_2) \tag{7.91}$$

The acceleration term σ_u has the units of meters per second per second. The parameter σ_2 is related to the change in angular velocity or bearing rate, however. To convert angular velocity into angular acceleration we must multiply by the scan time T, as with σ_1. But since σ_u represents linear acceleration, we must divide by the radial distance $R + r(k)$ to convert to the desired angular-acceleration form. As an approximation, however, assume that the estimated range deviation $r(k)$ is small compared with the average range R. We can then use R in all the calculations instead of changing the parameter at each iteration. It is left for the reader to justify the relation

$$\sigma_2^2 = \frac{\sigma_1^2}{R^2} = \frac{A^2T^2}{3R^2}(1 + 4P_1 - P_2) \tag{7.92}$$

As an example, for the ASR radar, with $T = 4$ s and $R = 100$ km (60 mi) with $A = \frac{1}{3}g$ and $P_1 = 0.1$, $P_2 = 0.3$, as before (note that these really play a negligible role in the calculations), we have $\sigma_1^2 = 62$ and $\sigma_2^2 = 62 \times 10^{-10}$. For the ARSR radar, with $T = 10$ s and $R = 320$ km (200 mi), we have for the same maximum acceleration $\sigma_1^2 = 390$ and $\sigma_2^2 = 38 \times 10^{-10}$. Other values of scan time T and radial distance R produce correspondingly different numbers.

Note that the standard deviations σ_1 and σ_2 increase with T, as pointed out earlier: since the random variables $u_1(k)$ and $u_2(k)$ represent *changes* in range *rate* and bearing *rate*, respectively, they must have the property of increasing with time for a fixed acceleration. Equation (7.92) for the variance of the change in bearing rate also says that this quantity decreases with range for a *fixed linear acceleration*. This is also to be expected. (Conversely, the closer in the aircraft is, the higher its angular velocity and acceleration for a fixed linear acceleration.)

With the two parameters σ_1^2 and σ_2^2 of matrix U_k specified, we now have all the information needed to completely define the Kalman filter operations of Fig. 7.7*c*, which provide, at the two output taps, the recursive estimates and one-step predictions of range $r(k)$, range rate $\dot{r}(k)$, bearing $\theta(k)$, and bearing rate $\dot{\theta}(k)$. All that remains to be done is to initialize the gain matrix K_k.

This initialization is carried out by specifying the first value of the covariance matrix $V(k \mid k)$ appearing in Eq. (7.76). For note, from the recursive equations (7.74) to (7.76) that form the basis for finding K_k algorithmically, that initialization is begun with Eq. (7.76). Once we find the first value of $V(k+1 \mid k)$, we can, from Eq. (7.74), find the first value of K_k. We then update $V(k \mid k)$, using Eq. (7.75), find the next value of $V(k+1 \mid k)$, and proceed recursively. To find the first value of $V(k+1 \mid k)$ we must in turn initialize the covariance matrix $V(k \mid k)$ of the filtered estimate. From its definition this matrix is given as

$$V(k|k) = E[[\mathbf{s}(k) - \hat{\mathbf{s}}(k)][\mathbf{s}(k) - \hat{\mathbf{s}}(k)]^T] \tag{7.93}$$

As noted earlier, the diagonal terms are just the mean-squared errors in the signal-vector estimates.

Filter initialization requires a first estimate as well as a first error-covariance matrix corresponding to the use of that first estimate. A first estimate can be found in several ways. In some estimation problems an optimal (least mean-squared error) estimator $\hat{s}(1 \mid 1)$ can be found using the orthogonality principle or, equivalently, by starting with a previous $\hat{s}(0 \mid 0) = 0$, which is indeed the optimal estimate of the zero-mean signal components when no observations are available. In such cases, the corresponding error-covariance matrix $V(0 \mid 0)$ would be simply the steady-state covariance matrix C of the signal vector since

$$\begin{aligned} V(0 \mid 0) &= E[[\mathbf{s}(0) - \hat{\mathbf{s}}(0 \mid 0)][\mathbf{s}(0) - \hat{\mathbf{s}}(0 \mid 0)]^T] \\ &= E[\mathbf{s}(0)\mathbf{s}^T(0)] = E[\mathbf{s}(k)\mathbf{s}^T(k)] = C \end{aligned} \tag{7.94}$$

This initialization approach cannot be used for the present air-traffic tracking example because the signal vector does not have a steady-state covariance matrix. The cumulative effect of many acceleration perturbations is to give $r(k)$ and $\theta(k)$ variances (about their nominal values) which grow with k. Another initialization approach is particularly suited for such problems in which the steady-state initialization is inappropriate. This second technique is based on getting a suboptimal but reasonable estimate using the first few observations. The Kalman recursion can then be used to process future observations optimally, as long as the error-covariance matrix for the initial ad hoc estimate is calculated correctly. This approach is often used in practice since it also works better when the signal model is actually a linearization of a nonlinear system: good linearization requires an initial estimate which is fairly accurate, and an ad hoc estimate can often provide such an estimate. (Remember that the Kalman filter is not directly applicable when the observations are nonlinear functions of the signal.)

A reasonable ad hoc initialization can be developed for this particular problem, using the two available sensor measurements, range and bearing. The first set of these two measurements, $\mathbf{x}(1)$ $(k = 1)$, can be used to estimate range and bearing but not range rate and bearing rate (unless we arbitrarily take these as zero). The second pair of measurements $\mathbf{x}(2)$ $(k = 2)$ provides the additional two numbers required to make independent estimates of the four parameters. We thus assume that we start the filter going only after obtaining the first two measurements $\mathbf{x}(1)$ and $\mathbf{x}(2)$. Using these, we produce the four-component signal-vector estimate $\hat{\mathbf{s}}(2)$. But this tells us that the first value of the covariance matrix $V(k \mid k)$ we can calculate is $V(2 \mid 2)$, since this involves the estimate $\hat{\mathbf{s}}(2)$. From Eqs. (7.76) and (7.74) the first gain matrix calculated will be K_3, the matrix used in finding the third signal estimate $\hat{\mathbf{s}}(3)$.

What form should the initial estimate $\hat{\mathbf{s}}(2)$ take in terms of the two sensor outputs $\mathbf{x}(1)$ and $\mathbf{x}(2)$? One particularly satisfying form is to say that the data pair $\mathbf{x}(2)$ should be used to estimate range and bearing directly, while both $\mathbf{x}(2)$ and $\mathbf{x}(1)$ are used to find the two rate estimates. Specifically, we take as the four estimates [the four components of $\hat{\mathbf{s}}(2)$]

$$\hat{s}_1(2) = \hat{r}(2) = x_1(2) \tag{7.95}$$

$$\hat{s}_2(2) = \hat{\dot{r}}(2) = \frac{1}{T}[x_1(2) - x_1(1)] \tag{7.96}$$

$$\hat{s}_3(2) = \hat{\theta}(2) = x_2(2) \tag{7.97}$$

$$\hat{s}_4(2) = \hat{\dot{\theta}}(2) = \frac{1}{T}[x_2(2) - x_2(1)] \tag{7.98}$$

Not only are these intuitively satisfying (the range and bearing estimates are simply the latest sensor readings themselves, while the rate estimates are the differences in the two sensor readings divided by T), but they are in keeping with the first-order system dynamics assumed for the aircraft [Eqs. (7.77) and (7.79): this corresponds to neglecting the effect of acceleration in the estimates above].

These simple estimates serve as the basis of the simplest filter that could be designed to provide running estimates of the four desired signal terms. This is the *two-point extrapolator* which continually uses the last data set to estimate the range and bearing and the last two sets of data to estimate range rate and bearing rate.† In this case sample 2 is replaced generally by k and 1 by $k - 1$. This filter is basically a nonrecursive filter. (It is used frequently, in another example, in the implementation

† Ibid., p. 105.

of data-compression algorithms for efficient communication because of its simplicity.) A comparison of this filter with the Kalman and other filters in the radar tracking case indicates that it performs substantially worse in terms of tracking accuracy, but its implementation in terms of computer processing time and memory requirements is correspondingly less complex.†

On the basis of the description of $\hat{\mathbf{s}}(2)$ in terms of the two sets of data $\mathbf{x}(1)$ and $\mathbf{x}(2)$, as given in Eqs. (7.95) to (7.98), we can calculate the initial covariance matrix $V(2 \mid 2)$ and from this the initial Kalman gain matrix K_3. For this we need the error vector $\mathbf{s}(2) - \hat{\mathbf{s}}(2)$. Recall from the definition of $\mathbf{s}(k)$ that its four components are $r(k)$, $\dot{r}(k)$, $\theta(k)$, and $\dot{\theta}(k)$. The first and third components of $\mathbf{s}(2) - \hat{\mathbf{s}}(2)$ are then $r(2) - x_1(2)$ and $\theta(2) - x_2(2)$, respectively. We also have from Eqs. (7.77) and (7.78)

$$\dot{r}(2) = \dot{r}(1) + u_1(1) = \frac{r(2) - r(1)}{T} + u_1(1)$$

with a similar expression for $\dot{\theta}(2)$. Subtracting from these Eqs. (7.96) and (7.98), respectively, we obtain the second and fourth components of $\mathbf{s}(2) - \hat{\mathbf{s}}(2)$. But we also have from our model of additive measurement noise [Eqs. (7.84) and (7.85)] $x_1(1) = r(1) + n_1(1)$ and $x_2(1) = \theta(1) + n_2(1)$. Similar expressions hold for $x_1(2)$ and $x_2(2)$, the set of data samples at the second time interval. We substitute these expressions for $x_1(1)$, $x_2(1)$, etc., into the expressions for the components of $\mathbf{s}(2) - \hat{\mathbf{s}}(2)$, and it is left for the reader to finally show that

$$\mathbf{s}(2) - \hat{\mathbf{s}}(2) = \begin{bmatrix} -n_1(2) \\ \dfrac{n_1(1) - n_1(2)}{T} + u_1(1) \\ -n_2(2) \\ \dfrac{n_2(1) - n_2(2)}{T} + u_2(1) \end{bmatrix} \tag{7.99}$$

The evaluation of $V(2 \mid 2)$ is now readily carried out. Specifically, note that this is a 4×4 matrix given by

$$V(2 \mid 2) = E[[\mathbf{s}(2) - \hat{\mathbf{s}}(2)][\mathbf{s}(2) - \hat{\mathbf{s}}(2)]^T]$$
$$= \begin{bmatrix} V_{11} & V_{12} & V_{13} & V_{14} \\ V_{21} & V_{22} & V_{23} & V_{24} \\ V_{31} & V_{32} & V_{33} & V_{34} \\ V_{41} & V_{42} & V_{43} & V_{44} \end{bmatrix} \tag{7.100}$$

The V_{ij} element is simply given by $E[[s_i(2) - \hat{s}_i(2)][s_j(2) - \hat{s}_j(2)]]$. Using Eq. (7.99), these elements are readily evaluated as follows:

$$V_{11} = E[n_1^2(2)] = \sigma_r^2 \tag{7.101}$$

† Ibid., pp. 106, 107.

using Eq. (7.88). Similarly,

$$V_{12} = V_{21} = E\left[n_1(2)\left[\frac{n_1(2) - n_1(1)}{T} + u_1(1)\right]\right] = \frac{\sigma_r^2}{T} \tag{7.102}$$

using the assumption that the measurement noise is white ($E[n_1(2)n_1(1)] = 0$) and uncorrelated with the maneuver noise $u_1(1)$. Proceeding in a similar manner, we find

$$V_{22} = \frac{2\sigma_r^2}{T^2} + \sigma_1^2 \tag{7.103}$$

$$V_{33} = \sigma_\theta^2 \tag{7.104}$$

$$V_{44} = \frac{2\sigma_\theta^2}{T^2} + \sigma_2^2 \tag{7.105}$$

$$V_{34} = V_{43} = \frac{\sigma_\theta^2}{T} \tag{7.106}$$

All other terms turn out to be zero. The matrix $V(2 \mid 2)$ can then equally well be written

$$V(2 \mid 2) = \begin{bmatrix} V_{11} & V_{12} & 0 & 0 \\ V_{21} & V_{22} & 0 & 0 \\ 0 & 0 & V_{33} & V_{34} \\ 0 & 0 & V_{43} & V_{44} \end{bmatrix} \tag{7.100a}$$

with the eight elements defined above in terms of the four variances discussed earlier and the scan time T. (The reader is urged, as a check, to test the dimensionality of each of the elements of the matrix.)

As an example, let the rms noise in the range sensor be equivalent to 1 km. Thus $\sigma_r = 1000$ m. Let the rms noise σ_θ in the bearing sensor be 1° or 0.017 rad. This defines the noise-covariance matrix R_k. [If the sensor measurements are gaussian-distributed about the values $r(k)$ and $\theta(k)$, respectively, there is thus a 0.68 probability that the measurements $x_1(k)$ and $x_2(k)$ will lie within ± 1000 m and $\pm 1°$ of $r(k)$ and $\theta(k)$, respectively. The range-sensor rms noise deviation is 1 percent of its output at $R = 100$ km.] For various values of average range R, scan time T, and maximum acceleration A, we can now calculate the two maneuver-noise variances σ_1^2 and σ_2^2, using Eqs. (7.91) and (7.92). This defines the maneuver-noise covariance matrix U_k. From Eqs. (7.101) to (7.106) we determine the elements of the initial covariance matrix $V(2 \mid 2)$ of the filtered estimate. Using Eq. (7.76), we find the initial value $V(3 \mid 2)$ of the covariance matrix of the one-step prediction, and from Eq. (7.74) we find K_3, the initial setting of the gain matrix, used to obtain the first Kalman estimate and prediction of the four desired parameters (Fig. 7.7c). We then repeat by finding $V(3 \mid 3)$, from Eq. (7.75), and so on.

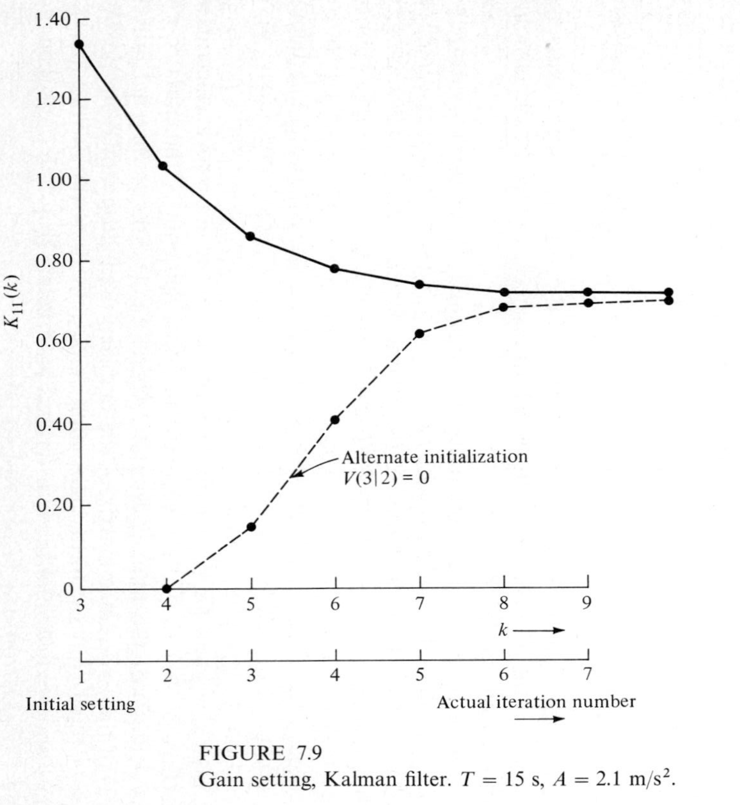

FIGURE 7.9
Gain setting, Kalman filter. $T = 15$ s, $A = 2.1$ m/s^2.

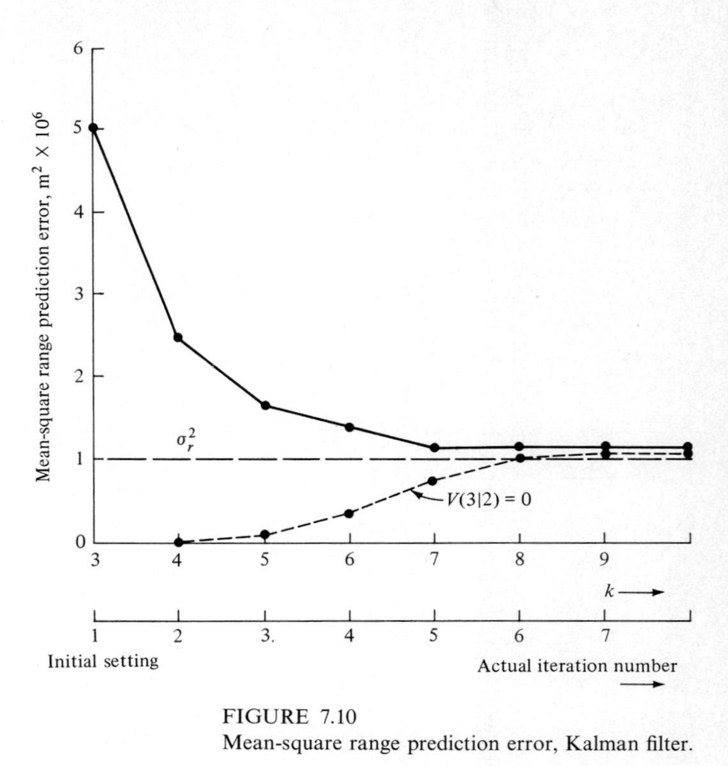

FIGURE 7.10
Mean-square range prediction error, Kalman filter.

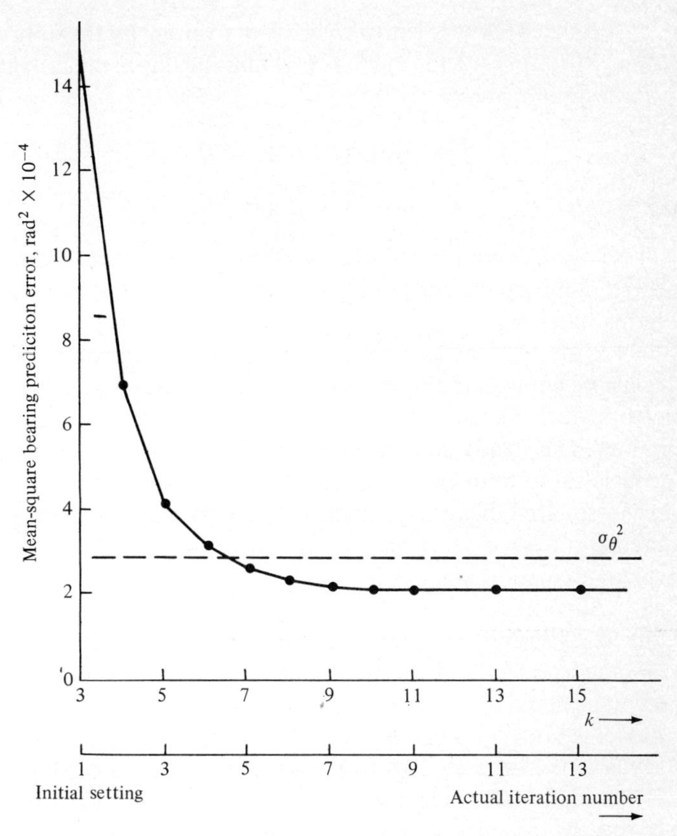

FIGURE 7.11
Mean-square bearing prediction error, Kalman filter.

In Figs. 7.9 to 7.11 we have plotted the results of a computer calculation of the appropriate matrices for range $R = 160$ km (100 mi), scan time $T = 15$ s, and a maximum acceleration $A = 2.1$ m/s^2. We have assumed the two parameters P_1 and P_2, the probabilities of maximum and zero acceleration, respectively (Fig. 7.8), to be zero for simplicity. For these numbers $\sigma_1^2 = 10^3/3$ and $\sigma_2^2 = 1.3 \times 10^{-8}$. We have plotted as solid curves only the first diagonal element, $K_{11}(k)$, of the gain matrix and the first and third diagonal elements of the one-step prediction covariance matrix $V(k+1|k)$. These correspond, respectively, to $E[[s_1(k+1) - \hat{s}_1(k+1|k)]^2] = E[[r(k+1) - \hat{r}(k+1|k)]^2]$ and

$$E[[s_3(k+1) - \hat{s}_3(k+1|k)]^2] = E[[\theta(k) - \hat{\theta}(k+1|k)]^2]$$

the mean-squared errors in the prediction of range and bearing. Similar results could be obtained for other components of the two matrices. (The dotted curves in Figs. 7.9 and 7.10 correspond to a different initialization which will be described shortly.)

As a check it is left to the reader to show by the appropriate matrix manipulation $[V(3 \mid 2) = \Phi V(2 \mid 2)\Phi^T + U_2]$ that the initial mean-squared errors in the prediction estimate are

$$E[[r(3) - \hat{r}(3|2)]^2] = 5\sigma_r^2 + \sigma_1^2 T^2 = 5.075 \times 10^6$$

and

$$E[[\theta(3) - \hat{\theta}(3|2)]^2] = 5\sigma_\theta^2 + \sigma_2^2 T^2 = 0.145 \times 10^{-2}$$

These agree of course with the two initial mean-squared errors plotted. (Note that for these numbers the mean-squared sensor errors swamp out the maneuver-variance terms initially.)

Note that the gain setting drops quickly to a final steady-state value; six iterations are enough, in this example, to reach steady state. The mean-squared prediction errors also decrease monotonically, as expected, to their minimum steady-state values. The steady-state mean-squared range-prediction error in this example is just a little more than the mean-squared measurement noise σ_r^2, while the steady-state mean-squared bearing-prediction error is somewhat less than σ_θ^2. The bearing-prediction error converges to its steady-state value in eight iterations, while the range-prediction error converges in five iterations.

Superimposed on the solid curves in Figs. 7.9 and 7.10 are dotted curves obtained by using an alternate procedure, in which the first prediction $\hat{\mathbf{s}}(3 \mid 2)$ is assumed perfect and the matrix $V(3 \mid 2)$ is therefore the zero matrix. Although the airplane could agree to fly over a known reference point at time t_3, this initialization model is not very meaningful for the air-tracking problem. We introduce it merely to see how the filter evolves from this alternate kind of start. (This model *is* useful for errors in inertial navigators on board the plane which are zero when the plane takes off but which grow as position estimates are updated from the plane's noisy radar measurements.) From Eq. (7.75) this initialization also makes $V(3 \mid 3) = 0$, so that $V(4 \mid 3) = U$, etc., and $K_3 = 0$. The set $\mathbf{x}(3)$ of *data* samples is not used since we assumed perfect knowledge of $\mathbf{s}(3)$. As shown by Figs. 7.9 and 7.10, the gain settings and mean-squared prediction errors grow with time and approach the same steady-state values as with the first initialization procedure, but more slowly.

Figures 7.12 to 7.14 show the variation of the Kalman filter gain setting (specifically the first diagonal element of the matrix again), the mean-squared range-prediction error, and the mean-squared bearing-prediction error, for two additional examples; the scan time in both cases is taken to be 5 s (comparable to the short-range air-traffic-control radar), but the maximum acceleration A has two widely different values. (Note that the values assumed, 6.4 and 35 m/s^2, are quite unrealistic, but they have been chosen to provide sufficiently wide variation in the plots obtained.) The gain setting and mean-squared errors for the example of Figs. 7.9 to 7.11 are included in Figs. 7.12 to 7.14 so that three cases in all can be compared.

Note that these curves follow trends that appear quite reasonable. When the scan time T is decreased from 15 to 5 s, for the same σ_1^2, the gain setting and mean-squared errors decrease. Since the time between filter iterations has been decreased, the variation in range and bearing estimates decreases as well (the perturbations in acceleration have *less* time over which to act). New estimates thus carry less

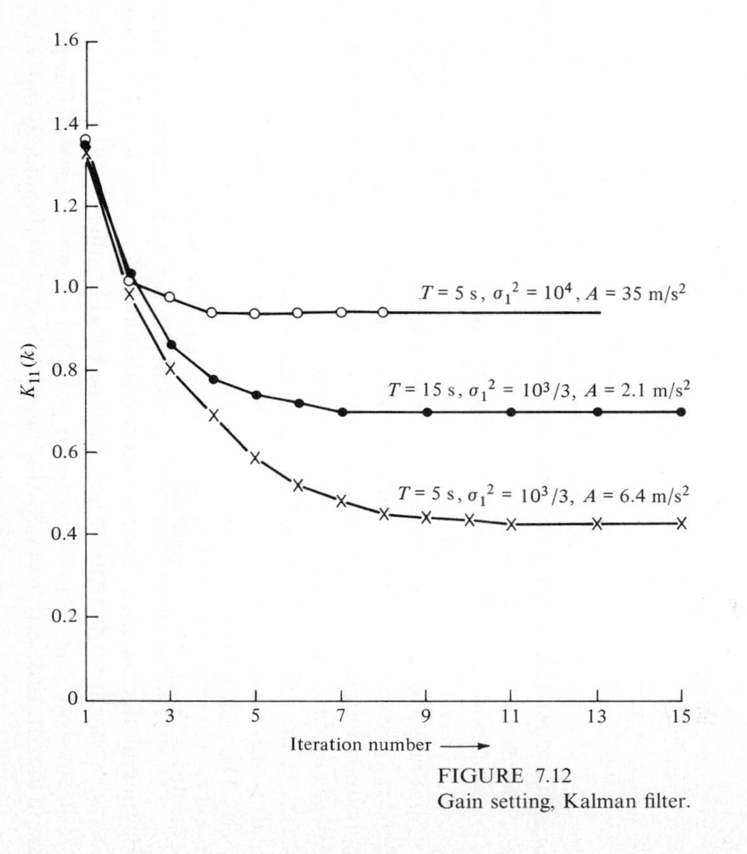

FIGURE 7.12
Gain setting, Kalman filter.

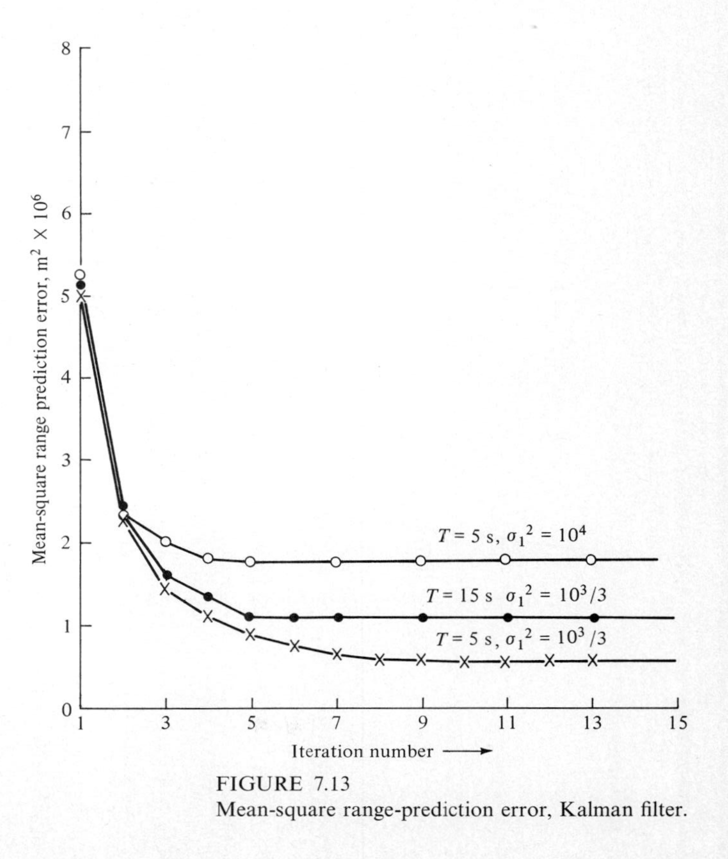

FIGURE 7.13
Mean-square range-prediction error, Kalman filter.

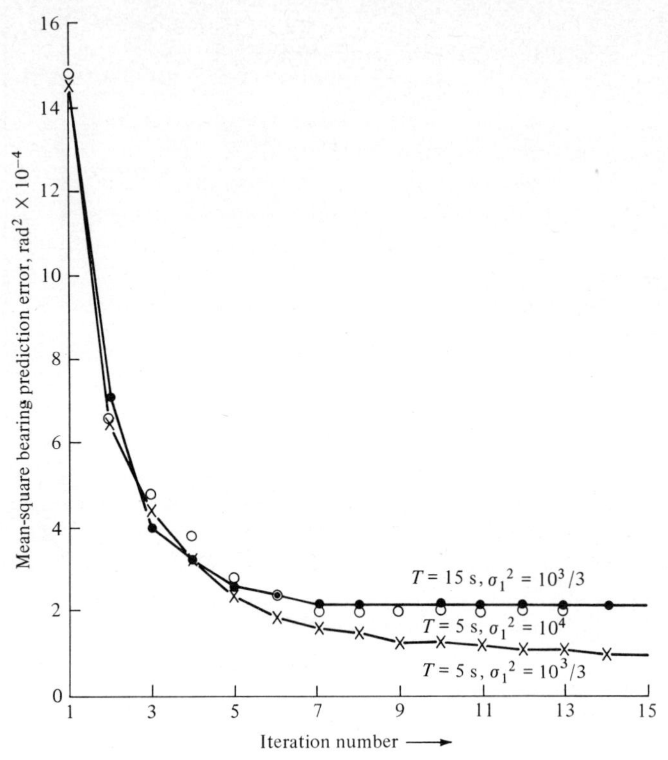

FIGURE 7.14
Mean-square bearing prediction error, Kalman filter.

information and can be weighted less. If the maximum acceleration A is now increased (increasing σ_1^2) for the same scan time T, the mean-squared errors and gain go up again, since range and bearing estimates will be expected to vary more from iteration to iteration. Note that although the three sets of curves (particularly the gain- and range-prediction error figures) show some dispersion, after a few iterations, as expected, they all begin very nearly at the same points at the first iteration. This is simply due to the numbers chosen. As noted earlier, the values of σ_r^2 and σ_θ^2 chosen are large enough to swamp out the effect of the maneuver noise, even for large σ_1^2 and large T, in the initial iterations. Smaller values of measurement noise would thus result in differences in the curves during the initial iteration period.

Augmented Kalman Filter Equations

In developing the material above on the application of Kalman filtering to the radar tracking problem, we assumed a very simple model for the aircraft flight dynamics. As indicated by Eqs. (7.78) and (7.80), we assumed the aircraft acceleration components in the r and θ directions each to be represented by a white-noise random process.

[Recall that the two acceleration terms are actually $u_1(k)/T$ and $u_2(k)/T$.] The aircraft maneuvers are therefore assumed to be such that the acceleration components are random and uncorrelated from one radar scan to the next. This may be a valid approximation in some situations, particularly for slow-scan radars, but for more rapidly scanning radars with relatively small T (the ASR radar has $T = 4$ s) this may no longer be valid. One would expect the aircraft acceleration, even though random, to remain correlated over successive scan intervals.

How does one take this more realistic situation into account? The simplest way of indicating that there is some memory to a random acceleration term $u(k)$ is to represent it as a first-order autoregressive process itself. Specifically, as we have already done several times in this book, let the acceleration component $u(k)$ be represented by the first-order dynamic equation

$$u(k+1) = \rho u(k) + w(k) \tag{7.107}$$

with $w(k)$ a zero-mean white-noise process of variance σ_w^2. The variance of $u(k)$ is thus $\sigma_u^2 = \sigma_w^2/(1-\rho^2)$. The parameter ρ is the correlation parameter we are free to adjust to make the $u(k)$ process fit the acceleration dynamics as best as possible; $R_u(j) \equiv E[u(k)u(k+j)] = \sigma_u^2 \rho^{|j|}$. If we let $\rho = 1$ and $\sigma_w^2 = 0$, we are left with a random acceleration variable u that does not vary from one time interval to the next. If, in the other extreme, $\rho = 0$, we are left with the white-noise acceleration-process model we have been discussing thus far in this section.

We can relate the correlation parameter ρ to both the aircraft and radar dynamics, as a first approximation, by letting

$$\rho = \begin{cases} 1 - \lambda T & T \le 1/\lambda \\ 0 & T > 1/\lambda \end{cases} \tag{7.108†}$$

with λ the inverse of the average aircraft-maneuver duration. If the radar scan interval is greater than $1/\lambda$, the maneuver resulting in a random change in acceleration is completed between scans and the radar sees an uncorrelated acceleration term from one interval to the next. If the radar scan interval is *less* than the maneuver duration, however, there is acceleration memory from one interval to the next, this memory increasing over more and more scan intervals as the scan interval is reduced relative to the maneuver duration.

Now assume that *each* of the acceleration components, in the r and θ directions, is representable by the same dynamic equation as Eq. (7.107). We use the same parameter ρ for both since each is a component of one acceleration vector. We can then write dynamic equations like Eq. (7.107) for both $u_1(k)$ and $u_2(k)$ in Eqs. (7.78) and (7.80). What does the Kalman filter processing sensor data subject to these modified signal dynamics look like now? Actually the results we have already obtained are still valid, at least conceptually, with a simple redefinition of the signal vector.

† Ibid., p. 109.

Recall that the Kalman filter was derived by assuming that the signal vector could be represented by the first-order difference equation

$$\mathbf{s}(k+1) = \Phi\mathbf{s}(k) + \mathbf{w}(k) \tag{7.109}$$

Here Φ is an appropriately defined signal-dynamics matrix, and $\mathbf{w}(k)$ is a white-noise matrix. Equations (7.77) to (7.80) were already in this form with $u_1(k)$ and $u_2(k)$ white-noise processes. In the present case they are not. But now consider rewriting the four signal-dynamics equations (7.77) to (7.80) with the *addition* of the two equations expressing the dynamic behavior of $u_1(k)$ and $u_2(k)$. We get as the complete set of six equations

$$r(k+1) = r(k) + T\dot{r}(k) \tag{7.110}$$

$$\dot{r}(k+1) = \dot{r}(k) + u_1(k) \tag{7.111}$$

$$u_1(k+1) = \rho u_1(k) + w_1(k) \tag{7.112}$$

$$\theta(k+1) = \theta(k) + T\dot{\theta}(k) \tag{7.113}$$

$$\dot{\theta}(k+1) = \dot{\theta}(k) + u_2(k) \tag{7.114}$$

$$u_2(k+1) = \rho u_2(k) + w_2(k) \tag{7.115}$$

Here $w_1(k)$ and $w_2(k)$ are both uncorrelated white-noise processes. These six equations are essentially of the same form as the original set of four [compare with Eqs. (7.77) to (7.80), $u_1(k)$ and $u_2(k)$ being white in that case]. If we define a *six-component* signal vector $\mathbf{s}(k)$, we in fact have identically the same problem handled earlier. Specifically, let the signal vector be defined as

$$\mathbf{s}(k) \equiv \begin{bmatrix} r(k) \\ \dot{r}(k) \\ u_1(k) \\ \theta(k) \\ \dot{\theta}(k) \\ u_2(k) \end{bmatrix} \tag{7.116}$$

The reader may then note that the six equations (7.110) to (7.115) are in fact represented by the single vector equation (7.109) if the signal-dynamics matrix Φ and the maneuver-noise vector $\mathbf{w}(k)$ are given respectively by

$$\Phi = \begin{bmatrix} 1 & T & 0 & 0 & 0 & 0 \\ 0 & 1 & 1 & 0 & 0 & 0 \\ 0 & 0 & \rho & 0 & 0 & 0 \\ 0 & 0 & 0 & 1 & T & 0 \\ 0 & 0 & 0 & 0 & 1 & 1 \\ 0 & 0 & 0 & 0 & 0 & \rho \end{bmatrix} \quad \text{and} \quad \mathbf{w}(k) = \begin{bmatrix} 0 \\ 0 \\ w_1(k) \\ 0 \\ 0 \\ w_2(k) \end{bmatrix} \tag{7.117}$$

The accelerations u_1 and u_2 were formerly in the input vector of a signal equation like (7.109). Here these accelerations are part of the augmented signal vector. As an example, u_1 enters the r equation via the coupling term of unity in the (2, 3) position of the Φ matrix.

The observations in this case are again noisy measurements of the two signal components $r(k)$ and $\theta(k)$, and so we use a 2×6 H matrix

$$\mathbf{x}(k) = \begin{bmatrix} r(k) + n_1(k) \\ \theta(k) + n_2(k) \end{bmatrix} = \underbrace{\begin{bmatrix} 1 & 0 & 0 & 0 & 0 & 0 \\ 0 & 0 & 0 & 1 & 0 & 0 \end{bmatrix}}_{H} \mathbf{s}(k) + \mathbf{n}(k) \tag{7.118}$$

This simple trick of augmenting the original signal-dynamics equations by adding the two equations representing the maneuver (acceleration) dynamics and then incorporating the additional variables $u_1(k)$ and $u_2(k)$ into the signal-vector definition has enabled us to handle this more complex situation. The Kalman filter solution is then the same as before, except that the dimensions of the matrices and vectors have been increased correspondingly. (The reader is encouraged to work out the Kalman filter for this case and show that it is of identically the same form as previously except for the fact that the H matrix is now 2×6, as already noted, the U matrix is 6×6, etc.) We have thus succeeded in setting up a more realistic signal model but at the cost of increased computational complexity.†

This technique of augmenting the signal equations, as well as the sensor equation, $\mathbf{x}(k) = H\mathbf{s}(k) + \mathbf{n}(k)$, is often used to include nonwhite noise as well as more complex signal dynamics in the Kalman filter development.

7.4 OTHER APPLICATIONS

In Sec. 7.3 we discussed at length the application of Kalman filtering to air traffic control. The ground radar tracking either an en route aircraft or one approaching an airport attempts to estimate the aircraft range and velocity from measurements taken once every scan.

In this section we discuss much more briefly several other applications of Kalman filtering. The first concerns itself with the estimation of the attitude angles of an orbiting spacecraft or satellite. Such orbiting spacecraft as weather satellites, environmental-sensing satellites, communication satellites, orbiting astronomical observatories, and many other satellites launched in recent years need this information to point or aim antennas, telescopes, cameras, etc., accurately. For the second application we have again chosen a space example, this one involving the in-flight determination of the position and velocity of a space vehicle on a circumlunar mission. This example is of particular interest because it represents a common estimation problem: the parameters to be estimated are nonlinearly related, and linearization must be used to set the equations into the form for which our previous analysis

† The comparative computer requirements for the two Kalman filters (the simpler and the more complex one) plus three other filters appropriate to the tracking problem are discussed in ibid., pp. 107 and 108. Also considered are the accuracies provided by the various filters for some specific tracking examples.

is applicable. As the third example we have chosen a ground-traffic problem, the estimation of the number of cars traversing a given section of a highway and their velocities. This information is of course needed in developing any automatic traffic-control system.

Determination of Satellite Attitude†

Assume that a satellite has been launched into an almost circular orbit about the earth. Under ideal conditions the orbital radius R, orbital frequency ω_0, and translational velocity v along the circular path are all related by the familiar equation of rotation

$$v = \omega_0 R \tag{7.119}$$

For a circular orbit the centrifugal force due to the circular motion must balance the force of gravity g (the centripetal force). This further enables us to relate the orbital radius and orbital frequency or angular velocity ω_0

$$g = \omega_0^2 R \tag{7.120}$$

Assume, for example, that the satellite is orbiting close to the earth's surface (its height above the earth is small compared to the earth's radius; a height of 200 km would be a typical example). R is then just the earth's radius. Using $g = 9.8\ \text{m/s}^2$ as the gravitational constant at the earth's surface and choosing $R = 6370$ km as the radius, we get from (7.120) $\omega_0 \approx 1.2$ mrad/s. The period of rotation is thus $2\pi/\omega_0 = 5200$ s, or 87 min. The translational velocity is, from (7.119), 7.8 km/s, or 28,000 km/h. All these numbers are of course familiar to regular readers of the daily newspaper. The assumption of a perfectly circular orbit is not good enough for the applications of such satellites, as noted earlier.

To carry out the necessary attitude control on board the satellite, accurate estimates of its attitude must be obtained. Three angles commonly used to define local attitude with reference to an assumed circular orbit are shown in Fig. 7.15. The angle φ measures roll about the axis $\bar{x}_0$ in the direction of motion, θ represents pitch (the up-down motion in the direction of motion), and ψ the so-called yaw angle or sidewise motion in the direction of travel, as seen from the satellite.

For small variations about the circular orbit three linear rotation equations relating the three angles may be written:‡

$$\dot{\varphi}(t) = \omega_0 \psi(t) + p \tag{7.121}$$

$$\dot{\theta}(t) = \omega_0 + q \tag{7.122}$$

$$\dot{\psi}(t) = -\omega_0 \phi(t) + r \tag{7.123}$$

† This example is taken from A. E. Bryson, Jr., and W. Kortüm, Estimation of the Local Attitude of Orbiting Spacecraft, *Automatica*, vol. 7, no. 2, pp. 163–180, March 1971.

‡ Ibid.

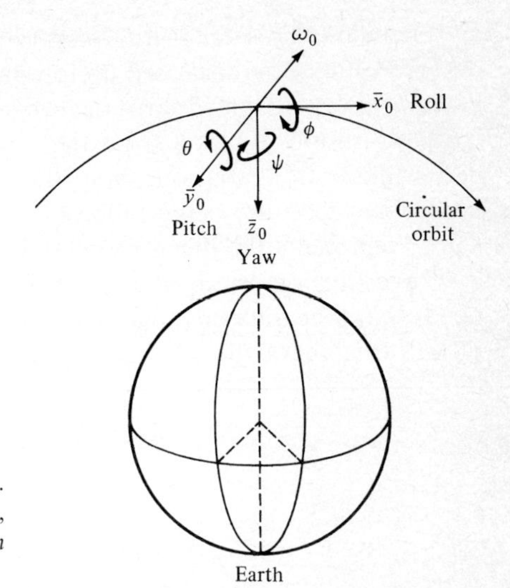

FIGURE 7.15
Definition of attitude angles. (*From A. E. Bryson and W. Kortüm, Automatica, vol. 7, no. 2, March 1971, Fig. 1, p. 164, with permission.*)

Here p, q, r represent the three components of vehicle angular velocity, measured with respect to the nominal angular velocity ω_0, as projected onto the three local axes of Fig. 7.15.

The pitch equation (7.122) appears uncoupled from the other two equations. It can thus be handled separately and simply and so is not included in the discussion following. We concentrate solely on estimation of the roll angle φ and yaw angle ψ.

If the two angular-velocity components p and r in (7.121) and (7.123) are known, a little thought will indicate that there is no estimation problem. One simply solves the two kinematic equations simultaneously using standard techniques to find $\varphi(t)$ and $\psi(t)$. The problem arises because they are *not* known exactly. More specifically, rate gyros on the roll-and-yaw axis are used to measure the angular-velocity components. Call the gyro outputs p_g (roll axis) and p_r (yaw axis). Because of random drift rates in the gyros, the gyro outputs are actually given by

$$p_g = p + D_g \tag{7.124}$$

and

$$p_r = r + D_r \tag{7.125}$$

respectively. D_g and D_r represent the random drift rates. The equations actually available for determining $\varphi(t)$ and $\psi(t)$ are

$$\dot{\varphi}(t) = \omega_0 \psi(t) + p_g - D_g \tag{7.126}$$

and

$$\dot{\psi}(t) = -\omega_0 \varphi(t) + p_r - D_r \tag{7.127}$$

The known gyro outputs p_g and p_r thus appear as forcing functions in the two equations (7.126) and (7.127) that have to be solved to find $\varphi(t)$ and $\psi(t)$. Because of superposition, their effect can always be added in separately in estimating $\varphi(t)$ and $\psi(t)$. The recursive-estimation technique is necessary to find an estimate of the response to the random forcing functions D_g and D_r. We shall thus suppress p_g and p_r

and include them later in the final estimate form. [Recall that in the linear mean-squared estimation discussed up to now, both in Chap. 6 and earlier in this chapter, we have always assumed zero-mean signals to be estimated. If the signals are not zero mean we simply subtract from the observations the mean value (assumed known) and estimate the variation about the mean. The mean value is then added back in later. The procedure being followed here is essentially the same: the response to p_g and p_r represents the mean value and is added back in later.]

We thus ignore p_g and p_r for the time being. Assume now that estimates are updated at time intervals spaced T s apart. To use the kinematic equations we should pick T small enough to have $\dot{\varphi}$ and $\dot{\psi}$ given by

$$\dot{\varphi} = \frac{\varphi(k+1) - \varphi(k)}{T} \tag{7.128}$$

and

$$\dot{\psi} = \frac{\psi(k+1) - \dot{\psi}(k)}{T} \tag{7.129}$$

Here we have introduced our previous notation $\varphi(t) = \varphi(kT) \equiv \varphi(k)$. For a 90-min orbit one would expect the updating to be carried out many times during one orbit, so that we should at least have $\omega_0 T \ll 2\pi$.

Introducing (7.128) and (7.129) into (7.126) and (7.127) and dropping p_g and p_r, as discussed above, we finally have for our state equations

$$\varphi(k+1) = \varphi(k) + \omega_0 T\psi(k) - TD_g(k) \tag{7.130}$$

and

$$\psi(k+1) = \psi(k) - \omega_0 T\varphi(k) - TD_r(k) \tag{7.131}$$

The dynamics of the random gyro drift terms $D_g(k)$ and $D_r(k)$ must now be considered in specifying the optimum Kalman filter. We shall assume them to be representable as white-noise processes. This will then give us a second-order Kalman filter as our optimum recursive estimator. Detailed considerations by Bryson and Kortüm in the paper from which this estimation example is taken indicate that the drift is more accurately modeled as a first-order autoregressive process.† One then must augment the two equations (7.130) and (7.131), as done in Sec. 7.3, and a fourth-order filter results. Details of this estimator with a numerical discussion appear in the paper, and the reader is referred to it for details. (The authors actually show that a third-order filter gives optimization accuracy comparable to the fourth-order one.) The second-order filter is accurate only for short periods of operation. For long periods one must use the more complex estimator. For our purpose, however, it suffices to stay with the simpler white-noise model.

As previously in this chapter, we now define the two-component vectors

$$\mathbf{s}(k) \equiv \begin{bmatrix} \varphi(k) \\ \psi(k) \end{bmatrix} \quad \text{and} \quad \mathbf{w}(k) \equiv \begin{bmatrix} -TD_g(k) \\ -TD_r(k) \end{bmatrix} \tag{7.132}$$

† Ibid., p. 166.

Equations (7.130) and (7.131) then appear in the usual vector form as

$$\mathbf{s}(k+1) = \Phi\mathbf{s}(k) + \mathbf{w}(k) \tag{7.133}$$

with the matrix Φ given by

$$\Phi = \begin{bmatrix} 1 & \omega_0 T \\ -\omega_0 T & 1 \end{bmatrix} \tag{7.134}$$

Equation (7.133) thus represents the dynamical equation of the vector signal [in this case $\varphi(k)$ and $\psi(k)$] to be estimated.

What measurements are available to carry out the estimation? A horizon sensor provides knowledge of the roll angle φ and is assumed to be the only instrument used. Instrument noise again precludes exact measurement of φ (otherwise estimation would not be necessary). The horizon-sensor output at each sampling interval is thus

$$x(k) = \varphi(k) + n(k) \tag{7.135}$$

We assume that the instrument noise $n(k)$ is white. Following the procedure of the previous sections, we write

$$x(k) = H\mathbf{s}(k) + n(k) \tag{7.135a}$$

with the measurement matrix H in this case just a row vector

$$H = [1 \quad 0] \tag{7.136}$$

Note that in this example with only one sensor used we get a scalar $x(k)$ representing the measurement data rather than the vector $\mathbf{x}(k)$ of previous sections. From (7.72) the optimum recursive estimate of $\varphi(k)$ and $\psi(k)$ is given by

$$\hat{\mathbf{s}}(k+1) = \Phi\hat{\mathbf{s}}(k) + K_{k+1}[x(k+1) - H\Phi\hat{\mathbf{s}}(k)] \tag{7.137}$$

with the K_k the Kalman gain matrix calculated, as previously, by equations like (7.74) to (7.76). This matrix depends of course on the variance of the gyro drifts and the instrument noise.

In this example K_{k+1} is the two-component column vector that may be written

$$K_{k+1} = \begin{bmatrix} K_\varphi \\ K_\psi \end{bmatrix} \tag{7.138}$$

In terms of these two Kalman gain terms, (7.137) can be expanded in terms of the two desired estimates $\hat{\varphi}(k)$ and $\hat{\psi}(k)$. It is left to the reader to show that the two recursive estimates are given respectively by

$$\hat{\varphi}(k+1) = \hat{\varphi}(k) + \omega_0 T\hat{\psi}(k) + K_\varphi[x(k+1) - \hat{\varphi}(k) - \omega_0 T\hat{\psi}(k)] \tag{7.139}$$

and $$\hat{\psi}(k+1) = -\omega_0 T\hat{\varphi}(k) + \psi(k) + K_\psi[x(k+1) - \hat{\varphi}(k) - \omega_0 T\hat{\psi}(k)] \tag{7.140}$$

These do not include the two gyro rate outputs p_g and p_r that should be added respectively to the two equations.

Two alternate interpretations of (7.139) and (7.140) are of interest. First recall from (7.73) that the predicted signal estimate one step in the future is related to the current estimate by the simple expression

$$\hat{\mathbf{s}}(k+1 \mid k) = \mathbf{\Phi}\hat{\mathbf{s}}(k) \tag{7.73}$$

Expanding this in terms of $\varphi(k)$ and $\psi(k)$, we have

$$\hat{\varphi}(k+1 \mid k) = \hat{\varphi}(k) + \omega_0 T\hat{\psi}(k) \tag{7.141}$$

and

$$\hat{\psi}(k+1 \mid k) = \hat{\psi}(k) - \omega_0 T\hat{\varphi}(k) \tag{7.142}$$

Hence (7.139) and (7.140) can be written

$$\hat{\varphi}(k+1) = \hat{\varphi}(k+1|k) + K_\varphi[x(k+1) - \hat{\varphi}(k+1|k)] \tag{7.139a}$$

and

$$\hat{\psi}(k+1) = \hat{\psi}(k+1|k) + K_\psi[x(k+1) - \hat{\varphi}(k+1|k)] \tag{7.140a}$$

The update term, multiplied by a Kalman gain coefficient, that must be added to the best predicted value of the signal to obtain the new estimate is the same in both cases. This is to be expected, since one sensor only, the horizon sensor, provides new data for updating the estimates. Since $x(k)$ is a noisy estimate of $\varphi(k)$, it is to be expected that the error term is of the form $x(k) - \hat{\varphi}(k+1 \mid k)$.

As the second interpretation, one that lends itself to a *continuously updating* estimator and to an analog implementation of it, let the update interval T become small enough for the changes in roll-and-yaw angles to be essentially differential in size. In the limit as $T \to 0$, it is left to the reader to show that (7.139) and (7.140), with p_g and p_r added back in, take on the continuous forms†

$$\dot{\hat{\varphi}}(t) = \omega_0 \hat{\psi}(t) + p_g + K'_\varphi[x(t) - \hat{\varphi}(t)] \tag{7.139b}$$

and

$$\dot{\hat{\psi}}(t) = -\omega_0 \hat{\varphi}(t) + p_r + K'_\psi[x(t) - \hat{\varphi}(t)] \tag{7.140b}$$

Here K'_φ and K'_ψ are new, continuous-form Kalman gain coefficients. The estimators are again similar in form to the original dynamic equations (7.126) and (7.127). [Note that if the update interval T is chosen small enough to satisfy $\omega_0 T \ll 1$, and if the roll estimate $\hat{\varphi}(k)$ and yaw estimate $\hat{\psi}(k)$ are comparable in size, the difference terms in both (7.139) and (7.140) are both of the form $x(k) - \hat{\varphi}(k)$, which is equivalent to the difference terms in (7.139*b*) and (7.140*b*). If $\omega_0 = 10^{-3}$, $T \sim 100$ s or less already provides this condition. It is thus apparent that it is the difference between the new horizon-sensor sample and the estimate of the roll angle $\hat{\varphi}(k)$ that is significant in providing updated estimates of *both* $\hat{\varphi}(k)$ and $\hat{\psi}(k)$. The $\omega_0 T\hat{\psi}(k)$ is small compared to $\hat{\varphi}(k)$ and may be neglected.] Since $\dot{\hat{\varphi}}(t)$ and $\dot{\hat{\psi}}(t)$ are respectively the time derivatives of $\hat{\varphi}(t)$ and $\hat{\psi}(t)$, the implementation of (7.139*b*) and (7.140*b*) suggests the analog implementation using integrators of Fig. 7.16. A digital version would involve using (7.139) and (7.140) with p_g and p_r added appropriately, which is essentially already

†Ibid.

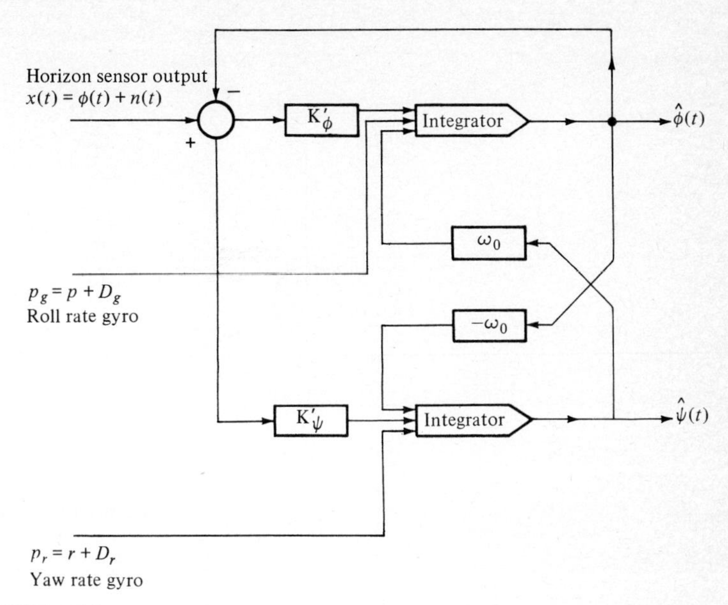

FIGURE 7.16
Roll-yaw filter using electronic integrators. (*From A. E. Bryson and W. Kortüm, Automatica, vol. 7, no. 2, March 1971, Fig. 2, by permission.*)

covered by the general Kalman filter implementation of Fig. 7.7. Bryson and Kortüm also discuss an implementation using rate-integrating gyros that eliminates the need for electronic integrators. They point out that this second implementation gives rise to a scheme called an *orbital gyrocompass.*

In-Flight Estimation of Position and Velocity of Circumlunar Vehicle†

In this example of the application of recursive-estimation techniques we consider the problem of midcourse guidance of a vehicle on a circumlunar mission. We visualize a space vehicle launched on a specified trajectory to take it around the moon and back. Unavoidable errors in the initial injection into the trajectory dictate a need for corrective maneuvers as the vehicle proceeds along its path. The on-board guidance system carries out the necessary control by using estimated position and velocity to determine the trajectory that would be followed in the absence of corrective maneuvers and from which the necessary corrections can be determined. We discuss here an on-board Kalman filter type of digital estimator for recursively estimating the position and velocity of the vehicle as it proceeds along the trajectory.

† G. L. Smith, S. F. Schmidt, and L. A. McGee, Application of Statistical Filter Theory to the Optimal Estimation of Position and Velocity on Board a Circumlunar Vehicle, *NASA Tech. Rep.* R-135, 1962.

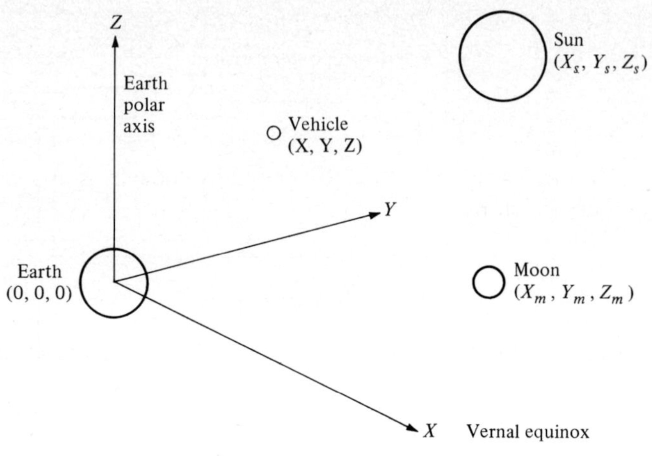

FIGURE 7.17
Geometry of midcourse guidance problem.

The geometry of the problem is indicated in Fig. 7.17.† The equations of the space vehicle, assuming a free-fall condition with the gravitational attraction of the earth, moon, and sun the only forces acting on the vehicle, are given by

$$\ddot{X} = -\frac{\mu_e X}{r^3}\left[1 + J\left(\frac{a}{r}\right)^2\left(1 - 5\frac{Z^2}{r^2}\right)\right] - \frac{\mu_m(X - X_m)}{\Delta_m^3} - \frac{\mu_m X_m}{r_m^3} - \frac{\mu_s(X - X_s)}{\Delta_s^3} - \frac{\mu_s X_s}{r_s^3} \tag{7.143}$$

$$\ddot{Y} = -\frac{\mu_e Y}{r^3}\left[1 + J\left(\frac{a}{r}\right)^2\left(1 - 5\frac{Z^2}{r^2}\right)\right] - \frac{\mu_m(Y - Y_m)}{\Delta_m^3} - \frac{\mu_m Y_m}{r_m^3} - \frac{\mu_s(Y - Y_s)}{\Delta_s^3} - \frac{\mu_s Y_s}{r_s^3} \tag{7.144}$$

and

$$\ddot{Z} = -\frac{\mu_e Z}{r^3}\left[1 + J\left(\frac{a}{r}\right)^2\left(3 - 5\frac{Z^2}{r^2}\right)\right] - \frac{\mu_m(Z - Z_m)}{\Delta_m^3} - \frac{\mu_m Z_m}{r_m^3} - \frac{\mu_s(Z - Z_s)}{\Delta_s^3} - \frac{\mu_s Z_s}{r_s^3} \tag{7.145}$$

The earth has been assumed oblate in shape. (The correction factor to a spherical earth model is the term inside the brackets of the first term in each equation.) Both the sun and moon have been assumed spherical. The various symbols and

† Ibid., p. 27.

coefficients of the three equations are defined as follows. r, r_m, and r_s represent respectively the distance of the vehicle, moon, and sun from the earth, and are given by

$$r = \sqrt{X^2 + Y^2 + Z^2} \qquad r_m = \sqrt{X_m^2 + Y_m^2 + Z_m^2} \qquad r_s = \sqrt{X_s^2 + Y_s^2 + Z_s^2}$$

Δ_m and Δ_s represent the distance of the vehicle from the moon and sun, respectively, and are given by

$$\Delta_m = \sqrt{(X - X_m)^2 + (Y - Y_m)^2 + (Z - Z_m^2)}$$

$$\Delta_s = \sqrt{(X - X_s)^2 + (Y - Y_s)^2 + (Z - Z_s)^2}$$

The three gravitational constants μ_e, μ_m, μ_s have the following values:

$$\mu_e = 3.986135 \times 10^{14} \text{ m}^3/\text{s}^2$$

$$\mu_m = 4.89820 \times 10^{12} \text{ m}^3/\text{s}^2$$

$$\mu_s = 1.3253 \times 10^{20} \text{ m}^3/\text{s}^2$$

The constant a is just the radius of the earth at the equator, $a = 6.37826 \times 10^6$ m, and $J = 1.6246 \times 10^{-3}$ is a measure of the earth's oblateness. (As a check, the gravitational force at the earth's surface in a perfectly spherical situation is $\mu_e/a^2 = 9.8 \text{ m/s}^2$, substituting in the values for a and μ_e above.)

The six variables to be estimated are the three vehicle-position components X, Y, and Z and the corresponding velocities $\dot{X}$, $\dot{Y}$, $\dot{Z}$. They define a six-component signal vector. To use the recursive-estimation approach of previous sections we must linearize these equations. To do this assume that at the time an observation is made (from which an estimate of the signal is to be determined) there is available a reference point (X_R, Y_R, Z_R) on the vehicle trajectory. Instead of estimating the actual vehicle-position coordinates (X, Y, Z) we need only estimate $X - X_R$, $Y - Y_R$, and $Z - Z_R$. On the assumption the difference terms are small, i.e., corrections required are small, each of the equations of motion (7.143) to (7.145) need only be expanded in a Taylor series about the reference point and the first terms in each series retained to have the desired linearized equations of motion.

Specifically, write the three equations of motion (7.143) to (7.145) in the shorthand form

$$\ddot{X} = f_1(X, Y, Z) \tag{7.143a}$$

$$\ddot{Y} = f_2(X, Y, Z) \tag{7.144a}$$

$$\ddot{Z} = f_3(X, Y, Z) \tag{7.145a}$$

with $f_1(\cdot)$, $f_2(\cdot)$, and $f_3(\cdot)$ each representing all the terms on the right-hand side of the respective equations. Now expand each in its Taylor series about X_R, Y_R, Z_R. The $\ddot{X}$ equation, for example, becomes

$$\ddot{X} = f_1(X_R, Y_R, Z_R) + \frac{\partial f_1}{\partial X}\bigg| (X - X_R) + \frac{\partial f_1}{\partial Y}\bigg| (Y - Y_R) + \frac{\partial f_1}{\partial Z}\bigg| (Z - Z_R) + \cdots \tag{7.143b}$$

FIGURE 7.18
On-board calculation of signal reference point.

where the three partial derivatives are all assumed evaluated at the reference point. The six components of the signal vector to be estimated are now written

$$s_1 = X - X_R \qquad s_2 = Y - Y_R \qquad s_3 = Z - Z_R$$

$$s_4 = \dot{s}_1(t) \qquad s_5 = \dot{s}_2(t) \qquad s_6 = \dot{s}_3(t)$$

The three linearized equations of motion now become very simply

$$\dot{s}_4 = \frac{\partial f_1}{\partial X} s_1 + \frac{\partial f_1}{\partial Y} s_2 + \frac{\partial f_1}{\partial Z} s_3 \tag{7.146}$$

$$\dot{s}_5 = \frac{\partial f_2}{\partial X} s_1 + \frac{\partial f_2}{\partial Y} s_2 + \frac{\partial f_2}{\partial Z} s_3 \tag{7.147}$$

$$\dot{s}_6 = \frac{\partial f_3}{\partial X} s_1 + \frac{\partial f_3}{\partial Y} s_2 + \frac{\partial f_3}{\partial Z} s_3 \tag{7.148}$$

(It is again understood that each derivative is evaluated at the appropriate reference coordinate.) It is apparent that the equations of motion are now in a form involving first-order linear differential equations from which the recursive estimates can be found. Before developing this, however, a word about the determination of the reference point (X_R, Y_R, Z_R) is in order.

If recursive estimation has been proceeding regularly, a previous estimate of the signal is available. Since the vehicle dynamics are known [the vehicle is assumed to be moving in a free-fall condition, obeying the nonlinear equations of motion (7.143) to (7.145)] and an on-board computer is assumed available, the *present* position of the vehicle based on the *previous estimate* as an initial condition can be found by solving (integrating) the equations of motion. This position found by using the previous estimate will be taken to be the reference point (X_R, Y_R, Z_R). This is diagrammed in Fig. 7.18. It is this calculation of a current point on the trajectory, based on a past estimate, that the current estimate is intended to correct. If the estimates are fairly accurate, relatively few estimates need be made and correspondingly few adjustments of the vehicle path by its on-board guidance and control system need be made.

To obtain the appropriate discrete-time Kalman filter for this example, one need only approximate the derivatives in Eqs. (7.146) to (7.148) by the usual difference terms. For example, we can write

$$\dot{s}_4(t) \approx \frac{s_4(k+1) - s_4(k)}{T} \tag{7.149}$$

with $s_4(k)$ again defined as the shorthand notation for $s_4(kT)$ and T again written as the interval between successive estimates. It is then left for the reader to show that the first-order vector equation corresponding to the three equations of motion (7.146) to (7.148) is given by

$$\mathbf{s}(k+1) = \Phi(k)\mathbf{s}(k) \tag{7.150}$$

with $\mathbf{s}(k)$ the six-component vector

$$\mathbf{s}(k) = \begin{bmatrix} s_1(k) \\ \vdots \\ s_6(k) \end{bmatrix}$$

and the *time-varying* Φ matrix given by

$$\Phi(k) = \begin{bmatrix} 1 & 0 & 0 & T & 0 & 0 \\ 0 & 1 & 0 & 0 & T & 0 \\ 0 & 0 & 1 & 0 & 0 & T \\ T\dfrac{\partial f_1}{\partial X}(k) & T\dfrac{\partial f_1}{\partial Y}(k) & T\dfrac{\partial f_1}{\partial Z}(k) & 1 & 0 & 0 \\ T\dfrac{\partial f_2}{\partial X}(k) & T\dfrac{\partial f_2}{\partial Y}(k) & T\dfrac{\partial f_2}{\partial Z}(k) & 0 & 1 & 0 \\ T\dfrac{\partial f_3}{\partial X}(k) & T\dfrac{\partial f_3}{\partial Y}(k) & T\dfrac{\partial f_3}{\partial Z}(k) & 0 & 0 & 1 \end{bmatrix}$$

Note that not only must the reference point (X_R, Y_R, Z_R) be found computationally by solving the three nonlinear equations of motion (7.143) to (7.145) as a separate task each time a new position and velocity estimate is to be made (actually in the interval T between the previous estimate and the current one) but the nine derivatives in the Φ matrix must be reevaluated as well.

Equation (7.150) here corresponds to the first-order vector model for the signal dynamics previously written as Eq. (7.56). There are of course two differences: as already noted, the Φ matrix is here time-varying, and there is no white-noise process $\mathbf{w}(k)$ driving the system of equations.

The lack of a white-noise driving process is due specifically to the midcourse free-fall conditions of motion assumed. The signal vector $\mathbf{s}(k)$ is still random, however, because of the random injection conditions specified in putting the vehicle into the initial point on its trajectory. This corresponds to random initial conditions $\mathbf{s}(0)$. We assume that $E[\mathbf{s}(0)] = 0$, to make our model correspond to those used previously, and that the covariance matrix $E[\mathbf{s}(0)\mathbf{s}^T(0)]$ of the initial conditions is specified. A little thought will indicate that this is equivalent to specifying the variance or mean-squared value of each of the six initial conditions, as well as any correlation between them that may exist. These can of course be determined from experience, from knowledge of the accuracy of placing a vehicle into its trajectory,

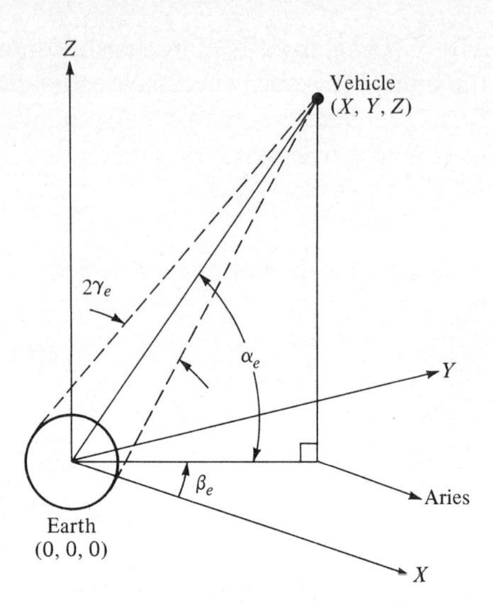

FIGURE 7.19
Example of angles measured by on-board instrumentation.

etc. This covariance matrix, which is assumed given, corresponds to U_0, the signal-driving noise-covariance matrix in Eqs. (7.71) and (7.76), as evaluated at $k = 0$. The successive U_k matrices, $k = 1, 2, \ldots$, are then taken to be zero in this example.

To complete the estimation example here we need the set of measurement data used to update the recursive estimates. One possible set of observations from which estimates of position can be made uses on-board optical instrumentation to measure various angles with respect to the earth.† (This then assumes that the vehicle at the time of observation is relatively close to earth; an alternate set would involve measuring angles with respect to various celestial bodies.) The three angles α_e, β_e, γ_e of Fig. 7.19 represent one such set from which a position fix can be made. The two angles α_e and β_e provide measurements of the direction of the vehicle-earth line of sight, and $2\gamma_e$ is the subtended earth angle. The angle β_e in the $Z = 0$ plane is measured counterclockwise from the X axis (the direction of Aries), while α_e is assumed positive if Z is negative. It is left for the reader to show that the following equations relate the angles to the vehicle position (X, Y, Z):

$$\alpha_e = -\sin^{-1}\frac{Z}{R} \tag{7.151}$$

$$\beta_e = \sin^{-1}\frac{Y}{(X^2 + Y^2)^{1/2}} \tag{7.152}$$

$$\gamma_e = \sin^{-1}\frac{R_0}{R} \tag{7.153}$$

Here R_0 is the radius of the earth, and $R = (X^2 + Y^2 + Z^2)^{1/2}$.

† Ibid., app. C.

Now consider a small perturbation about the reference point (X_R, Y_R, Z_R). The first terms of the Taylor series expansions of Eqs. (7.151) to (7.153) about the reference point produce linear approximations to these nonlinear equations. As an example, if $\Delta\alpha_e$ represents the difference in angle α_e evaluated at (X, Y, Z) and (X_R, Y_R, Z_R), it is readily shown that

$$\Delta\alpha_e \approx \frac{\partial\alpha_e}{\partial X} s_1 + \frac{\partial\alpha_e}{\partial Y} s_2 + \frac{\partial\alpha_e}{\partial Z} s_2 \tag{7.154}$$

The three partial derivatives are again to be evaluated at the reference point (X_R, Y_R, Z_R). Similar linear equations can be written relating $\Delta\beta_e$ and $\Delta\gamma_e$ to s_1, s_2, and s_3. These signal components are of course the quantities $X - X_R$, $Y - Y_R$, and $Z - Z_R$, respectively, as previously defined. As examples,

$$\frac{\partial\alpha_e}{\partial X} = \frac{XZ}{R^2\sqrt{X^2 + Y^2}} \qquad \frac{\partial\beta_e}{\partial Z} = 0$$

If we define $\mathbf{m}(k)$ to be the three-dimensional measurement vector

$$\begin{bmatrix} \Delta\alpha_e \\ \Delta\beta_e \\ \Delta\gamma_e \end{bmatrix}$$

with the angles measured at time kT, it is apparent that $\mathbf{m}(k)$ and the desired signal vector $\mathbf{s}(k)$ are related by the matrix equation

$$\mathbf{m}(k) = H(k)\mathbf{s}(k) \tag{7.155}$$

It is left for the reader to show that the *time-varying* $H(k)$ matrix in this case is the 3×6 matrix

$$H(k) = \begin{bmatrix} \frac{\partial\alpha_e}{\partial X}(k) & \frac{\partial\alpha_e}{\partial Y}(k) & \frac{\partial\alpha_e}{\partial Z}(k) & 0 & 0 & 0 \\ \frac{\partial\beta_e}{\partial X}(k) & \frac{\partial\beta_e}{\partial Y}(k) & \frac{\partial\beta_e}{\partial Z}(k) & 0 & 0 & 0 \\ \frac{\partial\gamma_e}{\partial X}(k) & \frac{\partial\gamma_e}{\partial Y}(k) & \frac{\partial\gamma_e}{\partial Z}(k) & 0 & 0 & 0 \end{bmatrix}$$

This is exactly the H matrix mentioned previously in Eq. (7.67), relating the desired signal vector to the measurement vector. In the previous examples H was time-invariant and of a particularly simple form involving 1s and 0s as entries.

As before, we assert that the instruments used to measure the three angles introduce instrument noise. The noises are again assumed to be zero mean and white, with their variances and possible correlations known.† We then write the data vector $\mathbf{x}(k)$ just as in Eq. (7.67):

$$\mathbf{x}(k) = \mathbf{m}(k) + \mathbf{n} = H(k)\mathbf{s}(k) + \mathbf{n}(k) \tag{7.156}$$

† If the instrument noises are nonwhite, they can be modeled as first-order autoregressive processes and the filter equations augmented, as in the previous section.

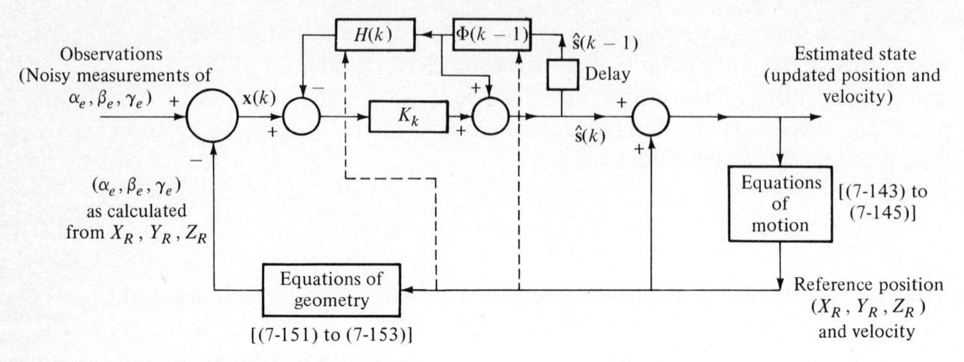

FIGURE 7.20
State estimation system, circumlunar vehicle; linearized version [$H(k)$ and $\Phi(k-1)$ depend on reference position].

The covariance matrix $R(k) = E[\mathbf{n}(k)\mathbf{n}^T(k)]$ has as its elements the variances and covariances (cross correlations) of the three instrument noises making up the instrument-noise vector $\mathbf{n}(k)$.

Since Eqs. (7.150) and (7.156) describing the signal dynamics and measurement characteristics, respectively, are in symbolic form identical with Eqs. (7.56) and (7.67), the recursive estimator for the six-component signal vector can be written down directly. From Eq. (7.72) it is just

$$\hat{\mathbf{s}}(k) = \Phi(k-1)\hat{\mathbf{s}}(k-1) + K_k[\mathbf{x}(k) - H(k)\Phi(k-1)\hat{\mathbf{s}}(k-1)] \tag{7.157}$$

The only difference here is that the Φ and H matrices are time-varying, as shown. It can be shown that this recursive form of the estimate would be optimum if the H and Φ matrices varied *independently* in time. Their variation here is a function of the state estimates. So Eq. (7.157) is not necessarily optimal. It is only a reasonable estimator to try in this nonlinear case. The Kalman gain matrix K_k is again calculated using the set of matrix equations (7.74) to (7.76) or an equivalent set. As noted above, for this example the signal-driving noise-covariance matrix U_k in Eq. (7.76) is zero, except for the initial matrix U_0.

This estimate is added to the stored reference position (X_R, Y_R, Z_R), as well as to the respective velocity components found by integrating the equations of motion, to obtain the updated estimate of position and velocity. These can be compared to a desired trajectory to see whether corrective control is needed. They are then integrated in turn to find the new reference point (X_R, Y_R, Z_R) to be used at the next estimation time.

Figure 7.20 is the block diagram of the entire state-estimation system. Note that the recursive calculation of the Kalman gain matrix K_k, the updating of the two matrices $\Phi(k-1)$ and $H(k)$, the trigonometric calculations prescribed by Eqs. (7.151) to (7.153), and the integration of the equations of motion (7.143) to (7.145) must all be done by an on-board computer.

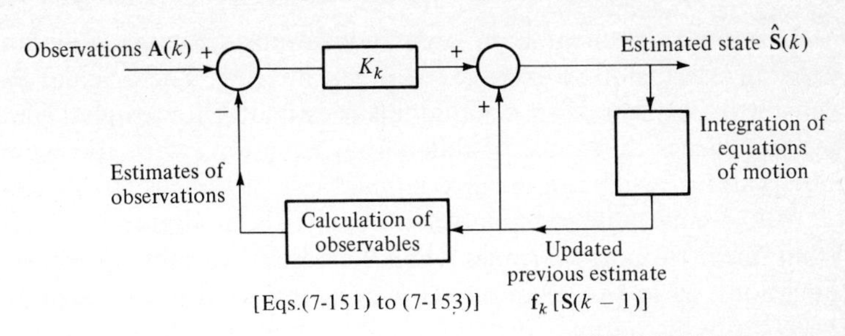

FIGURE 7.21
Extended Kalman filter.

The linearized estimator of Fig. 7.20 can be simplified considerably by some simple observations based on Eq. (7.157). The expression $\Phi(k-1)\hat{\mathbf{s}}(k-1)$ appearing in two places in that equation is the linear term in a Taylor series expansion [see Eqs. (7.143*b*), (7.146) to (7.148), and (7.150)]. This is the Taylor series representing the nonlinear solution of the equations of motion. Since the reference-position vector is itself obtained by integrating the equations of motion, we may just as well replace the $\Phi(k-1)\hat{\mathbf{s}}(k-1)$ term with the reference vector added to it by the nonlinear solution of the equations of motion directly. Note that the $H(k)$ matrix in Eq. (7.157) was also obtained by linearizing the nonlinear equations (7.151) to (7.153) relating the angles measured to the vehicle position vector. This too can be replaced by the nonlinear equations themselves, since, as seen from Fig. 7.20, they are needed anyway in estimating the angles.

Call the estimate of the state vector $\mathbf{S}(k)$. Then the nonlinear solution of the equations of motion can be written

$$\mathbf{S}(k) = \mathbf{f}_k[\mathbf{S}(k-1)]$$

with $\mathbf{f}_k$ representing the nonlinear functions. Let $\mathbf{h}_k$ represent the nonlinear equations of geometry (7.151) to (7.153). Then the entire process of estimation, using the *actual* state vector $\mathbf{S}(k)$ and not the deviation vector $\mathbf{s}(k)$, can be written in concise form

$$\hat{\mathbf{S}}(k) = \mathbf{f}_k[\mathbf{S}(k-1)] + K_k\{\mathbf{A}(k) - \mathbf{h}_k[\mathbf{f}_k[\mathbf{S}(k-1)]]\} \tag{7.158}$$

Here $\mathbf{A}(k)$ represents the *actual* observation of the angles. The nonlinear function $\mathbf{f}_k[\mathbf{S}(k-1)]$ can also be written by comparison with previous discussions, as the predicted value of $\hat{\mathbf{S}}(k)$ or $\hat{\mathbf{S}}(k \mid k-1)$ [see Eq. (7.73), for example].

Equation (7.158) is called the *extended* Kalman filter. The gain matrix K_k is the same matrix used in the linear Kalman filter and must still be calculated, as previously, by assuming small (linear) perturbations about each point estimated. K_k requires a $\Phi(k-1)$ calculated as a perturbation about $\mathbf{S}(k-1)$ and an H_k as a perturbation about $\mathbf{f}_k[\mathbf{S}(k-1)]$. The extended Kalman filter is diagrammed in Fig. 7.21.

A final comment is in order concerning ordinary Kalman filtering and the generalized version of extended Kalman filtering. The original linear form is a convenient realization of an optimal linear estimator for a signal which evolves according to a *linear* difference or differential equation, when the estimates are based on observations which are linear combinations of the desired signal-plus-noise.

It seems quite reasonable to change the updating and correcting operations from linear to nonlinear ones when the desired variables are described by nonlinear equations, as in this example. However, this kind of extended Kalman filter is not necessarily an optimal processor of the available data for computing least mean-squared error estimates. If the initial uncertainties are small, the linearization *approximations* will be relatively accurate and subsequent estimates should be fairly good. However, it is not unusual for extended Kalman filters to diverge from the true signal trajectory. Much recent research has been devoted to improving these filters and to finding alternative nonlinear estimators.

Application of Kalman Filtering to Traffic Systems†

The extended Kalman filter introduced in the last paragraph arises naturally in an application of recursive estimation to traffic control. We shall be brief in the presentation here, leaving it to the reader to study the article from which this material is taken if he is interested in more details.

Consider a section of highway L m long (Fig. 7.22). The object is to estimate at times $t = kT$ the number of vehicles $y(k)$ present in the section. Two sets of sensors are used: one at the entrance to the section counts the vehicles u_{1k} and their average velocity V_{1k} entering the section in the T-s interval between kT and $(k + 1)T$, and another counts the number u_{2k} and their average velocity V_{2k} leaving the section in the same T-s interval. Assume that the number-count sensors used introduce some inaccuracy representable by a composite white-noise source $w(k)$. It is then apparent that the relation between $y(k + 1)$ and $y(k)$, the number of vehicles at time $k + 1$ and k, respectively, is given by

$$y(k+1) = y(k) + u_{1k} - u_{2k} + w(k) \tag{7.159}$$

In this formulation the u_{1k} and u_{2k} counts are modeled as deterministic driving forces. Note how similar this is to the rate gyro outputs in the orbiting-satellite example discussed earlier [see Eqs. (7.126) and (7.127)].

The input and output average speeds V_{1k} and V_{2k} are used to find the average speed $\bar{V}(k)$ in the L-m section empirically. One approximation suggested in the original paper is to write

$$\bar{V}(k) = \alpha V_{1k} + (1 - \alpha)V_{2k} \tag{7.160}$$

For moderately heavy traffic the constant α is somewhere between 0.5 and 0.7.

† M. W. Szeto and D. C. Gazis, Application of Kalman Filtering to the Surveillance and Control of Traffic Systems, *Transp. Sci.*, vol. 6, no. 4, pp. 419–439, November 1972.

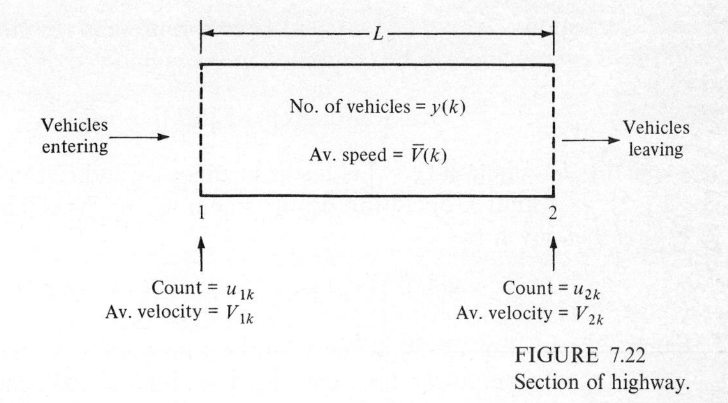

FIGURE 7.22
Section of highway.

It is well known (and almost intuitively obvious) in traffic studies that the mean velocity of a traffic stream and its density are inversely related: as the density of traffic increases, the average velocity decreases. In some studies a piecewise-linear approach has been utilized to model this effect. Szeto and Gazis used traffic data taken from Lincoln Tunnel† measurements and indicate that the data relating velocity and density fit a (nonlinear) bell-shaped curve of the form

$$\bar{V}(k) = b_k e^{-[y(k)/a_k L]^2} + n(k) \tag{7.161}$$

The density is of course given by $y(k)/L$ (the number of vehicles in the L-m section divided by L), a_k and b_k are two parameters to be estimated, and $n(k)$ is the usual white-noise term introduced to account for deviations away from the ideal model.

The quantities to be estimated in this example are the number of vehicles $y(k)$ and the two unknown parameters a_k and b_k. A signal vector $\mathbf{S}(k)$ can thus be defined with these three quantities as its components:

$$\mathbf{S}(k) = \begin{bmatrix} y(k) \\ a_k \\ b_k \end{bmatrix} \tag{7.162}$$

In terms of this vector the relation between the measured data $\bar{V}(k)$ [we have previously used $\mathbf{x}(k)$ to represent the data] and the signal vector $\mathbf{S}(k)$ is nonlinear, of the form

$$\bar{V}(k) = h[\mathbf{S}(k)] + n(k) \tag{7.163}$$

Here $h[\,\cdot\,]$ is of course the gaussian, or bell-shaped, function of Eq. (7.161).

To finish setting up the model assume that a_k and b_k are signal *parameters* that do not vary with time. These are exactly the signals discussed in Chap. 6. Then

$$a_{k+1} = a_k \tag{7.164}$$

and

$$b_{k+1} = b_k \tag{7.165}$$

† This tunnel under the Hudson River connects New York City and New Jersey and has been the subject of many studies.

Equations (7.159), (7.164), and (7.165) represent the model of signal dynamics in this example and can be combined in the signal-vector form

$$\mathbf{S}(k+1) = \mathbf{f}_k[\mathbf{S}(k)] + \mathbf{w}(k) \tag{7.166}$$

[Actually the function $\mathbf{f}_k[\,\cdot\,]$ is linear in this case and can be written explicitly, if so desired, as a matrix operating on the signal vector $\mathbf{S}(k)$.] The recursive estimate of $\mathbf{S}(k)$ is then given by

$$\mathbf{S}(k+1) = \mathbf{f}_k[\mathbf{S}(k)] + K_k(\bar{V}(k) - h\{\mathbf{f}_k[\mathbf{S}(k)]\}) \tag{7.167}$$

[compare with Eq. (7.158)]. The Kalman gain matrix K_k must again be found by *linearizing* the nonlinear function $h[\,\cdot\,]$ of Eqs. (7.163) and (7.162) about some reference point, from which the measurement matrix H can be evaluated. The function $\mathbf{f}_k[\mathbf{S}(k)]$ must also generally be expanded about the reference point to obtain the signal-dynamics matrix Φ. (In this case, as already noted, $\mathbf{f}_k[\,\cdot\,]$ is linear.)

The reader is referred to the original paper for a discussion of the application of the Kalman filtering technique to real traffic data, comparing estimates to actual counts of vehicles in a given section of the Lincoln Tunnel. The results are described as very good.† An extension of this technique to the estimation of density over several sections of highway for control purposes is considered in detail in the same paper.

In another approach to the use of recursive estimators for traffic estimation and control, joint estimates of speed and density are considered.‡ A simplified version of this approach appears as one of the problems at the end of this chapter.

PROBLEMS

7.1 *Recursive Estimator for Two Data Samples*
(*a*) Starting with the orthogonalty relation of Eq. (7.23), carry through the calculations indicated to first prove Eq. (7.22).
(*b*) Continue by checking Eqs. (7.28) and (7.30).
(*c*) Carry out the necessary calculations to obtain Eq. (7.31).

7.2 *Recursive Estimator Using the $k + 1$ Sample* Carry out the calculations necessary to obtain Eqs. (7.34) to (7.36). Show that the algebraic manipulations of Prob. 7.1 are appropriate in this case as well.

7.3 Refer to Prob. 6.3, relating to a velocity sensor mounted on a radar. Repeat the problem using a recursive-estimation technique to estimate the vehicle velocity, measured with respect to the average 10,000 mi/h velocity. Show that the fifth estimate, using data sample x_5, agrees with the result of part (*b*) in Prob. 6.3.

7.4 The number of vehicles per hour traversing a section of a highway during the peak of the rush-hour period on a given day is to be estimated using an automatic counter and timer. Four measurements are made, producing the values 5905, 5610, 6050, and 6220.

† Ibid.; see especially fig. 6, p. 428.

‡ N. E. Nahi, Freeway Traffic Data Processing, *Proc. IEEE*, vol. 61, no. 5, pp. 537–541, May 1973; N. E. Nahi and A. N. Trivedi, Recursive Estimation of Traffic Variables: Section Density and Average Speed, *Transp. Sci.*, August 1973.

(*a*) Find the least mean-squared linear estimate if it is known from previous measurements that the average vehicle-per-hour count is 5800, with a standard deviation of 200 vehicles per hour about that value. Instrument error may be assumed zero on the average, uncorrelated from sample to sample, and with an rms (standard deviation) value of 300 vehicles per hour. Compare the estimate with a sample-mean estimate.
(*b*) Find the new estimate if an additional measurement is made and found to be 5880 vehicles per hour. Compare the mean-squared errors in the two cases (four and five samples).
(*c*) Repeat part (*b*) if the fifth measurement gives 6280 vehicles per hour instead.

7.5 The number of vehicles per hour traversing a section of a highway is to be estimated from counter measurements. Measurements are spaced far enough apart in time so that the number of vehicles counted will be expected to vary with time. (This differs from Prob. 7.4, in which the measurements are presumed taken close together in time, so that the vehicle count being estimated is not expected to change over that interval.) Assume that the vehicle count can be modeled as a first-order autoregressive process like that of Eq. (7.37). Take $a = 0.5$; assume the average count is 6000 vehicles per hour, with the standard deviation about this value 300 vehicles per hour. The counting-instrument errors are uncorrelated from measurement to measurement, zero mean, with an rms error of 200 vehicles per hour. Find the successive estimates of the vehicle count, using linear recursive estimation, if the measurements made are, in order, 5900, 5800, 6050, 6100, and 5950, all in units of vehicles per hour.

7.6 Show that the vector equation of (7.56) corresponds to the N signal equations of (7.54).

7.7 Find the H matrix of Eq. (7.67) if $x_1(k) = 2s_1(k) + 3s_2(k) + n_1(k)$ with all the other $x_j(k)$'s as given in Eq. (7.66).

7.8 Let the signal and observation vectors in Eqs. (7.56) and (7.67) have one component only. Show that the three vector equations (7.68), (7.70), and (7.71) reduce to the three scalar equations (7.48), (7.49), and (7.50) if the Φ, H_k, $V(k \mid k-1)$, and R_k matrices are appropriately defined scalars.

7.9 Refer to Sec. 7.3 on the application of recursive estimation to radar tracking.
(*a*) Verify that the vehicle-acceleration model discussed preceding Eq. (7.90) is represented by Fig. 7.8 and that the variance of the acceleration is then described by Eq. (7.90).
(*b*) Show that the variance of the maneuver noise u_2 in Eq. (7.80) is given by Eq. (7.92).
(*c*) For initialization of the recursive estimator it is necessary first to evaluate the $V(2 \mid 2)$ matrix of Eq. (7.100), as discussed in the text. Show that for the tracker and dynamics of the vehicle assumed, this matrix can be written in the form of Eq. (7.100*a*), with the components given by Eqs. (7.101) to (7.106). In doing this first justify Eq. (7.99).
(*d*) Carry out a computer calculation to check the solid curves of Figs. 7.9 to 7.11. Use the same parameter values chosen for the curves.
(*e*) Verify the initial values of Figs. 7.10 and 7.11 by actual calculation of the $V(3 \mid 2)$ matrix.

7.10 A radar tracker has a scan time of 10 s and a range of 160 km. The aircraft being tracked has a maximum acceleration of $\frac{1}{3}g$. Assume that the vehicle dynamics can be modeled as in the first part of Sec. 7.3. Carry out a computer calculation of the different components of the Kalman gain matrix. Compare $K_{11}(k)$ with the plots of Fig. 7.12. Assume in this case that the sensor noises are substantially smaller; $\sigma_r = 0.25$ km, and $\sigma_\theta = 0.1°$.

7.11 Refer to the second-order roll-yaw filter discussed in the satellite-attitude example of Sec. 7.4. Let the variance of the instrument (horizon-sensor) noise be 10^{-6} rad^2. Take the rms drift rates of the roll- and yaw-rate gyros to be 10^{-5} rad/s each. Determine the Kalman gain coefficients K_ϕ and K_ψ of Eqs. (7.137) to (7.140) for a reasonable number of iterations.

7.12 Show that the optimum recursive estimator of Eq. (7.137) is given by the two equations (7.139) and (7.140) in the case of the second-order roll-yaw filter for satellite-attitude control discussed in Sec. 7.4.
(*b*) Show that these two equations take on the continuous-time forms of Eqs. (7.139*b*) and (7.140*b*) in the limit as the update interval T goes to zero.

7.13 *Fourth-Order Gyrocompass Filter for the Estimation of the Attitude of Orbiting Spacecraft*† Extend the second-order filter discussed in the satellite-attitude example of Sec. 7.4 by assuming each of the gyro drifts D_g and D_r in (7.130) and (7.131) to be modeled as a first-order autoregressive process. Use the approach discussed at the end of Sec. 7.3: define an augmented four-component signal vector, specify the 4×4 signal-dynamics matrix Φ, and obtain the fourth-order Kalman filter appropriate in this case. Compare with the second-order Kalman filter derived in the text.

7.14 Refer to the circumlunar-vehicle example discussed in Sec. 7.4.
(*a*) Show that the three linearized equations of motion (7.146) to (7.148) give rise to the first-order signal-vector equation (7.150).
(*b*) Show that the measurement angles of Fig. 7.19 are related to the vehicle coordinates by the nonlinear equations (7.151) to (7.153).
(*c*) Verify that the linearized versions of these equations are representable by the vector equation (7.155), with the H matrix the time-varying matrix shown in the text.

7.15 *Estimation of Speed and Number of Cars in a Section of Highway*‡ Consider a section of highway X m long carrying $N(k)$ cars at time kT. The average speed of the cars is $u(k)$ m/s. Sensors located at the beginning of this section measure the number $I_1(k)$ entering the X-m stretch every T s and their average speed $V_1(k)$. For control purposes it is desired to maintain a continually updated estimate of $N(k)$ and $u(k)$. In this problem we consider a very simplified approach for carrying this out. The reader is referred to the two references cited for details of the model noted here and various Kalman filter approaches for carrying out the desired estimation. Additional sensors located at the output of the section are considered in the papers as well.

A little thought will indicate that if the $N(k)$ cars are uniformly distributed along the X-m length and are all traveling at a velocity of $u(k)$ m/s, those leaving the section in T s must be located within a distance $Tu(k)$ m from the end of the section, the number leaving thus being $Tu(k)N(k)/X$. Actually the cars are not uniformly spread out and are not all traveling at the same speed. The number leaving will thus vary by a random factor $\varphi(k)$ about $Tu(k)N(k)/X$. We assume that $\varphi(k)$ is a zero-mean random variable and independent from one time interval to the next. From this discussion the number of cars at time $(k+1)T$ is readily written. It must be the number at time kT plus those entering in T s less those leaving. This gives the conservation equation

$$N(k+1) = \left[1 - \frac{Tu(k)}{X}\right] N(k) + I_1(k) - \varphi(k) \tag{i}$$

This is one of the equations used in the estimation approach. It couples the two variables $u(k)$ and $N(k)$ to be estimated.

† Bryson and Kortüm, loc. cit., p. 168.

‡ Nahi, loc. cit.; Nahi and Trivedi, loc. cit.

If it is now assumed that the speed $u(k)$ is derived from a first-order autoregressive process with parameter α $[u(k+1) = \alpha u(k) + \eta(k);\ E[u(k)] = \mu]$ it can be shown† that the following equation is obtained relating the variables and sensor outputs:

$$u(k+1) = \alpha u(k) - \frac{I_1(k)u(k)}{N(k)} + \frac{V_1(k)I_1(k)}{N(k)} + \eta(k) + \psi(k) \qquad \text{(ii)}$$

Here $N(k) \gg 1$ has been assumed, and $\psi(k)$ is another white-noise quantity.

Both Eqs. (i) and (ii) are coupled nonlinear equations. To simplify the estimation problem we now assume that we are interested in estimating variations $\Delta u(k)$ about the mean speed μ, that we measure input speeds $\Delta V_1(k)$ about the mean speed, that we measure the number of cars $\Delta I_1(k)$ entering, as a deviation away from a known average number $E[I_1(k)]$, and that we are interested in estimating *variations* $\Delta N(k)$ about the actual number of cars $N(k)$ in the section.

Equations (i) and (ii) then reduce to the linear equations

$$\Delta u(k+1) = \alpha\, \Delta u(k) + \frac{z_1(k)}{N(k)} + n(k) \qquad \text{(iii)}$$

$$\Delta N(k+1) = \left[\frac{1 - Tu(k)}{X}\right] \Delta N(k) - \frac{TN(k)}{X} \Delta u(k) \qquad \text{(iv)}$$

We have also assumed that $I_1(k) \ll N(k)$. The variable $z_1(k) = I_1(k)\, \Delta V_1(k)$. In deriving estimators using these two equations it is first assumed that $u(k)$ and $N(k)$ are *known* quantities. In the final implementation the estimates of these quantities may be used instead.

(*a*) Define a four-component signal vector $\mathbf{s}$ with $\Delta u(k)$, $\Delta N(k)$, $z_1(k)$, and $\Delta I_1(k)$ as the components. Set up the first-order vector equation incorporating Eqs. (iii) and (iv). What is the Φ matrix in this example?

(*b*) Let the two sensor outputs be

$$x_1(k) = \Delta I_1(k) + n_1(k) \qquad \text{and} \qquad x_2'(k) = \Delta V_1(k) + n_2'(k)$$

with $n_1(k)$ and $n_2'(k)$ two white-noise sources to account for sensor and measurement inaccuracies. Show that a noisy measurement of the quantity $z_1(k)$ defined in part (*a*) can be obtained by multiplying the two sensor outputs appropriately. Specifically, show that

$$x_2(k) \equiv \{E[I_1(k)] + x_1(k)\} x_2'(k) = z_1(k) + n_2(k)$$

Express $n_2(k)$ in terms of $n_1(k)$ and $n_2'(k)$. Find its mean value, variance, and other statistics in terms of the statistics of $n_1(k)$ and $n_2'(k)$.

(*c*) Set up the vector equation relating the measurement vector $\mathbf{x}(k)$ to the signal vector $\mathbf{s}(k)$ and find the 2×4 H matrix for this example.

(*d*) Draw block diagrams for the Kalman filter obtained in this example and describe the filter.

† Ibid.

INDEX

INDEX